Thermal History of Sedimentary Basins

Nancy D. Naeser Thane H. McCulloh
Editors

Thermal History of Sedimentary Basins

Methods and Case Histories

With 197 Illustrations

Springer-Verlag
New York Berlin Heidelberg
London Paris Tokyo

Nancy D. Naeser
United States Geological Survey
Denver Federal Center
Denver, Colorado 80225, USA

Thane H. McCulloh
Mobil Exploration and Producing Services, Inc.
Dallas, Texas 75265-0232, USA

Library of Congress Cataloging-in-Publication Data
Thermal history of sedimentary basins : methods and case histories /
edited by Nancy D. Naeser and Thane H. McCulloh.
p. cm.
Papers from the Society of Economic Paleontologists and Mineralogists Research Symposium on "Thermal History of Sedimentary Basins—Methods and Case Histories," held during the annual convention of the American Association of Petroleum Geologists in New Orleans, March 1985.
Includes bibliographies and index.
1. Earth temperature—Congresses. 2. Sedimentation and deposition—Congresses. 3. Organic geochemistry—Congresses.
I. Naeser, Nancy, Dearlen, 1944- . II. McCulloh, Thane Hubert, 1926- . III. Society of Economic Paleontologists and Mineralogists.
IV. Research Symposium on "Thermal History of Sedimentary Basins—Methods and Case Histories" (1985 : New Orleans, La.)
QE509.T48 1988
551.1′2—dc19 88-4024

Printed on acid-free paper.

Typeset by Publishers Service, Bozeman, Montana.
Printed and bound by Edwards Brothers, Inc., Ann Arbor, Michigan.
Printed in the United States of America.

9 8 7 6 5 4 3 2 1

ISBN 0-387-96702-8 Springer-Verlag New York Berlin Heidelberg
ISBN 3-540-96702-8 Springer-Verlag Berlin Heidelberg New York

Preface

The collection of papers in this volume is a direct result of the Society of Economic Paleontologists and Mineralogists Research Symposium on "Thermal History of Sedimentary Basins: Methods and Case Histories" held as part of the American Association of Petroleum Geologists Annual Convention in New Orleans in March 1985. The original goal of the symposium was to provide a forum where specialists from a variety of disciplines could present their views of methods that can be used to study the thermal history of a sedimentary basin or an important portion of a basin. An explicit part of that goal was to illustrate each method by presentation of a case history application. The original goal is addressed by the chapters in this volume, each of which emphasizes a somewhat different approach and gives field data in one way or another to illustrate the practical usefulness of the method.

The significance of our relative ignorance of the thermal conductivities of sedimentary rocks, especially shales, in efforts to understand or model sedimentary basin thermal histories and maturation levels is a major thrust of the chapter by Blackwell and Steele.

Creaney focuses on variations in kerogen composition in source rocks of different depositional environments and the degree to which these chemically distinct kerogens respond differently to progressive burial heating.

Molecular indicators of thermal maturity from kerogen and kerogen extract are the principal subject of the chapter by Curiale, Larter, Sweeney, and Bromley. They review this subject against a broad background of the more commonly measured bulk thermal maturity parameters such as Rock-Eval pyrolysis data and vitrinite reflectance. In particular, they examine specific aromatization reactions relative to measured "maturity levels" with appropriate concerns about reaction kinetics and heating rates.

Bulk organic matter maturities (mainly vitrinite reflectance) from many localities that are classified as burial diagenesis, geothermal system, or contact metamorphic environments are examined by Barker in terms of maximum temperature, exposure time, and reaction rates. He concludes that maximum temperature reached is the overwhelmingly dominant control and that reaction time spent at temperatures only slightly lower than the maximum can be neglected.

Empirical chemical thermometers based on the compositions of the dissolved substances in waters from oil wells, hot springs, and geothermal wells can be used to estimate the subsurface reservoir temperatures from the surface to depths corresponding to 350°C, according to Kharaka and Mariner. Estimates are within 10°C of measured values for reservoir temperatures higher than about 70°C. A new Mg-Li geothermometer is presented and recommended for all subsurface water (except those from gas wells).

Determination of paleotemperatures from measurements made on fluid inclusions in diagenetic minerals is the subject of an up-to-date review by Burruss, in which a pointed discussion is presented of the problems posed by reequilibration of early diagenetic fluid inclusions under deeper burial pressure-temperature conditions.

Pytte and Reynolds assemble and discuss data from six selected sites where duration of time near peak temperature can be estimated for the smectite to illite transformation. An empirical kinetic reaction model is fit to the data and adequately describes the smectite to illite reaction extent for conditions that range from volcanic contact alteration to long-term burial diagenesis under low-temperature conditions. Temperature is the dominant control on reaction progress (provided the chemistry of the system permits reaction).

The use of $^{40}Ar/^{39}Ar$ age spectrum analysis of detrital low-temperature potassium feldspar from buried clastic sequences as a means to trace the thermal evolution of a sedimentary basin is the subject of the chapter by Harrison and Burke. Data from the southernmost San Joaquin basin, California, the Albuquerque basin section of the Rio Grande Rift, New Mexico, and the southern Viking Graben of the North Sea basin are used to illustrate the technique.

Naeser, Naeser, and McCulloh discuss fission-track dating of detrital apatite and zircon from clastic sequences as a method of defining the overall temperature-time history of a basin. In addition, localized temperature anomalies, the sediment provenance, and the sedimentation record of the basin can be analyzed. They illustrate the method with studies in the San Joaquin basin, California, and the Green River basin, Wyoming.

Green, Duddy, Gleadow, and Lovering also discuss the application of fission tracks to basin studies, but they emphasize the importance of using the shortening of fission tracks in apatite, and the resulting variation in mean track length and track length distribution, with progressive annealing as a paleotemperature indicator. They illustrate this with data from wells in the Otway basin, Australia.

Feinstein, Kohn, and Eyal have combined vitrinite reflectance and fission-track data to reconstruct the thermal history recorded in folded rocks of a discontinuous succession of Early Permian to Tertiary age in southern Israel. According to the authors, following a brief intense Jurassic thermal episode, the coalification process "froze," despite progressive burial, reflecting a thermal decay since Early Cretaceous time.

Armagnac, Bucci, Kendall, and Lerche present a method of using the variation in vitrinite reflectance with depth in a well to estimate the thickness of sediment removed at unconformities. They illustrate use of the method in wells in the National Petroleum Reserve of Alaska and in a well in the Rharb basin, Morocco.

A one-dimensional finite-element modeling procedure is used by Issler and Beaumont to study the postrifting thermal and subsidence history of the Labrador continental margin, northeastern Canada. Model-derived temperatures are used to predict vitrinite reflectance and the progress of selected aromatization-isomerization biomarker compound reactions. Preliminary results from seven wells show overall good agreement with observations of maturity measures, crustal thickness (refraction), corrected bottom-hole temperatures, and paleobathymetry and stratigraphy.

McDonald, von Rosenberg, Jines, Burke, and Uhler use a two-dimensional, transient, finite-difference modeling procedure to derive the time-temperature history of sedimentary basins. The model simulates processes that occur during basin evolution, including subsidence, sedimentation, uplift and erosion, internal heat generation, surface temperature variations, magmatic activity, faulting, and changes in sediment thermal conductivity with burial. Use of the model is illustrated for a hypothetical Late Cenozoic basin.

A steady-state two-dimensional, finite-difference, numerical modeling procedure is used by Hagen and Surdam to examine the thermal evolution of the northern Bighorn basin, Wyoming and Montana. Integration of this thermal evolution model with time-temperature reconstructions derived from basin geology results in temperature histories for Cretaceous hydrocarbon source rocks.

Heasler and Surdam use their previously proposed thermal model for coastal California as a basis for modeling hydrocarbon maturation in Miocene Monterey Formation source rocks of the Pismo and Santa Maria basins. The authors use the model to predict the level of thermal exposure necessary to generate high API gravity crude oil in the rocks.

We would like to thank all of the authors and the reviewers of the chapters in this volume. Their conscientious work has made the volume possible. We thank Dale Issler and Christopher Beaumont for permission to use a modified version of one of their figures as the cover illustration. We would also like to thank Tom Kostick, U.S. Geological Survey, Denver, for preparing the cover illustration and for helping to modify a number of illustrations in the volume. Nancy Naeser wishes to thank Ian Mackenzie, New Zealand Department of Scientific and Industrial Research, for the early training in editing that made working on this volume so much easier.

NANCY D. NAESER
THANE H. MCCULLOH

Contents

Contributors

CHARLENE ARMAGNAC
Department of Geology, University of South Carolina, Columbia, South Carolina 29208, USA

CHARLES E. BARKER
United States Geological Survey, Denver, Colorado 80225, USA

CHRISTOPHER BEAUMONT
Oceanography Department, Dalhousie University, Halifax, Nova Scotia B3H 4J1, Canada

DAVID D. BLACKWELL
Department of Geological Sciences, Southern Methodist University, Dallas, Texas 75275, USA

BRUCE W. BROMLEY
Unocal Science and Technology Division, Unocal Corporation, Brea, California 92621, USA

JAMES BUCCI
Department of Geology, University of South Carolina, Columbia, South Carolina 29208, USA

KEVIN BURKE
Department of Geosciences, University of Houston, Houston, Texas 77004, USA

W. H. BURKE, JR.
Mobil Research and Development Corporation, Dallas, Texas 75381, USA

ROBERT C. BURRUSS
United States Geological Survey, Denver, Colorado 80225, USA

STEPHEN CREANEY
Esso Resources (Canada) Ltd., Calgary, Alberta T2P 0H6, Canada

Joseph A. Curiale
Unocal Science and Technology Division, Unocal Corporation, Brea, California 92621, USA

Ian R. Duddy
Geotrack International, Department of Geology, University of Melbourne, Parkville, Victoria 3052, Australia

Moshe Eyal
Department of Geology and Mineralogy, Ben Gurion University of the Negev, Beer Sheva 84105, Israel

Shimon Feinstein
Department of Geology and Mineralogy, Ben Gurion University of the Negev, Beer Sheva 84105, Israel

Andrew J.W. Gleadow
Department of Geology, La Trobe University, Bundoora, Victoria 3083, Australia

Paul F. Green
Geotrack International, Department of Geology, University of Melbourne, Parkville, Victoria 3052, Australia

E. Sven Hagen
Pecten International Company, Houston, Texas 77001, USA

T. Mark Harrison
Department of Geological Sciences, State University of New York at Albany, Albany, New York 12222, USA

Henry P. Heasler
Department of Geology and Geophysics, University of Wyoming, Laramie, Wyoming 82071, USA

Dale R. Issler
Institute of Sedimentary and Petroleum Geology, Calgary, Alberta T2L 2A7, Canada

W. R. Jines
Mobil Research and Development Corporation, Dallas, Texas 75381, USA

Christopher G. St. C. Kendall
Department of Geology, University of South Carolina, Columbia, South Carolina 29208, USA

Yousif K. Kharaka
United States Geological Survey, Menlo Park, California 94025, USA

Barry P. Kohn
Department of Geology and Mineralogy, Ben Gurion University of the Negev, Beer Sheva 84105, Israel

Stephen R. Larter
Institut for Geology, Department of Geology, University of Oslo, Blindern, N-0316 Oslo 3, Norway

Ian Lerche
Department of Geology, University of South Carolina, Columbia, South Carolina 29208, USA

John F. Lovering
Department of Geology, Flinders University, Bedford Park, South Australia 5042, Australia

Robert H. Mariner
United States Geological Survey, Menlo Park, California 94025, USA

Thane H. McCulloh
Mobil Exploration and Producing Services, Inc., Dallas, Texas 75265-0232, USA

A. E. McDonald
Mobil Research and Development Corporation, Dallas, Texas 75381, USA

Charles W. Naeser
United States Geological Survey, Denver, Colorado 80225, USA

Nancy D. Naeser
United States Geological Survey, Denver, Colorado 80225, USA

A. M. Pytte
Chevron Overseas Petroleum Inc., San Ramon, California 94583-0946, USA

R. C. Reynolds
Department of Earth Sciences, Dartmouth College, Hanover, New Hampshire 03755, USA

John L. Steele
Department of Geological Sciences, Southern Methodist University, Dallas, Texas 75275, USA

Ronald C. Surdam
Department of Geology and Geophysics, University of Wyoming, Laramie, Wyoming 82071, USA

Robert E. Sweeney
Unocal Science and Technology Division, Unocal Corporation, Brea, California 92621, USA

L. M. Uhler, Jr.
Mobil Research and Development Corporation, Dallas, Texas 75381, USA

D. U. von Rosenberg
Mobil Research and Development Corporation, Dallas, Texas 75381, USA

1
Thermal History of Sedimentary Basins: Introduction and Overview

Thane H. McCulloh and Nancy D. Naeser

Foreword

Interest in the thermal histories of sedimentary rocks and basins has grown rapidly since 1970 and is now intense. The main reason behind this acceleration is the increasing awareness that the natural processes responsible for generating oil and gas from kerogens of petroleum source rocks depend essentially on burial heating. Debates about the relative roles of other factors (time, heating rates, kerogen types, specific kerogen components, natural catalysis, and so forth) go on, but geochemists and virtually all petroleum geologists agree that heating of preserved sedimentary organic detritus is essential for oil and gas generation, and that burial in sedimentary basins or depocenters is required to achieve sufficient heating for commercial accumulations to occur.

Heat flows continuously from the interior of the earth toward the surface. This flow furnishes the bulk of energy that heats buried sediments and promotes the organic and mineralogical reactions that control the generation, migration, and accumulation of commercial accumulations of oil and gas. It has been estimated that the earth is losing heat into space at a mean value of between 3×10^{13} and 4×10^{13} watts for the entire surface area (Verhoogen, 1980, p. 13; Jaupart, 1984). However, this flow is very unevenly distributed, being higher through the sea floor and in tectonically active and geologically young regions (where volcanism may be evident) and lower in stable continental interiors and pre-Cambrian shield areas. There are two main sources of this heat. One is the heat transferred across the boundary from convecting upper mantle material into the overlying lithosphere. The other is a contribution from radioactive decay of uranium, thorium, and potassium within the lithosphere. Spatial and temporal lithospheric inhomogeneities, coupled with quasi-steady-state dynamism of the convecting upper mantle, account for the uneven flow of heat through the surface layers. Because of such inhomogeneities, each region or problem must be considered separately and from an empirical point of view (Blackwell, 1983).

Two fundamentally different approaches have evolved to study the thermal conditions and thermal evolution of sedimentary basins. One approach is fundamentally empirical, descriptive, and deductive. Observational data from wells and well samples, or in some instances from outcrops, seeps, or springs, are assembled to permit interpretations about the present thermal condition of the basin. The earlier thermal evolution of the basin is then deduced on the basis of the time-stratigraphic information available from surface geology, drilling, geophysical data, and lithological, stratigraphic, and structural models of the basin. The other approach is fundamentally theoretical and inductive. A physical model of the earth's outer layers (lithosphere and upper mantle) is assumed for an initial time prior to basin formation. A causative mechanism for basin inception is further assumed, generally invoking thinning of the lithosphere by tectonic stretching (Sleep, 1971; McKenzie, 1981), and stratigraphic data are used to constrain the timing of the basin-forming event. The thermal and structural evolution of the basin are then inductively modeled. Comparison of the modeling products with observational data should provide a test of the validity of the initial assumptions and the modeling procedures. In both approaches, or in any combination of the two, observational and analytical data are essential, either intrinsically or for validation. Hence the heavy emphasis, currently and during the past 20 years, on investigations aimed at quantifying thermal conditions in geologically young sedimentary basins (Kehle et al., 1970; Tanaka and Sato, 1977) or on identifying and quantifying natural consequences and products of thermal alteration of sedimentary materials that might provide measures of thermal exposure during burial heating (Hood and Castaño, 1974; Héroux et al., 1979).

Thermal Conditions and Heat Transport

Direct observation of subsurface temperature, geothermal gradient, and (to a lesser extent) thermal properties of rocks has become a generally accepted and widely practiced facet of petroleum exploration, especially in actively subsiding, geologically young sedimentary basins. A substantial and diverse literature attests to the extent of such work, a few selected examples of which are: Doebl and Teichmüller (1979), Catala (1984), Perrier and Raiga-Clemenceau (1984), and Thamrin (1985). Basic to all such work is the equation relating conductive heat flow through sedimentary (or other) rocks to their thermal conductivity (K) and the geothermal gradient (dT/dz):

$$Q_z = K(dT/dz)$$

where Q_z is the vertical component of heat flow. If the rocks are horizontally layered and heat transfer is by conduction only, the thermal conductivity and the temperature gradient are inversely proportional to one another for an unchanging heat flow. Where heterogeneous strata are steeply dipping, or where convective fluid movements complicate the transfer of heat, much more complex relationships pertain.

Seabed heat-flow measurements adjacent to active sea-floor spreading ridges are known to be strongly affected by hydrothermal circulation of cooling seawater through fracture networks in newly formed oceanic crust (Lister, 1980). Such effects persist for long time periods and for large distances from spreading axes. Convective transport and concentration of heat is also known from localized centers on land where geothermal systems in sedimentary rocks have been explored by drilling (Meinhold, 1967; Eckstein, 1979; Elders et al., 1980; Halfman et al., 1984; Rabinowicz et al., 1985). Such occurrences, apart from the cryptic influence of compaction dewatering flow (Bethke, 1985), strongly suggest that subsurface temperatures in sedimentary basins are not controlled everywhere only by conductive heat flow. Although episodes of pronounced convective heat transport might be expected to be localized, transient events of brief duration, the associated temperature increase may be significant (30° to 40°C above steady-state background for periods of 10^4 to 10^5 years). An excursion of such a magnitude would scarcely influence the reactions responsible for generation or thermal decomposition of oil. However, primary expulsion and migration of oil and gas already released from source rock kerogens would be drastically accelerated through such thermal exposure, mainly because of the added buoyancy induced in such hydrocarbon fluids by virtue of their very large coefficients of thermal expansion relative to water or brine. The timing of such events in relation to oil migration and trap availability would therefore seem to deserve attention.

Indicators of Thermal Exposure

New techniques and approaches, and refinements of old techniques, for indirectly gaining temperature or paleotemperature information from sedimentary rocks have grown in pace with the growth in awareness that such information is important in practical ways. A broad array of older and newer approaches is consequently now available (Tissot et al., 1971; McCulloh et al., 1978; Héroux et al., 1979; Naeser, 1979; Briggs et al., 1981; Hower, 1981; Harrison and Bé, 1983; Miknis and Smith, 1984; Roedder, 1984; and others). Most of these are analytical or observational in nature and require access to holes drilled into sedimentary rocks or, more commonly, to samples of rocks or pore fluids extracted from such exploratory holes. Where special circumstances provide outcrops or natural seeps or springs that afford exceptional sampling opportunities, some of these observational techniques may be applied to gain information that relates to thermal conditions in the depths of a basin, but this is not the usual case.

Solid Organic Indicators

The measurement of the optical reflectance of vitrinite occupies a unique place in the array of petrographic and geochemical approaches used to study the thermal history of sedimentary rocks. Profiles of vitrinite reflectance versus burial depth are a widely accepted and generally used measure of thermal maturity (Hood et al., 1975; Dow, 1977, 1978; Bostick et al., 1978; Teichmüller, 1979). This is so despite the fact that close examination of the data used to establish the most frequently cited correlation between calculated thermal maturities and measured vitrinite reflectances (Waples, 1980) shows an amount of scatter that many would deem unacceptable. This, plus the insensitivity of vitrinite reflectance at high values of TTI, should raise questions about the practical use of this approach. It has been repeatedly and independently demonstrated that the scatter is geologically meaningful and not just a statistical artifact of measurement or calculation errors (Hutton and Cook, 1980; Kalkreuth, 1982; Newman and Newman, 1982; Walker et al., 1983; Kalkreuth and Macauley, 1984, pp. 42–43; Price and Barker, 1984; Wenger and Baker, 1986). Factors other than temperature and time can clearly influence the reflectance of vitrinite under burial conditions up to at least the peak of oil generation from oil-prone kerogen ($R_o \approx 1.0$). The influence of these factors is substantially different from basin to basin or area to area within one basin, and even from bed to bed within a sequence of a single maturity. Shifts in reflectance of vitrinite of as much as 0.35% appear related to differences in hydrocarbon content of the total organic fraction of which the vitrinite is a part. Therefore, the preeminent standard in normal use for routine determinations and comparisons of thermal exposure seems to be on a shaky or not completely sound footing, perhaps especially in richly oil-prone basins (McCulloh, 1979). Recognition of the limitations of such a widely used analytical approach would seem an essential first step toward making full and proper use of it, as well as a prerequisite in the search for the most truthworthy thermal exposure parameter. Reliance should not be placed on vitrinite reflectance profiles alone for determination of paleogeothermal gradients or paleogeothermal heat flow without recognition of the inherent limitations of such data.

A number of solid-phase organic maturity measures have been used or advanced as supplements to, or substitutes for, the optical reflectance of vitrinite. Several of these are noteworthy in terms of the objectives of this review. For example, thermal alteration indices (TAI) of kerogen concentrates are frequently used in conjunction with vitrinite reflectance data (Staplin, 1969). Because of various preparation, analytical, and intrinsic problems, this technique yields data that are considerably less precise than vitrinite reflectance and correlate only in broad and general terms with both hydrocarbon-generation parameters and maturity (Powell et al., 1982; Petersen and Hickey, 1985). "As is the case with nearly all organic geochemical techniques, reliable interpretations can be made if the limitations of the method are considered and the results are cross-correlated with other methods" (Petersen and Hickey, 1985). Solid state ^{13}C cross-polarization/magic-angle spinning (CP/MAS) nuclear magnetic resonance (NMR) spectroscopy has been advanced as a nondestructive technique for identification of coal macerals and determination of coal rank or potential oil yields from oil shale (Miknis et al., 1981; Zilm et al., 1981; Dereppe et al., 1983; Hagaman et al., 1984). Although this technique offers promise, especially for developing fine relative maturity graduations in the very low maturity region (i.e., TTI [Waples, 1980] values between 1 and 20), absolute maturity cannot be judged without independent knowledge of kerogen composition. Rock-Eval pyrolysis data (Espitalié et al., 1977) are sometimes used to monitor or gauge the level of maturity of source rock kerogen despite numerous limitations and complications (Peters, 1986). At best, Rock-Eval data provide a measure of thermal maturity that is even less satisfactory and more problematic than measurements of vitrinite reflectance or thermal alteration index, especially at very low and very high hydrogen index values (Monthioux et al., 1985). They should be used to gauge thermal maturity only where more suitable parameters are not available.

Molecular Organic Indicators

A large and rapidly growing number of organic molecular indicators of thermal maturity have been suggested and used, mainly because of analytical instrumental advances, but partly because of the uncertainty or ambiguity about interpretation of bulk solid-phase organic maturity parameters. This procedure is based on analyses of crude oils and/or source rock extracts. Most or all of the organic molecular indicators have notable limitations, apart from the fundamental difficulty of telling indigenous from migrated extractable organic matter, especially where the two may be mixed. Nevertheless, even with limitations and problems, thermal maturity parameters based on crude oils and extracts play an important and increasingly significant role in discerning subsurface thermal conditions and history, particularly when used in conjunction with other methods.

One of the most firmly established molecular chemical indicators is the carbon preference index (CPI) from sequences of uniform and suitable (kerogen Type II-III) organic matter type (Bray and Evans, 1961; Allan et al., 1977; Tissot and Welte, 1978, pp. 460–463; Durand and Oudin, 1980). Even this parameter fails if the rocks contain hydrogen-rich kerogen from strongly reducing depositional environments or have been penetrated by migrating oil. A low CPI may not necessarily imply thermal maturity even though a high CPI generally indicates immaturity.

Early recognition of the possibility of using the equilibrium constants of an isomerization reaction involving hydrocarbon compounds occurred in Russia (Savina and Volikovskij, 1968). Shortly afterward, reductions in the ratios of benzothiophene to dibenzothiophene were attributed to increasing thermal exposure of high-sulfur crude oils (Ho et al., 1974), while tetracyclic aromatic hydrocarbon distributions in extracts from Early Jurassic shales of the Paris Basin were used as thermal maturity indicators (Tissot et al., 1974). In rapid succession, a large number of more-or-less useful aromatic and aliphatic molecular reactions have been investigated and proposed as indicators of thermal maturity (Seifert and Moldowan, 1978, 1980; Mackenzie et al., 1980, 1981; Mackenzie and Maxwell, 1981; Cornford et al., 1983; Suzuki, 1984; Alexander et al., 1985). Extensions and refinements of this work are continuing while applications to basin evaluation have already begun (Mackenzie and McKenzie, 1983; Beaumont et al., 1985). At this point it may be well to remember that:

> No one technique can be applied universally, and no simple equivalence can be drawn up between the various techniques. Vitrinite reflectance is the most widely used technique, while the rapidly developing field of molecular ratios has the important advantage that measurements can be made on both source-rocks and reservoired oil, thus allowing the establishment of a direct genetic link. (Cornford, 1984, p. 175)

Mineral Indicators

In contrast with most or all reactions involving organic chemicals or chemical mixtures, a few reactions or changes involving minerals are simple and relatively well understood. Fluid inclusions trapped in diagenetic overgrowths, intergranular pore-filling phases, or fracture-lining minerals furnish opportunities for the determination of maximum burial temperatures and fluid pressures (Mullis, 1979; Visser, 1982; and others). The fluids, trapped while migrating, record properties that make them akin in some cases to maximum recording thermometers.

Sufficient burial heating of detrital apatite crystals in coarse-grained clastic strata results in annealing of accumulated spontaneous fission tracks (Naeser, 1979, 1981; Briggs et al., 1981; Gleadow and Duddy, 1981; Gleadow et al., 1983). The kinetics of this relatively simple annealing process are understood well enough that major features of burial thermal history can be deciphered from suitable well profiles. The range of temperatures through which annealing of apatite fission tracks takes place spans the oil generation window.

Careful application of oxygen isotopic analyses to neoformed diagenetic silicate minerals in deeply buried shales and sandstones has been shown to provide a basis for estimating burial temperatures (Yeh and Savin, 1977; Savin and Lee, 1984). Underlying this approach is the presumption that diagenetic silicates crystallizing in low temperature subsurface environments do so in oxygen isotopic equilibrium with ambient water and that postcrystallization isotopic exchange is negligible. This approach focuses attention on the importance of the role of aqueous fluids in bringing about mineralogical changes as burial heating progresses, as does the approach based on fluid inclusion analyses.

Analysis of the $^{40}Ar/^{39}Ar$ age spectrum of detrital microcline was introduced by Harrison and Bé (1983) as a method to constrain the timing and thermal intensity of heating in sediments. For comparable durations of heating, Ar loss from detrital microline occurs at higher temperatures than fission-track annealing in apatite and is thus somewhat less sensitive to low-temperature events.

The mineral laumontite, which occurs in some mineralogically immature clastic strata, furnishes the last of the relatively straightforward inorganic means discussed here for gaining information about paleogeothermal conditions in sedimentary basins. Empirical evidence in actively subsiding basins and a few geothermal systems has shown that laumontite crystallizes according to strict subsurface temperature and fluid pressure (P_f) conditions through a temperature range from 32°C to >200°C (McCulloh et al., 1978, pp. 24–32; McCulloh et al., 1981). Crystallization occurs only where permeable rocks are invaded by moving (rising) pore waters of appropriate ionic strengths and compositions and at suitable conditions of T and P_f. If P_f for the time of crystallization can be estimated, T is fixed. Once again, the importance of the role of aqueous fluids and water-rock interactions is manifest.

Clay mineral assemblages in general and the smectite to illite transformation in particular have been used by many investigators as indicators of thermal history (Perry and Hower, 1970; Hoffman and Hower, 1979). Some authors emphasize the importance of detrital mineral composition and depositional environment as the determinant of final diagenetic products (Heling, 1969, 1974; Aoyagi et al., 1987). Others emphasize the role of temperature during burial as an overriding determinant (Pollastro and Barker, 1986). Both temperature and pressure are evidently important (Velde, 1984), together with pore-fluid chemistry, which in part reflects depositional environment. The degree of influence of these variables is still under investigation (Jennings and Thompson, 1986) and will ultimately dictate the extent to which profiles of clay mineral sequences are reliable indicators of thermal conditions or paleogeothermal history, either alone or when calibrated locally in terms of other thermal indicators.

Mineral-Water Interactions

The important role of mineral-fluid interactions during burial history of sedimentary sequences has been repeatedly mentioned, directly or by inference, in preceding paragraphs. It is fitting, therefore, to focus briefly on the "chemical geothermometers" based on concentrations of various dissolved constituents of subsurface waters collected from hot springs and wells, especially in geothermal systems (Fournier, 1981; Henley et al., 1984; Kharaka and Mariner, this volume). Various versions of such chemical geothermometers have been used with considerable success to estimate subsurface reservoir temperatures from chemical analyses of waters issuing from depths where temperatures are in the range from 30°C to about 350°C. The fact that at reservoir temperatures higher than $\approx 70°C$ such estimates seem to agree with observed temperatures within $\pm 10°C$ testifies eloquently to the dynamism and rapidity of water-rock interactions in the depth range of interest to petroleum geologists. Perhaps this is why such a close relationship seems to exist, both in carbonate rock sequences and in many clastic rock sequences, between porosity-depth gradients and temperature gradients, subsurface temperatures, or thermal maturity (time-temperature index or its integral) (Maxwell, 1964; McCulloh et al., 1978; Schmoker, 1984). If so, there may be some expectation that forward modeling of basin structural-sedimentological-thermal evolution might eventually progress to the level of providing a capability to predict expectable maximum reservoir porosities for different depths and locations about a depocenter.

Predictive Calculations of Thermal Maturity

The need for information in advance of drilling has been a powerful incentive to develop predictive approaches to assessing thermal conditions and histories in basins where the principal geological constraints derive mainly or only from reflection seismic data and regional geological analogies. Predictive modeling of thermal evolution of sedimentary basins and their contents proceeds from first principles and requires elaborate numerical simulations. Evolution of basin geometry and lithology, heat flow, thermal conductivity, surface temperatures, and time-stratigraphy are necessary input elements. Integration of computed time-temperature histories with some more-or-less elaborate kinetic model for petroleum generation from basin-specific source rocks (Sweeney et al., 1987) allows computation of thermal maturity, and oil and gas generation, as a function of time and position in the basin. Initial efforts to develop such models necessarily occur where results of exploration drilling provide "ground truth." The degree to which model results match the observations should be a test of the adequacy of a given model or modeling approach.

Forward (inductive) modeling of basin structural and thermal evolution was first clearly conceptualized by Sleep (1971) and independently was developed more fully in concept by Fischer (1975). A simple mathematical theory was then advanced for predictively relating the causative subsidence, basement (crustal) heat flow, and the evolution of the gravity field associated with crustal extension and thinning during rifting (McKenzie, 1978). Such theories were next employed to study the structural and thermal evolution of several kinds of sedimentary basins (Royden et al., 1980; McKenzie, 1981; and others). One of the latest chapters in this ongoing discourse (Barton, 1984, p. 227) concludes that "unfortunately, the thermal predictions made by the [crustal stretching] model cannot be tested directly using existing borehole temperature measurement technology" because of the presumed poor quality of the data obtained during oil-field operations.

Deductive deterministic modeling of the thermal history of sedimentary basins began at least as early as 1959 (Grossling, 1959). Models built around measured geothermal gradients (and/or heat flow and thermal conductivities) and observed time-stratigraphic and lithological and structural configurations were developed with successively greater refinement and complication to calculate thermal history and petroleum generation (Tissot, 1973; Sharp and Domenico, 1976; Yükler and Welte, 1980; Goff, 1983; Yükler and Kokesh, 1984; Heum et al., 1986). Vitrinite reflectance data provide the maturation yardstick for most of these efforts, which are proceeding with greater and greater elaboration (Eggen, 1984; Leadholm et al., 1985). Of substantial interest is the view expressed by Eggen (1984) that present-day high-temperature gradients (and corresponding high heat-flow values) on the flanks of the Viking Graben are excessive when used to calculate vitrinite reflectance gradients for comparison with those observed in deep wells. Eggen concludes that the currently high heat-flow regions are relatively recent developments and that there is a relationship between hydrocarbon accumulations and local thermal anomalies. He explicitly rules out deep crustal sources for the excess heat observed and thereby sounds a cautionary note for all who would model basin thermal history.

Coupling of inductive forward modeling procedures with deductive deterministic (backward) modeling efforts is rare. Important examples are the modeling of the thermal and maturation evolution of the Otway Basin, Australia (Middleton and Falvey, 1983) and of Jurassic source rocks in part of the northern North Sea (Goff, 1983). Vitrinite reflectance data are used in both areas as the yardstick for judging the validity of modeling results even though Middleton and Falvey recognize explicitly that " . . . organic diagenesis modeling is relatively insensitive to precise details of thermal history" (p. 271).

Postscript

The following chapters in this volume span the spectrum of the principles, techniques, approaches, and applications of importance to the study of the thermal history of sedimentary basins. Topics range from the significance of the thermal conductivity of sedimentary rocks, through discussions of the various organic and inorganic thermal maturity indicators, to proposed modeling procedures. We anticipate that several of the chapters in this volume may be somewhat controversial, but we include them in the hope that they will help stimulate discussion and further research into the thermal history of sedimentary basins.

References

Alexander, R., Kagi, R.I., Rowland, S.J., Sheppard, P.N., and Chirila, T.V. 1985. The effects of thermal maturity on distributions of dimethylnaphthalenes and trimethylnaphthalenes in some ancient sediments and petroleums. Geochimica et Cosmochimica Acta 49:385–395.

Allan, J., Bjorøy, M., and Douglas, A.G. 1977. Variation in the content and distribution of high molecular weight hydrocarbons in a series of coal macerals of different rank. In: Campos, R., and Goni, J. (eds.): Advances in Organic Geochemistry 1975. Madrid, Enadimsa, pp. 633–655.

Aoyagi, K., Chilingarian, G.V., and Yen, T.F. 1987. Clay mineral diagenesis in argillaceous sediments and rocks. Energy Sources 9:99–109.

Barton, P. 1984. Crustal stretching in the North Sea: Implications for thermal history. In: Durand, B. (ed.): Thermal Phenomena in Sedimentary Basins. Paris, Editions Technip, pp. 227–233.

Beaumont, C., Boutilier, R., Mackenzie, A.S., and Rullkötter, J. 1985. Isomerization and aromatization of hydrocarbons and the paleothermometry and burial history of Alberta foreland basin. American Association of Petroleum Geologists Bulletin 69:546–566.

Bethke, C.M. 1985. A numerical model of compaction-driven groundwater flow and heat transfer and its application to the paleohydrology of intracratonic sedimentary basins. Journal of Geophysical Research 90: 6817–6828.

Blackwell, D.D. 1983. Heat flow in the northern Basin and Range province. Geothermal Resources Council Special Report 13, pp. 81–92.

Bostick, N.H., Cashman, S.M., McCulloh, T.H., and Waddell, C.T. 1978. Gradients of vitrinite reflectance and present temperature in the Los Angeles and Ventura Basins, California. In: Oltz, D.F. (ed.): A Symposium in Geochemistry: Low Temperature Metamorphism of Kerogen and Clay Minerals. Los Angeles, Pacific Section, Society of Economic Paleontologists and Mineralogists, pp. 65–96.

Bray, E.E., and Evans, E.D. 1961. Distribution of n-paraffins as a clue to recognition of source beds. Geochimica et Cosmochimica Acta 22:2–15.

Briggs, N.D., Naeser, C.W., and McCulloh, T.H. 1981. Thermal history of sedimentary basins by fission-track dating (abst.). Nuclear Tracks 5:235–237.

Catala, G. 1984. Technical aspects of measurements of temperatures in boreholes: Some aspects of their interpretation. In: Durand, B. (ed.): Thermal Phenomena in Sedimentary Basins. Paris, Editions Technip, pp. 37–46.

Cornford, C. 1984. Source rocks and hydrocarbons of the North Sea. In: Glennie, K.W. (ed.): Introduction to the Petroleum Geology of the North Sea. London, Blackwell Scientific Publications, pp. 171–204.

Cornford, C., Morrow, J.A., Turrington, A., Miles, J.A., and Brooks, J. 1983. Some geological controls on oil composition in the U.K. North Sea. In: Brooks, J. (ed.): Petroleum Geochemistry and Exploration of Europe. London, The Geological Society, pp. 175–201.

Dereppe, J.-M., Boudou, J.-P., Moreaux, C., and Durand, B. 1983. Structural evolution of a sedimentologically homogeneous coal series as a function of carbon content by solid state ^{13}C n.m.r. Fuel 62:575–579.

Doebl, F., and Teichmüller, R. 1979. Zur Geologie und heutiger Geothermik im mittleren Oberrhein-Graben. In: Inkohlung und Geothermik. Fortschritte in der Geologie von Rheinland und Westfalen 27, pp. 1–17.

Dow, W.G. 1977. Kerogen studies and geological interpretation. Journal of Geochemical Exploration 7: 79–99.

Dow, W.G. 1978. Petroleum source beds on continental slopes and rises. American Association of Petroleum Geologists Bulletin 62:1584–1606.

Durand, B., and Oudin, J.L. 1980. Examples of migration of hydrocarbons in a deltaic series, the Mahakam Delta, Kalimantan, Indonesia. Proceedings of the Tenth World Petroleum Congress, Bucharest, 1979, vol. 2, pp. 3–11.

Eckstein, Y. 1979. Heat flow and the hydrologic cycle: Examples from Israel. In: Cermák, V., and Rybach, L. (eds.): Terrestrial Heat Flow in Europe. New York, Springer-Verlag, pp. 88–97.

Eggen, S. 1984. Modelling of subsidence, hydrocarbon generation and heat transport in the Norwegian North Sea. In: Durand, B. (ed.): Thermal Phenomena in Sedimentary Basins. Paris, Editions Technip, pp. 271–283.

Elders, W.A., Hoagland, J.R., and Williams, A.E. 1980. Hydrothermal alteration as an indicator of temperature and flow regime in the Cerro Prieto geothermal field of Baja California. Geothermal Resources Council Transactions 4:121–124.

Espitalié, J., Madec, M., Tissot, B.P., Mennig, J.J., and Leplat, P. 1977. Source rock characterization method for petroleum exploration. Proceedings of the Ninth

Annual Offshore Technology Conference, Houston, TX, vol. 3, pp. 439–448.

Fischer, A.G. 1975. Origin and growth of basins. In: Fischer, A.G., and Judson, S. (eds.): Petroleum and Global Tectonics. Princeton, NJ, Princeton University Press, pp. 47–78.

Fournier, R.O. 1981. Application of water chemistry to geothermal exploration and reservoir engineering. In: Rybach, L., and Muffler, L.J.P. (eds.): Geothermal Systems: Principles and Case Histories. New York, Wiley, pp. 109–143.

Gleadow, A.J.W., and Duddy, I.R. 1981. A natural long-term track annealing experiment for apatite. Nuclear Tracks 5:169–174.

Gleadow, A.J.W., Duddy, I.R., and Lovering, J.F. 1983. Fission track analysis: A new tool for the evaluation of thermal histories and hydrocarbon potential. Australian Petroleum Exploration Association Journal 23:93–102.

Goff, J.C. 1983. Hydrocarbon generation and migration from Jurassic source rocks in the E Shetland Basin and Viking Graben of the northern North Sea. Journal of the Geological Society of London 140:445–474.

Grossling, B.F. 1959. Temperature variations due to the formation of a geosyncline. Geological Society of America Bulletin 70:1253–1282.

Hagaman, E.H., Schell, F.M., and Cronauer, D.C. 1984. Oil-shale analysis by CP/MAS-^{13}C n.m.r. spectroscopy. Fuel 63:915–919.

Halfman, S.E., Lippmann, M.J., Zelwer, R., and Howard, J.H. 1984. A geological interpretation of the geothermal fluid movement in the Cerro Prieto Field, Baja California, Mexico. American Association of Petroleum Geologists Bulletin 68:18–30.

Harrison, T.M., and Bé, K. 1983. $^{40}Ar/^{39}Ar$ age spectrum analysis of detrital microclines from the southern San Joaquin Basin, California: An approach to determining the thermal evolution of sedimentary basins. Earth and Planetary Science Letters 64:244–256.

Heling, D. 1969. Relationship between initial porosity of Tertiary argillaceous sediments and paleosalinity in the Rheintalgraben (SW Germany). Journal of Sedimentary Petrology 39:246–254.

Heling, D. 1974. Diagenetic alteration of smectite in argillaceous sediments of the Rhinegraben (SW Germany). Sedimentology 21:463–472.

Henley, R.W., Truesdell, A.H., and Barton, P.B., Jr. 1984. Fluid-mineral equilibria in hydrothermal systems. Reviews in Economic Geology: Vol. 1. Chelsea, MI, Society of Economic Geologists, 267 pp.

Héroux, Y., Chagnon, A., and Bertrand, R. 1979. Compilation and correlation of major thermal maturation indicators. American Association of Petroleum Geologists Bulletin 63:2128–2144.

Heum, O.R., Dalland, A., and Meisingset, K.K. 1986. Habitat of hydrocarbons at Haltenbanken (PVT-modelling as a predictive tool in hydrocarbon exploration). In Spencer, A.M., et al. (eds.): Habitat of Hydrocarbons on the Norwegian Continental Shelf. London, Graham and Trotman, pp. 259–274.

Ho, T.Y., Rogers, M.A., Drushel, H.V., and Koons, C.B. 1974. Evolution of sulfur compounds in crude oils. American Association of Petroleum Geologists Bulletin 58:2338–2348.

Hoffman, J., and Hower, J. 1979. Clay mineral assemblages as low grade metamorphic indicators: Application to the thrust faulted disturbed belt of Montana, U.S.A. In: Scholle, P.A., and Schluger, P.K. (eds.): Aspects of Diagenesis. Society of Economic Paleontologists and Mineralogists Special Publication 26, pp. 55–79.

Hood, A., and Castaño, J.R. 1974. Organic metamorphism: Its relationship to petroleum generation and application to studies of authigenic minerals. United Nations Economic Commission for Asia and Far East, Committee for Coordination of Joint Prospecting for Mineral Resources in Asian Offshore Areas, Technical Bulletin 8:85–118.

Hood, A., Gutjahr, C.C.M., and Heacock, R.L. 1975. Organic metamorphism and the generation of petroleum. American Association of Petroleum Geologists Bulletin 59:986–996.

Hower, J. 1981. Shale diagenesis. In: Longstaffe, F.J. (ed.): Clays and the Resource Geologist. Mineralogical Society of Canada Short Course Handbook 7, pp. 60–80.

Hutton, A.C., and Cook, A.C. 1980. Influence of alginite on the reflectance of vitrinite from Joadja, N.S.W., and some other coals and shales containing alginite. Fuel 59:711–714.

Jaupart, C. 1984. On the thermal state of the earth. In: Durand, B. (ed.): Thermal Phenomena in Sedimentary Basins. Paris, Editions Technip, pp. 5–9.

Jennings, S., and Thompson, G.R. 1986. Diagenesis of Plio-Pleistocene sediments of the Colorado River delta, southern California. Journal of Sedimentary Petrology 56:89–98.

Kalkreuth, W. 1982. Rank and petrographic composition of selected Jurassic-Lower Cretaceous coals of British Columbia, Canada. Canadian Petroleum Geology Bulletin 30:112–139.

Kalkreuth, W., and Macauley, G. 1984. Organic petrology of selected oil shale samples from the lower Carboniferous Albert Formation, New Brunswick, Canada. Canadian Petroleum Geology Bulletin 32:38–51.

Kehle, R.O., Schoeppel, R.J., and Deford, R.K. 1970. The AAPG geothermal survey of North America. Geothermics (Special Issue): 358-367.

Leadholm, R.H., Ho, T.Y., and Sahai, S.K. 1985. Heat flow, geothermal gradients and maturation modelling on the Norwegian continental shelf using computer methods. In: Thomas, B.M., et al. (eds.): Petroleum Geochemistry in Exploration of the Norwegian Shelf. London, Graham and Trotman, pp. 131–143.

Lister, C.R.B. 1980. Heat flow and hydrothermal circulation. Annual Reviews of Earth and Planetary Science 8:95–117.

Mackenzie, A.S., Hoffman, C.F., and Maxwell, J.R. 1981. Molecular parameters of maturation in the Toarcian shales, Paris Basin, France: III. Changes in aromatic steroid hydrocarbons. Geochimica et Cosmochimica Acta 45:1345–1355.

Mackenzie, A.S., and Maxwell, J.R. 1981. Assessment of thermal maturation in sedimentary rocks by molecular measurements. In: Brooks, J. (ed.): Organic Maturation Studies and Fossil Fuel Exploration. New York, Academic Press, pp. 239–254.

Mackenzie, A.S., and McKenzie, D.P. 1983. Isomerization and aromatization of hydrocarbons in sedimentary basins formed by extension. Geological Magazine 120:417–528.

Mackenzie, A.S., Patience, R.L., Maxwell, J.R., Vandenbroucke, M., and Durand, B. 1980. Molecular parameters of maturation in the Toarcian shales, Paris Basin, France: I. Changes in configuration of acyclic isoprenoid alkanes, steranes and triterpanes. Geochimica et Cosmochimica Acta 44:1709–1721.

Maxwell, J.C. 1964. Influence of depth, temperature, and geologic age on porosity of quartz sandstone. American Association of Petroleum Geologists Bulletin 48:697–709.

McCulloh, T.H. 1979. Implications for petroleum appraisal. In: Cook, H.E. (ed.): Geologic Studies of the Point Conception Deep Stratigraphic Test Well OCS-CAL 78-164 No. 1, Outer Continental Shelf, Southern California, United States. U.S. Geological Survey Open-File Report 79-1218, pp. 26–42.

McCulloh, T.H., Cashman, S.M., and Stewart, R.J. 1978. Diagenetic baselines for interpretive reconstructions of maximum burial depths and paleotemperatures in clastic sedimentary rocks. In: Oltz, D.F. (ed.): A Symposium in Geochemistry: Low Temperature Metamorphism of Kerogen and Clay Minerals. Los Angeles, Pacific Section, Society of Economic Paleontologists and Mineralogists, pp. 18–46.

McCulloh, T.H., Frizzell, V.A., Jr., Stewart, R.J., and Barnes, I. 1981. Precipitation of laumontite with quartz, thenardite, and gypsum at Sespe Hot Springs, western Transverse Ranges, California. Clays and Clay Minerals 29:353–364.

McKenzie, D.P. 1978. Some remarks on the development of sedimentary basins. Earth and Planetary Science Letters 40:25–32.

McKenzie, D.P. 1981. The variation of temperature with time and hydrocarbon maturation in sedimentary basins formed by extension. Earth and Planetary Science Letters 55:87–98.

Meinhold, R. 1967. Über den Zusammenhang geothermischer, hydrodynamischer und geochemischer Anomalien und deren Bedeutung für die Klärung der Entstehung von Erdöllagerstätten. Zeitschrift für angewandte Geologie 14:233–240.

Middleton, M.F., and Falvey, D.A. 1983. Maturation modeling in Otway Basin, Australia. American Association of Petroleum Geologists Bulletin 67:271–279.

Miknis, F.P., and Smith, J.W. 1984. An NMR survey of United States oil shales. Organic Geochemistry 5:193–201.

Miknis, F.P., Sullivan, M., Bartuska, V.J., and Maciel, G.E. 1981. Cross-polarization magic-angle spinning ^{13}C NMR spectra of coals of varying rank. Organic Geochemistry 3:19–28.

Monthioux, M., Landais, P., and Monin, J.-C. 1985. Comparison between natural and artificial maturation series of humic coals from the Mahakam delta, Indonesia. Organic Geochemistry 8:275–292.

Mullis, J. 1979. The system methane-water as a geologic thermometer and barometer from the external part of the Central Alps. Bulletin de Minéralogie 102:526–536.

Naeser, C.W. 1979. Thermal history of sedimentary basins: Fission-track dating of subsurface rocks. In: Scholle, P.A., and Schluger, P.R. (eds.): Aspects of Diagenesis. Society of Economic Paleontologists and Mineralogists Special Publication 26, pp. 109–112.

Naeser, C.W. 1981. The fading of fission tracks in the geologic environment: Data from deep drill holes (abst.). Nuclear Tracks 5:248–250.

Newman, J., and Newman, N.A. 1982. Reflectance anomalies in Pike River coals: Evidence of variability in vitrinite type, with implications for maturation studies and "Suggate rank". New Zealand Journal of Geology and Geophysics 25:233–243.

Perrier, J., and Raiga-Clemenceau, J. 1984. Temperature measurements in boreholes. In: Durand, B. (ed.): Thermal Phenomena in Sedimentary Basins. Paris, Editions Technip, pp. 47–54.

Perry, E.A., and Hower, J. 1970. Burial diagenesis in Gulf Coast peletic sediments. Clays and Clay Minerals 18:165–177.

Peters, K.E. 1986. Guidelines for evaluating petroleum source rocks using programmed pyrolysis. American

Association of Petroleum Geologists Bulletin 70:318–329.

Petersen, N.F., and Hickey, P.J. 1985. Visual kerogen assessment of thermal history (abst.). American Association of Petroleum Geologists Bulletin 69:296.

Pollastro, R.M., and Barker, C.E. 1986. Application of clay-mineral, vitrinite reflectance, and fluid inclusion studies to the thermal and burial history of the Pinedale anticline, Green River basin, Wyoming. In: Gautier, D.L. (ed.): Roles of Organic Matter in Sediment Diagenesis. Society of Economic Paleontologists and Mineralogists Special Publication 38, pp. 73–83.

Powell, T.G., Creaney, S., and Snowdon, L.R. 1982. Limitations of use of organic petrographic techniques for identification of petroleum source rocks. American Association of Petroleum Geologists Bulletin 66:430–435.

Price, L.C., and Barker, C.E. 1984. Suppression of vitrinite reflectance in amorphous rich kerogen: A major unrecognized problem. Journal of Petroleum Geology 8:59–84.

Rabinowicz, M., Dandurand, J.-L., Jaknbowski, M., Schott, J., and Cassan, J.-P. 1985. Convection in a North Sea oil reservoir: Inferences on diagenesis and hydrocarbon migration. Earth and Planetary Science Letters 74:387–404.

Roedder, E. 1984. Fluid inclusions. Reviews in Mineralogy: Vol. 12. Washington, DC, Mineralogical Society of America, 644 pp.

Royden, L., Sclater, J.G., and Von Herzen, R.P. 1980. Continental margin subsidence and heat flow: Important parameters in formation of petroleum hydrocarbons. American Association of Petroleum Geologists Bulletin 64:173–187.

Savin, S.M., and Lee, M. 1984. Estimation of subsurface temperatures from oxygen isotope ratios of minerals. In: Durand, B. (ed.): Thermal Phenomena in Sedimentary Basins. Paris, Editions Technip, pp. 65–70.

Savina, J.D., and Volikovskij, A.S. 1968. Possibilités d'utilisation des constantes d'équilibre des réactions d'isomérisation des hydrocarbures pour évaluer les températures de formation du pétrole et du gaz. Gazovaya Promyshlennost 13(6):8–11.

Schmoker, J.W. 1984. Empirical relation between carbonate porosity and thermal maturity: An approach to regional porosity prediction. American Association of Petroleum Geologists Bulletin 68:1697–1703.

Seifert, W.K., and Moldowan, J.M. 1978. Applications of steranes, terpanes and monoaromatics to the maturation, migration and source of crude oils. Geochimica et Cosmochimica Acta 42:77–95.

Seifert, W.K., and Moldowan, J.M. 1980. The effect of thermal stress on source-rock quality as measured by hopane stereochemistry. In: Douglas, A.G., and Maxwell, J.R. (eds.): Advances in Organic Geochemistry 1979. Oxford, Pergamon Press, pp. 229–237.

Sharp, J.M., Jr., and Domenico, P.A. 1976. Energy transport in thick sequences of compacting sediments. Geological Society of America Bulletin 87:390–400.

Sleep, N.H. 1971. Thermal effects of the formation of Atlantic continental margins by continental breakup. Geophysical Journal of the Royal Astronomical Society 24:325–350.

Staplin, F.L. 1969. Sedimentary organic matter, organic metamorphism, and oil and gas occurrence. Canadian Petroleum Geology Bulletin 17:47–66.

Suzuki, N. 1984. Estimation of maximum temperature of mudstone by two kinetic parameters; epimerization of sterane and hopane. Geochimica et Cosmochimica Acta 48:2273–2282.

Sweeney, J.J., Burnham, A.K., and Braun, R.L. 1987. A model of hydrocarbon generation from Type I kerogen: Application to Uinta basin, Utah. American Association of Petroleum Geologists Bulletin 71:967–985.

Tanaka, T., and Sato, K. 1977. Estimation of subsurface temperature in oil and gas producing areas, northeast Japan. Japanese Association of Petroleum Technologists Journal 42(4):229–237.

Teichmüller, M. 1979. Die Diagenese der kohligen Substanzen in dem Gesteinen des Tertiärs und Mesozoihums des mittleren Oberrhein-Grabens. Fortschritte in der Geologie von Rheinland und Westfalen 27:19–49.

Thamrin, M. 1985. An investigation of the relationship between the geology of sedimentary basins and heat flow density. Tectonophysics 121:45–62.

Tissot, B.P. 1973. Vers l'évaluation quantitative du pétrole formé dans les bassins sédimentaires. Revue Association Francaise des Techniciens du Pétrole 222:27–31.

Tissot, B.P., Califet-Debyser, Y., Deroo, G., and Oudin, J.L. 1971. Origin and evolution of hydrocarbons in early Toarcian shales, Paris Basin, France. American Association of Petroleum Geologists Bulletin 55: 2177–2193.

Tissot, B.P., Durand, B., Espitalié, J., and Combaz, A. 1974. Influence of the nature and diagenesis of organic matter in the formation of petroleum. American Association of Petroleum Geologists Bulletin 58:499–506.

Tissot, B.P., and Welte, D.H. 1978. Petroleum Formation and Occurrence. New York, Springer-Verlag, 538 pp.

Velde, B. 1984. Transformations of clay minerals. In: Durand, B. (ed.): Thermal Phenomena in Sedimentary Basins. Paris, Editions Technip, pp. 111–116.

Verhoogen, J. 1980. Energetics of the Earth. Washington, DC, National Academy of Sciences, 139 pp.

Visser, W. 1982. Maximum diagenetic temperature in a petroleum source-rock from Venezuela by fluid inclusion geothermometry. Chemical Geology 37:95–101.

Walker, A.L., McCulloh, T.H., Petersen, N.F., and Stewart, R.J. 1983. Anomalously low reflectance of vitrinite, in comparison with other petroleum source-rock maturation indices, from the Miocene Modelo Formation in the Los Angeles Basin, California. In: Isaacs, C.M., and Garrison, R.E. (eds.): Petroleum Generation and Occurrence in the Monterey Formation, California. Los Angeles, Pacific Section, Society of Economic Paleontologists and Mineralogists, pp. 185–190.

Waples, D.W. 1980. Time and temperature in petroleum formation: Application of Lopatin's method to petroleum exploration. American Association of Petroleum Geologists Bulletin 64:916–926.

Wenger, L.M., and Baker, D.R. 1986. Variations in organic geochemistry of anoxic-oxic black shale-carbonate sequences in the Pennsylvanian of the Mid-continent, U.S.A. Organic Chemistry 10:85–92.

Yeh, H.-W., and Savin, S.M. 1977. Mechanism of burial metamorphism of argillaceous sediments: 3. O-isotope evidence. Geological Society of America Bulletin 88:1321–1330.

Yükler, M.A., and Kokesh, F. 1984. A review of models used in petroleum resource estimation and organic geochemistry. In: Brooks, J., and Welte, D. (eds.): Advances in Petroleum Geochemistry: Vol. 1. London, Academic Press, pp. 69–113.

Yükler, M.A., and Welte, D. 1980. A three-dimensional deterministic dynamic model to determine geologic history and hydrocarbon generation, migration, and accumulation in a sedimentary basin. In: Fossil Fuels. Paris, Editions Technip, pp. 267–285.

Zilm, K.W., Pugmire, R.J., Larter, S.R., Allan, J., and Grant, D.M. 1981. Carbon-13 CP/MAS spectroscopy of coal macerals. Fuel 60:717–722.

2
Thermal Conductivity of Sedimentary Rocks: Measurement and Significance

David D. Blackwell and John L. Steele

Abstract

The thermal histories of sedimentary basins and their effect on organic maturation are topics of active study. The focus of these studies is on large-scale thermal events, such as an initial rifting event, that affect temperatures in a basin. Events of less global significance, however, are more important to the internal temperatures of a sedimentary basin. Such effects as internal thermal events (magma intrusion, diaparism), contrasts in heat production of U, Th, and K in the sediments and underlying basement, large- and small-scale flow of fluid, and thermal conductivity variations, both vertical and horizontal, can raise or lower temperatures much more than lithospheric-scale events. The nature and effect of such thermal effects are briefly discussed in this chapter. The most basic effect, but one of the least well known, is the thermal conductivity of the rocks in the basin. If the mean thermal conductivity cannot be accurately predicted, even the most sophisticated and appropriate modeling techniques for analyzing thermal histories and organic maturation levels may fail when applied to real basins. Temperature variations related to thermal conductivity variations are illustrated using precision temperature-gradient logs from various sedimentary basin settings. Different ways of determining the thermal conductivity of sedimentary rocks are discussed, including laboratory measurements on cuttings and core samples, in situ direct measurements, inference from well log measurements of travel time, gamma-ray activity and so forth, conversion of seismic reflection travel time to thermal resistance, and inversion of detailed temperature logs. Laboratory measurements are in some cases unreliable, especially for shales, one of the most abundant sedimentary lithologies. Actual shale thermal conductivities appear to be 25 to 50% lower than the literature values and do not appear to vary as a function of compaction in the expected way. Thus, some sort of in situ technique of thermal conductivity determination is needed. The use of precision temperature logs with spot sampling for laboratory comparison is favored and several examples of this technique from the Midcontinent, Gulf Coast, and Rocky Mountains are illustrated. The detailed temperature log from the Gulf Coast demonstrates high gradients in shale sections at 2 km depth because of the low thermal conductivity. The thermal properties of shale have implications for interpretation of the thermal effects of geopressuring.

Introduction

This chapter discusses thermal properties in sedimentary basins and techniques of their measurement. The use of precision temperature-gradient logs in holes where conductive heat flow predominates and that have reached thermal equilibrium is emphasized as the most satisfactory way to obtain in situ information. Studies of the thermal properties of sedimentary rocks in the laboratory are numerous; however, as is often the case in geology, there are sampling problems, and the completeness and applicability of the laboratory results may be questioned. It does not appear, in fact, that there is enough information available to estimate mean thermal conductivity effectively for a section of sedimentary rock. If the mean thermal conductivity cannot be accurately predicted, even the most sophisticated and appropriate modeling techniques for analyzing thermal histories and maturation levels may fail when applied to real basins. The object of this chapter is to investigate the

TABLE 2.1. Controls on temperatures in sedimentary basins.

Description	Scale of effect
Initial thermal event	10–20%
Transient external thermal events	5–10%
Radioactivity of basement and sediments	20–50%
Fluid flow	50–100%
Thermal properties of basin	50–100%
Internal thermal events (intrusion, diapirism, etc.)	Large

thermal conductivity of sedimentary rocks and to illustrate its importance in determining the temperature structure of sedimentary basins.

In the past few years a lot of attention has been given to study of the thermal history of sedimentary basins. Most of these studies emphasize the effects of large-scale tectonic events, particularly the basin-initiating event, as a major influence on thermal history. Although these large-scale events are significant, static and dynamic phenomena within the basin will be more significant contributors to the thermal history of a particular basin. In the first part of this chapter factors that might affect temperatures in sedimentary basins are briefly reviewed. Techniques of measurement of thermal conductivity are summarized and some investigations of detailed temperature gradient and heat-flow studies carried out in carefully selected holes in sedimentary rocks are described. The information that these studies give on the thermal properties of sedimentary rocks in situ is highlighted.

General aspects of the application of temperature/heat-flow techniques and information to exploration are discussed by Roberts (1981), Gretener (1981), and Blackwell (1986). The discussion in this chapter will focus on the more specific aspect of thermal conductivity.

Controls on Temperatures

There are many factors that affect temperatures in sedimentary basins. A list of the major effects is shown in Table 2.1. The details of how each of these factors affects temperatures in a given basin depend on many variables, so this discussion must be oversimplified. In a general way the division of basins into passive margin, cratonic, or active tectonic types reflects the relative influences of the different factors listed in Table 2.1. For example, in a passive margin or a cratonic sedimentary basin, subsequent internal or external thermal events are generally not significant. In an active tectonic setting such as a rift zone or an accretionary prism, internal thermal effects may dominate the thermal history, at least locally.

The basic equation for calculating temperature in sedimentary basins depends on the thermal properties, the thermal conductivity, of the rock and the heat flow. The relationship between these quantities is

$$Q_z = K\frac{dT}{dz} \qquad (1)$$

where Q_z is the vertical component of heat flow, K is the thermal conductivity, and dT/dz is the temperature gradient. The units of thermal conductivity typically used are mcal/cms°C (10^{-3}cal/cms°C) in the CGS system, whereas in the SI system the units are $Wm^{-1}K^{-1}$. Units of heat flow are μcal/cm²s (10^{-6} cal/cm²s) in the CGS system, and mWm^{-2} in the SI system. Conversion factors between the two systems of units are 41.84 mWm^{-2} = 1 μcal/cm²s, and 0.4184 $Wm^{-1}K^{-1}$ = 1 mcal/cms°C. The average global heat flow is 60 mWm^{-2} (1.5 μcal/cm²s). Units of temperature gradient most commonly used are numerically the same in both the SI and CGS systems (mKm^{-1} and °C/km, respectively). In much of the exploration literature, temperature gradient is assigned a confusing array of units. The most useful units are °F/100 ft (1°F/100 ft = 18.2°C/km), but the units of ft/°F and °F/1,000 ft are sometimes encountered.

Equation 1 is written for vertical heat flow in a homogeneous medium. In a horizontally layered medium with heat transfer by conduction only, the thermal conductivity and the temperature gradient are inversely proportional because the heat flow does not change with depth. In this case over several lithological layers, Equation 1 can be rewritten as

$$Q_z = K_1\left(\frac{dT}{dz}\right)_1 = K_2\left(\frac{dT}{dz}\right)_2 = \cdots = K_n\left(\frac{dT}{dz}\right)_n. \qquad (2)$$

The major effects operative in sedimentary basins that affect the internal temperature either directly or by affecting the heat flow are listed in Table 2.1. Since Sleep (1971) initially pointed out that the subsidence history in many sedimentary basins is similar to the predicted subsidence due to the thermal cooling of a lithospheric slab, initial-event thermal effects have received the most attention. Such thermal models based on the papers of McKenzie (1978, 1981) and others have been extensively applied. An example of this type of analysis is the paper by Sclater and Christie (1980) on the thermal history of the central North Sea basin. As shown in Table 2.1, however, the effect of initial thermal events on temperatures in a sedimentary basin is small and the primary effect is on the synrift sediments. This lack of influence is illustrated in Figure 2.1 (after McKenzie, 1981). The base of the sediments deposited after the end of the rift activity is highlighted as is the depth range of organic maturation. It is clear that by the time the first-deposited sediments have subsided deep enough to begin thermal maturation, the initial thermal effect of formation of the sedimentary basin is insignificant, and the background heat flow and the assumed thermal conductivity in this model are the important factors in determining the basin temperatures. Thus unless the source rocks for hydrocarbon generation are within the rift sediments, the initial thermal event will have little effect on subsequent thermal maturation of hydrocarbons in the basin.

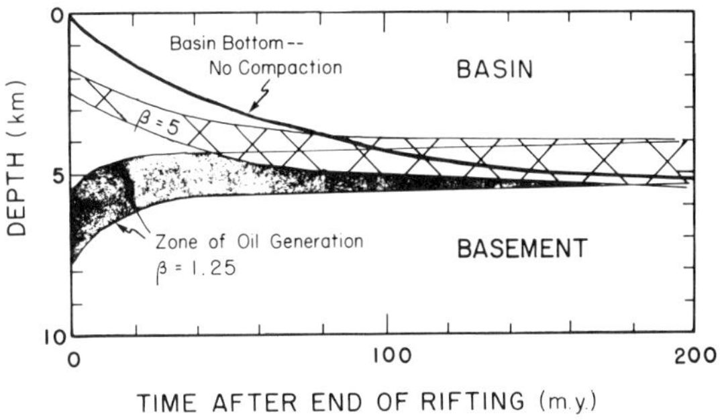

FIGURE 2.1. Zone of oil generation and basin sediment thickness as a function of time for a sedimentary basin formed by an instantaneous stretching event of modest (beta = 1.25) and major (beta = 5) magnitude. Base of sediments deposited after basin rifting is labeled "basin bottom." The figure is after McKenzie (1981, Fig. 4) and β is the stretching factor (final length divided by original length).

Once a sedimentary basin forms, sublithospheric heat-flow changes may also affect the temperatures. Short of a reinitiation of rifting, however, the effect of a transient thermal event will be quite small because the thermal event would be active at the base of the lithosphere at a depth of 50 to 100 km. The delay and attenuation associated with conduction of heat through 40 to 90 km of the lithosphere would act to limit reasonable effects to less than 5 to 10% of the background heat flow. Only if the presence of an outside event is quite obvious from the geological history of the region could the influence be larger.

In contrast, internal transient thermal events may have a huge effect, at least locally, on the thermal history of the sedimentary basin. Intrusion of a dike or sill into sedimentary rocks can locally change the temperatures and heat flow by orders of magnitude. Particularly in active tectonic regions, the location and history of transient thermal events are important in understanding the thermal evolution of sedimentary rocks. For example, in an area such as the Basin and Range province of Nevada, the same stratigraphic horizon may have a vastly different maturation history in different areas depending on its local thermal history. Studies by Poole and Claypool (1984) in the eastern Great Basin show major contrasts in the maturation of organic material of individual units throughout the region. In the Railroad Valley area of eastern Nevada, Duey (1983) has suggested that there is a correspondence between hydrocarbon accumulations and temperature and heat-flow anomalies. Considering the multifaceted nature of transient thermal events, highly site-specific studies are required.

Another internal thermal event of possible significance is diapirism. If a shale or salt diapir rises rapidly enough to depart from thermal equilibrium, heat may be transported to shallow depths. Even when the rise of the diapir ceases and thermal equilibrium is reached, the diapir may have an effect on the thermal field if its thermal properties differ from the surrounding terrain. Salt domes have higher thermal conductivity than typical sedimentary rocks; therefore, heat is refracted into the salt domes and heat-flow values are typically higher over salt domes than over the surrounding terrain. On the other hand, a shale diapir would have lower thermal conductivity and would refract heat in the surrounding medium, resulting in lower heat flow over the diapir than adjacent to it. A recent discussion of the thermal effect of salt domes is given by O'Brien and Lerche (1984).

Lateral changes in heat flow within a sedimentary basin will change the temperature in direct proportion to the heat-flow variations (see Equation 1). One cause of the heat-flow variations within a sedimentary basin is variations in the radioactivity (uranium, thorium, and potassium) of the underlying basement. Over large areas of continents, the surface heat flow is linearly related to heat production of the upper crust (Birch et al., 1968; Lachenbruch, 1968; Roy, Blackwell, et al., 1968). For example, within the continental interior of the United States, Combs and Simmons (1973) have related geographic variations in heat flow to variations in lithology (and by inference, heat production) in the basement. In particular, areas of granitic basement have higher-than-average heat flow, whereas areas of mafic basement have lower-than-average heat flow. Variations in heat flow due to variations of basement radioactivity can be large enough to cause regional variation in the maturation levels of hydrocarbons. Thus, one variable affecting the distribution of oil and gas within the continental interior of the United States might be variation in the basement heat flow. Rahman and Roy (1981) (see also Coates et al., 1983) documented extremely high radioactivity in some basement rocks in northern Illinois. If such high-heat-production rocks were present beneath a thick section of sedimentary rock, the resulting high-temperature zones could cause local increase in the maturation level of hydrocarbons. The extreme heat-flow (and consequent temperature) variations are a factor of two or three to one, but the typical scale of such effects is on the order of 25 to 50% over a large area (hundreds of square kilometers) and only 10 to 25% over a smaller area (tens of square kilometers).

Uranium, thorium, and potassium also are found within the sedimentary package and the heat production from these rocks contributes to the overall heat flow. In a thick package of sedimentary rocks, the heat-flow contribution from heat production within the sediments themselves may be considerable and lead to higher heat-flow and temperature values than would be predicted on the basis of basement radioactivity alone. Keen and Lewis (1982) documented this effect for sedimentary rocks in the eastern North America passive continental margin. The typical heat production of these rocks is sufficient to add a heat-flow increment of 5 mWm^{-2} for every kilometer of sedimentary rock. If similar values apply to the Gulf Coast where the basin thickness exceeds 10 km, the heat-flow increment due to the internal heat production could be 25 to 50% of the background heat flow. Local exothermic or endothermic chemical reactions in the sedimentary package could also change the temperatures, although the effect would be localized and in general relatively small.

In an actively subsiding sedimentary basin, the dumping of large quantities of sediment at the surface temperature into the basin has a significant effect on the overall temperature history. The thermal effects of sedimentation are discussed by Sharp and Domenico (1976). In addition to the effect of sedimentation itself, there are many dynamic effects, such as the expulsion of pore water, which come into play in a basin that is actively subsiding. The overall thermal structure is complicated and the consequences of the many possible effects as yet unknown. Much work remains to determine the overall effect of sedimentation on the thermal history of a basin. The fact that so little work has been done is surprising since one of the major sedimentary basins of hydrocarbon interest, the Gulf Coast, certainly falls in the category of an actively subsiding basin where many dynamic processes are in action.

Local and regional water flow may have a significant effect on the thermal history of many sedimentary basins. Some models of the migration and trapping of hydrocarbons are directly associated with groundwater flow in sedimentary basins (Roberts, 1980; Toth, 1980). Water movement has a significant effect on temperature because convective heat transfer is so efficient. Such effects may be on a basin-wide scale or on a local scale. Large-scale water flow has been associated with heat-flow variations in the Prairies basin in Alberta (Lam et al., 1985) and in the Denver basin in the Midcontinent (Gosnold, 1984, 1985). Theoretical studies of fluid flow in sedimentary basins have been discussed by Domenico and Palciauskas (1973), Garven and Freeze (1984), and Smith and Chapman (1983), among others. Local water flow also may have significant effects on temperatures and may be associated with local hydrocarbon entrapment (see, for example, the studies in southern Louisiana by Harrison [1980] and Gatenby [1980]). Regional-scale water flow effects may be on the order of 10 to 50% of the background temperature heat flow, whereas local effects can be much greater, particularly if fluid flow associated with high-temperature geothermal systems is included (see Barker, 1983).

The most prosaic but ultimately most basic parameter affecting subsurface temperature is the thermal conductivity of the sedimentary rocks. Since the thermal gradient is inversely proportional to the thermal conductivity for a given heat flow, there is obviously a one-to-one correlation between mean thermal conductivity and mean temperature. Thermal conductivity must have a major effect on the temperatures, in conjunction with all the other effects discussed above. These thermal conductivity effects may be due to vertical variations in a horizontal layered section or due to lateral juxtapositions of contrasts in thermal conductivity associated with folds, faults, or salt or shale diapirs (see the earlier discussion).

If the thermal conductivity values of sedimentary rocks were known or could be predicted from other information, such as lithology or porosity, there would be no need for the following discussion. One of the important conclusions reached in this chapter is that we do not, in fact, have enough information to estimate mean thermal conductivity effectively for a section of sedimentary rock. If the mean conductivity cannot be accurately predicted, even the most sophisticated and appropriate modeling techniques applied to the various phenomena discussed in the previous paragraphs are not sufficient for accurate temperature predictions. Consequently, the remainder of the chapter deals with the thermal conductivity, its measurement and causes of its variation in sedimentary rocks.

Correlation of Temperature Gradients and Thermal Conductivity

Accuracy of Temperature Measurements

In the early phases of the petroleum industry a significant amount of attention was given to temperature studies in exploration wells, and the development of accurate temperature logging devices was discussed in early papers (Hawtof, 1930; Van Ostrand, 1934, 1937). Temperatures in these early studies were measured using maximum reading mercury thermometers at several depths in holes that had reached thermal equilibrium. The development of continuous downhole temperature logging tools more or less corresponded with the wide conversion from cable tool to rotary drilling techniques. Because most wireline logging is done immediately following completion of drilling, the temperatures are still disturbed by the massive circulation of surface fluid during rotary drilling. As a result, temperature logs taken at the time other wireline logs are run do not represent equilibrium conditions and are difficult to interpret. Hence, there are many papers that discuss attempts to develop correction schemes for the nonequilibrium temperature data. To this day no detailed study of the complete return to thermal equilibrium of a deep hole has been published. Without such studies, it cannot be inferred at what point during the recovery the subtle correlations of temperature and lithology discussed in the following section appear in the nonequilibrium data. Cable tool holes, because of the minimal fluid circulation, would be more informative but are rarely available.

The technical ability to obtain accurate, precise temperature logs has been available for the last 30 to 40 years. This technical ability has not been tapped in hydrocarbon exploration, however. Logging systems are available that can record temperatures with a precision of 0.001 °C over a digitizing interval of a few centimeters (a temperature difference of 0.001 °C corresponds to 1 °C/km for a 1-m digitizing interval, 2 °C/km for 0.5-m digitizing interval, and 4 °C/km for a 0.25-m digitizing interval). Blackwell and Spafford (1987) give an up-to-date discussion of the techniques and equipment used for temperature logging in continental heat-flow measurements. Logs are run with the tool going down the hole so that undisturbed temperatures are obtained and slowly enough so that thermal equilibrium in the sensor is maintained (or nonequilibrium corrections must be made). Logging velocities are typically 10–15 m/min. The use of filtering in correcting for the temperature response of logging tools has been discussed by Costain (1970), Conaway (1977), and Nielsen and Balling (1984). The limitation on the accuracy of temperature measurements of the rock is not asso-

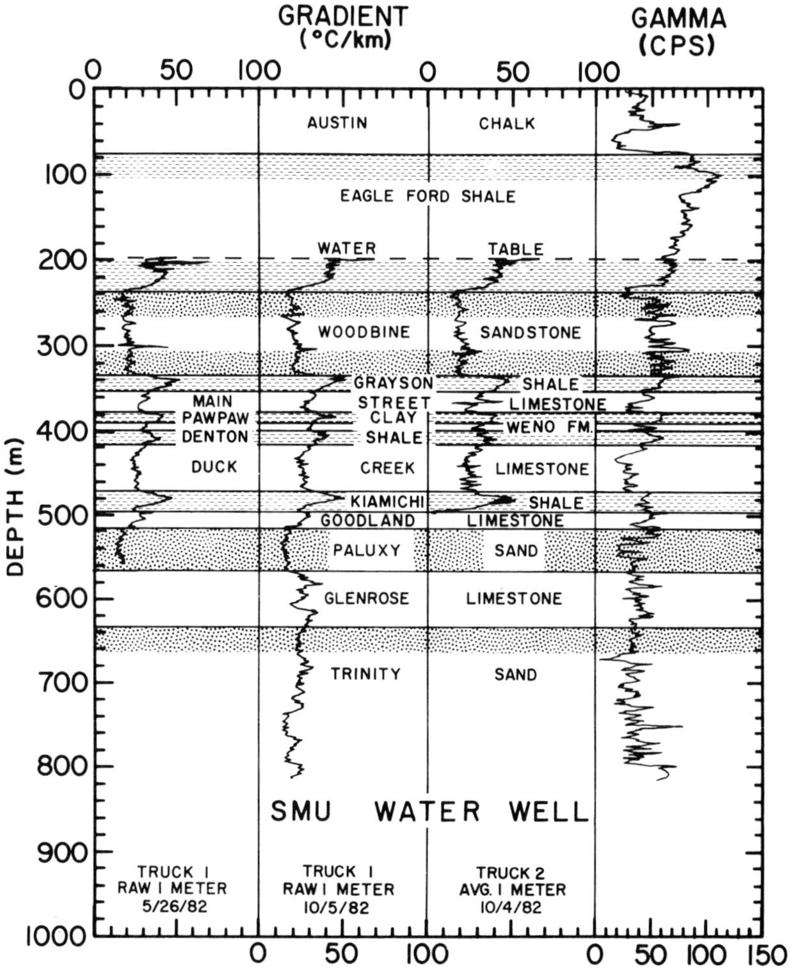

FIGURE 2.2. Temperature-gradient and total gamma-ray logs for a well on the Southern Methodist University campus. The section includes Upper and Lower Cretaceous rocks. Formation names are shown. Sand units are shown by the dot pattern and shale units are shown by a line pattern. The remainder of the units are composed of marl or limestone. Temperatures were digitized at 1-m or 0.5-m intervals to a precision of 0.001 °C and are not shown above the water table. Truck 1 logs were made with a conventional thermistor probe, the Truck 2 log was made with a semiconductor sensor (see Blackwell and Spafford, 1987).

ciated with instrumentation (it could be made at least one to two orders of magnitude better) but with thermal noise in the well, even at thermal equilibrium, associated with small convection cells. Detailed studies of the convection phenomenon have been published by Diment (1967), Gretener (1967), and Sammel (1968).

Correlation of Temperature Gradient and Lithology

The relationship of temperature to lithology in a conductive setting is illustrated in Figure 2.2. This figure shows three high-precision equilibrium temperature-gradient curves calculated from the difference of temperature over either a 0.5-m or a 1-m interval. This type of "gradient" log is used for interpretation because it displays correlations with geology and hole effects better than a temperature-depth curve. The three logs shown in Figure 2.2 were made with two completely different sets of instrumentation: a conventional thermistor probe and a probe containing a semiconductor sensor with an output of 1 mA/°C. The logs were run at a speed of 12 m/min and digitally recorded at either 1.0-m or 0.5-m depth intervals to a precision of 0.001 °C.

A comparison of the first two curves on the left shows the reproducibility of the gradient log using the same tool at different times. A comparison of the first two curves to the third curve demonstrates the reproducibility using completely different instrumentation. The lithology and formations encountered in the hole are given in the figure. The rocks are Upper and Lower Cretaceous shales, marls, and limestones, with some sand (the Woodbine, Paluxy, and Trinity Sandstones). The very close and reproducible correspondence of high temperature gradients with shale and low temperature gradients with sand and limestone is clear. Gradients are 40 to 50°C/km (2.2 to 2.7°F/100 ft) in the former units, and 15 to 20°C/km (0.8 to 1.1°F/100 ft) in the latter units. One interesting result is that the temperature-gradient log shows the contact between the Woodbine Sandstone and the Eagle Ford Shale more clearly than does the total gamma-ray log often used to separate sands and shales.

The kind of detailed information available from an equilibrium temperature log may be compared to the information available from a set of bottom-hole temperature (BHT) data of the type often used in attempts to correlate thermal data with hydrocarbon accumulations. Figure 2.3 shows a comparison of BHT data and precision temperature logs for eight sites in western Nebraska (Gosnold et al., 1982). Each log and its associated BHT data (from holes within 10 km of the log) are shown by the same symbol. It is obvious that even with the correction of the BHT for drilling disturbance (if it were possible) the information content would not compare to the temperature logs. For example, BHT information is usually only available over a limited depth range in a given area because most of the holes bottom, and are logged, at similar depths. Nonetheless, BHT data are widely available and may be effectively used in some cases.

An extreme example of the effect of lithology on subsurface temperatures (from eastern Kansas near Big Springs) is shown in Figure 2.4. The temperature gradient in the upper part of the hole (in Pennsylvanian shale and limestone) is about 50°C/km (2.5°F/100 ft), whereas the temperature gradient in the lower part of the hole (in Mississippian and older carbonate rocks) is about 18°C/km (1°F/100 ft). The errors that arise from using a single BHT and the surface temperature to calculate a gradient, and then using that gradient to calculate

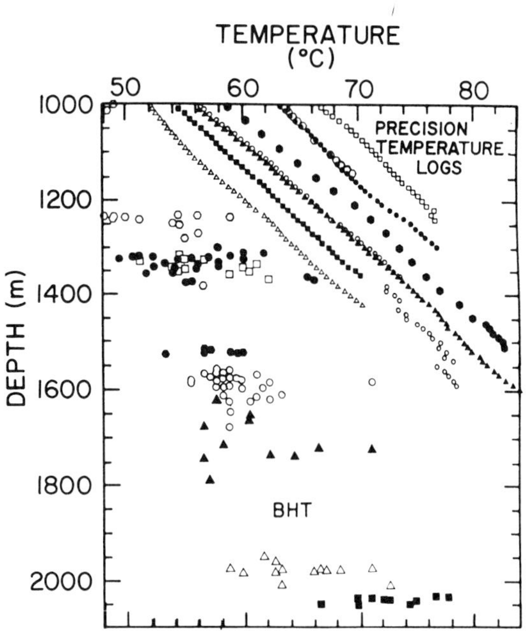

FIGURE 2.3. Comparison of BHT and precision temperature logs for some holes in southwestern Nebraska (after Gosnold et al., 1982). BHT data within a 10-km radius of a particular deep well with a precision temperature log are shown by the same symbol used for plotting the temperature log for that well.

temperatures at other depths by extrapolation or interpolation are illustrated on the figure. For accurate understanding of the subsurface thermal conditions, a knowledge of the thermal properties of the rocks, as well as temperature, is necessary. If temperature data are available as a function of depth in several wells, then vertical as well as horizontal variations can be explored. Thermal conductivity variations are generally the cause of vertical and lateral temperature variations.

Techniques of Thermal Conductivity Measurement

Laboratory Techniques

Most values of thermal conductivity for rocks come from measurements made on samples in the laboratory. Characteristic values of the thermal conductivity of sedimentary rocks from laboratory measurements are presented by Clark (1966) and by Roy et al. (1981). The various laboratory techniques are summarized in more detail in discus-

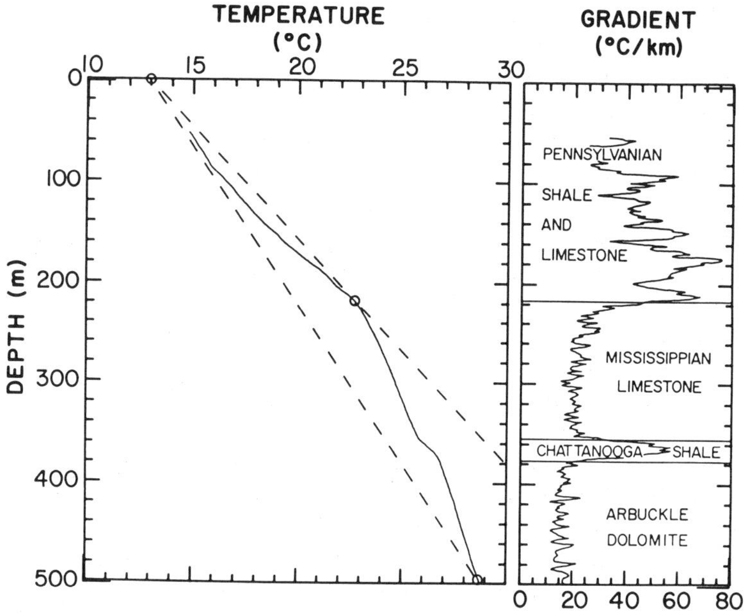

FIGURE 2.4. Temperature-depth and temperature gradient-depth plots for USGS Watson No. 1 well in eastern Kansas near Big Springs (SE¼SW¼SE¼ sec. 18, T. 18 S., R. 23 E.). The generalized stratigraphy encountered in the hole is shown. Dashed lines show the results that would be obtained by calculating an average gradient for the well using the mean annual surface temperature and a BHT from 220 m depth or from 500 m depth.

sions by Gretener (1981) and Blackwell and Spafford (1987). Typical measurement techniques include the divided-bar steady-state measurements suitable for use on core and some cuttings samples and the needle-probe transient measurement suitable for use on very soft materials and cuttings. The divided-bar technique as typically used was described by Birch (1950). This technique is designed to work on machined cylinders of rock. An axial load is applied to a saturated sample and a temperature drop is imposed across the sample. The temperature drop is compared to a temperature drop across standard materials such as quartz and silica glass which are exchanged for the sample, so that a relative measurement is obtained. Divided-bar measurements can be made under realistic conditions of temperature and pore pressure (Somerton, 1975), although most measurements are made at room temperature on saturated samples with a few hundred bars axial load. For anisotopic samples several cylinders cut in different orientations may be measured.

Sass et al. (1971) described a procedure to extend the divided-bar type technique to measurements on cuttings samples. In this modification of the technique, cuttings samples are placed in a plastic cylinder with water, and the plastic cylinder is substituted for a rock cylinder so that the thermal conductivity of the mixture is obtained. The thermal conductivity of the rock from which the cuttings come is then calculated using a mixing formula. This cuttings technique is applicable to only isotropic samples because there is no way to control the orientation along which the thermal conductivity is determined for anisotropic samples. In addition, sampling problems are common. Small samples are generally used and it is not easy to relate the measurements to a specific type of rock unless the unit is quite thick and the samples are not contaminated by caving.

The needle-probe technique involves inserting a long needle into (soft) rock or mud. The needle contains a heater wire and a thermistor (temperature sensor). The heater is turned on and a temperature versus time history is measured from which the thermal conductivity can be deduced. This technique is usually used for heat-flow measurements in the oceans because the needle will penetrate the soft muds in which the gradient measurements are made.

Vacquier (1985) has recently described a modification of the transient needle-probe technique. This technique involves placing a needle probe on the surface of the rock and performing a transient measurement. This surface technique has advantages because the standard needle-probe technique is not suitable for most (lithified) samples. Neither of the needle-probe techniques is simple in the case of anisotropic samples because measurements in two or three directions and solution of a tensor equation are required to determine the thermal conductivity. A major limitation with any laboratory technique is related to sampling difficulties. In the case of certain rock types the sampling difficulties are very significant, as will appear in the subsequent discussion.

Temperature has a significant effect on the thermal conductivity of rocks. The quartz content of the rock is the primary variable because the temperature coefficient of thermal conductivity of quartz is the highest of the common rock-forming minerals. Robertson (1979) prepared a very useful table with contours of thermal conductivity as a function of temperature based on measurements from room temperature to about 300°C. Houbolt and Wells (1980) included a temperature correction to thermal conductivity in their indirect technique (see below) by using a polynomial approximation to thermal conductivity data as a function of temperature. This technique is sufficient if the conductivity at room temperature is known; however, if the lithological section varies in composition (quartz content) with depth, then this type of approach is approximate at best.

In Situ Techniques

Direct Measurements

An ideal solution to the difficulties of sampling and measurement of thermal conductivity values in the laboratory would be to obtain in situ measurements of thermal conductivity. In the past much effort has been expended in an attempt to develop such techniques (Beck et al., 1971). Most of the techniques tested involve inserting a tool into a drill hole and locking it into place. Some sort of a transient thermal conductivity measurement based on the rate of temperature rise during a heating cycle or the rate of temperature drop following the end of a heat pulse is attempted. To date, the results of such experiments have been unsatisfactory because of difficulties associated with hole size and shape, induced water convection by heating, time required for measurements, and so forth. Thus, whereas the concept seems useful, and some suitable techniques are available, those techniques are not often used.

Therefore, in situ techniques are generally restricted to indirect methods. A number of different types of indirect methods have been proposed. These include correlation of various well-log parameters with thermal conductivity, correlation of reflection two-way travel times with thermal resistance, and calculation of thermal conductivity from temperature-gradient logs. Each of these techniques is discussed in the remainder of this section of the chapter.

Well-Log Correlations

Techniques of determining thermal conductivity using well-log properties such as electrical conductivity or resistivity, P-wave travel time or velocity, gamma-gamma density, or natural gamma-ray activity have been discussed in a number of papers. The determinations are certainly feasible because the primary factors that affect thermal conductivity are the composition and the porosity of the rock. Therefore, if rock type and porosity can be determined from well logs, thermal conductivity can be estimated. The general determination of lithology and porosity, however, requires a complete suite of well logs and detailed analysis, and as yet no complete discussion of a well-log technique for thermal conductivity determination has been described.

An example of the type of information that might be used is illustrated in Figure 2.5, which shows a detailed temperature-gradient log and associated total gamma-ray and P-wave velocity logs from a hole in Kansas. There is a reasonably close correlation between variations in gradient, presumably related to variations in thermal conductivity, and changing character of the natural gamma and velocity logs. In particular, the shales typically show slow velocity (high travel time) and high gamma-ray activity, whereas the limestones typically show low gamma-ray activity and high velocities (low travel times). The Pennsylvanian section is composed of thin beds of sandstone, limestone, and shale. Due to the large number of thin-bedded units, the problem of aliasing of the sampling by the well logs is a major limitation for the application of in situ techniques to this section of the hole.

Hagedorn (1985) has recently summarized literature results dealing with the use of well-log measurements to determine in situ thermal con-

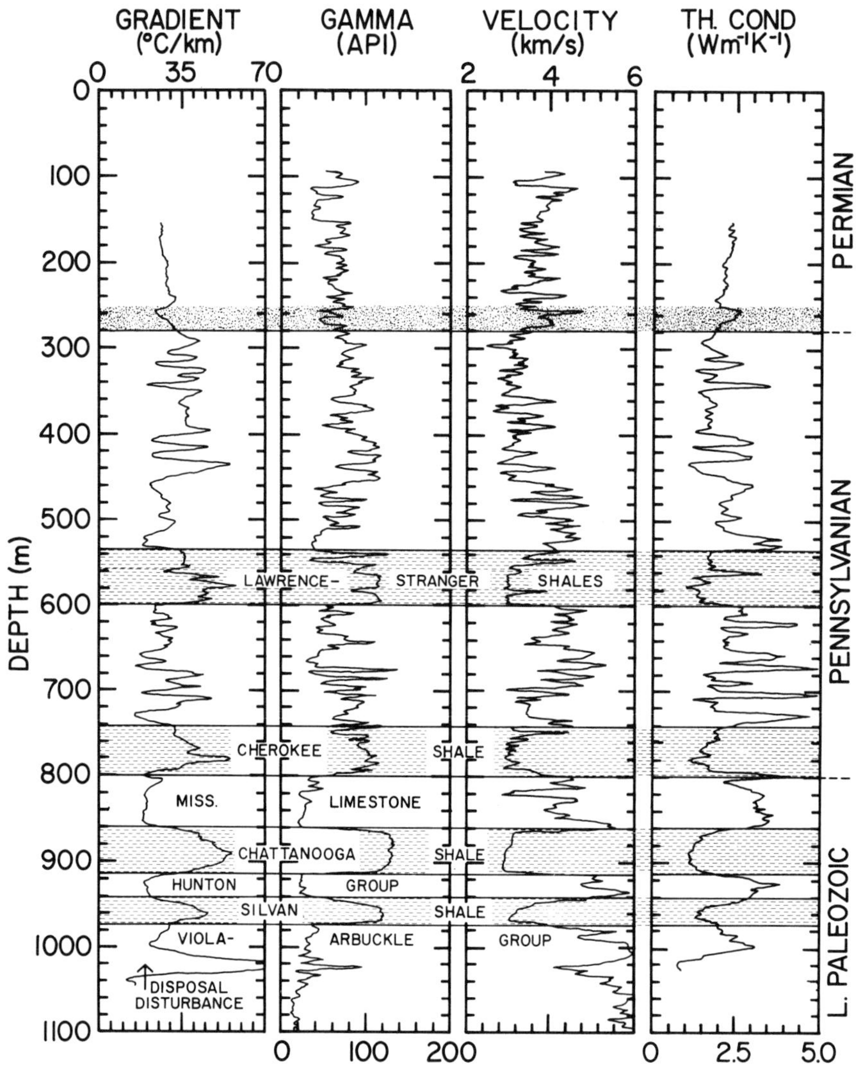

FIGURE 2.5. Temperature gradient-depth, total gamma-ray, interval velocity, and calculated thermal conductivity logs for USGS GEIS No. 1 well near Salina, Kansas (SW¼SW¼SW¼ sec. 32, T. 13 S., R. 2 W.). Several specific formations as well as generalized ages of the rocks are shown.

ductivity. The following review is summarized from his discussion. Anand et al. (1973) investigated the correlation of physical properties and thermal conductivity. They worked with the Berea, Boise, Bandera, and several other unnamed sandstones. They obtained a relationship between thermal conductivity, density, permeability, porosity, and formation resistivity factor. Although they obtained a reasonably good fit, the work was on dry samples and permeability is difficult to measure in situ. In order to relate the formula to saturated samples, they used a nonlinear multiple regression analysis on literature data.

Goss and Combs (1976) approached the problem by looking at many different rock properties and discarding the least significant. They worked on sandstones from the Imperial Valley, California, and included a few samples of siltstones and limestones, one sample of dolomite, and one sample of shale. All of these samples were hydrothermally altered and cemented. A linear approach was taken due to the small size of the data set. They obtained a relationship between bulk density, porosity, permeability, formation factor, sonic velocity, free-fluid index, and thermal conductivity, which gave a good fit with a correlation coefficient of $R = 0.926$. By discarding in sequence the least-significant variables, they obtained the following practical formula, depending only on porosity and sonic velocity, with an $R = 0.966$,

$$K = 0.84 - 0.040\Phi + .000695\, V_p \tag{3}$$

where
K = conductivity in $Wm^{-1}K^{-1}$
Φ = porosity (%)
V_p = sonic velocity in m/s

Relations were also derived using both velocity and porosity alone, but sign problems arose when using only one variable. A better fit was found when electrical conductivity was included. Goss and Combs (1976) stated, however, that since electrical conductivity is a difficult parameter to measure accurately in situ, it would not be practical to include this property. The above equation gave a prediction accuracy of $\pm 10\%$, or about the error expected from conventional divided-bar measurements on cuttings samples. They concluded that this equation would probably not directly apply to other geological settings.

Merkle et al. (1976) worked with data from a well drilled through the Edwards Limestone near San Antonio, Texas. They related thermal conductivity to sonic velocity, density, and the neutron response coefficient, each multiplied by its respective percentage of each mineral constituent. This procedure was used to give the approximate mineralogy of the rock. The in situ thermal conductivity was determined by assuming a standard thermal conductivity for each mineral constituent and using the mixing formula developed by Sass et al. (1971),

$$K_{\text{bulk}} = \Sigma\, K_i^{C_i} \tag{4}$$

where

K = conductivity in $Wm^{-1}K^{-1}$
C = decimal fraction of constituent i

Evans (1977) worked with an extensive suite of sedimentary rocks in the North Sea. The lithologies represented in his data set encompass most typical sedimentary rock types. From these data, using multiple linear regression techniques, he obtained the empirical formula,

$$K = -0.049\,\Phi - 0.160\, V_p + 3.60\rho - 5.50 \tag{5}$$

where
K = bulk thermal conductivity in $Wm^{-1}K^{-1}$
Φ = porosity (%)
V_p = sonic velocity in m/s
ρ = bulk density in g/cm^3.

A fit with an $R = 0.9$ was obtained. This correlation coefficient translates to $\pm 15\%$ of values as measured on a divided bar, which are taken to be ± 5 to 10% of the true conductivities. Evans (1977) therefore inferred that the accuracy of this method is similar to the accuracy of conventional techniques.

Although there have been several equations developed for the relationship of well-log parameters to thermal conductivity, the studies so far have been confined in area or lithology. There is obviously additional room for study, because the results so far are of limited geographic and/or lithological scope.

Other studies not specifically related to laboratory measurements are discussed in the literature. Empirical correlations of well-log parameters and temperature gradient are discussed by Reiter et al. (1980), Blackwell et al. (1981), and Hagedorn (1985). Reiter et al. (1980) found a good correlation between temperature gradient and the induction conductivity log. Blackwell et al. (1981) and Hagedorn (1985) used cross-plot and multivarient linear correlations and found a best relation between temperature gradient and the natural gamma-ray and travel time logs.

Beck (1976) presented an example of a detailed temperature-depth log as an indicator of relative thermal conductivity. He also showed a negative correlation between electrical resistivity and thermal resistivity (the inverse of thermal conductivity). He emphasized the increased lithological resolution of using temperature logs in conjunction with more conventional well logs.

Travel Time

Houbolt and Wells (1980) suggested a novel method for obtaining the mean thermal conductivity (or its inverse, the thermal resistivity): the relation of travel time from reflection profiles to mean thermal properties of the section sampled by the travel time information. In their technique the travel time data from a reflection profile are used together with temperature data from a nearby control well to solve for the constants in the equation

$$Q = \ln[(c + T_z)/(c + T_0)]/[a(t_z - t_0)] \quad (6)$$

where

c, a = empirically determined constants

In the case of the BU (see below),

$$a = 1.039 \text{ and } c = 80.031$$

The variables are:

T_0 = intercept or surface temperature (°C)
T_z = temperature measured at depth z (°C)
t_0 = intercept travel time corresponding to intercept temperature
t_z = one-way travel time to depth z (s)
Q = heat flow ($\mathrm{mWm^{-2}}$).

For multiple layers the equation is

$$T_{(i,j)} = T_0 \, e^{aQ(t_j - t_0)} + ce^{aQ[(t_j - t_0)^{-1}]} \quad (7)$$

They concluded that this equation worked well for siliclastic and carbonate sequences. They also argued that an estimate of thermal conductivity could then be obtained by the relationship,

$$K = [V_p * Q] \,/\, [a(c + T_z)] \quad (8)$$

which holds true if the heat flow is a constant throughout the sedimentary sequence.

They used data from the well, Bolderij-1, in the North Sea to create a "standard" heat-flow unit called the Bolderij Unit (BU). They suggested that 1 BU was equal to 77 $\mathrm{mWm^{-2}}$. They then used their procedure and the BU calibration to evaluate relative heat flow in other areas. More will be said about this technique in following sections of this chapter.

Temperature-Gradient Logs

The final way to obtain in situ relative and absolute thermal conductivity values is to use a high-precision temperature-depth log. The close correlation between temperature and lithology in a conductive setting has been illustrated in Figures 2.2, 2.4, and 2.5. If no structural or fluid flow problems are present in or near a hole, it can be assumed that Equation 2 applies and the heat flow is equal to the product of the temperature gradient times the thermal conductivity. *If the heat flow is known and is constant and the geothermal gradient is known*, the thermal conductivity can be directly determined. To start the process, thermal conductivity measurements or accurate estimates are required of at least one of the major rock types encountered in the hole.

Data from the well in Kansas shown in Figure 2.5 can be used to illustrate this technique. The limestones and dolomites of the Mississippian, Hunton, and Viola-Arbuckle units are relatively isotropic and homogeneous, and cuttings samples of these units can be obtained from which accurate in situ thermal conductivity values can be measured in the laboratory. Once these measurements are obtained, the temperature log can be converted to a thermal conductivity log using as a calibration the heat flow calculated using the known intervals of thermal conductivity. This process is illustrated as the thermal conductivity curve on Figure 2.5. The values on this curve are calculated directly from the temperature gradient, based on the measured thermal conductivities on cuttings samples of the carbonate rocks in the hole and a calculated heat flow of 57 $\mathrm{mWm^{-2}}$. The inferred values of thermal conductivity range from 1.09 to 1.25 $\mathrm{Wm^{-1}K^{-1}}$ for the predominently shale units to the measured values of 2.2 to 2.9 $\mathrm{Wm^{-1}K^{-1}}$ for the carbonate units. The temperature gradient spike at 1,040 m is related to a fluid disposal effect (nonconductive) caused by use of the well for fluid disposal before logging. This technique relies primarily on the assumption of conductive heat transfer within the hole and should give the least-biased relative values. Nonetheless, the lack of high-precision temperature logs in equilibrium holes in sedimentary rocks limits the application of this technique.

Thermal Conductivity and Heat Flow in Kansas

In 1981 four holes were drilled in the state of Kansas by the U.S. Geological Survey for aquifer studies. These holes were drilled to or slightly into basement rocks and were designed to study aquifer characteristics in the Arbuckle Group. The holes were extensively studied, cutting samples were obtained, and a complete suite of well logs was run, including precision temperature logs collected by personnel from Southern Methodist University. Blackwell et al. (1981) attempted to determine heat-flow values for these holes and encountered some surprises. Log results from two of these wells have already been illustrated (Figs. 2.4 and 2.5). Heat-flow analysis proceeded in a conventional manner in that cuttings samples were obtained and divided-bar measurements of the thermal conductivity were made. Averages of the measured thermal conductivity values were then calculated for various depth intervals and compared to the temperature gradients to calculate the heat flow.

A peculiar result was obtained for all holes and is illustrated in Figure 2.6 for the hole whose temperature-depth curve and lithology are shown in Figure 2.4. The geological section in this hole includes, in order, 220 m of Pennsylvanian rocks consisting predominantly of shale and limestone, 130 m of Mississippian limestone, 20 m of Chattanooga Shale, and 100+ m of carbonate rocks of the Arbuckle Group. The measured thermal conductivity of cutting samples is shown as the solid-line bar graph in the second panel of Figure 2.6. The measured thermal conductivity is almost constant from a depth of 100 m to the top of the Arbuckle dolomite at 375 m; the average value is about 2.1 $Wm^{-1}K^{-1}$. The measured thermal conductivity of the Chattanooga Shale is slightly lower, approximately 1.9 $Wm^{-1}K^{-1}$. Upon multiplication of these average thermal conductivity values times the observed temperature gradients shown in the first panel, the heat flow shown by the solid line in the third panel was calculated. It would be expected that this heat flow would be essentially constant with depth in this hole, because there is no evidence for significant heat transport by aquifer flow or thermal convection, or for structural complexities that might cause heat-flow refraction. What is seen, however, is that the calculated heat flow is proportional to lithology and ranges from values of over 90 mWm^{-2} in the Pennsylvanian shales and the Chattanooga Shale to approximately 65 mWm^{-2} in the carbonate sections. It is interesting to note that the heat flow is essentially the same in the Mississippian limestone and in the Arbuckle dolomite, even though there is a significant difference in the measured thermal conductivity.

If there were systematic aquifer effects, then the heat flow should vary in a uniform way with depth; however, the calculated apparent heat flow is, as a sequential function of increasing depth, high, low, high, and low again. This interleaved pattern of contrasting heat-flow values is not explained by any reasonable convective effect. The same effect was observed in the calculated heat flow (not illustrated) for the hole shown in Figure 2.5. In that hole, there are three major shales within the carbonate section and the calculated heat-flow values in the shales were 50 to 100% above the calculated heat-flow values in the carbonate units.

The inescapable conclusion is that the cuttings measurements of the thermal conductivity of shales are inaccurate and invalid. This result is not surprising because shales are anisotropic. The thermal conductivity of a sheet silicate is very high in the direction parallel to the sheet, but quite low perpendicular to the sheet. Thermal conductivity variations of a factor of two to three due to orientation are to be expected. Because the techniques used to measure thermal conductivity of cuttings do not allow an orientation of the shale fragments in a proper manner, thermal conductivity values intermediate in value between the perpendicular-to and parallel-to foliation values will be obtained. Because the shale units in Kansas are nearly flat, the measured cuttings thermal conductivity values will be biased toward the high side. It might be expected that the situation would be improved if core samples of the shales were available for thermal conductivity determination; however, the thermal conductivity values inferred for the shales in Kansas (1.18 ± 0.03 $Wm^{-1}K^{-1}$; see Fig. 2.5) are lower than values of shale reported in the literature. Since these Kansas shales are Paleozoic in age and have quite low porosities and high densities (2.52–2.57 g/cm^3), these low values, which are typical of clays and muds, were unexpected.

An attempt to estimate indirectly the thermal conductivity of the noncarbonate sections was made using correlations between the temperature gradient and the well-log information. A good correspondence was found between temperature gradient and total gamma-ray activity or velocity (travel time). Cross plots for these three parameters from the Kansas holes are shown in Figure 2.7. The holes intersect the same lithology, so the same general relationships are shown with the exception that one of the holes (12S) shows a slightly differ-

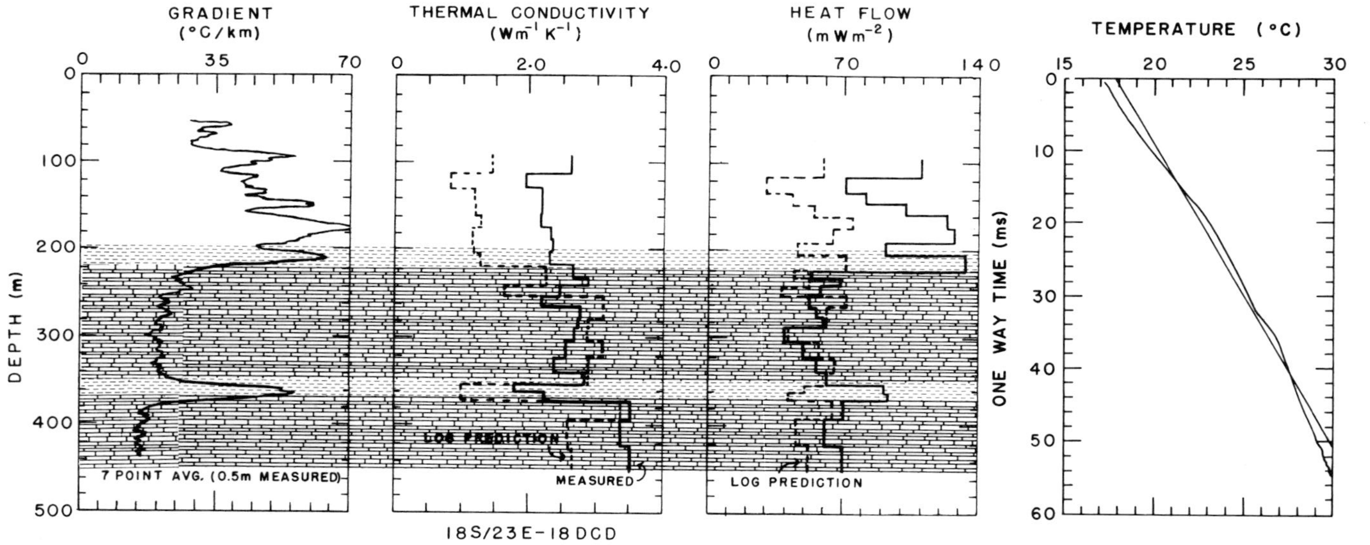

FIGURE 2.6. Plot of temperature gradient, thermal conductivity, and heat flow as a function of depth for the Watson No. 1 well (Big Spring, Kansas, Fig. 4). The solid line in the thermal conductivity column shows the thermal conductivity calculated from measurements of cuttings samples in the interval shown. The solid line in the heat-flow column indicates the heat flow calculated from the averaged temperature gradient and the measured thermal conductivity value within each interval. The dashed lines in both the thermal conductivity and heat-flow columns indicate the thermal conductivities and heat-flow values obtained by measurement/ inference for the carbonates and clastic units, respectively, from well-log cross-plots and the assumption of conductive heat transfer in the hole. The dashed curves are preferred as a realistic estimate of the in situ values. A plot of summed travel time as a function of temperature is shown in the final panel.

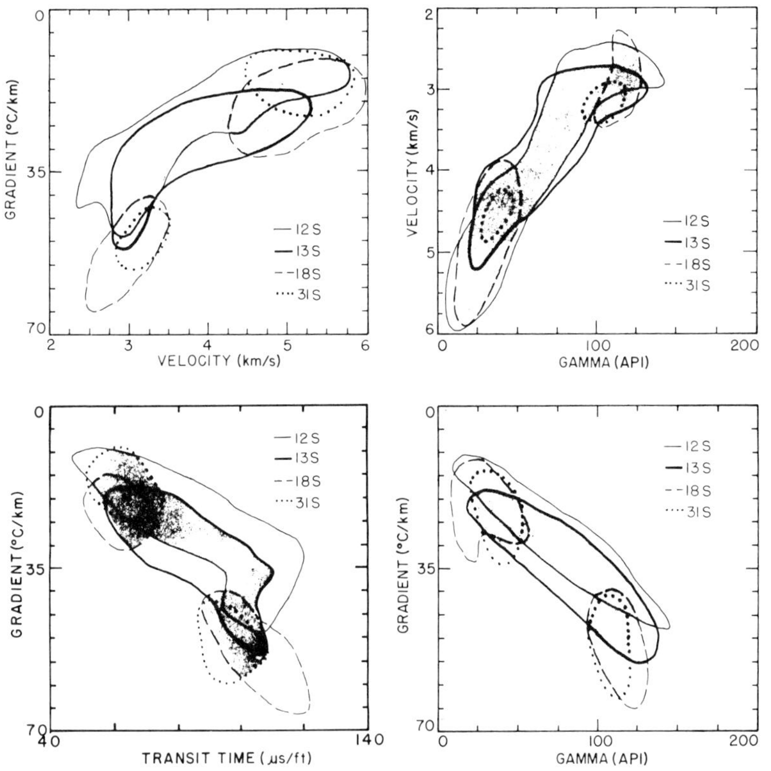

FIGURE 2.7. Cross-plots of temperature gradient-velocity, velocity-natural gamma-ray activity, temperature gradient-travel time, and temperature gradient-natural gamma-ray activity for several holes in eastern Kansas. The general area of the individual cross-plot points is outlined and overlap areas are shaded.

ent relationship between gradient and the other parameters. This difference is presumed to be related to a difference in heat flow at that site. It is to be expected empirically that these log techniques would work for the section in Kansas because of the strong contrast in properties between the carbonate rocks and the shales. As discussed in the previous section, the log technique is more complicated if additional lithologies are included. The correlation of log properties with thermal conductivity at the present time is empirical at best.

Using the well-log technique to approximate an overall relationship between thermal gradient and thermal conductivity for the holes, interval thermal conductivities were then calculated. Interval thermal conductivities calculated from the well logs are shown as the dashed curves in Figure 2.6. The predicted thermal conductivity values for shales are on the order of 1.09 to 1.25 $Wm^{-1}K^{-1}$ and average 1.18 $\pm$ 0.03 $Wm^{-1}K^{-1}$. The predicted heat-flow values average 60 $\pm$ 3 mWm^{-2} with no significant variation between lithologies.

In this geological setting the travel time technique suggested by Houbolt and Wells (1980) would apply reasonably well. In Figure 2.6 a plot is shown of the temperature as a function of the summed travel time. The relationship is reasonably linear and if a calibration between travel time and thermal conductivity were available, the heat flow could be calculated from the slope.

This example illustrates the various ways in which thermal conductivity can be determined. Once reliable information on the in situ values is obtained from a hole or series of holes in a sedimentary sequence, that information can be used in other thermal studies such as thermal history modeling and analysis of bottom-hole temperature information.

Discussion

Examples of Temperature-Gradient–Depth Curves

There have not been many descriptions of the effect of thermal conductivity on temperature-depth curves or interval temperature-gradient values in sedimentary basins. The best examples

TABLE 2.2. Typical order of magnitude in situ thermal conductivity values of rocks at 20°C.

Lithology	$Wm^{-1}K^{-1}$
Claystone and siltstone	0.80–1.25
Shale	1.05–1.45
Sand	1.70–2.50
Sandstone	2.50–4.20
Quartzite	4.20–6.30
Lithic sand	1.25–2.10
Graywacke	2.70–3.35
Limestone	2.50–3.10
Dolomite	3.75–6.30
Salt	4.80–6.05
Anhydrite	4.90–5.80
Coal	<0.5
Water	0.59
Granite	2.50–3.35
Basalt and andesite	1.45–2.10
Rhyolite glass	1.25–1.45
Rhyolite ash	0.60–1.05
Rhyolite welded tuff	1.70–2.10

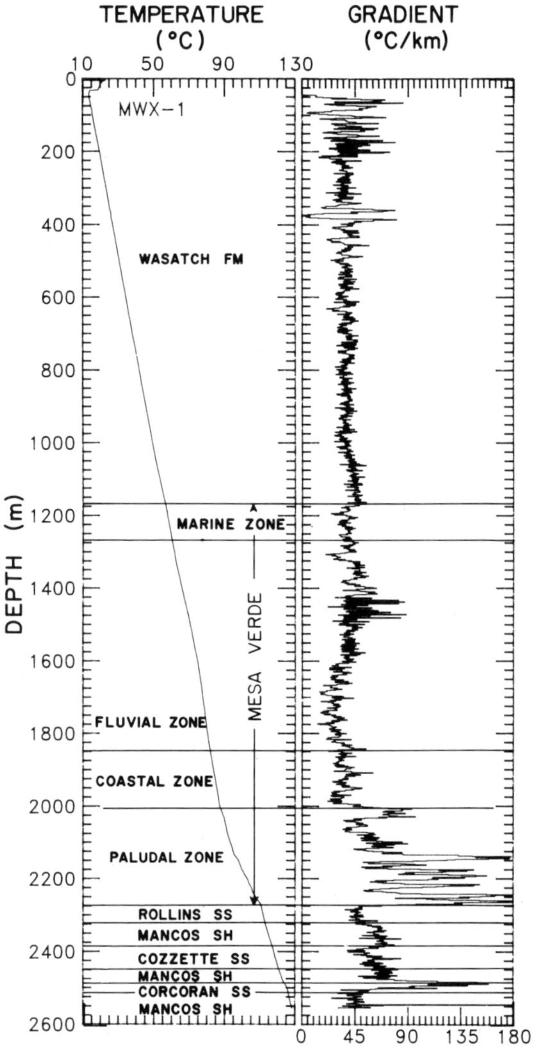

FIGURE 2.8. Precision temperature-depth curve for Sandia MWX-1 well near Rifle, Colorado. The geological section encountered in the hole is shown. A rapid change in temperature of approximately 15°C (27°F) over a depth interval of 200 m between 2,100 and 2,300 m is associated with a coal section. Geothermal gradients range as high as 180°C/km (10°F/100 ft) within the coal intervals. The average temperature gradient value for the hole above 2,000 m is approximately 40°C/km (2.2°F/100 ft).

conveniently available, such as those cited by Gretener (1981) in his discussion of the relationship of temperature, gradient, and thermal conductivity, are of relatively poor quality. Single detailed temperature-gradient–depth logs are described by Decker, Roy et al. (1968), Conaway and Beck (1977; also Beck, 1982), and Reiter et al. (1980), but these examples do not appear to be familiar to explorationists. As a further demonstration of the usefulness of accurate temperature logs in illustrating lithology and thus relative thermal conductivity in sedimentary basins, several other examples are discussed in this section.

The most obvious effects of thermal conductivity on temperature-gradient logs occur when there are significant contrasts in thermal conductivity between thick geological units. For example, in the hole at Salina, Kansas (Fig. 2.5), the geological units, while they have strong contrasts in thermal conductivity, are thin enough so that the overall temperature-depth curve is quite linear. On the other hand, the temperature-depth curve from the hole near Big Springs, Kansas (Fig. 2.4) is very nonlinear because the rocks in the upper and lower parts of the hole have contrasting thermal conductivity values.

Generalized thermal conductivity values of different rock types at 20°C are listed in Table 2.2. The contrast in values will decrease at higher temperatures. Rocks with exceptionally high thermal conductivity include quartzite, dolomite, salt, and anhydrite. Rocks with low thermal conductivity include shale, tuff, and especially coal. Coals have the lowest thermal conductivity of typical rocks that occur in sedimentary basins (see Kayal and Christoffel, 1982) and can have a profound effect on temperature versus depth curves. This effect is illustrated by the temperature-depth curve from the MWX-1 well near Rifle, Colorado. This well was drilled by Sandia Laboratories as a tight gas sands research well. After the hole had reached equilibrium, a precision temperature log was made by Southern Methodist University (Fig. 2.8). The well is in the Eocene Wasatch Formation and underlying Cretaceous rocks including the Mesa Verde Group. Below 2,000 m in the hole, a series of coal beds are encountered. The coals are too thin to be resolved individually on the log; therefore the gradients that are plotted in Figure 2.8 include some lithologies in addition to the coals. Nonetheless, the contrast between the typical gradients in the sandstones and siltstones in the upper part of the hole (ranging from 20° to 40°C/km) compared to the temperature gradients of 150° to 180°C/km in the coal sections suggests a thermal conductivity contrast of four or more. The effect on the temperature-depth curve is spectacular as a temperature change of almost 15°C occurs over a depth interval of approximately 250 m. Projections of temperature versus depth using bottom-hole temperature information from above the coal section would significantly underestimate the temperature in the lower part of the sedimentary section. Since this temperature range is within the maturation window, the maturation effects would be incorrectly evaluated as well. In a study of organic maturation level within the well, Bostick and Freeman (1984) noted anomalously high vitrinite reflectance values in the bottom part of the hole as compared to the trend above 2,000 m. The temperature-depth data clearly illustrate that this rapid change in the vitrinite reflectance is associated with a rapid change in temperature due to the low thermal conductivity associated with the coals.

Another example of the relationship of temperature gradient to lithology is shown in Figure 2.9. This figure shows temperature gradient versus depth and formation names for the Shell Chapman No. 1 well near Hempstead, Texas, drilled in a typical Gulf Coast sand-shale section. The sands can be identified on the temperature-gradient log by the lowest gradients, which average approximately 25°C/km. The shale sections, particularly thick shales of the Cook Mountain and Sparta Formations, stand out prominently because of the high temperature gradients within these intervals (45° to 50°C/km). Although no thermal conductivity measurements have been made for samples in this well, the close correspondence between gradient and lithology and the approximately constant gradients within these lithologies as a function of

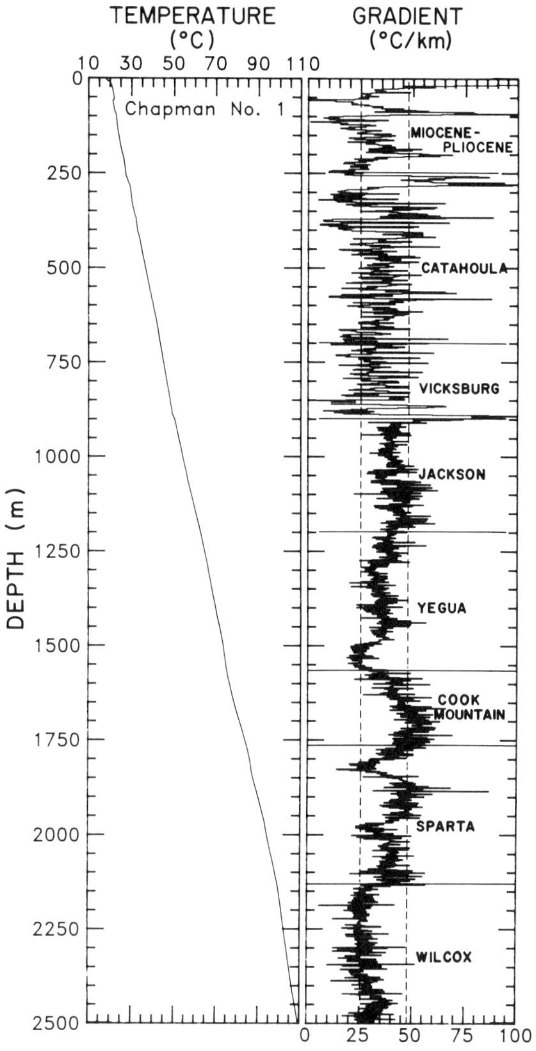

FIGURE 2.9. Temperature-depth and temperature gradient versus depth plots for the Shell Chapman No. 1 well near Hempstead, Texas. The various units are shown as are the approximate locations of a "sand line" and a "shale line." The sand line corresponds to a temperature gradient of 25°C/km (1.4°F/100 ft), whereas the shale line corresponds to a temperature gradient of 45°C/km (2.5°F/100 ft).

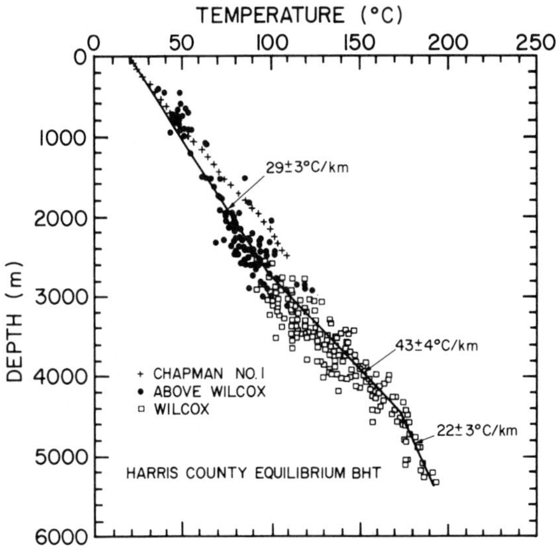

FIGURE 2.10. Plot of "equilibrium" bottom-hole temperatures versus depth for the Wilcox geothermal fairway in Harris County, Texas. Data are from Bebout et al. (1979). Average gradients for different segments of the data are shown. Temperature versus depth data from the Chapman No. 1 well in nearby Hempstead County are shown by the plus symbols.

depth, suggest that heat flow is essentially constant with depth and that the variations are due to contrasts in lithology. These values are consistent with a sand thermal conductivity on the order of 2.2 ± 0.2 $Wm^{-1}K^{-1}$ and shale thermal conductivities on the order of 1.1 to 1.2 $Wm^{-1}K^{-1}$ (compare to Figures 2.5 and 2.6).

This observed contrast in thermal conductivity for sand and shale can be used to investigate a problem in the thermal structure of the Gulf Coast: the reason for the high geothermal gradients associated with top-of-geopressure. Figure 2.10 shows a summary of "equilibrium" bottom-hole temperatures from Harris County assembled by Bebout et al. (1979). The top of the geopressure zone is near the top of the Wilcox Formation. The mean temperature gradient (based on bottom-hole temperatures) to the top of the geopressure zone at about 3 km is 29 ± 3°C/km. Within the geopressure zone between 3 km and 5 km, temperature gradients are on the order of 43 ± 4°C/km. At greater depths gradients decrease to values of 22 ± 3°C/km.

Temperatures from the Chapman No. 1 well are also plotted in Figure 2.10 to illustrate their relationship to the bottom-hole temperature data from Harris County. The temperatures in the Chapman No. 1 well in adjoining Waller County are higher than the bottom-hole temperatures in Harris County. Apparently either the bottom-hole temperatures above 2.5 km are systematically lower in Harris County due to a higher proportion of sand, or the bottom-hole temperatures are in fact not equilibrium temperatures and are 5° to 10°C low. The contrast in temperature gradient across the geopressure zone is almost exactly equal to the value that would be predicted by the contrast in temperature gradient between a predominantly sand and a predominantly shale section, and the contrast in temperature gradients compares closely to that observed between shales and sandstones on a finer scale in the Chapman No. 1 well a few kilometers to the north. Thus, thermal conductivity contrasts seem adequate to explain the change in temperature gradient across the top of the geopressure zone.

It has been suggested that upward flow of fluids through the geopressure zone might contribute to the observed high temperature gradients. The vertical transport of heat by one-dimensional fluid flow through a porous medium was originally discussed by Bredohoeff and Papadopulos (1965), and their model can be used to investigate whether the apparent upward bow of the temperature versus depth relation between 3 km and 6 km is related to a component of upward fluid flow. Calculations using the formulation of Mansure and Reiter (1979) suggest that upward velocities on the order of 0.3 to 0.6 cm/year would be required to raise the gradient from 25 to 30°C/km to 40 to 45°C/km. Assuming reasonable volumes of fluid within and below the geopressure zone, velocities of this order of magnitude would require a volume of fluid that would be sufficient to deplete the high pressure fluids within a time scale on the order of 100,000 to 200,000 years. Therefore, the maintenance of high gradients through the top-of-geopressure is not likely to be due exclusively to fluid flow but must be partly or predominantly due to the contrast in thermal conductivity.

The temperature data from the Gulf Coast also illustrate a major limitation with the Houbolt and Wells (1980) technique applied to the relatively unconsolidated sand and shale section of the Gulf Coast. Sonic velocity is not a reliable lithological indicator in this section. Since the temperature gradients vary by a factor of two depending on the lithology, the travel time through this section cannot be used to determine average thermal conductivity or thermal resistance to better than ±25 to 50% without independent lithological knowledge.

The similarity of sand and shale velocities points out a possible use of temperature-depth logs. Since sand/shale ratios are quite important to the interpretation of seismic reflection data, much effort has been spent to determine ways to measure sand/shale ratios best. Conventional well logs (gamma-ray, for example) often fail to discriminate sands and shales (see Fig. 2.2, for example). Temperature-gradient logs, however, show major contrasts between the two types of lithologies and thus can be used for accurate separation of sands and shales.

This point can also be illustrated using data from a hole in western Nebraska (Fig. 2.11). Detailed temperature-gradient and total gamma-ray logs are shown for this well. The well penetrates Cretaceous units including the Pierre, Niobrara, Carlisle, and Greenhorn Formations and the Dakota Group (Omandi, Skull Creek, and Lakota Formations) as well as a Jurassic section and a Permian section (980 m to total depth) with some interbedded salt. The temperature-gradient log is much more unit diagnostic in the Cretaceous section than is the gamma-ray log. The Niobrara Formation in particular has a distinctive gradient signature characterized by an almost constant temperature gradient with an interval of very low temperature gradient immediately below and an interval of very high temperature gradient immediately above. The unit above the Niobrara is an organic-rich member of the Pierre Shale. The contrast in temperature gradient in the Dakota Group between the Omandi and Lakota Sandstones on the one hand and the Skull Creek Shale on the other is also more obvious than are contrasts in the gamma-ray log response. The lowest gradients observed in the hole are coincident with salt layers within the Permian section, so salt would have the highest thermal conductivity of the units encountered (see Table 2.2).

The final example is from the Williston Basin in North Dakota (Fig. 2.12). The section encountered in this well is very similar to the section encountered in the Nebraska well; in fact, many of the characteristic lithological (temperature gradient) units carry through all the way across this part of the Midcontinent. The temperature data for the North Dakota well are from Combs (1970; see also Combs and Simmons, 1973). Also shown are gamma-ray and velocity logs from a nearby well obtained from a log library. The potential use of accurate temperature-depth curves for lithological correlation is obvious.

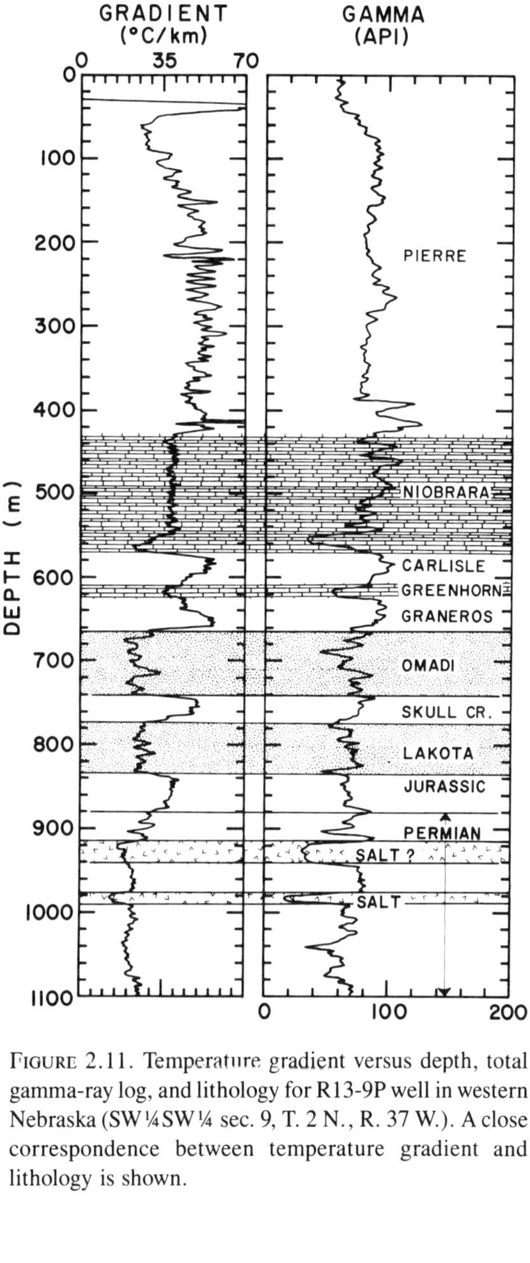

FIGURE 2.11. Temperature gradient versus depth, total gamma-ray log, and lithology for R13-9P well in western Nebraska (SW¼SW¼ sec. 9, T. 2 N., R. 37 W.). A close correspondence between temperature gradient and lithology is shown.

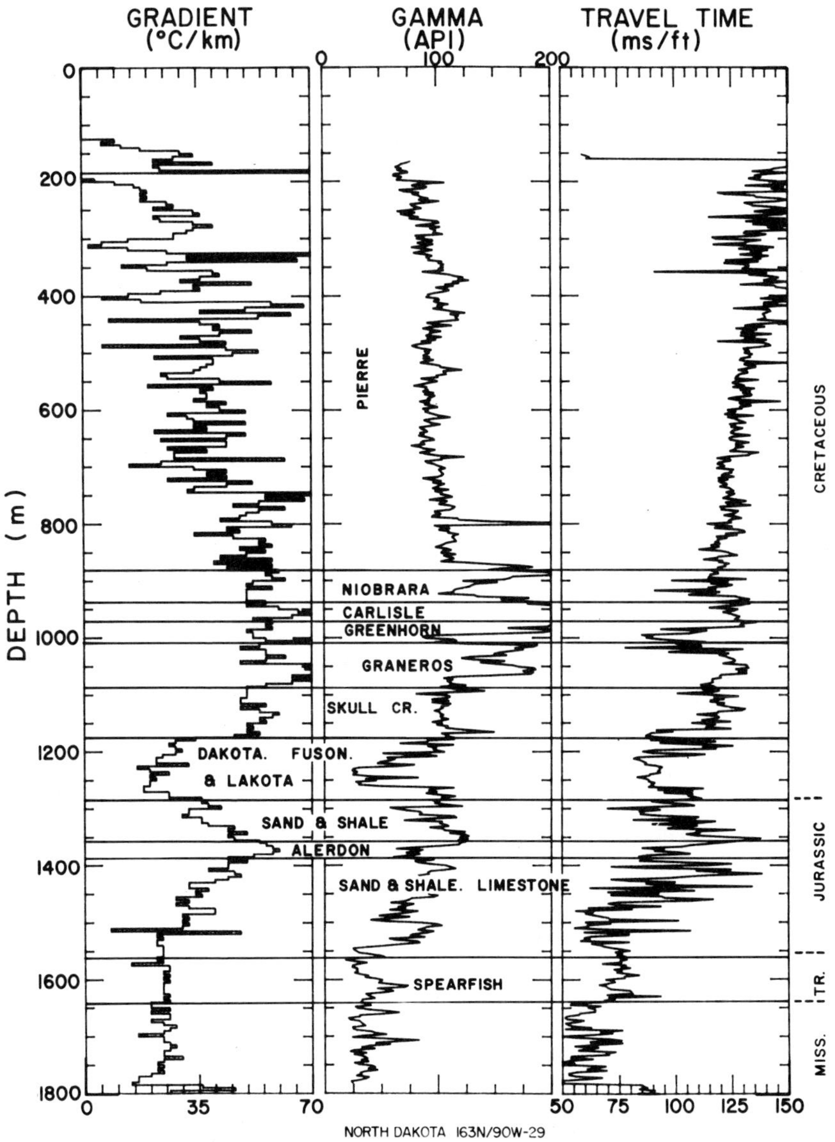

FIGURE 2.12. Temperature gradient-depth data from the Carrie Hovland No. 1 well in the Williston Basin, North Dakota (temperature data from Combs, 1970). Total gamma-ray and travel time versus depth information for a nearby well (1-B Anderson) are also shown. Units encountered in the wells are shown in the figure.

Previous Estimates of Shale Thermal Conductivity

In the examples presented here, shales ranging in age from Cenozoic to Paleozoic appear to have thermal conductivity values in the range 1.1 to 1.3 $Wm^{-1}K^{-1}$. This value is in contrast to careful laboratory measurements. For example, Judge and Beck (1973) discussed a detailed study of heat flow in the western Ontario sedimentary basin. As part of this study, they made detailed thermal conductivity measurements on an extensive suite of core samples from the lower Paleozoic section encountered there. They then calculated interval heat-flow values for each of the lithological units using the measured thermal conductivity values. Consistent results were obtained in all formations up and down each hole and from hole to hole with the exception of the Ordovician Meaford-Dundas and Collingwood shale intervals. Interval heat flow calculated for the Ordovician shale averaged 60% above the heat flow calculated for the other intervals within each hole. Judge and Beck (1973) attributed such variations to anomalous heat production in the shale or some nonconductive effect. If, however, the ratio is taken of the gradients in the shale compared to the gradients in the other intervals of the holes, assuming constant heat flow, then the predicted thermal conductivity value for the shale is 1.2 $Wm^{-1}K^{-1}$. This value is precisely that predicted for the shales in Kansas compared to the carbonate units! In situ measurements of shale thermal conductivity in the Salina and Cabot Head Formations (Beck et al., 1971) in a hole on the campus of Western Ontario University agreed with laboratory measurements (Conaway and Beck, 1977). The difficulty in working with the shale lithology is thus clearly outlined.

Similarly, Hyndman et al. (1979) discussed an abrupt change in gradient in a hole on Sable Island, Canada. In spite of this change, which had all the characteristics of a thermal conductivity change, measurements of thermal conductivity on cutting samples showed no contrast up and down the hole. The lithology indicated by samples associated with the anomalously high heat flow and high gradient interval was shale. In this case also the ratio of gradients is approximately the same as that between shales and sands in the sections discussed here.

Sass and Galanis (1983) discussed measurements by the needle-probe technique of one core sample of the Pierre Shale from a hole in South Dakota. They obtained values of 1.25 to 1.49 $Wm^{-1}K^{-1}$ for the needle probe inserted parallel to the axis of the core and 1.22 to 1.31 $Wm^{-1}K^{-1}$ for the needle probe inserted perpendicular to the axis of the core. These values are among the lowest in the literature but still appear 10 to 20% high relative to inferred in situ values.

Significance of Results

The importance of understanding the thermal conductivity of the rock units in a sedimentary basin before attempting to understand its temperature structure is illustrated by the various examples presented in this chapter. For this purpose accurate, detailed temperature logs are needed. The necessary density of such logs would depend on facies changes and an extensive number of such logs might not be needed for every basin. A few wells can be used for calibration of the thermal conductivity of a particular geological section. The much more accessible bottom-hole temperature data that exist may then be used to extend and interpolate the isolated points of detailed temperature study.

Inclusion of thermal conductivity measurements in a thermal analysis package is not a wasted exercise. Published estimates of the thermal conductivity of shale appear to be in error by 50 to 100%. Since shale forms a significant fraction of the section in sedimentary basins, accurate subsurface temperature-depth predictions *cannot* be made if the thermal properties of shale are not well understood. For example, Sclater and Christie (1980) have done a very careful thermal analysis of the central North Sea basin. In the basin approximately 75% of the section is composed of clay and shale. They assumed a thermal conductivity versus porosity curve for shale that predicts thermal conductivity values on the order of 2 $Wm^{-1}K^{-1}$ over most of the depth range. If the thermal conductivity of the shales in the North Sea is similar to the thermal conductivity of shales in the areas discussed here, then the value assumed by Sclater and Christie (1980) overestimates the actual thermal conductivity of shale by approximately 75%. Since shale represents 75% of the sedimentary section, the resulting error in estimation of temperatures is on the order of 50%. Because the thermal maturation indices double for every 10°C change in temperature, an error of 50% in the predicted temperature versus depth is obviously very serious if those temperatures are going to be used to predict the thermal maturation rank in a sedimentary basin.

Shales are anomalous in that the thermal conductivity appears to be a weak or nonexistent function of compaction. The reason for this anomalous behavior is related to the mineralogical nature of shale. At shallow depths the clay particles are more or less randomly oriented so that the mineralogical contribution to the overall thermal conductivity is high. Porosity is high and water content great, however, so the overall thermal conductivity value is low, 0.7 to 1.0 $Wm^{-1}K^{-1}$. As compaction proceeds, the clay particles are progressively oriented until their direction of good thermal conduction becomes horizontal and their direction of poor thermal conduction becomes vertical. This effect almost offsets loss of water due to compaction with the result that shale conductivity does not vary much with depth. Many other phenomena such as structural variations associated with transformation to kaolinite may have a significant effect on shale conductivity. These questions remain to be investigated.

It is clear that much remains to be learned about in situ thermal conductivities in sedimentary basins and that such information may not be easily obtained. The significance of obtaining such information and completely understanding thermal conductivity variations, however, is important if adequate thermal analyses of sedimentary basins are to be made. The full interplay of lithology, porosity, structure, and other variables in the determination of the overall thermal conductivity of sedimentary basins must be understood.

Acknowledgments. Don Steeples of the Kansas Geological Survey acquainted us with the availability of the wells in Kansas from which the shale thermal conductivity discrepancy was most obvious. Shari Kelley carried out calculations of the fluid flow velocity through the geopressured region based on the possible temperature-gradient effect. Dan Hagedorn summarized results of well logging and thermal conductivity techniques. Mobil Dallas Research Laboratories and Shell Research Laboratories furnished support for collection of some of the thermal data discussed in this paper. Helpful comments on the manuscript were made by Nancy Naeser, Alan Beck, and Robert Roy.

References

Anand, J., Somerton, W.H., and Gomaa, E. 1973. Predicting thermal conductivities of formations from other known properties. Journal of the Society of Petroleum Engineers 13:267–273.

Barker, C.E. 1983. Influence of time on metamorphism of sedimentary organic matter in liquid-dominated geothermal systems, western North America. Geology 11:384–388.

Bebout, D.G., Weise, B.D., Gregory, A.R., and Edwards, M.B. 1979. Wilcox sandstones in the deep subsurface along the Texas Gulf Coast, their potential for production of geopressured energy. Texas Bureau of Economic Geology DOE Report ET28461, 219 pp.

Beck, A.E. 1976. The use of thermal resistivity logs in stratigraphic correlation. Geophysics 41:300–309.

Beck, A.E. 1982. Climatically perturbed temperature gradients and their effect on regional and continental heat-flow means. Tectonophysics 41:17–39.

Beck, A.E., Anglin, F.M., and Sass, J.H. 1971. Analysis of heat-flow data—*in situ* thermal conductivity measurements. Canadian Journal of Earth Sciences 8:1–20.

Birch, F. 1950. Flow of heat in the Front Range, Colorado. Geological Society of America Bulletin 61:567–630.

Birch, F., Roy, R.F., and Decker, E.R. 1968. Heat-flow and thermal history in New England and New York. In: Zen, E., White, W.S., Hadley, F.B., and Thompson, J.B. (eds.): Studies of Appalachian Geology: Northern and Maritime. New York, Interscience, pp. 437–451.

Blackwell, D.D. 1986. Use of heat-flow/temperature measurements, including shallow measurements, in hydrocarbon exploration. In: Davidson, M. (ed.): Unconventional Methods in Exploration for Petroleum and Natural Gas, IV. Dallas, Southern Methodist University Press, pp. 321–351.

Blackwell, D.D., and Spafford, R.E. 1987. Experimental methods in continental heat-flow. In: Sammis, C.G., and Henyey, T.L. (eds.): Geophysics Field Measurements. Methods of Experimental Physics: Vol. 24, Part B. New York, Academic Press, pp. 189–226.

Blackwell, D.D., Steele, J.L., and Steeples, D.W. 1981. Heat-flow determination in Kansas and their implications for midcontinent heat-flow patterns. EOS, Transactions of the American Geophysical Union 62:392.

Bostick, N.H., and Freeman, V.L. 1984. Vitrinite reflectance and paleotemperature models tested at DOE's multiwell experiment site in the Piceance Basin, Colorado. U.S. Geological Survey Report EMG-OGR, 10 pp.

Bredohoeff, J.D., and Papadopulos, I.S. 1965. Rates of vertical groundwater movement estimated from the earth's thermal profile. Water Resources Research 1:325–328.

Clark, S.P., Jr. 1966. Thermal conductivity. In: Clark, S.P., Jr. (ed.): Handbook of Physical Constants. Geological Society of America Memoir 97, pp. 459–482.

Coates, M.S., Haimson, B.C., Hinze, W.J., and Van Schmus, W.R. 1983. Introduction to the Illinois deep hole project. Journal of Geophysical Research 88:7267–7275.

Combs, J.B. 1970. Terrestrial heat-flow in north central United States. Ph.D. dissertation, Massachusetts Institute of Technology, Cambridge, 317 pp.

Combs, J.B., and Simmons, G. 1973. Terrestrial heat-flow determinations in the northcentral United States. Journal of Geophysical Research 78:441–461.

Conaway, J.G. 1977. Deconvolution of temperature gradient logs: Geophysics 42:823–837.

Conaway, J.G., and Beck, A.E. 1977. Fine-scale correlation between temperature gradient logs and lithology. Geophysics 42:1401–1410.

Costain, J.K. 1970. Probe response and continuous temperature measurements. Journal of Geophysical Research 75:3969–3975.

Diment, W.H. 1967. Thermal regime of a large diameter borehole: Instability of the water column and comparison of air- and water-filled conditions. Geophysics 32:720–726.

Domenico, P.A., and Palciauskas, V.V. 1973. Theoretical analysis of forced convective heat transfer in regional ground-water flow. Geological Society of America Bulletin 84:3803–3814.

Duey, H.D. 1983. Oil generation and entrapment in Railroad Valley, Nye County, Nevada. Geothermal Resources Council Special Report 13, pp. 199–206.

Evans, T.R. 1977. Thermal properties of North Sea rocks. The Log Analyst 18(2):3–12.

Garven, G., and Freeze, R.A. 1984. Theoretical analysis of the role of groundwater flow in the genesis of stratabound ore deposits. American Journal of Science 284:1085–1174.

Gatenby, G.M. 1980. Exploration ramifications of subsurface fluid migrations in the Lake Borgne-Valentine area of Southeastern Louisiana. Transactions of the Gulf Coast Association of Geological Societies 30:91–104.

Gosnold, W.D., Jr. 1984. Heat-flow and groundwater movement in the Central Great Plains. In: Jorgensen, D.G., and Signor, D.C. (eds.): Proceedings of the Geohydrology Dakota Aquifer Symposium. Worthington, OH, Water Well Journal Publishing Company, pp. 70–75.

Gosnold, W.D., Jr. 1985. Heat-flow and groundwater flow in the Great Plains of the United States. Journal of Geodynamics 4:247–264.

Gosnold, W.D., Jr., Eversoll, D.A., and Carlson, M.P. 1982. Three years of geothermal research in Nebraska. In: Ruscetta, C.A. (ed.): Geothermal Direct Heat Program Roundup Technical Conference Proceedings: Vol. 1. U.S. Department of Energy Report ID12079-79, pp. 142–157.

Goss, R., and Combs, J. 1976. Thermal conductivity measurement and prediction from well log parameters with borehole application. In: Second United Nations Symposium on the Development and Use of Geothermal Resources. Washington, DC, U.S. Government Printing Office, pp. 1019–1027.

Gretener, P.E. 1967. On the thermal instability of large diameter wells: An observational report. Geophysics 32:727–738.

Gretener, P.E. 1981. Geothermics: Using temperature in hydrocarbon exploration. American Association of Petroleum Geologists Education Course Note Series 17, 170 pp.

Hagedorn, D.N. 1985. The calculation of synthetic thermal conductivity logs from conventional geophysical well logs. M.S. thesis, Southern Methodist University, Dallas, TX, 110 pp.

Harrison, F.W., III. 1980. The role of pressure, temperature, salinity, lithology, and structure in hydrocarbon accumulation in Constance Bayou, Deep Lake, and Southeast Little Pecan Lake Fields, Cameron Parish, Louisiana. Transactions of the Gulf Coast Association of Geological Societies 30:113–129.

Hawtof, E.M. 1930. Results of deep well temperature measurements in Texas. American Petroleum Industry Production Bulletin 205:62–108.

Houbolt, J.J.H.C., and Wells, P.R.A. 1980. Estimation of heat-flow in oil wells based on a relation between heat conductivity and sound velocity. Geologie en Minjnbouw 59:215–224.

Hyndman, D.D., Jessop, A.M., Judge, A.S., and Rankin, D.S. 1979. Heat-flow in the Maritime Provinces of Canada. Canadian Journal of Earth Sciences 16:1154–1165.

Judge, A.S., and Beck, A.E. 1973. Analysis of heat-flow data: Several bore holes in a sedimentary basin. Canadian Journal of Earth Sciences 10:1494–1507.

Kayal, J.R., and Christoffel, D.A. 1982. Relationship between electrical and thermal resistivities for differing grades of coal. Geophysics 47:127–129.

Keen, C.E., and Lewis, T. 1982. Measured radiogenic heat production in sediments from continental margin of eastern North America: Implications for petroleum generation. American Association of Petroleum Geologists Bulletin 66:1402–1407.

Lachenbruch, A.H. 1968. Preliminary geothermal model of the Sierra Nevada. Journal of Geophysical Research 73:6977–6989.

Lam, H.L., Jones, F.W., and Majorowicz, J.A. 1985. A statistical analysis of bottom-hole temperature data in southern Alberta. Geophysics 50:677–684.

Mansure, A.J., and Reiter, M. 1979. A vertical groundwater movement correction for heat-flow. Journal of Geophysical Research 84:3490–3496.

McKenzie, D.P. 1978. Some remarks on the development of sedimentary basins. Earth and Planetary Science Letters 40:25–32.

McKenzie, D.P. 1981. The variation of temperature with time and hydrocarbon maturation in sedimentary

basins formed by extension. Earth and Planetary Science Letters 55:87–98.
Merkle, R.H., Maccary, L.M., and Chico, R.S. 1976. Computer techniques applied to formation evaluation. The Log Analyst 17(3):3–10.
Nielsen, S.B., and Balling, N. 1984. Accuracy and resolution in continuous temperature logging. Tectonophysics 103:1–10.
O'Brien, J.J., and Lerche, I. 1984. The influence of salt domes on paleotemperature distributions. Geophysics 49:2032–2043.
Poole, F.G., and Claypool, G.E. 1984. Petroleum source-rock potential and crude-oil correlation in the Great Basin. In: Woodward, J., Meissner, F.F., and Clayton, J.L. (eds.): Hydrocarbon Source Rocks of the Greater Rocky Mountain Region. Denver, Rocky Mountain Association of Geologists, pp. 179–229.
Rahman, J.L., and Roy, R.F. 1981. Preliminary heat-flow measurements at the Illinois deep drill hole (abst.). EOS, Transactions of the American Geophysical Union 62:388.
Reiter, M., Mansure, A.J., and Peterson, B.K. 1980. Precision continuous temperature logging and comparison with other types of logs. Geophysics 45: 1857–1868.
Roberts, W.H., III. 1980. Design and function of oil and gas traps. In: Roberts, W.H., III, and Cordell, R.J. (eds.): Problems of Petroleum Migration. American Association of Petroleum Geologists Studies in Geology: Vol. 10. American Association of Petroleum Geologists, pp. 217–240.
Roberts, W.H., III. 1981. Some uses of temperature data in petroleum exploration. In: Gottlieb, B.M. (ed.): Unconventional Methods in Exploration for Petroleum and Natural Gas, II. Dallas, Southern Methodist University Press, pp. 8–48.
Robertson, E.C. 1979. Thermal conductivities of rocks. U.S. Geological Survey Open-File Report 79-356, 31 pp.
Roy, R.F., Beck, A.E., and Touloukian, Y.S. 1981. Thermophysical properties of rocks. In: Touloukian, Y.S., Judd, W.R., and Roy, R.F. (eds.): Physical Properties of Rocks and Minerals: Vol. II-2. New York, McGraw-Hill Cindus, pp. 409–502.
Roy, R.F., Blackwell, D.D., and Birch, F. 1968. Heat generation of plutonic rocks and continental heat-flow provinces. Earth and Planetary Science Letters 5:1–12.
Roy, R.F., Decker, E.R., Blackwell, D.D., and Birch, F. 1968. Heat-flow in the United States. Journal of Geophysical Research 73:5207–5222.
Sammel, E.A. 1968. Convective flow and its effect on temperature logging in small diameter wells. Geophysics 33:1004–1012.
Sass, J.H., and Galanis, S.P., Jr. 1983. Temperatures, thermal conductivity, and heat-flow from a well in Pierre Shale near Hayes, South Dakota. U.S. Geological Survey Open-File Report 83-25, 10 pp.
Sass, J.H., Lachenbruch, A.H., and Munroe, R.J. 1971. Thermal conductivity of rocks from measurements on rock fragments and its application to heat-flow determinations. Journal of Geophysical Research 76:3391–3401.
Sclater, J.C., and Christie, P.A.F. 1980. Continental stretching: An explanation of the post-mid-Cretaceous subsidence of the Central North Sea Basin. Journal of Geophysical Research 85:3711–3739.
Sharp, J.M., Jr., and Domenico, P.A. 1976. Energy transport in thick sequences of compacting sediment. Geological Society of America Bulletin 87: 390–400.
Sleep, N.H. 1971. Thermal effects of the formation of Atlantic continental margins by continental break up. Geophysical Journal of Royal Astronomical Society 24:325–350.
Smith, L., and Chapman, D.S. 1983. On the thermal effects of groundwater flow: I. Regional scale systems. Journal of Geophysical Research 88:593–608.
Somerton, W.H. 1975. Thermal properties of partially liquid saturated rocks at elevated temperatures and pressures. American Petroleum Institute Research Project Report 155, 35 pp.
Toth, J. 1980. Cross-formational gravity-flow of groundwater: A mechanism of the transport and accumulation of petroleum (the generalized hydraulic theory of petroleum migration). In: Roberts, W.H., III, and Cordell, R.J. (eds.): Problems of Petroleum Migration. American Association of Petroleum Geologists Studies in Geology: Vol. 10. American Association of Petroleum Geologists, pp. 121–167.
Vacquier, V. 1985. The measurement of thermal conductivity of solids with a transient linear heat source on the plane surface of a poorly conducting body. Earth and Planetary Science Letters 74:275–279.
Van Ostrand, C.E. 1934. Temperature gradients. In: Wrather, W.E., and Lahee, F.H. (eds.): Problems in Petroleum Geology. Tulsa, OK, American Association of Petroleum Geologists, pp. 989–1021.
Van Ostrand, C.E. 1937. On the estimation of temperatures at moderate depths in the crust of the earth. Transactions of the American Geophysical Union 18 (pt. 1):21–33.

3 Reaction of Organic Material to Progressive Geological Heating

Stephen Creaney

Abstract

The generation of oil is a process which begins at some point during the burial history of a source rock. This "onset of maturation" is dictated largely by temperature and residence time. However, the nature of the source rock itself also influences the hydrocarbon product being expelled from the source rock. The vast majority of the world's oil can be ascribed to source rocks of the following types:

A. Marine mudrocks that were deposited under anoxic conditions and dominated by planktonic organisms and anaerobic bacteria. This type of source rock can have a carbonate or clay inorganic matrix and total organic carbon values from 1-30 weight % (commonly 4-10% when immature). Examples of this classical oil source rock are the source rocks of Western Canada, the Middle East, and the Malm of the North Sea.

B. Specific coal facies such as torbanites and cannel coals containing a mixture of hydrogen rich plant detritus (for example, spores, pollen, cuticle, resin, and algae). Deposition was probably in open water areas of an overall coal swamp environment. Examples of these source rocks can be found in the Gippsland Basin, Canadian Beaufort Sea, and Southeast Asia.

C. Lacustrine organic-rich deposits that are rich in freshwater algae, which ultimately result in high wax crude oils. Examples are relatively rare but include major source rocks in the Uinta Basin and China. This type of source rock will not be discussed further in this chapter.

The effect of increasing maturity on marine mudrocks of the Devonian Duvernay Formation of Alberta will be described in detail to illustrate oil generation from the Type A (marine) source rocks. The data base for this study consisted of 47 conventional cores, ranging from immature to completely overmature, and 93 oils from separate accumulations presumed to have been sourced from the Duvernay.

An illustration of oil generation in a Type B (coaly) source rock is provided by a single core from the Lower Cretaceous of the Beaufort-Mackenzie Basin plus many of the oils and condensates reservoired in that area.

Introduction

The process of hydrocarbon generation in the earth's crust has been studied for many years (see Hunt, 1979, for historical review) and the general consensus appears to be that the following are required to generate oil and gas: 1) a source rock, and 2) exposure to elevated temperatures (maturity).

The subsequent secondary migration and accumulation of oil and gas is complex and will not be dealt with in this chapter. The present volume is a discussion of the methods currently available to determine the thermal history of a sedimentary basin and therefore deals with the maturity aspect of oil and gas generation.

Hydrocarbon generation models have been presented by many authors (e.g., Shibaoka et al., 1973; Tissot et al., 1974; Vassoevich et al., 1974; Hood et al., 1975; Dow, 1977; Snowdon and Powell, 1982). Most of these models describe three stages of source rock evolution: 1) an immature stage—source rocks have not achieved sufficient time/temperature exposure to generate oil but may have produced biogenic methane, 2) a mature

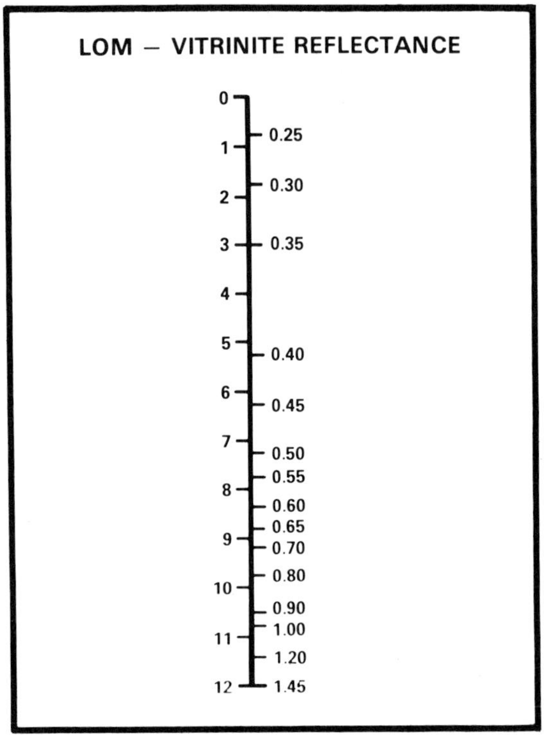

FIGURE 3.1. Correlation of level of organic metamorphism (LOM) and vitrinite reflectance. (From Hood et al., 1975.)

stage – source rocks begin to generate oil and thermogenic gas, and 3) an overmature stage – source rocks are exposed to a thermal regime that allows only gaseous hydrocarbons to be generated and preserved. Several authors have studied the effect of organic matter variability on the nature and amount of hydrocarbons capable of being generated (Tissot et al., 1974; Hunt, 1979). Frequently this type of study is chemical in nature and assumes that all types of organic material can be present in any depositional environment. The geological aspects of source rock accumulation are not used to constrain the possible organic matter variability in source rock analysis.

The present chapter is an attempt to introduce the reader to the process of hydrocarbon generation via a discussion of the interplay of source rock composition and thermal maturity. Throughout this discussion, maturity is referred to in terms of the LOM (Level of Organic Metamorphism) scale (from Hood et al., 1975). This scale provides a convenient means of discussing maturity, and can be used to correlate maturity data derived from a variety of different analyses. Figure 3.1 illustrates the correlation of LOM to vitrinite reflectance.

Oil Source Rock Types

Potential oil source rocks are organic-rich sediments containing organic material that is sufficiently hydrogen rich (Tissot et al., 1974) to convert mainly to oil during thermal maturation. The organic materials that are generally richest in hydrogen are marine plankton, freshwater algae, spores, pollen, leaf cuticle, resin, and anaerobic bacteria (Van Krevelen, 1961; Tissot et al., 1974). All the above materials are oil prone and would be classified as Type I or II kerogen (Tissot et al., 1974). However, they are usually deposited in very different depositional environments. Hedberg (1964, 1968) discussed the subdivision of oil source rocks into "marine" and "nonmarine" based largely on the geological differences between these depositional environments plus the differences in oils presumed to have been sourced from marine and nonmarine rocks. Some more recent studies have concentrated on characterizing the geochemical "signatures" of marine and nonmarine hydrocarbons (e.g., Grantham et al., 1981; Lijmbach et al., 1981; Rohrback, 1981; Thomas, 1982; Durand and Parratte, 1983).

Figure 3.2 outlines a simplified classification of oil source rocks with examples of basins in which they occur. The remainder of this chapter discusses the effect of increasing maturity on a marine oil source rock and a nonmarine oil source. Both are illustrated with examples from Canadian basins. The imbalance in detail and sample coverage in favor of the marine source rock case study reflects the current knowledge level associated with the nonmarine type.

A Marine Source Rock: A Case Study from the Devonian of Western Canada

Geological Setting

The Upper Devonian of east-central Alberta, Canada, contains a marine, oil-prone source rock – the Duvernay Formation (Fig. 3.3). The stratigraphic

TYPE	WATER	ORGANIC MATTER	EXAMPLES
TYPE A	MARINE	TYPE II KEROGEN PLANKTON + ANAEROBIC BACTERIA	MIDDLE EAST WESTERN CANADA NORTH SLOPE NORTH SEA (MALM) PARIS BASIN
TYPE B	FRESH BRACKISH	TYPE II/III KEROGEN SPORES, POLLEN, CUTICLE, FRESHWATER ALGAE, VITRINITE, INERTINITE	BEAUFORT BASIN (CANADA) SCOTIAN SHELF (CANADA) GIPPSLAND BASIN (AUSTRALIA)
TYPE C	FRESH	TYPE I KEROGEN FRESHWATER ALGAE	GREEN RIVER SHALE (INTERIOR U.S.) CHINA

FIGURE 3.2. General source rock classification based on depositional environment.

relationships of this formation are depicted in Figure 3.4.

The Duvernay Formation comprises a sequence of dark-brown to black, bituminous, slightly argillaceous carbonates interbedded with grey-green, calcareous shales. These sediments are characteristically organic rich, with total organic carbon (TOC) values of 2.0 to 17%, and exhibit plane-parallel, millimeter lamination (Fig. 3.5). In the East Shale Basin (see Figs. 3.3 and 3.4) the Duvernay Formation overlies platform carbonates of the Cooking Lake Formation with minor discontinuity, and is the basinal equivalent of surrounding lower Leduc Formation reefs. The Duvernay thickens northward and eastward up a depositional slope, passing into lithologies more typical of the overlying lower Ireton Formation, which conformably overlies it elsewhere. Time stratigraphic markers in the basinal sequence cut obliquely through the top of the Duvernay Formation, indicating the facies-controlled nature of this upper contact (Stoakes, 1980). Basin filling proceeded from east to west, and the top of the Duvernay Formation also rises stratigraphically in this direction. Where the Cooking Lake Formation is absent, as in the West Shale Basin, the Duvernay conformably overlies rocks of similar basinal aspect, referred to as the Majeau Lake Member of the Duvernay Formation. The Majeau Lake Member is the basinal equivalent of the Cooking Lake Platform and, as such, predates reef growth. Lithologically it is identical to the Duvernay Formation. Undoubtedly, the Majeau Lake Member contributed hydrocarbons to the Leduc Formation reservoirs, and for convenience it has been grouped with the Duvernay to the west of the Rimbey-Morinville Leduc reef trend.

The Duvernay and Majeau Lake represent accumulations under marine, deep-water, low-energy, basinal conditions. Oxygen-starved, euxinic conditions are suggested by the absence of fauna, preservation of organic material (Type II oil-prone kerogen), and color of the sediment. Evidence suggests that euxinic conditions existed in water depths on the order of 100 m in the East Shale Basin (Stoakes, 1980). Undoubtedly the presence of anoxic conditions, combined with slow sedimentation rates within this depositional basin, are the main reasons for preservation of abundant organic material in this rich source rock.

Euxinic laminites show the highest TOCs (up to 17% by weight), with bioturbated dysaerobic or aerobic sediments exhibiting markedly lower organic content (less than 1.0% by weight). Compositional analysis of the laminites reveals that they are, indeed, true carbonate source rocks with very low clay contents (Fig. 3.5). The requirement for an anoxic sediment/water interface for the preservation of abundant organic matter has been well documented by Demaison and Moore (1980). Further details of the deposition of the Duvernay source rock are outlined in Stoakes and Creaney (1984).

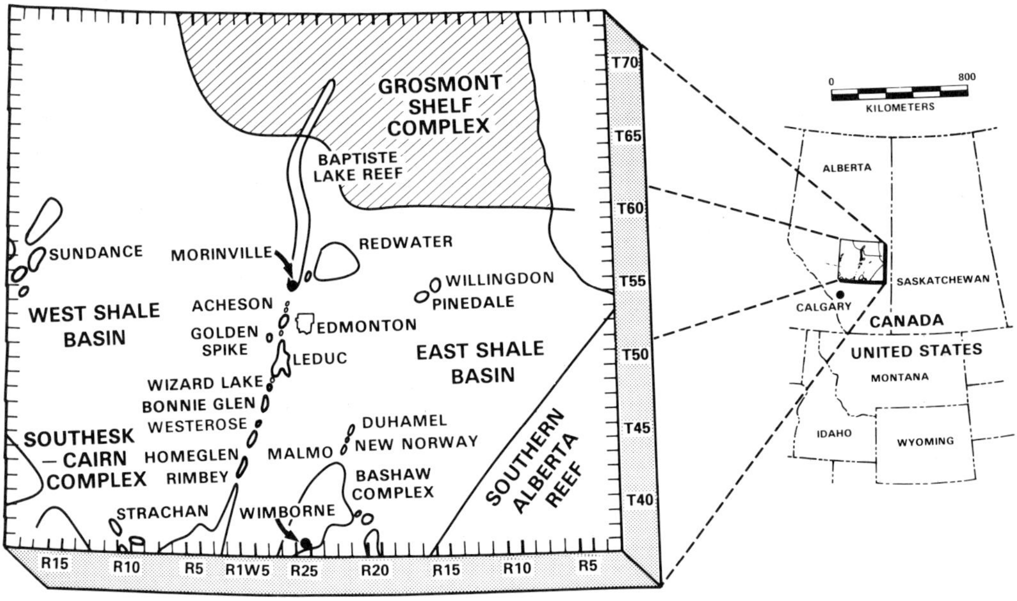

FIGURE 3.3. Location map of east-central Alberta, Canada. Significant Leduc age (Upper Devonian) features are identified.

Figure 3.6 indicates the locations of the 47 cores analyzed for present study. This large data base combined with the degree of stratigraphic control, the availability of oils presumed to have been sourced from the Duvernay Formation, and the homogeneity of organic type (Fig. 3.7) make the Duvernay an excellent example of the response of a marine source rock to maturity increase.

Effect of Increasing Maturity

The effect of increasing maturity on a source rock can be observed as progressive changes in the solid organic residue (kerogen) or as changes in the

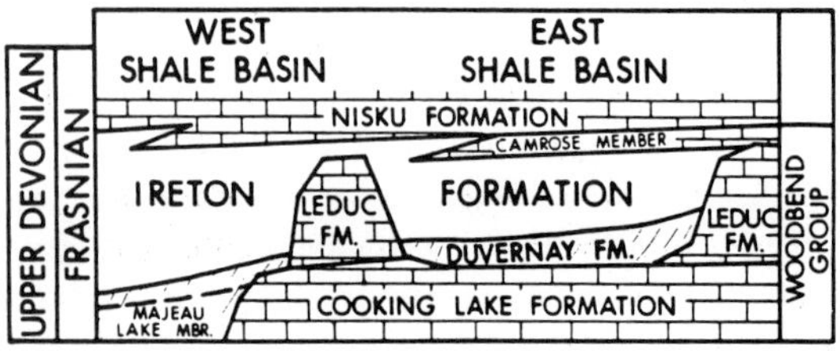

FIGURE 3.4. Generalized stratigraphy of the Devonian Woodbend Group, east-central Alberta, Canada. (From Stoakes and Creaney, 1984.)

hydrocarbons being released into source rock porosity from the solid organic matter.

Figure 3.8 is a LOM map for the Duvernay Formation in east-central Alberta. LOM data are derived from a combination of Rock-Eval pyrolysis (T_{max}) data on the source interval itself, plus vitrinite reflectance information projected from overlying Lower Cretaceous coals. The tectonic and thermal histories of this basin are relatively simple (Deroo et al., 1977) and, hence, strong correlations between maturity and present burial depth are possible (Fig. 3.9) and can be used to map maturity across broad areas of the basin.

The most convenient way to express the various stages of source rock maturity is through an analysis of the contents of source rock porosity. The onset of maturity can be defined as the first point when thermogenic hydrocarbons appear in source rock porosity. Analyses that sample pore contents include total solvent extraction, cuttings gas analysis, and Rock-Eval pyrolysis (S_1). These analyses may also sample the solid organic matter, to some extent, but in oil-prone source rocks this contribution is very rapidly overwhelmed by hydrocarbons appearing in source rock porosity. Figure 3.10 is a composite plot of total solvent extract yield (normalized to total organic carbon) and S_1 (in milli-

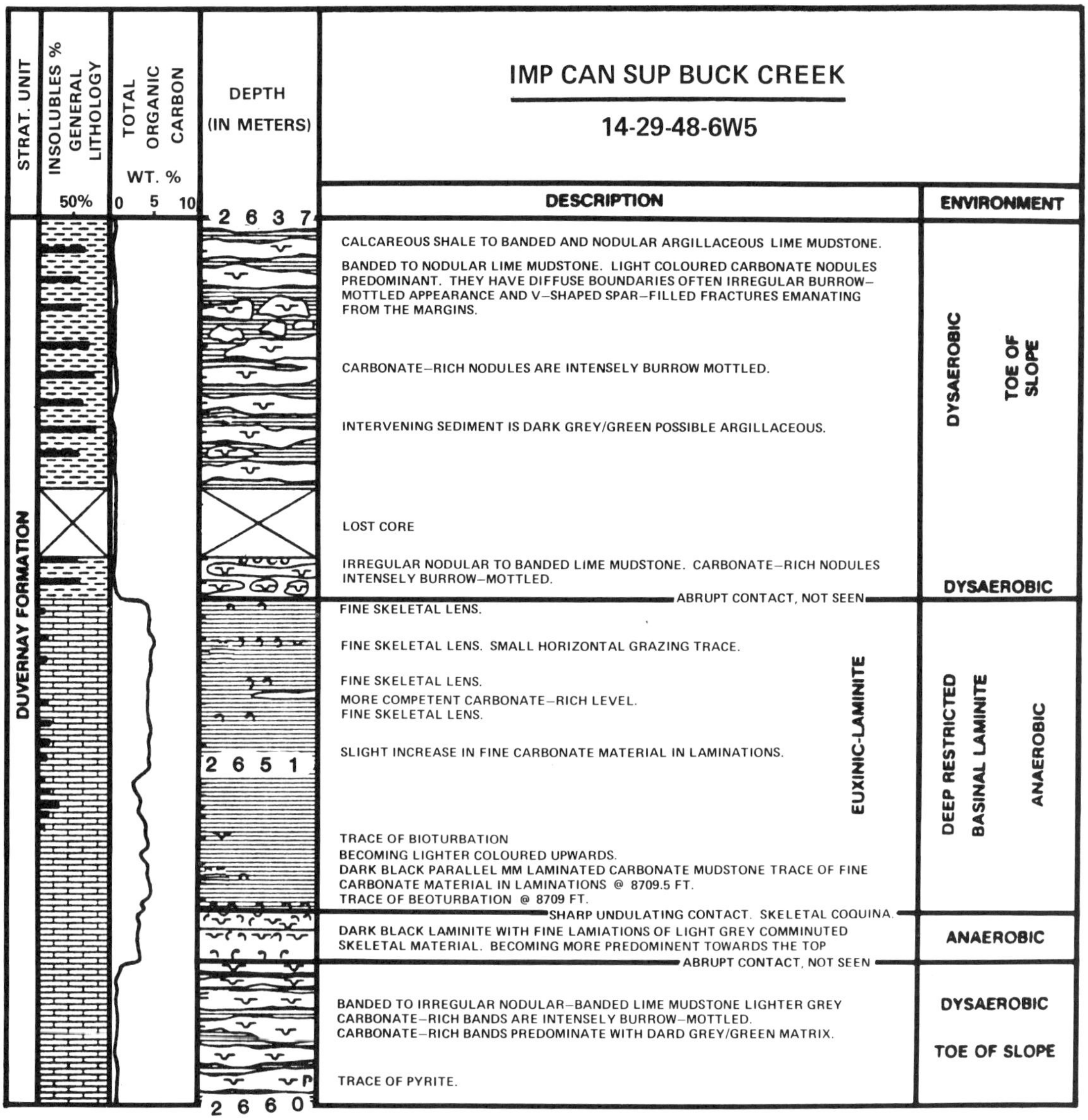

FIGURE 3.5. Example of detailed lithology, facies, total organic carbon, and acid-insoluble residue log for the Duvernay Formation. (From Stoakes and Creaney, 1984.)

grams of hydrocarbons per gram of rock) against LOM. It is apparent that hydrocarbons begin to appear in source rock porosity at a LOM of 6 to 8 (vitrinite reflectance, 0.43 to 0.57). A maximum occurs in the curves at a LOM of 10 to 10.5 (vitrinite reflectance, 0.82 to 0.92) indicating maximum oil saturation in the source rock. It is at this maturity that maximum oil expulsion from this type of source rock might be expected. As maturity increases, the total amount of liquid hydrocarbon in porosity decreases due to a combination of porosity reduction during compaction, oil expulsion, and conversion of oil to gas.

Liquid hydrocarbons become virtually undetectable at a LOM of 12 to 12.25. A LOM of 12.25 is considered here to be the onset of overmaturity, and is characterized by the absence of liquid hydrocarbons in porosity with a switch to gaseous hydrocarbon preservation only. This is better observed using cuttings gas analysis, because extract and pyrolysis analyses do not observe hydrocarbons in the C_1-C_4 range. Bailey et al.

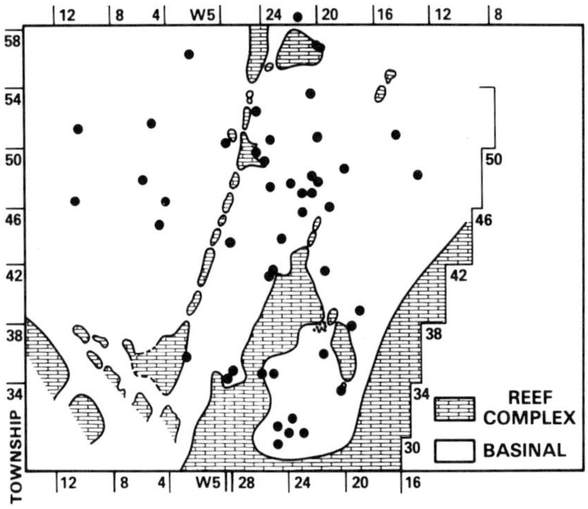

FIGURE 3.6. Location map of Duvernay Formation cores used in this study.

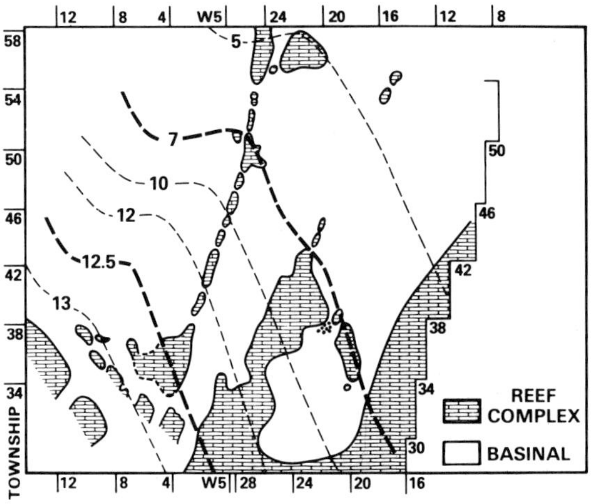

FIGURE 3.8. Regional maturity map of the Duvernay Formation in east-central Alberta, Canada (contours are LOM units).

(1974) summarized the effect of overmaturity on marine source rocks. Briefly, the percentage of components with more than one carbon atom decreases across the mature/overmature boundary until only methane remains. Figure 3.11 shows the trend in compound depletion with passage into the zone of overmaturity, based on analyses of many wells in the Western Canada Basin. Obviously, the ability to retain liquid hydrocarbons in source rock porosity is absent after a LOM of 12.5 (vitrinite reflectance, 1.65) and wet gas

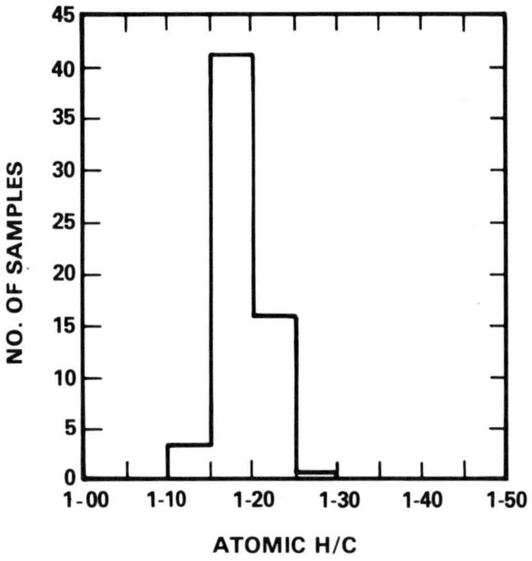

FIGURE 3.7. Distribution of atomic H/C analyses from the Duvernay Formation in two wells (Esso Redwater 10-27-57-21w4 and Esso Redwater 16-28-57-21w4). The narrow range attests to the lack of variation through time in organic matter type within this source rock. Both cores are at a LOM of 5.

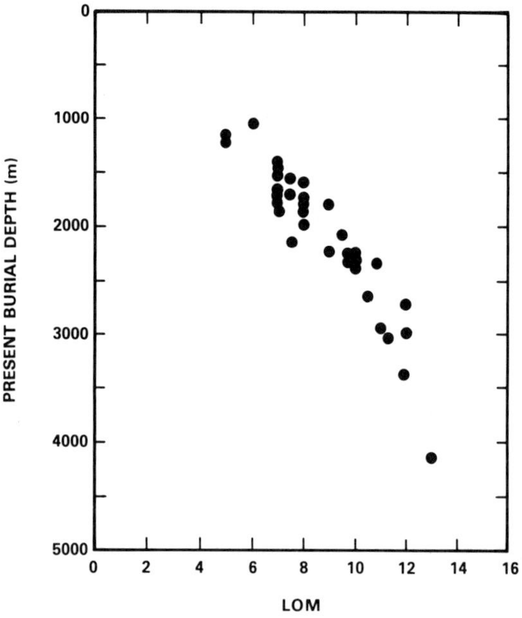

FIGURE 3.9. Relationship between maturity (expressed as LOM) and present burial depth for the Duvernay Formation in east-central Alberta, Canada.

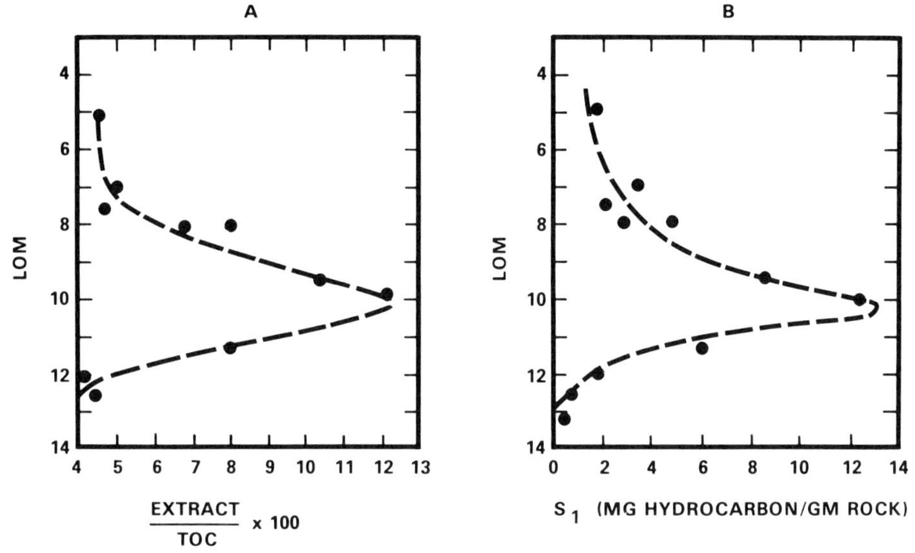

FIGURE 3.10. A, Relationship between total extract yield (expressed as a percent of total organic carbon) and maturity (LOM) for the Duvernay Formation. B, Relationship between Rock-Eval pyrolysis S_1 yield (in milligrams of hydrocarbon per gram of rock) and maturity (LOM) for the Duvernay Formation.

(C_2+) is absent by a LOM of 13 (vitrinite reflectance, 1.85).

Distribution of Reservoired Hydrocarbon Maturity

In recent years a few authors have published work that attempted to assess the maturity of reservoired hydrocarbons (Lijmbach et al., 1981; James, 1983). This approach provides the possibility of increasing our insight into the secondary migration pathways taken by oil and gas from source to trap. In the present study, the same approach has been used to test the oil generation model outlined above for a marine source rock. If the total extract components removed by solvent extraction from a source rock are representative of oil retained in source rock porosity, then they will reflect the

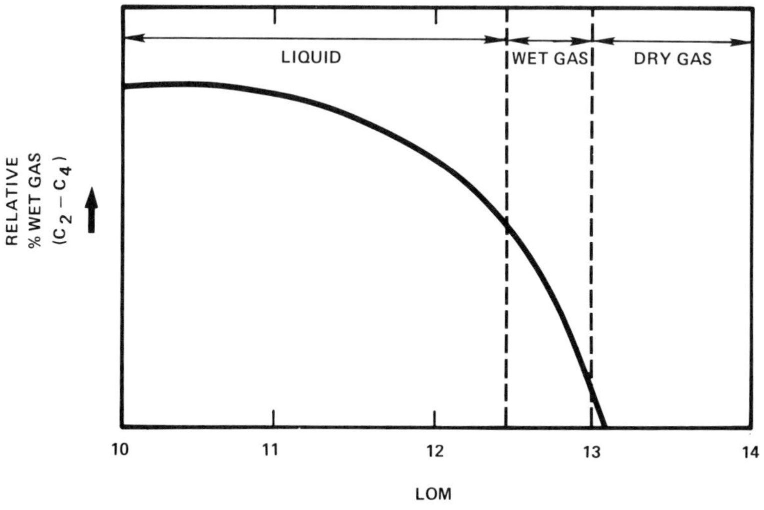

FIGURE 3.11. Schematic relationship between cuttings gas composition and advanced maturity (LOM) (based on analyses of many wells in western Canada; see Bailey et al. [1974] for an example of an individual well profile).

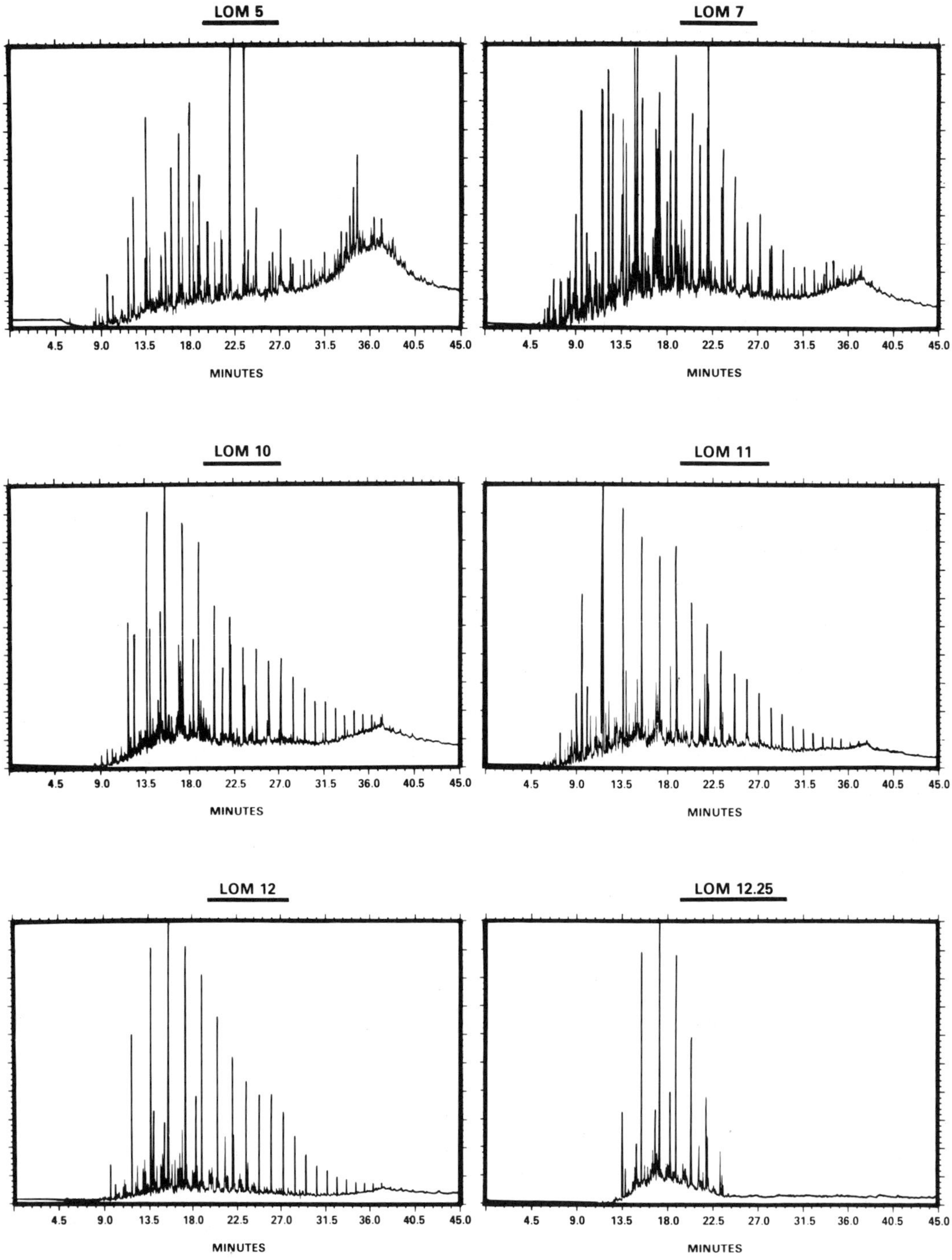

FIGURE 3.12. Total extract gas chromatographs from Duvernay Formation core samples at a range of maturity values (expressed as LOM). GC conditions: *GC*, Hewlett Packard Model 5710 fitted with an auxiliary liquid CO_2 cryogenic attachment; *Split*, 250:1 or 300:1; *Column*, 15 m SE-30 WCOT (0.2 mm I.D.) fused silica; *Sample*, 5 μL of 5% solution; *Program*, 0°C–300°C at 8°C/min, hold at 300°C for 8 min.

maximum maturity attained by that source rock at that location. Figure 3.12 is a composite of whole extract gas chromatograms (GC) from a series of Duvernay Formation cores of increasing maturities. Certain progressive changes can be observed:

1. Virtually no n-alkanes are present in the immature extract and the n-alkane spectrum develops with increasing maturity.
2. The unresolved "hump" (steranes and so forth) decreases progressively with increased maturity.
3. At LOM values greater than 12 (vitrinite reflectance, >1.45) the heavy n-alkanes diminish extremely rapidly as thermal degradation "cracks" the long-chain hydrocarbons.

Overall comparison of these total extract GC profiles with whole oil gas chromatograms of oils reservoired in Leduc Formation reefs (and sourced from the Duvernay) allows the assignment of an approximate LOM to the reservoired oil. This assignment is made based on n-alkane/isoprenoid ratios, overall n-alkane profile, and the relative magnitude of the sterane/triterpane hump. The resulting LOM value is obviously somewhat subjective and, as with all attempts to apply a maturity value to a mobile phase, subject to a number of sources of error. The prime sources of error and an assessment of their applicability would be:

1. Mixing of oils from different sources (No other sources are documented that could contribute significantly to Leduc Formation reservoirs.)
2. Lateral organic facies changes within the source (The lack of clastic input in this carbonate depositional system and the Devonian age of the rock largely preclude this.)
3. Mixing of oils from various maturities through time (Realistically, this must happen in all traps, and early, low-maturity oil will be subsequently overwhelmed with later, higher maturity oil. However, in general the resulting oil mixture will still have a bias toward the peak maturity values (LOM, 10 to 11) because of the exponentially increasing volumes of oil being generated from the source at this maturity [Waples, 1980].)

Figure 3.13 shows the whole oil GC traces for two oils from opposite ends of the maturity spectrum, as examples of the assignment of an oil LOM

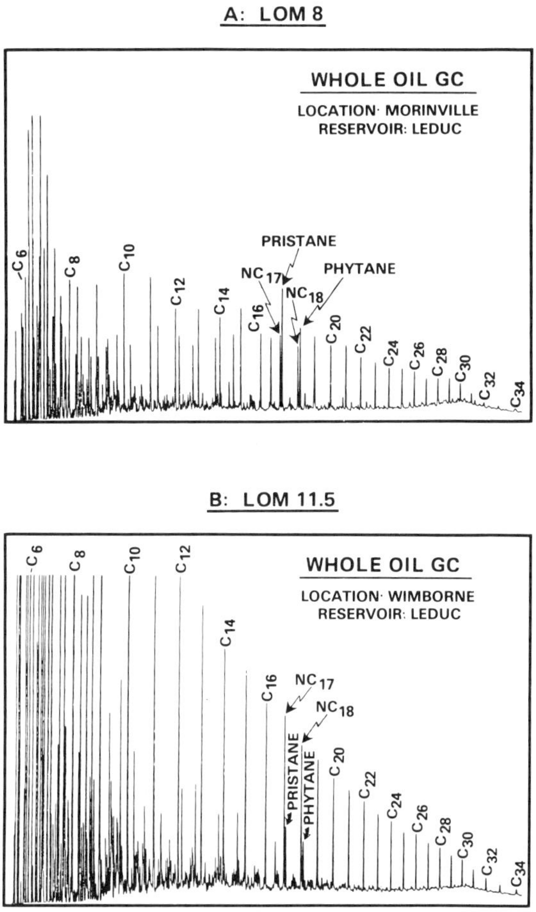

FIGURE 3.13. Whole oil gas chromatograms of two Leduc Formation reservoired oils sourced from the Duvernay Formation source rock. A, Morinville oil–LOM 8. B, Wimborne oil–Lom 11.5. GC conditions as in Figure 3.12 except sample = 0.5 μL.

(see Fig. 3.3 for oil sample locations). This process has been repeated on oils from 93 separate pools in Devonian rocks of the Western Canada Basin. All oils are presumed to have been sourced from the Duvernay Formation marine source rock. Figure 3.14 is a histogram of the total volume of reservoired oil in place (samples of which were used in this analysis) versus the assigned LOM value. It is apparent that the resulting distribution confirms the generation model established for the Duvernay marine source rock (see Fig. 3.10). Small amounts of liquid hydrocarbons occur which have LOMs <8 or >12 and the maximum in the distribution occurs at a LOM of 9 to 11.

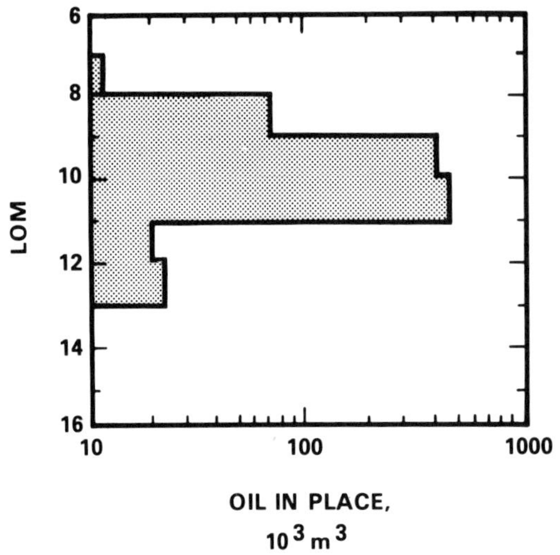

FIGURE 3.14. Distribution of Duvernay Formation sourced oil reserves according to the average maturity (LOM) of each pool/field. Ninety-three oil samples were involved in this analysis.

Terrigenous Source Rocks: A General Model

Hedberg (1964, 1968) recognized the association of high-wax, low-sulfur crude oils with nonmarine, paralic host sediments; and many subsequent authors have pointed to the possibility of "coals" and "coaly shales" sourcing nonmarine crude oils (Powell and McKirdy, 1973; Connan, 1974; Lijmbach et al., 1981; Durand and Parratte, 1983).

Coal is a complex mixture of plant-derived debris that accumulates in anaerobic swamp environments. The exinite or liptinite (hydrogen-rich) component is the fraction that is most likely to source oil upon maturation and the vitrinite (hydrogen-poor) is most likely to source gas. Inertinite (the charcoal component) is probably completely nonvolatile during metamorphism. Since an individual coal seam contains a mixture of all of the above components, both gaseous and liquid hydrocarbons may be generated. Furthermore, the proportion of exinite in an average coal is low (i.e., $< 10\%$), and hence the volume of liquid hydrocarbons generated may be small relative to the volume of gas generated. The vitrinite component probably assimilates a significant proportion of these liquids prior to expulsion, while some of the remainder has been observed as solid hydrocarbon filling fractures and voids in coals (Teichmüller, 1974). However, significant volumes of oil do occur in several basins around the world that have been ascribed to "coal" or coaly sediment source rocks. Examples are the Gippsland Basin (Threlfall et al., 1976), Beaufort-Mackenzie Basin (Snowdon and Powell, 1982), and South China Sea (Lijmbach et al., 1981).

Teichmüller (1962) introduced an organic facies model that defined the facies variations possible in a coal swamp. The facies that was considered to facilitate the enrichment of the exinite component was the lagoonal/open water facies. Figure 3.15 is a slight modification of this coal facies model indicating the overall source potential of these sediments. Thus the lagoonal/embayment environments (described as "Lake" on Fig. 3.15) are probably where "terrigenous" oil-prone source rocks accumulate. The detailed stratigraphic and areal distribution of this type of environment is considerably less predictable than the widespread marine source rock.

This potential for lateral organic facies changes in the "terrigenous" environment (see Fig. 3.15) is further complicated by the diversity of chemical moieties included under the term *exinite/liptinite*, for example, resinite (tree resin), cutinite (leaf cuticle), sporinite (spores), alginite (aquatic algae), and liptodetrinite (fragmental, small pieces of exinite/liptinite of unknown affinity). All of the above have different chemical compositions, and probably respond differently to increasing maturity. The process of their incorporation into lagoonal muds (floating and by wind) probably creates an almost random distribution in the sediment. Thus a variety of oil types may be recognized that have an overall terrigenous aspect but considerable difference in detail. Snowdon and Powell (1982) proposed a model of hydrocarbon generation from terrigenous organic matter that distinguished resin-rich from resin-poor, high-exinite sediments. However, the model was somewhat schematic with reference to maturity, because only samples of the reservoired oils and condensates were available with no data on recognized source rocks.

Example

As mentioned above, the present data base is extremely limited for use as a case study of a non-

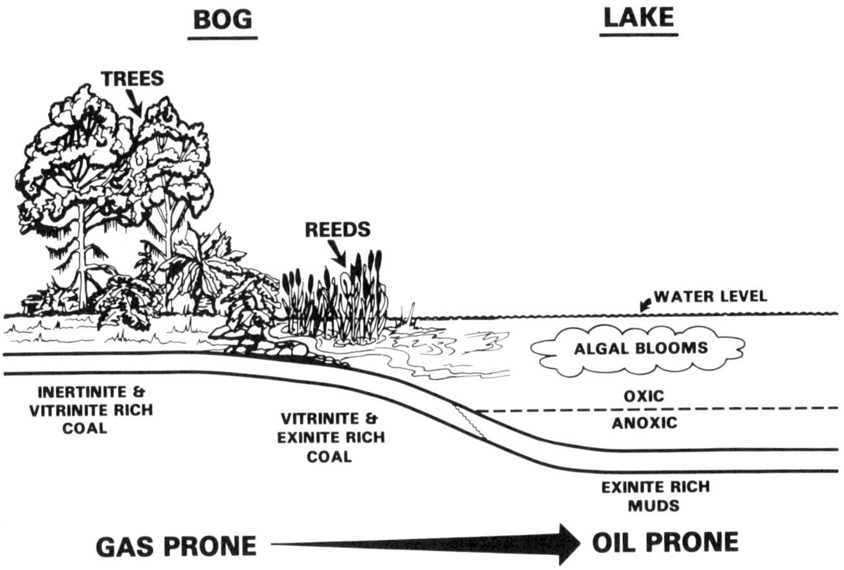

FIGURE 3.15. A general model of terrestrial, oil-prone source rock accumulation. (Modified from Teichmüller, 1962.)

marine source rock. The data presented here are from a single core taken in the Lower Cretaceous of the Beaufort-Mackenzie Basin in Arctic Canada. The core was taken through a potential reservoir unit but also intersected a series of coal swamp-related lagoonal mudstones. The organic richness (TOC) and hydrogen indices (S_2/TOC) of this section are plotted on Figure 3.16. The material is obviously very rich and contains oil-prone organic matter. Organic petrographic analysis of three samples from this section indicated that up to 28% of the organic material was exinitic and that this fraction was normally resinite and sporinite in variable proportions.

The vitrinite reflectance of these samples is 0.55% (LOM, 7.5 to 8.0). Thus, the sediments are organically rich, oil prone, and contain land plant material representing a single maturity level. Examination of the total extract chromatograms from this section (Fig. 3.17) confirms the anticipated chemical variability of these rocks. n-Alkanes vary from almost absent (at 2670.3 m depth) to dominating the chromatogram (2680.4 m). A common feature of all the chromatograms is the distribution of aromatic compounds (as identified on Fig. 3.17, sample from 2630.1 m).

The same features can be observed in reservoired liquid hydrocarbons from the Beaufort-Mackenzie Basin and elsewhere. Figure 3.18 shows a series of whole oil chromatograms from oils and condensates in the Beaufort-Mackenzie Basin, Arctic Canada, and the Scotian Shelf, East Coast Canada which the author believes to have been generated from source facies similar to that described above. The presence of an aromatic profile that is extremely similar to the Lower Cretaceous source facies (Fig. 3.17) suggests that this is potentially a rapid diagnostic "fingerprint" for this type of source. The critical conclusion for the purpose of this study is that all the oils and condensates in Figure 3.18 could have been generated from a variety of terrigenous exinitic source facies. Without further samples of this type of source facies, at a variety of maturities, it is not yet possible to define the response to maturity and the features of the generated product. However, the facies dependence of the generated hydrocarbons probably ceases to apply during the transition into overmaturity. This phase of hydrocarbon generation is dominated by cracking reactions and would be expected to be very similar to the process described earlier for the Duvernay Formation. This suggestion is confirmed somewhat by the observation that in coals the exinite macerals become optically indistinguishable from vitrinite at a LOM of 12.25 (vitrinite reflectance, 1.60%)

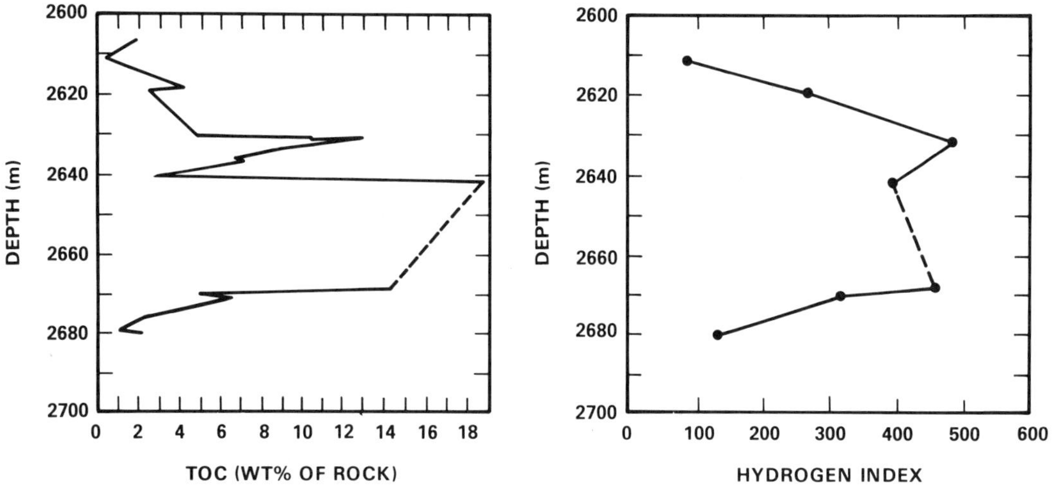

FIGURE 3.16. Plot of total organic carbon and hydrogen index verses depth in coal swamp, lagoonal sediments, Lower Cretaceous, Beaufort-Mackenzie Basin, Canada.

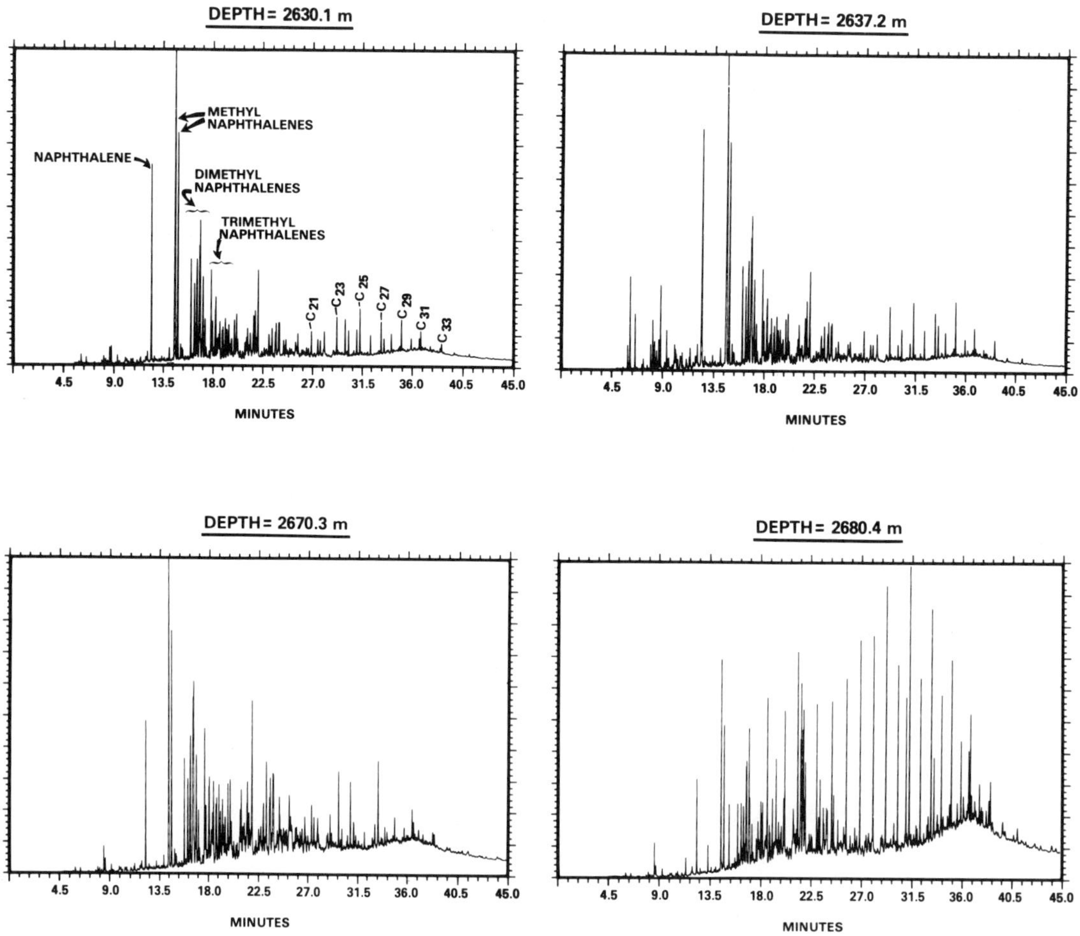

FIGURE 3.17. Total extract gas chromatograms from samples of the core illustrated in Figure 3.16. GC conditions as in Figure 3.12.

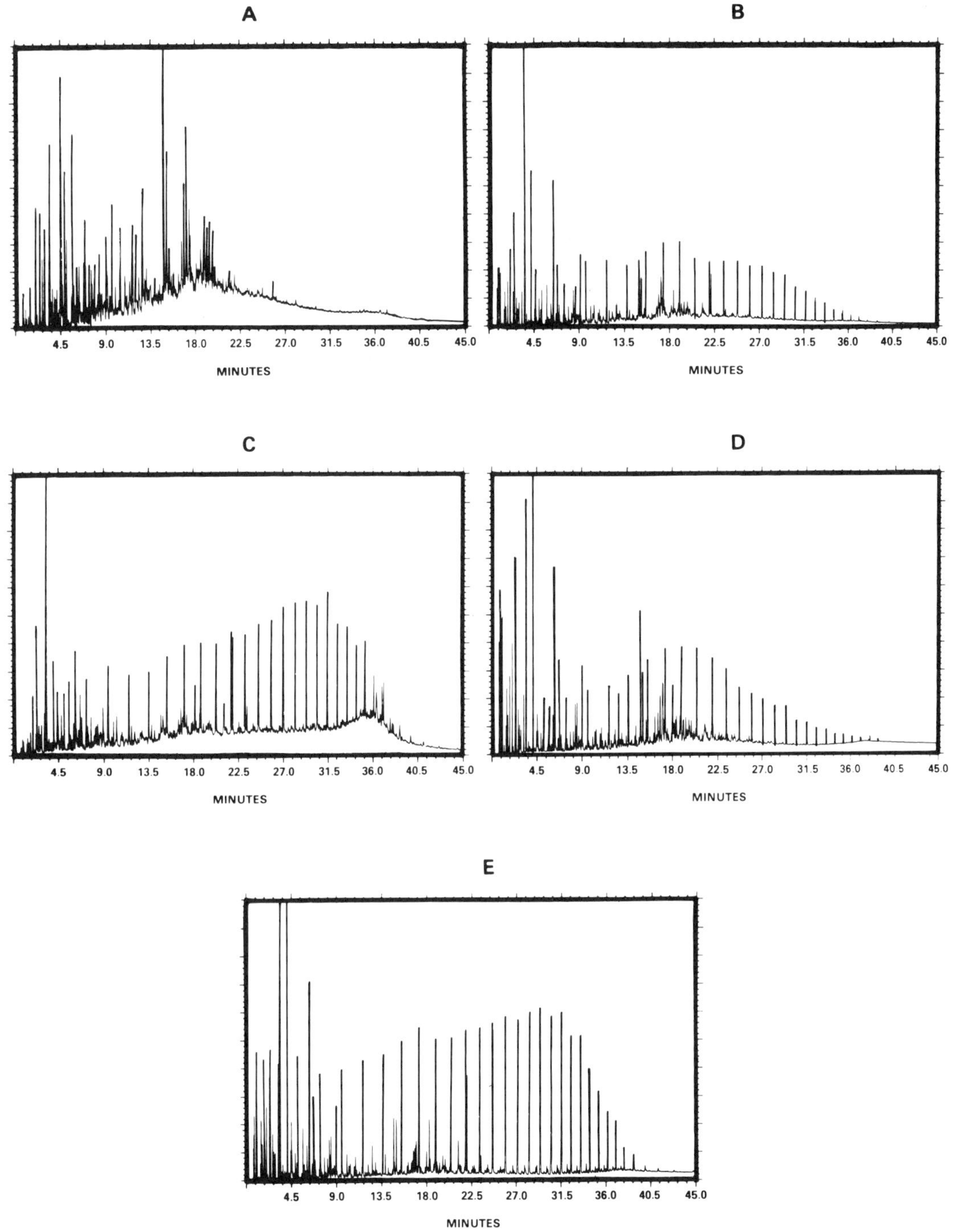

FIGURE 3.18. Whole oil gas chromatograms from a series of oils and condensates known to be sourced from terrestrial organic matter. GC conditions as in Figure 3.12 except sample = 0.5 μL. A, Biodegraded oil, Shell Niglintgak M-19 well, DST #16, 1,746.5 m–1,752.6 m, Beaufort-Mackenzie Basin, Canada. B, Oil, Shell Niglintgak M-19 well, DST #9, 2,174.7 m–2,180.8 m, Beaufort-Mackenzie Basin, Canada. C, Oil, Imperial Pullen E-17 well, F.I.T. #1, 3,590.5 m, Beaufort-Mackenzie Basin, Canada. D, Oil/condensate, Shell Primrose N-50 well, 1,610.4 m–1,650.5 m, Scotian Shelf, Canada. E, Oil, Gippsland Basin, Australia.

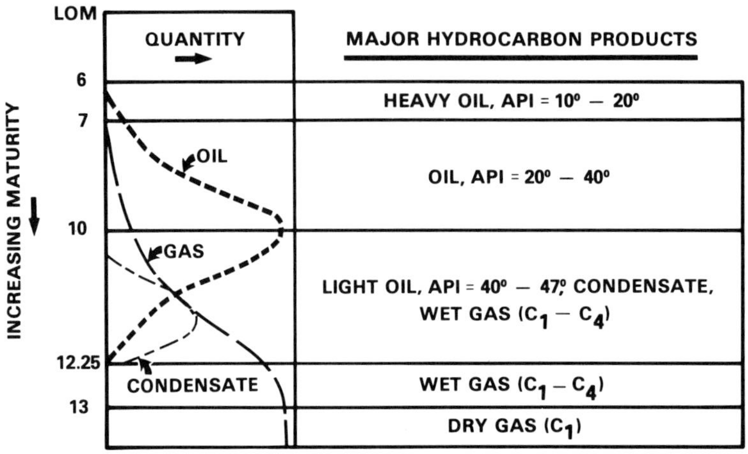

FIGURE 3.19. Schematic maturity (LOM) versus hydrocarbon type and quantity plot for marine, Type II kerogen source rocks.

Creaney, 1980). The suggestion by Snowdon and Powell (1982) that resin-rich terrestrial organic matter can generate light naphthenic oil and condensate at LOM levels as low as 7 remains poorly tested although the thermogravimetric data quoted by Snowdon (1980) support early hydrocarbon generation from pure resin samples.

The total extract chromatograms shown in Figure 3.17 appear less mature than any of the oils or condensates in Figure 3.18 (excluding the biodegraded Beaufort Basin example), which suggests that a LOM of 8 is insufficient for significant "terrigenous" oil generation.

Conclusions

The following observations and conclusions can be made from the above discussion:

1. Two types of source rock depositional environment exist: marine and nonmarine. Both have distinctly different stratigraphic and areal distributions within a basin.
2. The organic matter accumulating in the marine realm is less diverse and complex in space and time than the variety available in the nonmarine, coal swamp environment. Thus, it is relatively simple to distinguish hydrocarbons of marine origin from those of nonmarine origin, but extremely difficult to distinguish hydrocarbons from source rocks of the same type. Variations in degree of anoxia can produce chemical changes in either source rock environment, but such changes can occur within a single source as well as between sources.
3. The chemical homogeneity of organic matter in marine source rocks results in a relatively simple response to increasing maturity. Figure 3.19

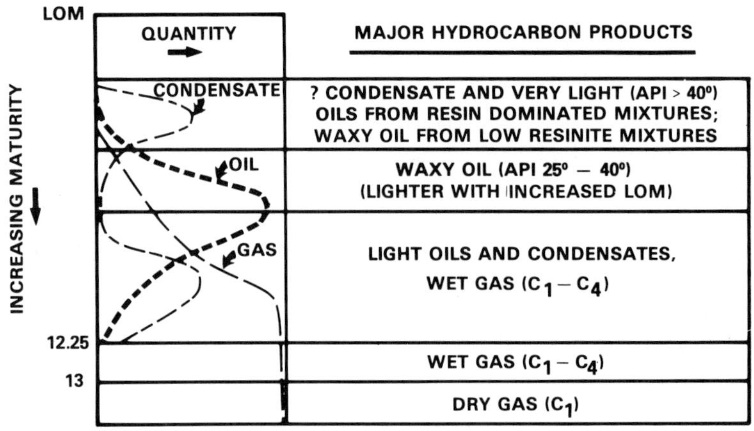

FIGURE 3.20. Schematic maturity (LOM) versus hydrocarbon type and quantity plot for terrestrial/nonmarine, Type II kerogen source rocks. (In part after Snowdon and Powell, 1982.)

is a schematic summary of the response of marine (Type II) organic matter to increased maturity, as developed in a study of the Devonian Duvernay Formation of the Western Canada Basin.

Liquid hydrocarbons begin to be generated at a LOM of 6 to 7 (vitrinite reflectance, 0.44 to 0.48%) and peak at around LOM 10 to 10.5 (vitrinite reflectance, 0.82 to 0.92%). Gas is an increasing part of the product at LOM 10 and above. The liquid hydrocarbons increase in API gravity and their average molecular weight decreases until only compounds less than C_4 are preserved at LOM 12.5 (vitrinite reflectance, 1.65%). Methane (C_1) is the only remaining hydrocarbon at LOMs >13 (vitrinite reflectance, 1.85%).

4. The diverse number of organic entities that can accumulate as Type I or II (oil-prone) organic matter in the nonmarine, coal swamp lagoonal environment includes spores, pollen, leaf cuticle, resin, and latex. These materials have variations in chemical signature and presumably different responses to increasing maturity. Figure 3.20 is a tentative schematic summary of the response of terrestrial, Type II organic matter to maturity based on analyses from a single core in the Canadian Beaufort-Mackenzie Basin plus information available in the literature (Snowdon, 1980; Snowdon and Powell, 1982).

   The onset of hydrocarbon generation in this model is obscure, and hence the lack of information on the LOM axis of Figure 3.20. An indication that the type of organic matter dictates the product of maturity has also been included at the lower maturity levels, with resin-derived light oils and condensates occurring in this region. The cracking reactions that dominate the trend to overmaturity probably occur at the same maturity levels as for marine organic matter, that is, no liquids above LOM 12.5 and primarily dry gas above LOM 13.

Acknowledgments. The author wishes to thank Esso Resources (Canada) Ltd. and Exxon Production Research Company for permission to publish this work. The portion of this work dealing with the Devonian of Western Canada could not have been completed without the input of F.A. Stoakes. The following individuals are also acknowledged for technical discussion and/or assistance: C.W.D. Milner, D.L. Layton, J.D. Stroud, S. Triffo, K. Wedin, R. Adolphe, and C.A. Price. The manuscript was improved by the reviews of F.A. Stoakes, Q.R. Passey, and A.T. James. Dr. J.M. Jones of Geo-Optics Ltd., U.K. performed the organic petrography.

References

Bailey, N.J., Evans, C.R., and Milner, C.D. 1974. Applying petroleum geochemistry to search for oil: Examples from Western Canada Basin. American Association of Petroleum Geologists Bulletin 58: 2284–2294.

Connan, J. 1974. Diagenese naturelle et diagenese artificielle de la matiere organique a elements vegetaux predominants. In: Tissot, B., and Bienner, F. (eds.): Advances in Organic Geochemistry 1973. Paris, Editions Technip, pp. 453–462.

Creaney, S. 1980. Petrographic texture and vitrinite reflectance variation on the Alston Block, north-east England. Proceedings of the Yorkshire Geological Society 42:553–580.

Demaison, G.J., and Moore, G.T. 1980. Anoxic environments and oil source bed genesis. American Association of Petroleum Geologists Bulletin 64:1179–1209.

Deroo, G., Powell, T.G., Tissot, B.P., and McCrossan, R.G. 1977. The origin and migration of petroleum in the Western Canadian Sedimentary Basin, Alberta: A geochemical and thermal maturation study. Geological Survey of Canada Bulletin 262, 136 pp.

Dow, W.G. 1977. Kerogen studies and geological interpretations. Journal of Geochemical Exploration 7:79–99.

Durand, B., and Parratte, M. 1983. Oil potential of coals: A geochemical approach. In: Brooks, J. (ed.): Petroleum Geochemistry and Exploration of Europe 1983. London, Blackwell, pp. 255–265.

Grantham, P.J., Posthuma, J., and Baak, A. 1981. Triterpanes in a number of far-eastern crude oils. In: Bjoroy, M., et al. (eds.): Advances in Organic Geochemistry 1981. New York, Wiley, pp. 675–683.

Hedberg, H.D. 1964. Geological aspects of origin of petroleum. American Association of Petroleum Geologists Bulletin 48:1755–1803.

Hedberg, H.D. 1968. Significance of high wax oils with respect to genesis of petroleum. American Association of Petroleum Geologists Bulletin 52:736–750.

Hood, A., Gutjahr, C.C.M., and Heacock, R.L. 1975. Organic metamorphism and the generation of petroleum. American Association of Petroleum Geologists Bulletin 59:986–996.

Hunt, J.M. 1979. Petroleum Geochemistry and Geology. San Francisco, Freeman, 617 pp.

James, A.T. 1983. Correlation of natural gas by use of carbon isotopic distribution between hydrocarbon components. American Association of Petroleum Geologists Bulletin 67:1176–1191.

Lijmbach, G.W.M., van der Veen, F.M., and Engelhardt, E.D. 1981. Characterization of crude oils and source rocks using field ionisation mass spectrometry. In: Bjoroy, M., et al. (eds.): Advances in Organic Geochemistry 1981. New York, Wiley, pp. 788–798.

Powell, T.G., and McKirdy, D.M. 1973. Relationship between ratio of pristane to phytane, crude oil composition and geological environment in Australia. Nature 243:37–39.

Rohrback, B.G. 1981. Crude oil geochemistry of the Gulf of Suez. In: Bjoroy, M., et al. (eds.): Advances in Organic Geochemistry 1981. New York, Wiley, pp. 39–48.

Shibaoka, M., Bennet, A.J.R., and Gould, K.W. 1973. Diagenesis of organic matter and occurrence of hydrocarbons in some Australian sedimentary basins. Australian Petroleum Exploration Association Journal 13:73–80.

Snowdon, L.R. 1980. Resinite—a potential petroleum source in the Upper Cretaceous/Tertiary of the Beaufort-Mackenzie basin. In: Miall, A.D. (ed.): Facts and Principles of World Petroleum Occurrence. Canadian Society of Petroleum Geologists Memoir 6:509–521.

Snowdon, L.R., and Powell, T.G. 1982. Immature oil and condensate: Modification of hydrocarbon generation model for terrestrial organic matter. American Association of Petroleum Geologists Bulletin 66: 775–788.

Stoakes, F.A. 1980. Nature and control of shale basin fill and its effect on reef growth and termination: Upper Devonian Duvernay and Ireton Formations of Alberta, Canada. Bulletin of Canadian Petroleum Geology 28:345–410.

Stoakes, F.A., and Creaney, S. 1984. Sedimentology of a carbonate source rock: Duvernay Formation of central Alberta. In: Eliuk, L. (ed.): Carbonates in Subsurface and Outcrop: Proceedings of the 1984 Canadian Society of Petroleum Geologists Core Conference, Calgary, Canada, pp. 132–147.

Teichmüller, M. 1962. Die Genese der Kohle: 4th Congrés de Stratigraphie et de Geologie du Carbonifere, vol. 3, pp. 699–722.

Teichmüller, M. 1974. Generation of petroleum-like substances in coal seams as seen under the microscope. In: Tissot, B., and Bienner, F. (eds.): Advances in Organic Geochemistry 1973. Paris, Editions Technip, pp. 380–407.

Thomas, B.M. 1982. Land-plant source rocks for oil and their significance in Australian basins. Australian Petroleum Exploration Association Journal 22:164–178.

Threlfall, W.F., Brown, B.R., and Griffith, B.R. 1976. Gippsland Basin, offshore. In: Leslie, R.B., Evans, H.J., and Knight, C.L. (eds.): Economic Geology of Australia and Papua, New Guinea. Australasian Institute of Mining and Metallurgy Monograph 7:41–67.

Tissot, B.P., Durand, B., Espitalie, J., and Combaz, A. 1974. Influence of nature and diagenesis of organic matter in formation of petroleum. American Association of Petroleum Geologists Bulletin 58:499–506.

Van Krevelen, D.W. 1961. Coal. New York, Elsevier, 514 pp.

Vassoevich, N.B., Akramkhodzhaev, A.M., and Geodekyan, A.A. 1974. Principal zone of oil formation. In: Tissot, B., and Bienner, F. (eds.): Advances in Organic Geochemistry 1973. Paris, Editions Technip, pp. 309–314.

Waples, D.W. 1980. Time and temperature in petroleum formation: Application of Lopatin's method to petroleum exploration. American Association of Petroleum Geologists Bulletin 64:916–926.

4
Molecular Thermal Maturity Indicators in Oil and Gas Source Rocks

Joseph A. Curiale, Stephen R. Larter,
Robert E. Sweeney, and Bruce W. Bromley

Abstract

Many detailed chemical parameters have been proposed as indicators of thermal maturity in oil and gas source rocks. Certain classical maturity parameters involving carbon preference indices and compound class ratios, such as hydrocarbons/extract yield and extract yield/total organic carbon, are less commonly used today, having been complemented with detailed molecular parameters. Among these parameters, the molecular distributions of metalloporphyrins, cyclic hydrocarbons, low-molecular-weight hydrocarbons, and gases are most commonly used. Recent instrumental advances have allowed the routine measurement of molecular ratios in geochemical organic matter, stimulating the development and use of biological markers, such as steranes, hopanes, and metallated tetrapyrroles, as thermal maturity indicators. Increased chromatographic resolution of source rock hydrocarbons has also led to the use of low-molecular-weight hydrocarbons, methylphenanthrenes, and aromatized steranes as maturity indicators. In this paper, we discuss these developments, emphasizing the applications and the pitfalls of using molecular maturity indicators.

Measurements of source rock thermal maturity attempt to describe the progress of the sum of the chemical reactions that convert sedimentary organic matter into oil and gas. Such measurements can be made on a molecular basis, using both soluble organic matter (C_{15+} hydrocarbons) and gases/gasoline range hydrocarbons (C_{15-}). The biological markers, including both aliphatic and aromatic hydrocarbons, are probably the most "ideal" molecular maturity indicators currently available, although their range of applicability is generally limited. Gases and gasoline range molecules (particularly their isotopic characteristics) also have great potential as thermal indicators, although an understanding of migrational fractionation is critical. While no single thermal maturity indicator will ever be the ultimate "ideal" parameter, molecular indicators will undoubtedly increase in importance in coming years, owing largely to their specificity and to the increasing ease of obtaining this type of data.

Introduction

The presence of thermally mature source rocks is a necessary element for the existence of a petroliferous basin. The maturity of an organic rich sediment is commonly used as a necessary criterion for distinguishing *actual* oil and gas source rocks from merely *potential* source rocks. Recognizing the need for thermal input in the generation of commercial quantities of petroleum from fine-grained sediments, source rock geochemists have developed numerous methods for assessing thermal maturity levels in a basin. Recently, the traditional "bulk" techniques involving compound class distributions, vitrinite reflectance, and spore coloration, to name a few, have been complemented by detailed molecular parameters, involving ratios and distributions of specific compounds. As instrumental technology for analytical chemistry advances, use of such molecular approaches to the problem of thermal maturity measurement will undoubtedly increase, hopefully leading to increasingly refined maturity indicators.

The purpose of this chapter is to present a synthesis of current thinking concerning molecular thermal maturity indicators that derive from the extractable organic matter of petroleum source rocks, in light of the strengths and weaknesses of

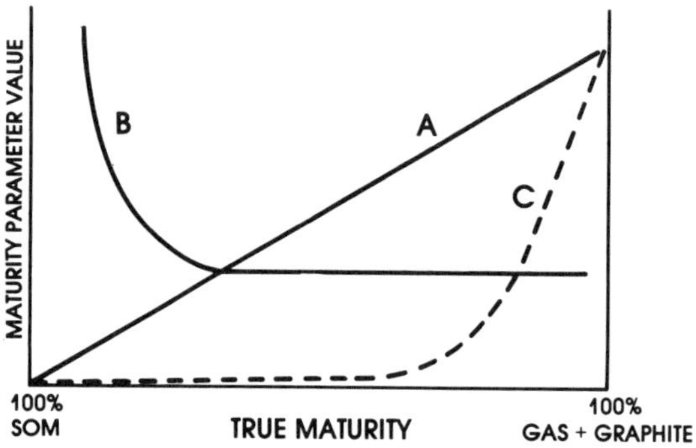

FIGURE 4.1. Values of maturity parameters versus true "maturity" (SOM = sedimentary organic matter). Line A illustrates an ideal maturity parameter—potential values are proportional to maturity over the entire possible range. Lines B (e.g., biomarker ratios) and C (e.g., graphite crystallinity) are typical of real maturity parameters, where the parameter's value is proportional to "maturity" over only a limited range.

the classical approaches using bulk measurements. Although practical applications are discussed to illustrate various points, the emphasis is mainly conceptual. Consequently, no attempt is made to document exhaustively every molecular indicator used to date. Rather, specific molecular indicators are reviewed in terms of their strengths and weaknesses, with an eye toward infusing some fresh ideas into the discussion of those thermal maturity parameters that are currently in vogue. We concentrate on hydrocarbon molecular indicators, including gases and gasoline range hydrocarbons as well as the C_{15+} fraction. Kerogen, because of its role as a progenitor of most of the hydrocarbons, is discussed also. We begin with an overview of the use of thermal maturity as a conceptual approach to petroleum exploration.

An adequate definition of thermal maturity, as the term is used by petroleum geochemists, is unavailable in the literature. Maturity, as used in this chapter, describes the position or extent of the chemical reactions that convert sedimentary organic material (SOM) into gas and graphite. One end member of this evolutionary pathway is bacterial and plant debris (mostly kerogen and bitumen, following diagenetic conversion), which are generally low in hydrocarbons. At the other extreme is dry thermal gas and hydrogen-deficient carbonaceous residue (graphite). "Maturity" is the extent to which this reaction complex has proceeded. Part way along this reaction coordinate axis, liquid and gaseous hydrocarbons are produced, then destroyed.

Figure 4.1 illustrates this concept of maturity. The x axis varies from 100% SOM on the left to 100% gas and graphite on the right. The objective of measuring chemical maturity parameters is to determine where on this reaction coordinate scale a kerogen, bitumen, oil, or gas sample lies. In this regard, the choice of maturity parameter is critical.

The ideal maturity parameter would derive from a reaction of the form

$$A \xrightarrow{k} B$$

where k is the rate constant in units of 1/time. In order to use such a reaction to derive an ideal maturity parameter, the following conditions must be satisfied:

1. A and B should be single compounds, and the reaction should be irreversible.
2. A should have no fate other than transformation to B.
3. B should be stable and have no source other than A.
4. The reaction should not proceed to completion in less than the full range of maturity of interest.

The best parameter for assessing maturity using such a reaction would be the ratio of the concentration of B to the sum of the concentrations of A and B, as this ratio would be independent of the initial concentration of A. Ideally, this ratio would be a linear function of maturity, varying from zero to one, and would represent the previously discussed reaction coordinate.

Many reactions related to the genesis of oil and gas from kerogen are conceptually of form $A \rightarrow B$ but have significant problems as indicators of maturity. In reality, most reactions are more similar to that shown in Figure 4.2. In this case, B is neither stable nor the unique product of A. Also, A is not conserved in reacting to form B, but has sources and sinks that are external to the reaction of interest.

In contrast to the irreversible $A \rightarrow B$ reaction systems characteristic of kerogen-based maturity parameters, equilibrium reaction systems are also commonly employed in maturity assessment, particularly those reactions involving biological marker compounds (discussed in more detail later).

$$A \underset{k_2}{\overset{k_1}{\rightleftharpoons}} B$$

An inherent disadvantage of equilibrium reactions is that, upon reaching equilibrium, no further information on maturity is available. Equilibrium reactions also suffer from the same problem as other maturity-related reactions, namely they do not proceed in a closed system (Fig. 4.2). However, with certain equilibria, such as those that involve epimers, this problem is not severe, insomuch as the reactants and products behave in very similar fashion during, for example, migration.

Another difficulty present in the search for the ideal maturity parameter involves kinetic considerations. At a given concentration the rate of conversion of A to B is solely a function of the kinetic constant k, and time. Unfortunately in geological situations k is not a constant but depends on such things as temperature, pressure, catalysis, and heating rate. This relationship of the kinetic constant to environmental factors may be quite different for different reactions. Even though two reactions may meet the above criteria and provide "ideal" maturity indicators, the two reactions will not necessarily yield the same measured maturity, and therefore will not be fully applicable to a full-range maturity study unless they can be calibrated to the reaction series of interest.

Although "ideal" maturity parameters that satisfy the conditions listed above are unavailable, it is critical to assess those parameters currently in use in terms of how closely they approach these conditions. Compilations of the many geochemical parameters that have been used as maturity indicators have been constructed by several workers (Tissot and Welte, 1978; Heroux et al., 1979, and references therein). It is our intent to discuss the more useful molecular maturity indicators of both historical and present-day importance, in light of the inherent weaknesses involved with their use.

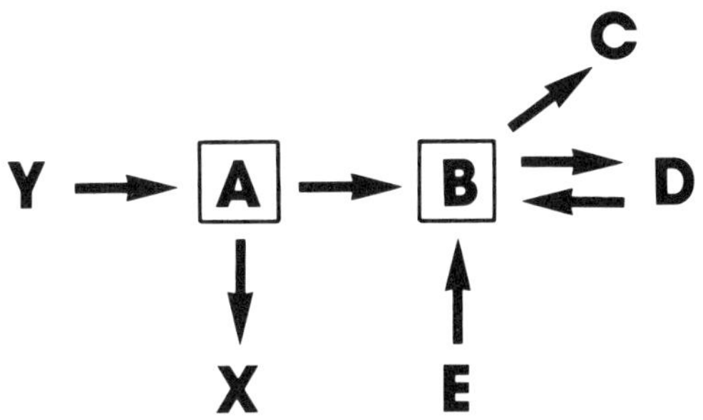

FIGURE 4.2. Typical nonideal reaction used to assess thermal maturity. Note that Reactant A has Source Y and Sink X, while Product B has Sources A and E, Sink C, and is in equilibrium with D. In such an open system, the assumptions for the "ideal" maturity parameter do not apply.

Soluble Organic Matter

Historical Approaches and Practical Difficulties

Thermal maturity in a petroleum geochemical context is best described in terms of the reaction coordinate concept discussed above. In discussing molecular maturity parameters in soluble organic matter, it is critical to distinguish between parameters that rely on the determination of the reaction coordinate of "complex systems" (i.e., those that involve multiple reactions), and true molecular maturation parameters, which (ideally) assess the extent of a single concerted reaction. In practice, the closest we come to true molecular maturation parameters are the stereochemical epimerization reactions involving the cyclic aliphatic biological markers. However, even these are not simple systems, insomuch as both reactant and product may have multiple sources and sinks. However, the complexity of these systems is far lower than that of most reaction systems commonly used for maturation determination.

Three main types of molecular maturation parameters involving soluble organic matter are currently in use. These are: 1) parameters based on the carbon number distribution of a single type of compound (e.g., carbon preference index); 2) parameters based on relative abundances of different types of specific hydrocarbons (e.g., isopre-

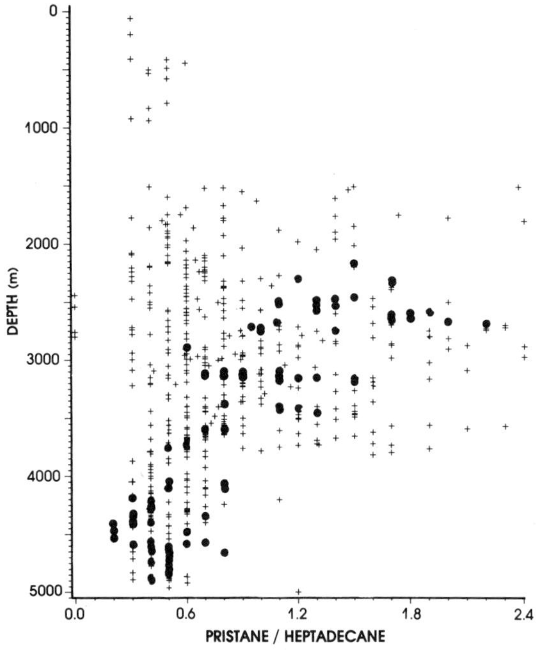

FIGURE 4.3. Depth versus pristane/heptadecane ratios for samples from a single province (plus signs and solid circles). Note that, for a single organic facies (shown as solid circles), a maturity trend is evident. This trend is obscured when samples from the entire province are considered.

noid/n-alkane ratios); and 3) more general "bulk-type" parameters that involve the gross organic matter within the sediment (e.g., hydrocarbon content as a percentage of total organic carbon). Several conceptual problems are associated with all three types of bitumen-related maturity parameters, arising from the fact that the typical source rock system is not closed. For example, in the case of the carbon preference index (CPI) for sequences of uniform organic matter type, increasing maturity results in a decrease in the CPI to near one in the oil window maturity range (Durand and Oudin, 1980). This decrease is due both to: 1) thermal degradation of existing normal alkanes, and (probably to the greatest extent) 2) dilution of preexisting alkane distributions having a high (inherited) CPI with more abundant generated n-alkanes having a near-unity CPI (Allan et al., 1977).

However, low-efficiency hydrocarbon expulsion from a source bed of nonunity CPI may sometimes result in expelled fluids with a near-unity CPI. With the ubiquitous occurrence of fluid movement in the subsurface, the use of CPI as an accurate source rock maturity indicator is therefore questionable. For example, if primary migration from a source bed took place by a solution mechanism where the ratio of solvent to solute (in this case n-alkanes) is low, then the n-alkane composition of the expelled phase will be independent of the composition and thermal maturity of the source bitumen, and be controlled solely by the molecular weight/solubility relationships of the solvent-solute system. (Solution expulsion mechanisms involving low-polarity solvents may be active when more efficient processes such as oil phase expulsion do not function efficiently.) In practice, for most feasible geological solvents, a log normal distribution of concentrations would be observed as a function of carbon number in a single homologous series (Bromley, 1983), with the CPI of this mixture being very close to one. Experiments using hydrous pyrolysis at relatively low temperatures (less than 300°C) sometimes reveal a minor expelled phase with a much lower CPI than the retained phase. Furthermore, it is not unusual to find sand/shale sequences in which the CPI of the often classically immature shale is greater than one, while that of the adjacent sand extract is nearer to one. Such migration-related problems may severely limit the usefulness of the CPI as a maturity parameter. A high CPI may indicate immaturity, but a low CPI need not indicate thermal maturity in all cases. Migration-induced fractionation may also affect isoprenoid/n-alkane ratios.

In general, however, it is facies variations that prove to be the downfall of most of the conventional thermal maturity parameters. Figure 4.3 shows the plot of the ratio of pristane/n-C17 versus depth for a series of sediments of uniform organic facies in a single stratigraphic unit, revealing a clear trend. However, if a mixed facies set of sediments is examined from the same geographic area, no such trend emerges (Fig. 4.3). A parameter such as an isoprenoid/n-alkane ratio depends for its effectiveness on the differing rates of generation and/or survival of two different hydrocarbon species. It has been shown that isoprenoid alkanes are released mainly during the early portion of the catagenic process (Van Graas et al., 1981; Burnham et al., 1982), whereas normal alkanes are

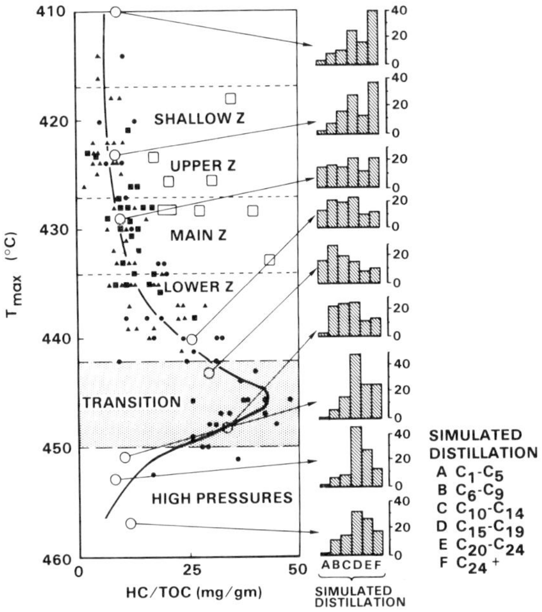

FIGURE 4.4. Rock-Eval T_{max} values versus HC/TOC ratios in coals. The classical bell-shaped thermal maturity curve is seen here. Also shown is the carbon number distribution (in percent) at various depths. (Modified after Durand and Oudin, 1980.)

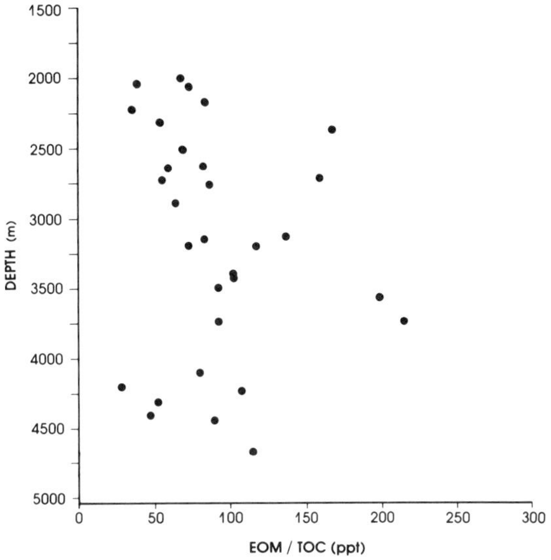

FIGURE 4.5. Depth versus mean EOM/TOC ratios for a Mesozoic source unit. Note the absence of the classical bell-shaped distribution with depth.

generated throughout the oil and gas window. However, the initial total abundances of, for example, pristane precursors in a sediment are strongly influenced by factors such as organic matter type, oxidation-reduction potential of the sediment, and degradation to smaller chain length isoprenoid hydrocarbons. In addition, it is implied by recent work of Goossens et al. (1984) that the extent of the bacterial biomass impact on the sedimentary organic matter can also effect initial abundances of pristane precursors. Thus both organic matter type and migratory overprinting can severely limit the usefulness of the classical maturity parameters.

The use of "bulk-type" maturity parameters, such as the ratio of normalized hydrocarbon to total organic carbon (HC/TOC) or normalized extractable organic matter to total organic carbon (EOM/TOC), is a clear example of problems attributable to migratory overprinting. In many classic cases (e.g., Fig. 4.4) a bell-shaped curve is seen if one plots, for example, HC/TOC versus either depth or a common maturity parameter such as vitrinite reflectance or Rock-Eval T_{max} (Fig. 4.4, after Durand and Oudin, 1980; Tissot, Espitalie, et al., 1974). This classic trend in EOM/TOC or HC/TOC is, however, often invoked even where the data only loosely support such a trend, and is generally interpreted as maturity related. However, such an interpretation could be groundless in cases where efficient migration of hydrocarbons from a source bed occurs. In these instances, no bell-shaped depth trend would develop. Figure 4.5 shows the normalized C_{15+} hydrocarbon content of a Mesozoic source bed as a function of depth for several wells. If one were ambitious, a bell-shaped maturity-related trend could be proposed for these data, although in reality we feel that such a trend would be, at best, tenuous. This is because efficient expulsion of hydrocarbons in a continuous oil phase is believed to be occurring in this sedimentary sequence. Thus no substantial buildup of hydrocarbons occurs in this section. In cases such as these we do not feel that HC/TOC or EOM/TOC ratios alone are reliable indicators of thermal maturity.

The decrease in the EOM/TOC or HC/TOC ratios in the higher rank part of the section (e.g., base of oil window) shown in Figure 4.4 could generally be considered to be monitoring the thermal cracking of larger molecules to smaller gaseous hydrocarbons and, thus, be an indicator of thermal maturity. However, when the molecular composition of the hydrocarbon suite is considered (Groups A–F in Fig. 4.4), it becomes obvious that the remaining hydrocarbons are depleted in the low-molecular-weight components. Since continuous generation and migration must be considered to occur, at some level, then it is more correct analytically to interpret the turnover in the HC/TOC ratio (at 445°C in Fig. 4.4) to indicate that at higher rank, migrational processes may dominate over generation processes in some instances. If generation of low-molecular-weight molecules was the dominant process, then the higher rank zone would be enriched in these molecules (i.e., Groups A–C; see Huc and Hunt, 1980).

It is apparent then, that a maturity-related trend in a geochemical parameter does not necessarily render that parameter a worthwhile or usable tool for the geochemist. The parameter must be reliably related in a quantitative manner to the series of chemical reactions associated with the generation of hydrocarbons over a wide range of maturity, before it can be recommended for use as a maturity indicator.

Aromatic Hydrocarbons as Thermal Maturity Indicators

Aromatic hydrocarbons are ubiquitous components of recent and ancient sediments. The precursor biochemicals for aromatic hydrocarbons in sediments and petroleum include terpenoids (Albrecht et al., 1976; Hyatsu et al., 1978; Baset et al., 1980; White and Lee, 1980; Allan and Larter, 1983), carotenoids (Day and Erdman, 1963; Ikan et al., 1975; Gallegos, 1980), quinones and phenols (Blumer, 1965; Allan and Larter, 1983), and unsaturated isoprenoid hydrocarbons (de Leeuw et al., 1977; Larter et al., 1983). Simple derivation of aromatic hydrocarbons from these biochemical precursors undoubtedly occurs during diagenesis/catagenesis. A major source of alkylated aromatic hydrocarbons is also from aromatization and isomerization of nonspecific skeletons formed during catagenesis. Strong evidence for this is provided by the general occurrence of all isomers of, for example, n-alkyl toluenes in sediments and crude oils (Solli et al., 1980). In addition, recent studies have claimed that isomeric distributions of two- and three-ring aromatic hydrocarbons are influenced more by maturity than by facies considerations, within the classic oil window range (Radke et al., 1980; Radke, Welte, et al., 1982; Radke, Willsch, et al., 1982; Alexander et al., 1985). These observations all indicate that aromatic hydrocarbons have a definite use as maturity tools.

Radke and associates (Radke et al., 1980; Radke, Willsch, et al., 1982; Radke and Welte, 1983) have recently attempted to introduce a quantitative approach to maturity determination, using tricyclic aromatic hydrocarbons. The method depends on the determination of phenanthrene and its four naturally occurring monomethyl homologues (1-, 9-, 2-, and 3-methylphenanthrenes) in the triaromatic fraction of sedimentary extracts. The assumption is made that 2- and 3-methylphenanthrene are derived from methylation of free phenanthrene as well as from the maturation of the less stable 1- and 9-methylphenanthrenes. The ratio of the abundances of the catagenically produced 2- and 3-methylphenanthrenes to the sum of phenanthrene, 1- and 9-methylphenanthrene is a parameter (the methylphenanthrene index [MPI]) which appears empirically to follow a maturity-related trend. While it is questionable if rearrangement of methylphenanthrenes and methylation of phenanthrene are the only sources of 2- and 3-methylphenanthrene (insomuch as phenanthrenes in general are produced by dehydrogenation of hydrophenanthrene precursors), and while the facies independence of the method is yet to be fully verified, the general approach is conceptually powerful. With refinement it may become a valuable tool in the petroleum geochemist's arsenal.

Other general trends in aromatic hydrocarbon distribution with rank level have been studied for coals (Radke, Willsch, et al., 1982; Allan and Larter, 1983; Radke and Welte, 1983) and other sediments containing nonmarine organic matter (Radke, Welte, et al., 1982). With increasing maturity, sediment extracts are observed to usually become progressively dominated by lower molecular weight, short-chain, substituted (dominantly methyl-substituted) alkylaromatic hydrocarbons. The proportion of carbon atoms in substituent groups decreases with increasing maturity (Allan and Larter, 1983). In addition, quantitative data relating different aromatic fractions are also useful as a maturity parameter, although problems associated with migration fractionation tend to complicate direct quantitative comparison of aromatic species having different structures, and consequently different polarities. Thus while the ratio of phenanthrene species to dihydrophenanthrene species has been found to change systematically with maturity for vitrinite and sporinite extracts (Allan and Larter, 1983), it is questionable whether such parameters will find general applicability to maturity determination in oils or source beds. In general, these types of trends are either too imprecise or observed too infrequently for use in routine exploration applications.

In addition to tricyclic compounds, the tetracyclic aromatics have received much attention in recent years. Modern application of tetracyclic aromatic hydrocarbon distributions as thermal maturity indicators began with Tissot, Durand, et al. (1974), who studied extracts of Early Toarcian (Early Jurassic) shales in the Paris Basin. Although specific aromatic compounds were not identified, it was observed that deeper samples showed a prominent relative decrease in C_{27}–C_{30} aromatic molecules, accompanied by an increase in C_{19}–C_{21} aromatics. This shift in carbon number distribution, later attributed to carbon-carbon bond cleavage in the side chain of C-ring monoaromatic steranes (Siefert and Moldowan, 1978; Mackenzie, Hoffman, et al., 1981), was the first hint of the usefulness of multiring aromatic hydrocarbons as maturation tools. Their use as thermal maturity indicators has expanded greatly in the past 10 years, due in large part to the widespread availability of computerized gas chromatography-mass spectrometry (Mackenzie et al., 1982).

The utility of aromatized steranes as maturity indicators became apparent even prior to specific structural elucidation of the molecules involved (Mackenzie, Hoffman, et al., 1981; Siefert et al., 1983). Recent studies have focused on the conversion of C-ring monoaromatic steranes to their ABC-ring triaromatic counterparts (I to II, Fig. 4.6). Four possible monoaromatic isomers (5α, 5β, 20R, and 20S; C-24 hydrogen configuration unknown) may dehydrogenate to yield two possible triaromatic isomers (20R and 20S; C-24 configuration unknown). Mackenzie, Hoffman, et al. (1981) have proposed such an aromatization reaction to explain increasing triaromatic/(monoaromatic + triaromatic) sterane ratios (from 0 to 1) with increasing maximum burial depth in Toarcian shales of the Paris Basin. The aromatization of monoaromatic steroid hydrocarbons to triaromatics is shown in Figure 4.7 for extracts from Jurassic shales of East Shetland Basin, North Sea (Mackenzie, 1984); in less than 1,000 m, the monoaromatic to triaromatic conversion is virtually complete. Generally, at maturity levels equivalent to peak hydrocarbon generation, triaromatic steranes are dominant. This observation usually limits the maturity indicator capabilities of aromatized steranes to vitrinite reflectance equivalent levels of less than 1.1%. This indicator is further limited by source rock age; Paleozoic shales often contain very low quantities of aromatized steroids relative to Tertiary shales (J.A. Curiale, unpublished results).

Realistically it must be considered that aromatic hydrocarbons in general have been used as routine maturity parameters for less than 10 years. There are many far older and far less effective nonaromatic hydrocarbon-based parameters still in use in petroleum geochemistry. It is anticipated that further development in this area, particularly in analytical capabilities, will produce useful and valuable methods. Until then, cautious use of aromatic hydrocarbons as maturity parameters is recommended.

Aliphatic Hydrocarbon Biological Markers as Thermal Maturity Indicators

Tetra- and pentacyclic aliphatic hydrocarbon biological marker measurements are now commonly used in organic geochemical studies (Mackenzie and Maxwell, 1981). Various ratios of these biomarkers have been utilized as empirical indicators of thermal maturity (Seifert and Moldowan, 1978; Mackenzie, 1984, and references therein). The great majority of the changes in these indicator ratios involve reactions taking place during early diagenesis, often involving unsaturated molecules, at a pregenerative maturity level for most oil and gas source rocks. We refer here to only those indicators that have been found most useful to the petroleum explorationist.

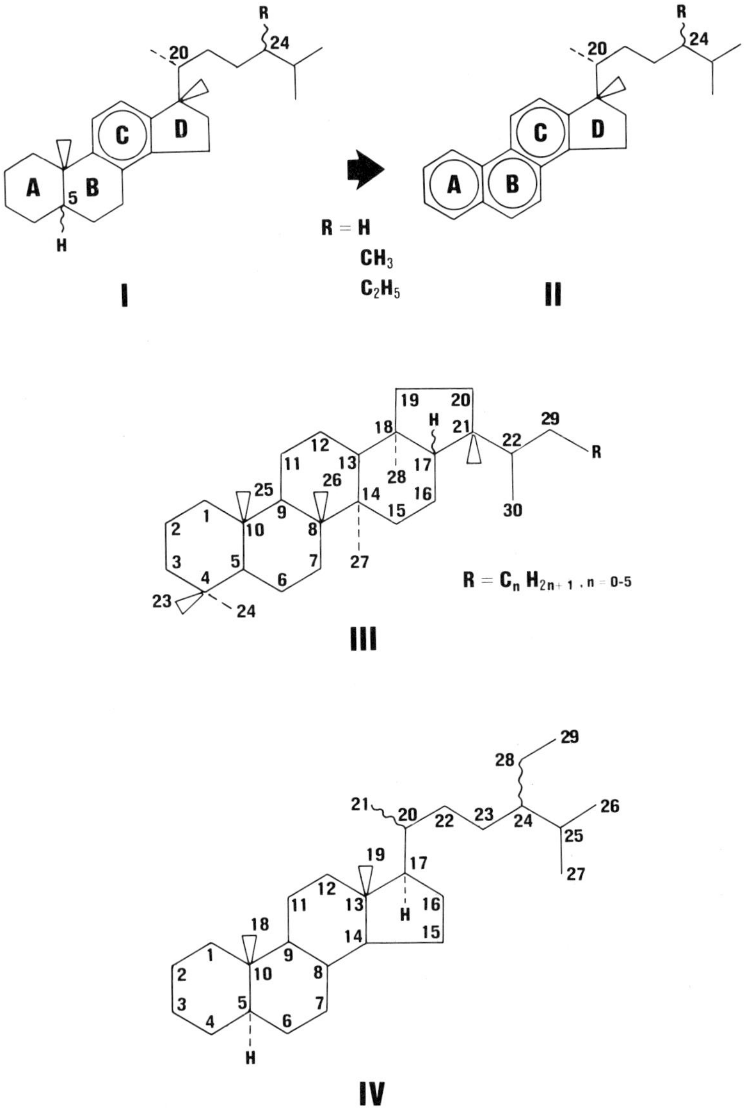

FIGURE 4.6. Molecular structures for monoaromatic steroid hydrocarbons (I), triaromatic steroid hydrocarbons (II), hopanoid hydrocarbons (III), and ethylcholestane (IV).

The recognition of aliphatic hydrocarbon biomarker compounds as some of the most sophisticated of the current molecular thermal maturity indicators in source rocks arises from observed correlations between stereoisomeric molecular ratios and commonly accepted nonbiomarker thermal maturity indicators (e.g., vitrinite reflectance). The biomarker indicators can be grouped into: 1) those resulting from a change in the number of carbon atoms in the molecule (noted earlier for alkylaromatic hydrocarbons), and 2) those resulting from isomeric changes at certain key chiral centers. While both types of indicators are commonly used, the first-discovered, and currently the most reliable, are those resulting from isomerization, and it is these that are discussed here. Applications of two of these epimerization reactions, those involving the hydrogen atoms at the C-22 position in hopanes and the C-20 position in steranes, are reviewed. The initial (biological) form of the molecules at these key chiral centers has the R configuration; subsequent isomerization occurs with increasing thermal input, leading to equilibrated mixtures of R and S epimers.

The hopanes found in geological systems are pentacyclic hydrocarbons having four six-membered

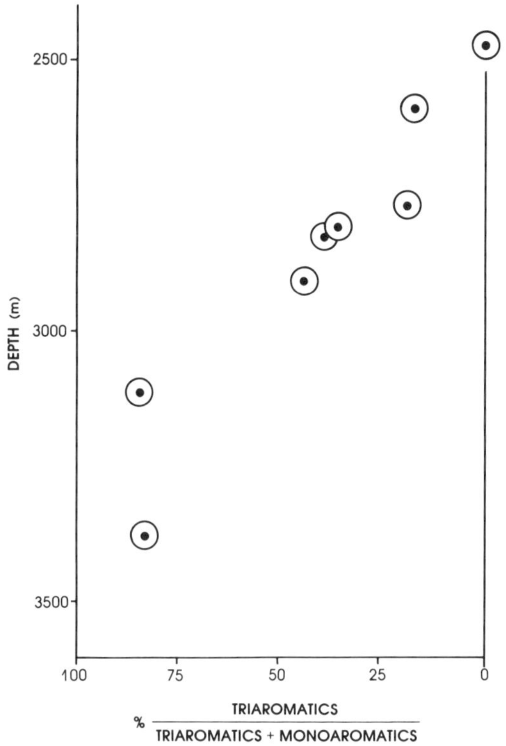

FIGURE 4.7. Depth versus percentage triaromatic steroid hydrocarbons. Triaromatic/(triaromatic + monoaromatic) ratio increases from 0 to 100%, in about 1,000 m. Data are for the Jurassic of East Shetland Basin, North Sea. (From Mackenzie, 1984, reprinted by permission of Academic Press.)

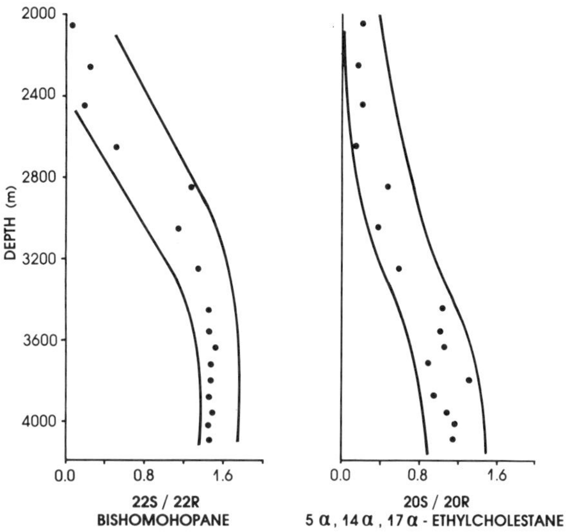

FIGURE 4.8. Depth versus 22S/22R-bishomohopane (left) and depth versus 20S/20R-5α,14α,17α-ethylcholestane (right). Equilibration points for both maturity indicators are below 3,400 m. Note stability of the hopane parameter following equilibration. Data for Mahakam Delta samples. (From Schoell et al., 1983, reprinted by permission of John Wiley & Sons.)

rings and one five-membered ring (Fig. 4.6, III, 17α[H]). Members of the hopane family possessing 30 or less carbon atoms have chiral centers at C-21 and all ring juncture locations (C-5, C-8, C-9, C-10, C-13, C-14, C-17, and C-18). Hopanes with greater than 30 carbon atoms contain one further chiral center, at C-22 (Fig. 4.6, III, $n>0$). The conversion of the initial 22R C_{31+} hopanes to an equilibrium mixture of 22R + 22S has been monitored in several petroliferous basins (e.g., Seifert and Moldowan, 1980). Extremely immature source shales have initial 22S/22R ratios of near zero, with the biological R epimer dominant. With increasing thermal maturity, the 22S/22R ratio increases monotonically to approximately 1.4 to 1.5 (Seifert and Moldowan, 1980; Schoell et al., 1983). Because this equilibrium value is generally reached early in the maturation sequence (Fig. 4.8) it can be used to determine whether a potential source rock has already generated petroleum. Furthermore, insomuch as almost all oils have 22S/22R ratios near 1.2 to 1.4, plots of the variation of this ratio versus a nonbiomarker thermal maturity indicator in source rocks may be useful in assessing whether or not thermal generation has begun, as measured by the nonbiomarker indicators, on a basinwide scale. For example, data in Figure 4.9 indicate that thermal generation has begun, as measured by vitrinite reflectance, by about $R_o = 0.6\%$ for this basin.

In addition to the C-22 epimers of C_{31+} hopanes, the ethylcholestane C-20 epimers (Fig. 4.6, IV) are also thermal maturity indicators. The initial biological C-20 configuration of ethylcholestane (R epimer) converts to an R + S mixture with increasing thermal input (Fig. 4.8), ultimately yielding a 20S/20R ratio of approximately 1.0 to 1.1 at equilibrium (Seifert and Moldowan, 1981; Schoell et al., 1983). While 20S/20R-ethylcholestane ratios can also be utilized as an indicator of the onset of thermal generation of hydrocarbons, their applicability is not as universal as that of the

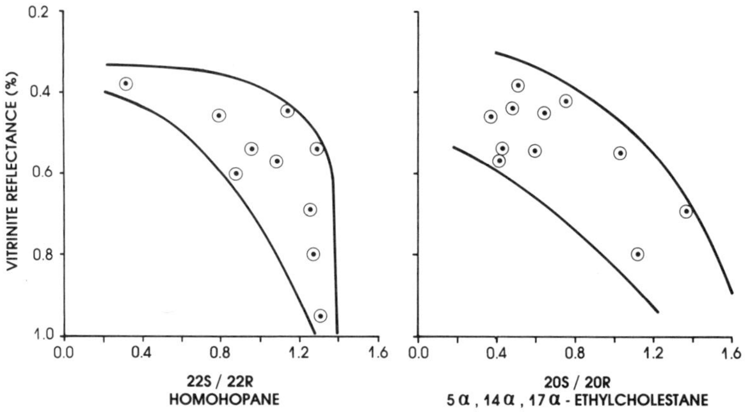

FIGURE 4.9. Vitrinite reflectance (R_0) versus 22S/22R-homohopane (left) and versus 20S/20R-5α,14α,17α-ethylcholestane (right). Samples analyzed are Mesozoic and Paleozoic source rocks of varying age and lithology, from a single, petroliferous basin. Plots show that, despite the large scatter in sample type, these two biomarker thermal maturity indicators can be useful indicators of the onset of thermal generation on a basin-wide scale as well as in a single well (see Fig. 4.8). Again, note stability of hopane parameter after thermal maturity has been reached, $R_0 > 0.6\%$.

22S/22R-C_{31+} hopane ratios. Typically, the onset of generation occurs at a 20S/20R ratio of approximately 0.6 to 0.8 (e.g., Fig. 4.9). However, several examples of crude oils are known in which this ratio is below 0.6, including mid-Miocene oils of the San Joaquin Basin, California, whose 20S/20R ratios are as low as 0.29 (Seifert and Moldowan, 1981; Curiale et al., 1985), for reasons that are currently unknown. Potentially, low 20S/20R sterane epimer ratios may denote early hydrocarbon generation. Although unusual, such low values highlight the problems that currently arise in utilizing sterane epimer ratios as thermal generation onset indicators, problems that have not yet been observed for hopane epimer ratios. Consequently, the use of a simple sterane ratio cutoff for any of the possible biomarker epimers is not recommended in basinal evaluations of thermal maturity.

As a supplement to the configurational isomers at the C-22 position in hopanes and the C-20 position in steranes, several other biological marker thermal maturation indicators are available in organic geochemical applications (Mackenzie et al., 1983). These include moretane/hopane (17β[H], 21α[H]/17α[H],21β[H]) ratios, 18α(H)/17α(H)-trisnorhopane ratios, and ratios of C_{27}–C_{29} steranes or monoaromatized steranes to their side chain-cleaved counterparts (Mackenzie, 1984, and references therein). The use of these and other molecular maturity parameters, considered to be exotic a decade ago, is now commonplace. Their future use as maturity indices in source rocks will depend in large part on the need for maturation data in regions where other maturity indicators are either unavailable or unreliable.

The rates of the biological marker aromatization and isomerization reactions discussed above have been investigated by McKenzie et al. (1983) and Mackenzie and McKenzie (1983) for extensional basins. Their laboratory results (Mackenzie, Lewis, et al., 1981) suggest that these reaction rates will vary significantly, depending upon source rock temperatures. Young, hot basins will experience monoaromatic-to-triaromatic sterane conversion much more rapidly than sterane or hopane isomerization, while the reverse situation obtains for old, cool basins. Thus biological marker measurements provide not only the capability of estimating thermal maturity, but also the possibility of deducing the thermal history of a basin as well. The ramifications of this possibility are discussed in more detail later in this chapter.

Tetrapyrroles as Thermal Maturity Indicators

Tetrapyrrolic organometallic molecules (Fig. 4.10, V and VI) are common constituents of source rock

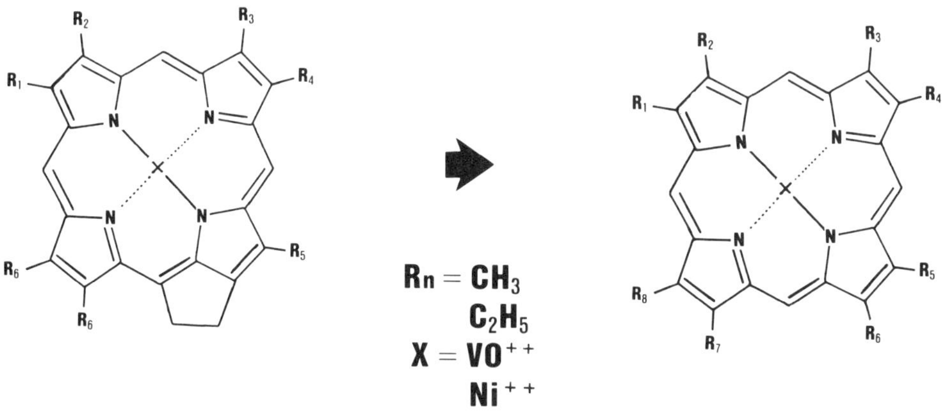

V VI

FIGURE 4.10. Molecular structures of DPEP metalloporphyrins (V) and etio metalloporphyrins (VI). X is any metal-containing divalent cation of adequate ionic radius.

extractable organic matter. Alkylated metalloporphyrins, specifically the deoxophylloerythroetioporphyrin (DPEP) and etioporphyrin structures, are the most common identified tetrapyrroles in source rock extracts. Both forms are structurally similar, the major difference being the presence of an isocyclic ring in the DPEP form (Fig. 4.10, structure V). Didyk et al. (1975) were among the first to investigate in detail the relative distributions of DPEP and etio forms in petroleum source rocks (using the Cretaceous La Luna Formation of Venezuela). They showed that the DPEP/etio ratio decreases with increasing thermal maturity, presumably due to cleavage of the isocyclic ring of the DPEP molecule. Mackenzie, Patience, et al. (1981) confirmed these results using extracts of the Toarcian shales of the Paris Basin, where the DPEP/etio ratio decreases rapidly with increasing maximum depth (Fig. 4.11). This ratio is presently recognized as a useful trend indicator of the onset of thermal hydrocarbon generation in the sedimentary column. The application of metalloporphyrins, and tetrapyrroles in general, to source rock thermal maturation studies will undoubtedly increase as more sophisticated maturity indicators become necessary. Unfortunately, current efforts are greatly restricted because of the involved procedures necessary in the quantitative extraction and isolation of metalloporphyrins from shales.

Gas and Gasoline Range Molecules

Our discussion to this point has involved molecules obtained via solvent extraction of sediments, which deals quantitatively with only the C_{15+} fraction. Lower molecular weight hydrocarbons, collectively termed *light hydrocarbons*, are lost during this type of processing. Alternative techniques, such as headspace analysis (Hachenberg, 1977) or distillation from the sediment (Philippi, 1975; Thompson, 1979; Huc and Hunt, 1980) have been used to study the absolute or carbon-normalized amount of light hydrocarbon molecules in a sediment. These amounts are related to the thermal maturity of the sediment if factors such as migration fractionation, leakage through reservoir seals, or biodegradation (James and Burns, 1984) do not cause significant secondary alteration of the light hydrocarbon fraction. The topics emphasized in this section are: 1) the distribution of light hydrocarbons in a sedimentary section, as a direct measure of catagenic hydrocarbon generation from SOM; and 2) the relationship between thermal maturity and stable isotope ratios of gases. The significance that light hydrocarbon abundances play in evaluating the dynamic interplay of generation and migration in the lower part of the oil generation zone has been referred to previously (Fig. 4.4). Again the discussion is conceptual in nature, and stresses molecular parameters.

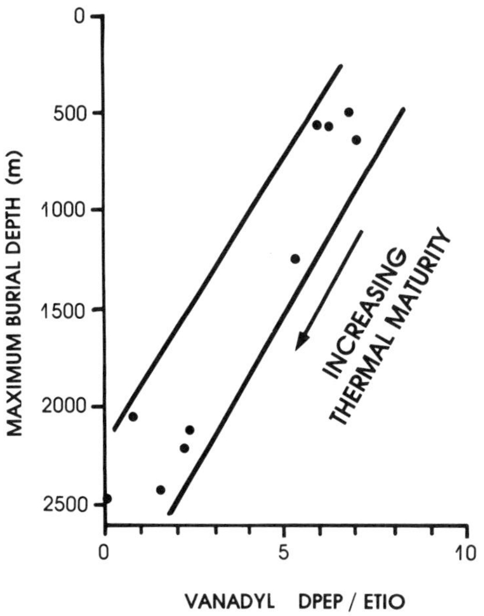

FIGURE 4.11. Burial depth versus DPEP/etio ratio of vanadyl petroporphyrins, for Toarcian shales of the Paris Basin. Note decreasing relative DPEP content with increasing burial depth/thermal maturity. See Figure 4.10 for structures involved. (From Mackenzie, Patience et al., 1981, reprinted by permission of Pergamon Journals.)

The continuing detailed light hydrocarbon analyses of young sediments collected from the JOIDES cruises (e.g., Jasper et al., 1982) substantiate earlier observations that the amount of gasoline range hydrocarbons present in recent sediments is miniscule. In contrast, the amount of light hydrocarbons present in buried shales is often orders of magnitude higher in concentration than recent sediment counterparts. This observation alone led to opinions such as that of Philippi (1975) that "... there is no escape from the conclusion that gasoline-range hydrocarbons are generated somewhere (deeper) in the subsurface" (p. 1354). This conclusion is supported by several detailed studies of the distribution of light hydrocarbons in well profiles (Fig. 4.12). A review of numerous plots of light hydrocarbon contents versus depth for subsiding basins suggests the following conclusions:

1. With the exception of methane, and to a lesser extent ethane, the concentration of light hydrocarbons is consistently low in the sediment section to a certain depth and then increases rapidly over an interval of 2,000 m (Leythaeuser et al., 1979).
2. Methane content is generally high at the top of the sedimentary section of a subsiding basin, due usually to biogenic derivation (Claypool and Kaplan, 1974), and either increases or decreases with depth, depending on the relative rates of generation and migration.
3. Certain alicyclic light hydrocarbons exhibit a consistent maximum at subsurface temperatures close to 77 °C (Thompson, 1979). A continuous increase in relative alkane content follows at higher temperatures. Ratios of normal alkanes to other molecules, such as the iso/normal butane and pentane ratios (Durand and Espitalie, 1972; Huc and Hunt, 1980; Snowdon and Powell, 1982) or normal hexane/methylcyclopentane ratios (Jonathan et al., 1975) change due to the preferential generation of normal alkanes, as opposed to cyclic alkanes, with increasing temperature. Temperature-dependent ratios of groups of gasoline range hydrocarbons have been used by Thompson (1979, 1983) to devise "heptane and isoheptane" indices for sediments and oils. By comparing the temperature distribution of the heptane and isoheptane ratios between sediments and oils, Thompson (1983) concluded that normal oils are generated at subsurface temperatures of the order of 138° to 149 °C.
4. The stable isotope ratio of methane is very negative (−90 o/oo to −60 o/oo) for shallow gas of biogenic origin, and becomes progressively more positive with depth. This isotope trend with depth was initially considered to be the result of either migration or generation processes. However, current opinion (Schoell, 1983) is that the methane carbon isotope ratio is generally a function of thermal maturity as well as source type. This concept has been further extended by James (1983) to the isotope ratios of the other low-molecular-weight hydrocarbons. He demonstrated that the carbon isotope ratio difference between gaseous hydrocarbon molecules can be used to estimate the thermal maturity of the source sediment.

The application of gases and gasoline range hydrocarbons to thermal maturity determinations

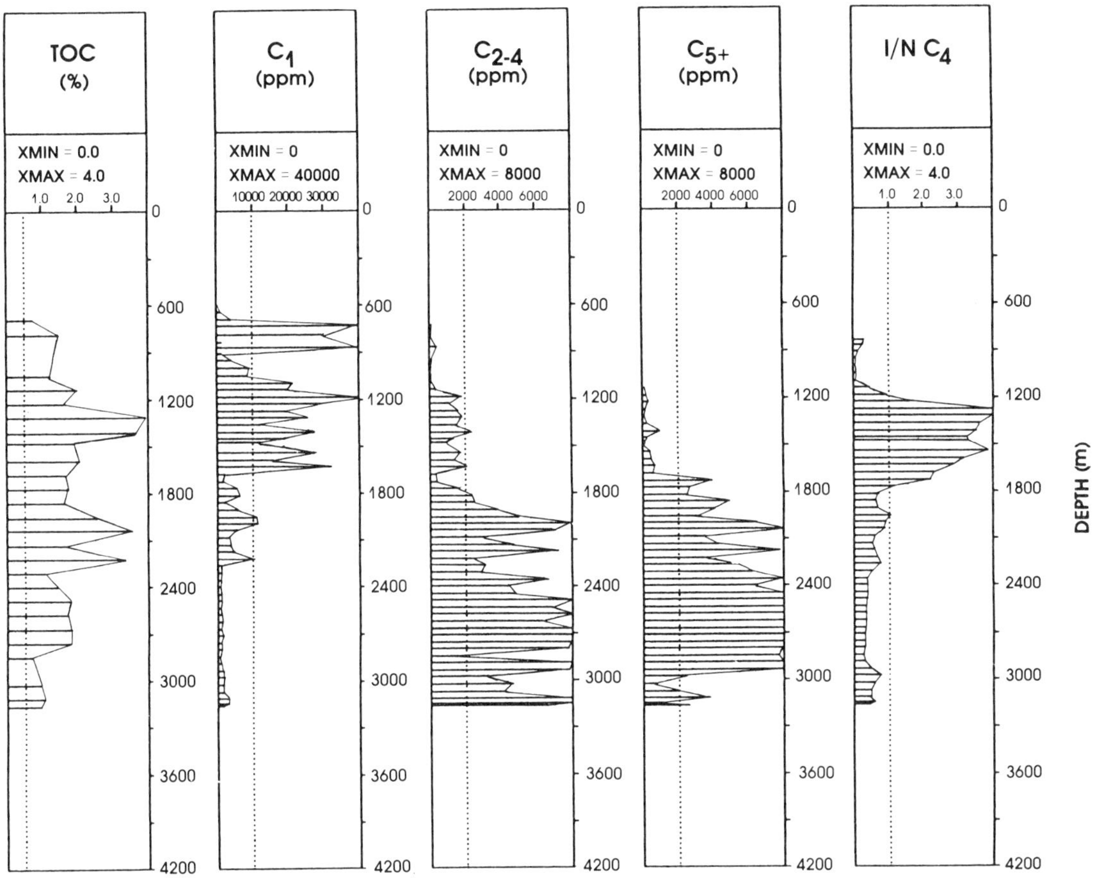

FIGURE 4.12. Depth distribution of total organic carbon (TOC), methane (C_1), ethane-butanes (C_{2-4}), gasoline range molecules (C_{5+}), and iso/normal butane ratio (I/N C_4) for a well from a subsiding basin.

is most apparent in hydrocarbon generation versus depth plots. It is often considered that the depth of significant hydrocarbon generation can be denoted by an exponential increase in hydrocarbon content in the sediment. For excellent source sections, the ratio of extractable HC/TOC can be used to document this depth (Fig. 4.13). However, as discussed earlier, for many well profiles a consistent depth trend such as this does not occur. Yet, for almost all well profiles from sedimentary sections from subsiding basins, a marked change with depth is noted in either the absolute or carbon-normalized amounts of the light hydrocarbon constituents. This depth interval could be used, in the absence of migration phenomena, to define the top of the hydrocarbon generation zone for the particular geological setting represented in the well. Thus, concentrations of low-molecular-weight hydrocarbons could theoretically be used as indicators of maturity. Other thermal maturity indicators can then be normalized to the light hydrocarbon-indicated "onset of generation" datum.

Methane concentration gradients that decrease with depth are also observed in many basins (as, for example, Fig. 4.12). This frequently results from migration of this small molecule, due to buoyancy and solution effects. Furthermore, the proportion of the ^{13}C isotope often increases with increasing depth in many basins (Schoell, 1984). This is often attributed to mixing of isotopically light biogenic methane ($\delta^{13}C = -70$ o/oo) with an isotopically heavier thermal component ($\delta^{13}C = -30$ o/oo to -45 o/oo). Figure 4.14 shows the methane $\delta^{13}C$ (isotopic) gradient through a Qua-

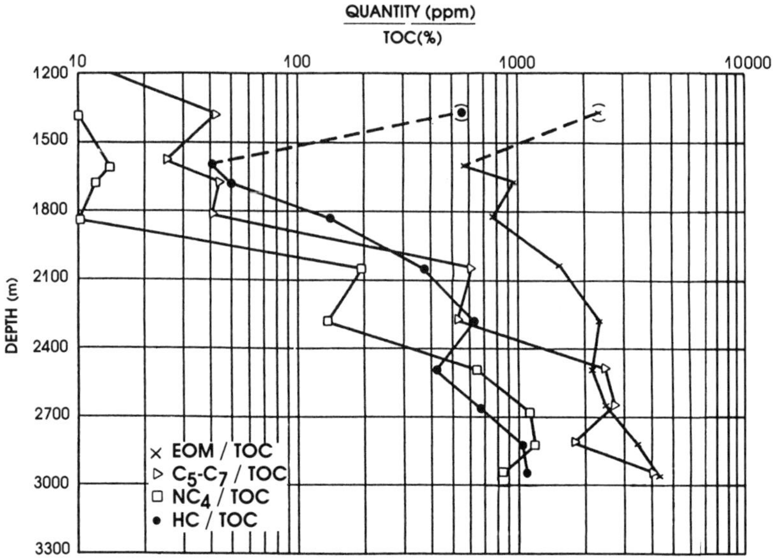

FIGURE 4.13. Depth versus ratios of normalized quantities of extractable organic matter (EOM), gasoline-range hydrocarbons (C_5–C_7), n-butane, and C_{15+}-extractable hydrocarbons (HC) to total organic carbon (TOC). Data for well profile are shown in Figure 4.12.

ternary sequence in a well where the R_o (vitrinite reflectance) value at 4,300 m is only 0.35%. The lower 1,000 m of the well is composed predominantly of lignite-rich sediment. A 30 o/oo carbon isotope ratio difference is observed for methane across a 1,200-m depth interval at low rank level. This isotope ratio gradient may be the result of isotopic fractionation related to migrational loss of methane from the lignites. It is difficult to retain isotopic mass balance and explain this isotope gradient shown in Figure 4.14 without loss of isotopically light methane. Though literature exists that indicates isotopic fractionation may not be important during secondary migration processes, adsorption/desorption processes (enhanced in the lignitic section of this example) common in fine-grained sediment may indeed induce isotopic fractionation.

It is clear that, while light hydrocarbon concentration measurements can provide a valuable tool in assessing the upper part of the oil generation zone, migration phenomena can also substantially affect any data obtained from within this zone. While valuable in providing an additional dimension in which to review geochemical processes, isotopic data are frequently complicated by factors other than thermal maturity.

Kerogen-Based Maturity Parameters

The precursor-product relationship between kerogen and most of the extractable hydrocarbons in source rocks has been generally accepted by most petroleum geochemists. Consequently, understanding of molecular maturity indicators in oil and gas source rocks requires an understanding of the composition of kerogen. At the present time, kerogen-based molecular maturity parameters are derived from nonequilibrium complex reaction systems, leading to difficulties in data interpretation. For example, although biomarkers are found in kerogens (Seifert, 1978; Mackenzie, 1984), no practical method of using their epimerization states as maturity indicators has yet been proposed. Thus no true equilibrium-based molecular maturity parameters are currently publicly available.

The simplest approach to the kerogen maturity problem is the production index (reaction coordinate) approach, which is the conceptual basis for several current geochemical approaches to maturation and generation. In Rock-Eval terminology, the production index (also called transformation ratio) is defined as the ratio of the low-temperature pyrolytic yield (S1) to the total pyrolytic yield (S1 + S2). Simplistically, an increasing production index, in the hypothetical instance of no actual expulsion of hydrocarbons from a source bed, represents an approximation to the true reaction coordinate for the reaction SOM → oil + gas + carbonaceous residue (Fig. 4.1). In fact, partial loss of S1 due to weathering and the drilling process, as well as migration of hydrocarbons from a sediment, complicates this picture. A more practical approximation is based on the Rock-Eval S2 or Hydrogen Index (HI) parameters. Then the reaction coordinate approximates, for the generation of oil and gas from kerogen, $M = (HI(i) - HI)/HI(i)$, where HI(i) is the Hydrogen Index just prior to the onset of hydrocarbon generation, and HI is the present-day Hydrogen Index. In this equation, the reaction coordinate, M, ranges from zero to a theoretical maximum of one. Utilizing this Hydrogen Index concept, it can be broadly concluded that in many cases, at least in terms of hydrocarbon yield if not composition, the Rock-Eval analysis system provides a good first-order estimate of the actual catagenic hydrocarbon yields. This is especially so for source rock systems containing abundant hydrogen-rich organic matter where mineral matrix phenomena are relatively unimportant in the analysis procedure (Espitalie, 1984; Larter, 1984).

It should be noted that any chemical property (and thus physiochemical property) of a kerogen is a function of organic matter type, in addition to maturity. All conventional kerogen-based maturation parameters depend for their efficacy on the degree to which the type of material being studied can be calibrated and allowed for. For example, reflectance of kerogen macerals does not become a maturity parameter unless the measurements are always made on the same type of material. Vitrinite reflectance (essentially vitrinite kerogen reflectance) is a successful maturity parameter because to a substantial extent vitrinitic type kerogens can be recognized in many sediment types, and the range of variation of chemical type (Larter and Senftle, 1985) is generally smaller for the vitrinitic kerogens than for the amorphous kerogen species found in sediments. In summary, kerogen-based molecular maturity parameters depend entirely for their success on the power of current kerogen typing procedures.

Current pyrolytic kerogen typing methods and their extent of evolution have been reviewed many times recently (Horsfield, 1984; Larter, 1984). In general our ability to type kerogens accurately and quantitatively is very poor. At least eight general kerogen types can be recognized in terms of their total hydrocarbon and normal hydrocarbon yields on pyrolysis (Larter and Senftle, 1985), and detailed variation within these gross types can at present be allowed for only crudely (Fig. 4.15). However, once kerogen type can be reliably and quantitatively estimated, then any change in composition for a monotypic kerogen set becomes potentially a maturation indicator. In addition, it is important to remember that in order for a single kerogen type to be useful in assessing maturation, the kinetics of maturation of the chosen kerogen type should be similar to that of the kerogen type sourcing the bulk of the hydrocarbons. Unfortunately, this point has been considered too infrequently in the literature.

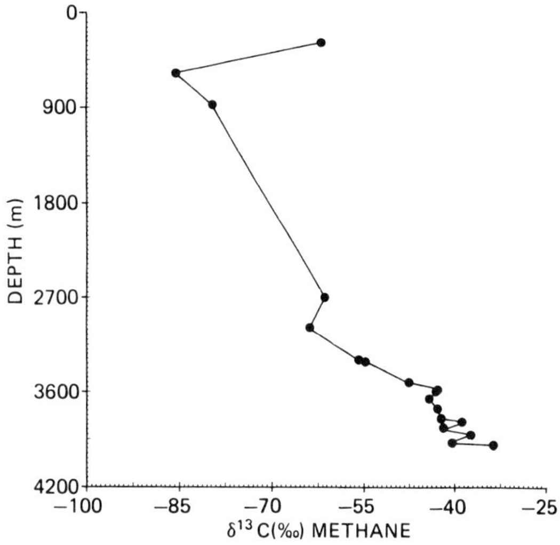

FIGURE 4.14. Carbon isotope ratios of methane in a rapidly deposited Quaternary sedimentary section. The interval below 3,500 m is composed predominantly of lignite, with the vitrinite reflectance at the bottom of the section being about 0.35%.

The Future

While bulk maturity parameters, such as the reflectance of vitrinite and the color of spores, will continue to be heavily used, it is probable that the use of molecular maturity indicators will increase

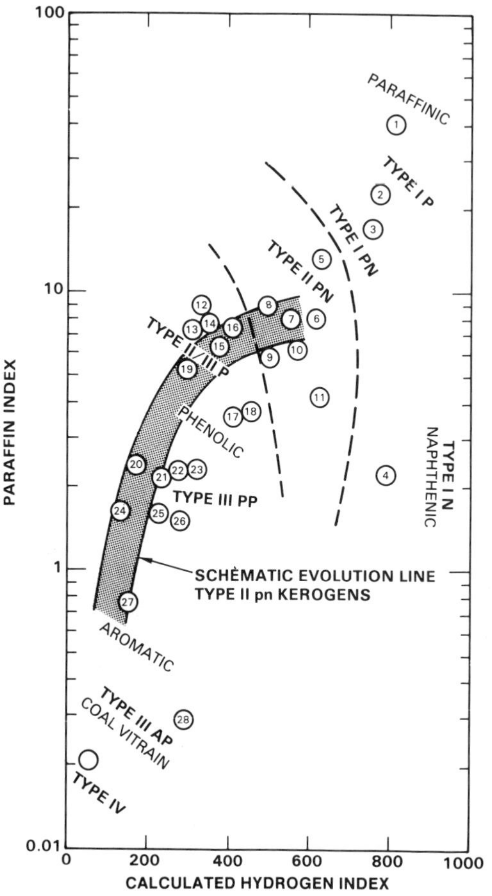

REPRESENTATIVE KEROGEN TYPES

TYPE Ip	— TORBANITE (BOTRYOCOCCUS) KEROGENS RICH IN ALGINITE.
TYPE Ipn	— TASMANITE (ALGINITE) RICH KEROGENS.
TYPE In	— RESINITE DERIVED MATERIALS.
TYPE II pn	— AMORPHINITE RICH KEROGENS FROM SOURCE ROCKS CONTAINING MARINE ORGANIC MATTER.
TYPE II/III p	— EXINITE RICH KEROGENS
TYPE II/IIIpp	— CARBONIFEROUS SPORINITES
TYPE III pp	— VITRINITE RICH KEROGENS FROM MARINE/DELTAIC SHALES.
TYPE III ap	— VITRINITE RICH KEROGENS FROM COAL SWAMPS.
TYPE IV	— INERTINITE RICH KEROGENS.

FIGURE 4.15. Paraffin Index (TOC-normalized C_9–C_{30} normal hydrocarbon yield) versus Hydrogen Index. At least eight distinct kerogen types can be discerned. Data obtained from 800°C kerogen flash pyrolysis of immature kerogens. The Paraffin Index (y axis) is calibrated such that 1 unit = 0.25% by weight. (After Larter and Senftle, 1985, reprinted by permission of Macmillan Magazines.)

dramatically in the future. This increase will not be without its problems. Of these, the most pressing is the still-unresolved question of reaction kinetics. Applying our early definition of thermal maturity (Fig. 4.1), it is apparent that a satisfactory assessment of the maturity of sedimentary organic matter requires a monitorable reaction system with kinetics similar, if not identical, to those of hydrocarbon generation.

The importance of reaction kinetics to the estimation of maturity is made clear if we compare two reactions discussed earlier, each of quite different mechanism, and involving differing reaction kinetics. These are the aromatization of C-ring monoaromatized steroid hydrocarbons to ABC-ring triaromatics (Figs. 4.6 and 4.7), and the epimerization of 5α,14α,17α-ethylcholestane (Figs. 4.6, 4.8, and 4.9). Use of reactions of this general type will be among the principal directions that molecular maturity parameter research will evolve. Furthermore, these two systems nicely illustrate some of the major problems associated with using molecular maturity parameters in general.

Figure 4.16 shows the reaction coordinates for the two reactions plotted as a function of depth of burial for three hypothetical basins (kinetic data from Mackenzie, 1984). The three basins all have equal thermal gradients (30°C/km) and temperatures of 10°C at the surface. All the basins are simple subsiding systems with linear subsidence

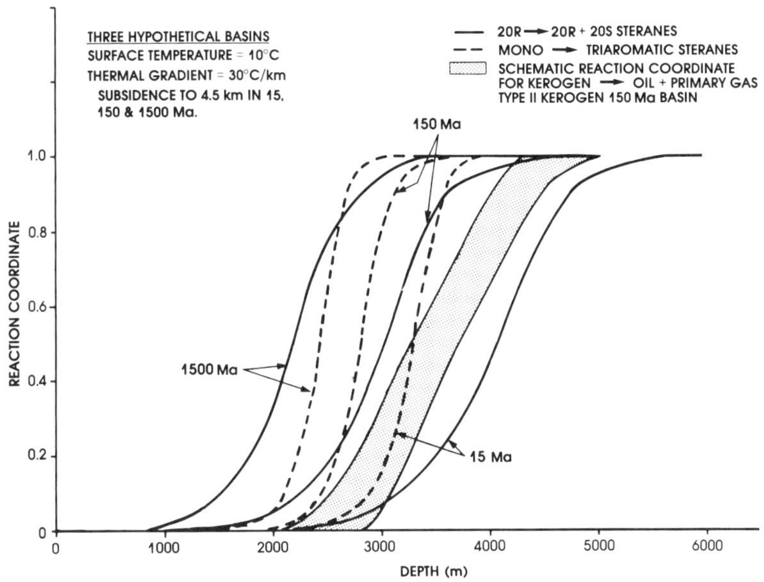

FIGURE 4.16. Modeling of biomarker epimerization and aromatization reactions using data of Mackenzie (1984) and equations after Juntgen and Van Heek (1968). The separation of the two reactions varies with heating rate.

rates, having ages (at 4.5 km) of 15 Ma, 150 Ma, and 1,500 Ma, respectively. Thermal hysteresis in the section is ignored. Using the equations of non-isothermal reaction kinetics derived by Juntgen and Van Heek (1968), the reaction coordinate for the biomarker aromatization and epimerization reactions (e.g., the "maturity") can be assessed as a function of basin heating rate.

Figure 4.16 shows that, whereas for the slowly subsiding basins the aromatization and epimerization reactions occur at similar temperatures, in the rapidly subsiding 15-Ma basin, where the sediments are heated at a much faster rate, the epimerization reaction (with lower activation energy) lags substantially. It is therefore clear that, without a knowledge of, and allowance for, variable heating (subsidence) rates in different basins, the direct application of these molecular parameters to maturity evaluation is hazardous (although see Suzuki, 1984). As the heating rate increases then, the reaction coordinates of the maturity parameters with different reaction mechanisms will diverge from one another and, more importantly, from the reaction coordinate system related to oil generation. Figure 4.16 also shows a schematic reaction coordinate plot for hydrocarbon generation from a real basin similar to that modeled as the 150-Ma basin. The oil generation curve plots in the position expected for a complex reaction system that has a net summed kinetic effect unlike that for the individual reactions involved in the system. It is clear that application of these molecular maturity parameters without regard for the complexities of the reaction assemblage involved can lead to major misinterpretations.

The reaction kinetics involved with molecular maturity parameters represent only one of the problems that must, and will, be addressed before such parameters can be utilized to their fullest. In addition, system closure questions must also be answered, in order to establish the uniqueness of the reactions involved. After these and other difficulties are overcome, it is entirely possible that reactions will emerge that satisfy more completely our initial definition of the ideal maturity parameter.

Acknowledgments. We are indebted to several colleagues, including Drs. J.R. Fox, E.C. Copelin, G.H. Smith, and J.T. Senftle for influencing our ideas concerning thermal maturity. We also wish to thank Dr. J.R. Fox for his review of an early draft of this manuscript, W.D. Ahlborn and M.A. Jicha for drafting and secretarial assistance, and Union

Oil Company for permission to publish. We appreciate excellent reviews by J.M. Moldowan, G.E. Claypool, and N.D. Naeser.

References

Albrecht, P., Vandenbrouke, M., and Mandenque, M. 1976. Geochemical studies on the organic matter from the Doula Basin (Cameroon): 1. Evolution of the organic matter and the formation of petroleum. Geochimica et Cosmochimica Acta 40:791–799.

Alexander, R.J., Kagi, R.I., Rowland, S.J., Sheppard, P.N., and Chirila, T.V. 1985. The effects of thermal maturity on distributions of dimethylnaphthalenes and trimethylnaphthalenes in some ancient sediments and petroleums. Geochimica et Cosmochimica Acta 49:385–395.

Allan, J., Bjoroy, M., and Douglas, A.G. 1977. Variation in the content and distribution of high molecular weight hydrocarbons in a series of coal macerals of different rank. In: Campos, R., and Goni, J. (eds.): Advances in Organic Geochemistry 1975. Madrid, Enadimsa, pp. 633–655.

Allan, J., and Larter, S.R. 1983. Aromatic structures in coal macerals. In: Bjoroy, M., et al. (eds.): Advances in Organic Geochemistry 1981. London, Wiley, pp. 534–545.

Baset, Z.H., Pancirov, R.J., and Ashe, T.R. 1980. Organic compounds in coal, structure and origins. In: Douglas, A.G., and Maxwell, J.R. (eds.): Advances in Organic Geochemistry 1979. Oxford, Pergamon Press, pp. 619–631.

Blumer, M. 1965. Organic pigments, their long term fates. Science 188:53–55.

Bromley, B.W. 1983. Observations on petroleum compositions – migration effects? Abstracts, Advances in Organic Geochemistry 1983, Hague, p. 117.

Burnham, A.K., Clarkson, J.E., Singleton, M.F., Wong, C.M., and Crawford, R.W. 1982. Biological markers from Green River kerogen decomposition. Geochimica et Cosmochimica Acta 46:1243–1253.

Claypool, G.E., and Kaplan, I.R. 1974. The origin and distribution of methane marine sediments. In: Kaplan, I.R. (ed.): Natural Gases in Marine Sediments. New York, Plenum Press, pp. 99–139.

Curiale, J.A., Cameron, D., and Davis, D.V. 1985. Biological marker distribution and significance in oils and rocks of the Monterey Formation, California. Geochimica et Cosmochimica Acta 49:271–288.

Day, W.C., and Erdman, J.G. 1963. Ionene, a thermal decomposition product of Beta-Carotene. Science 141:808–809.

De Leeuw, J.W., Simoneit, B.R., Boon, J., Rijpstra, W.I., De Lange, F., Van De Leeden, J.C.W., Correia, V.A., Burlingame, A.L., and Schenk, P.A. 1977. Phytol derived compounds in the geosphere. In: Campos, J., and Goni, R. (eds.): Advances in Organic Geochemistry 1975. Madrid, Enadimsa, pp. 61 79.

Didyk, B.M., Alturki, Y.I.A., Pillinger, C.T., and Eglinton, G. 1975. Petroporphyrins as indicators of geothermal maturation. Nature 256:563–565.

Durand, B., and Espitalie, J. 1972. Formation et evolution des hydrocarbures de C1 a C15 et des gaz permanents dans les argiles due toarcien du bassin de Paris. In: v. Gaertner, H.R., and Wehner, H. (eds.): Advances in Organic Geochemistry 1971. New York, Pergamon Press, pp. 455–468.

Durand, B., and Oudin, J.L. 1980. Examples of migration of hydrocarbons in a deltaic series, the Mahakem Delta, Kalimantan, Indonesia (Exemple de migration des hydrocarbures dans une series deltaique; Le Delta de la Mahakam, Kalimantan, Indonesie). Proceedings of the Tenth World Petroleum Congress, Bucharest, 1979, vol. 2, pp. 3–11.

Espitalie, J. 1984. Geochemical logging. In: Voorhees, K.J. (ed.): Analytical Pyrolysis: Techniques and Applications. London, Butterworths, pp. 276–304.

Gallegos, E.J. 1980. Alkylbenzenes derived from carotenoids in coals, by GC-MS. Journal of Chromatographic Science 119:177–185.

Goossens, H., de Leeuw, J.W., Schenk, P.A., and Brassell, S.C. 1984. Tocopherols as likely precursors of pristane in ancient sediments and crude oils. Nature 312:440–442.

Hachenberg, E.A. 1977. Gas Chromatographic Headspace Analysis. London, Heyden and Son, 127 pp.

Heroux, Y., Chagnon, A., and Bertrand, R. 1979. Compilation and correlation of major thermal maturation indicators. American Association of Petroleum Geologists Bulletin 63:2128–2144.

Horsfield, B. 1984. Pyrolysis studies and petroleum exploration. In: Brooks, J., and Welte, D.H. (eds.): Advances in Petroleum Geochemistry: Vol. 1. London, Academic Press, pp. 247–298.

Huc, A.Y., and Hunt, J.M. 1980. Generation and migration of hydrocarbons in offshore South Texas Gulf Coast sediments. Geochimica et Cosmochimica Acta 43:657–672.

Hyatsu, R., Winans, R.E., Scott, R.G., Moore, L.P., and Studier, M.H. 1978. Trapped organic compounds in coal. Fuel 57:547–554.

Ikan, R., Aizenstat, Z., Baedecker, M.J., and Kaplan, I.R. 1975. Thermal alteration experiments on organic matter in recent marine sediments: I. Pigments. Geochimica et Cosmochimica Acta 39:173–185.

James, A.T. 1983. Correlation of natural gas by use of carbon isotopic distribution between hydrocarbon components. American Association of Petroleum Geologists Bulletin 67:1176–1191.

James, A.T., and Burns, B.J. 1984. Microbial alteration of subsurface natural gas accumulations. American Association of Petroleum Geologists Bulletin 68:957–960.

Jasper, J.P., Whelan, J.K., and Hunt, J.M. 1982. Migration of C1 to C8 volatile organic compounds in sediments from the Deep Sea Drilling Project, Leg 75, Hole 530A. Initial Reports of the Deep Sea Drilling Project 75:1001–1008.

Jonathan, D.L., Hote, G., and Du Rouchet, J. 1975. Analyses geochemiques des hydrocarbures legers par thermovaporisation. Revue Institut Francais du Petrole 30:65–88.

Juntgen, H., and Van Heek, K.H. 1968. Gas release from coal as a function of heating rate. Fuel 47:103–117.

Larter, S.R. 1984. Application of analytical pyrolysis techniques to kerogen characterisation and fossil fuel exploration/utilization. In: Voorhees, K. (ed.): Analytical Pyrolysis: Techniques and Applications. London, Butterworths, pp. 212–275.

Larter, S.R., and Senftle, J.T. 1985. Improved kerogen typing for petroleum source rock analysis. Nature 318:277–280.

Larter, S.R., Solli, H., and Douglas, A.G. 1983. Phytol containing melanoidins and their bearing on the fate of isoprenoid structures in sediments. In: Bjoroy, M., et al. (eds.): Advances in Organic Geochemistry 1981. London, Wiley, pp. 513–523.

Leythaeuser, D., Schaefer, R.G., and Weiner, B. 1979. Generation of low molecular weight hydrocarbons from organic matter in source beds as a function of temperature. Chemical Geology 25:95–108.

Mackenzie, A.S. 1984. Applications of biological markers in petroleum geochemistry. In: Brooks, J., and Welte, D. (eds.): Advances in Petroleum Geochemistry: Vol. 1. London, Academic Press, pp. 115–214.

Mackenzie, A.S., Brassell, S.C., Eglinton, G., and Maxwell, J.R. 1982. Chemical fossils: The geologic fate of steroids. Science 217:491–504.

Mackenzie, A.S., Hoffman, C.F., and Maxwell, J.R. 1981. Molecular parameters of maturation in the Toarcian shales, Paris Basin, France: III. Changes in aromatic steroid hydrocarbons. Geochimica et Cosmochimica Acta 45:1345–1355.

Mackenzie, A.S., Lewis, C.A., and Maxwell, J.R. 1981. Molecular parameters of maturation in the Toarcian shales, Paris Basin, France: IV. Laboratory thermal maturation studies. Geochimica et Cosmochimica Acta 45:2369–2376.

Mackenzie, A.S., and Maxwell, J.R. 1981. Assessment of thermal maturation in sedimentary rocks by molecular measurements. In: Brooks, J. (ed.): Organic Maturation Studies and Fossil Fuel Exploration. New York, Academic Press, pp. 239–254.

Mackenzie, A.S., and McKenzie, D. 1983. Isomerization and aromatization of hydrocarbons in sedimentary basins formed by extension. Geological Magazine 120:417–470.

Mackenzie, A.S., Patience, R.L., and Maxwell, J.R. 1981. Molecular changes and the maturation of sedimentary organic matter. In: Atkinson, G., and Zuckerman, J.J. (eds.): Origin and Chemistry of Petroleum. New York, Pergamon Press, pp. 1–32.

Mackenzie, A.S., Ren-Wei, L., Maxwell, J.R., Moldowan, J.M., and Seifert, W.K. 1983. Molecular measurements of thermal maturation of Cretaceous shales from the Overthrust Belt, Wyoming, USA. In: Bjoroy, M., et al. (eds.): Advances in Organic Geochemistry 1981. London, Wiley, pp. 496–503.

McKenzie, D., Mackenzie, A.S., Maxwell, J.R., and Sajgo, C.S. 1983. Isomerization and aromatization of hydrocarbons in stretched sedimentary basins. Nature 301:504–506.

Philippi, G.T. 1975. The deep subsurface temperature-controlled origin of the gaseous and gasoline-range hydrocarbons of petroleum. Geochimica et Cosmochimica Acta 39:1353–1373.

Radke, M., Schaffer, R.G., and Leythauser, D. 1980. Composition of the soluble organic matter in coals and its relation to rank and liptinite fluorescence. Geochimica et Cosmochimica Acta 44:1787–1799.

Radke, M., and Welte, D.H. 1983. The methylphenanthrene index (MPI): A maturity parameter based on aromatic hydrocarbons. In: Bjoroy, M., et al. (eds.): Advances in Organic Geochemistry 1981. London, Wiley, pp. 504–512.

Radke, M., Welte, D.H., and Willsch, H. 1982. Geochemical study on a well in the Western Canada Basin: Relation of the aromatic distribution pattern to maturity of organic matter. Geochimica et Cosmochimica Acta 46:1–10.

Radke, M., Willsch, H., Leythauser, D., and Teichmüller, M. 1982. Aromatic components of coal: Relation of distribution pattern to rank. Geochimica et Cosmochimica Acta 46:1831–1848.

Schoell, M. 1983. Genetic characterisation of natural gases. American Association of Petroleum Geologists Bulletin 67:2225–2238.

Schoell, M. 1984. Stable isotopes in petroleum research. In: Brooks, J., and Welte, D.H. (eds.): Advances in Petroleum Geochemistry: Vol. 1. London, Academic Press, pp. 215–246.

Schoell, M., Teschner, M., Wehner, H., Durand, B., and Oudin, J.L. 1983. Maturity related biomarker and stable isotope variations and their application to oil/source rock correlation in the Mahakam Delta, Kalimantan. In: Bjoroy, M., et al. (eds.): Advances in

Organic Geochemistry 1981. London, Wiley, pp. 156–163.

Seifert, W.K. 1978. Steranes and terpanes in kerogen pyrolysis for correlation of oils and source rocks. Geochimica et Cosmochimica Acta 42:173–484.

Seifert, W.K., Carlson, R.M.K., and Moldowan, J.M. 1983. Geomimetic synthesis, structure assignment, and geochemical correlation application of monoaromatized petroleum steroids. In: Bjoroy, M., et al. (eds.): Advances in Organic Geochemistry 1981. London, Wiley, pp. 710–724.

Seifert, W.K., and Moldowan, J.M. 1978. Applications of steranes, terpanes and monoaromatics to the maturation, migration and source of crude oils. Geochimica et Cosmochimica Acta 42:77–95.

Seifert, W.K., and Moldowan, J.M. 1980. The effect of thermal stress on source-rock quality as measured by hopane stereochemistry. In: Douglas, A.G., and Maxwell, J.R. (eds.): Advances in Organic Geochemistry 1979. Oxford, Pergamon Press, pp. 229–237.

Seifert, W.K., and Moldowan, J.M. 1981. Paleoreconstruction by biological markers. Geochimica et Cosmochimica Acta 45:783–794.

Snowdon, L.R., and Powell, T.G. 1982. Immature oil and condensate: Modification of hydrocarbon generation model for terrestrial organic matter. American Association of Petroleum Geologists Bulletin 66:775–778.

Solli, H., Larter, S.R., and Douglas, A.G. 1980. Analysis of kerogens by pyrolysis-gas chromatography-mass spectrometry: II. Alkylnaphthalenes. Journal Analytical Applications in Pyrolysis 1:231–242.

Suzuki, N. 1984. Estimation of maximum temperature of mudstone by two kinetic parameters: Epimerization of sterane and hopane. Geochimica et Cosmochimica Acta 48:2273–2282.

Thompson, K.F.M. 1979. Light hydrocarbons in subsurface sediments. Geochimica et Cosmochimica Acta 43:657–672.

Thompson, K.F.M. 1983. Classification and thermal history of petroleum based on light hydrocarbons. Geochimica et Cosmochimica Acta 47:303–316.

Tissot, B., Durand, B., Espitalie, J., and Combaz, A. 1974. Influence of the nature and diagenesis of organic matter in the formation of petroleum. American Association of Petroleum Geologists Bulletin 58:499–506.

Tissot, B., Espitalie, J., Deroo, G., Tempere, C., and Jonathan, D. 1974. Origin and migration of hydrocarbons in the eastern Sahara (Algeria). In: Tissot, B., and Bienner, F. (eds.): Advances in Organic Geochemistry 1973. Paris, Editions Technip, pp. 315– 334.

Tissot, B., and Welte, D.H. 1978. Petroleum Formation and Occurrence: A New Approach to Oil and Gas Exploration. Berlin, Springer-Verlag, 538 pp.

Van Graas, G., de Leeuw, J.W., Schenk, P.A., and Haverkamp, J. 1981. Kerogen of Toarcian shales of the Paris Basin: A study of its maturation by flash pyrolysis techniques. Geochimica et Cosmochimica Acta 45:2465–2475.

White, C.M., and Lee, M.L. 1980. Identification and geochemical significance of aromatic components of coal. Geochimica et Cosmochimica Acta 44:1825–1832.

5

Temperature and Time in the Thermal Maturation of Sedimentary Organic Matter

Charles E. Barker

Abstract

Models that consider temperature alone or a combination of temperature and heating duration have been defined for the thermal maturation of sedimentary organic matter (OM). Both types of models appear to give adequate maximum paleotemperature estimates, but these estimates are imprecise due to problems in measuring thermal maturity (rank), heating duration, and maximum temperature (T_{max}). Must temperature and functional heating duration (t) be considered or is T_{max} alone sufficient to characterize thermal maturation to the precision level now possible in sedimentary environments? This question is addressed by developing a temperature-heating duration model for OM thermal maturation based on a broad geological data base and comparing it to other empirical models that consider T_{max} alone or T_{max} and t.

If a first-order reaction can be assumed for OM thermal maturation, then its reaction rate constant, k, is equal to $-\ln(f)\cdot(1/t)$, where t is the functional heating duration and f is the fraction of reactable OM. The fraction of reactable OM (f) is the complement of transformation ratio (r) estimated from Tissot and Espitalié's 1975 model of vitrinite reflectance (R_m) evolution. The functional heating duration for burial diagenesis is calculated by determining the elapsed time as temperature increases within 15°C of T_{max} without exceeding the time necessary for the controlling reactions to approach completion. Geologic field data and OM thermal maturation experiments extrapolated to a geologic time and temperature range suggest OM thermal maturation reactions approach completion in about 10^6 to 10^7 years during burial diagenesis. The elapsed time near T_{max} is used as the reaction time in geothermal systems and contact metamorphism by intrusive sheets.

The calculated reaction rate when plotted on an Arrhenius diagram [ln k versus absolute temperature $(1/T)$] falls along three subparallel straight-line segments. These segments correspond to OM thermal maturation in three different environments: burial diagenesis, geothermal systems, and contact metamorphism by intrusive sheets. These environments were individually analyzed by linear regression of $\ln(k)$ and $1/T$ data which for a single reaction mechanism conform to a straight line, $\ln(k) = (-E_a/R)1/T + \ln A$, where E_a is the activation energy for the reaction, R is the gas constant, and A is the Arrhenius factor. OM thermal maturation in burial diagenesis shows a strong linear relationship (correlation coefficient, $r = -0.84$) with an E_a of 9 kcal/mol (38 kJ/mol) and an A of 7×10^{-11}/s. OM thermal maturation in geothermal systems shows a moderate linear relationship ($r = -0.74$), an E_a of 11 kcal/mol (46 kJ/mol), and an A of 2.5×10^{-7}/s. OM thermal maturation from contact metamorphism by intrusive sheets is modeled by a linear relationship ($r = -0.64$), with an E_a of 12 kcal/mol (50 kJ/mol), and an A of 1.5×10^{-3}/s. In summary, these data conform to lines with a uniform and similar slope, indicating that OM thermal maturation consists of a pseudoreaction that has a similar E_a in each environment.

OM thermal maturation differs in burial diagenesis, geothermal systems, and contact metamorphism by intrusive sheets primarily in the reaction rate at a given temperature. This may be due to differences in pressure between these environments. The different reaction rate in each environment is also expressed in the time required for OM thermal maturation to stabilize. The regression line calculated for geothermal systems overlies a T_{max}-R_m model that is based on geologic evidence that indicates t was not important after 10^4 years. The regression line from contact metamorphism by igneous sheets is in fair agreement with data from laboratory experiments that indicate stabilization after 10^{-1} to 10^0 years.

Comparison of these calculated regression curves to published models of OM thermal maturation indicates that precision is not increased by considering heating duration. The data can be adequately modeled by considering T_{max} alone.

Introduction

Models describing the thermal maturation of sedimentary organic matter (OM) generally consider temperature alone or a combination of temperature and heating duration. Paleotemperature estimates from these models are imprecise due to problems in measuring thermal maturity (rank), heating duration, and maximum temperature (T_{max}). These problems make it difficult to distinguish which model type is the simplest effective explanation of thermal maturation in sedimentary environments. The simplest models only consider temperature, but is T_{max} alone adequate to characterize thermal maturation to the precision level now possible in sedimentary environments? This problem is approached by developing a temperature-heating duration model of OM thermal maturation based on a broad geological data base and by comparing it to other empirical models that consider T_{max} alone, or T_{max} and heating duration. Models that consider T_{max} alone are compared to those based on T_{max} and heating duration by using an estimate of the heating duration necessary for OM thermal maturation to stabilize. In any chemical system, reactions will stabilize when all the material reactible at that temperature is expended. If OM thermal maturation is relatively rapid, thermal stabilization can occur in the heating duration normally available in sedimentary systems. Experimental and geological evidence, presented later, suggests that stabilization at T_{max} does occur in sedimentary environments.

OM thermal maturation is modeled from published studies of temperature and rank in burial diagenesis, geothermal systems, and contact metamorphism by intrusive igneous sheets. A first-order kinetic equation is used to calculate the rate constant (k) of OM maturation as a function of reaction temperature. Regression analysis of these data plotted on an Arrhenius diagram establishes a baseline to compare the models. These models of OM thermal maturation are graphically analyzed using a linear-linear plot (Arrhenius diagram) of the logarithm of reaction rate (ln k) versus the reciprocal of the absolute temperature ($1/T$). The slope of the regression line is proportional to the activation energy of the reaction (E_a) and its y-axis intercept is equal to ln A (where A is the Arrhenius factor).

The advantage of this approach is that the data are compiled from numerous independent measurements in diverse sedimentary environments. The limitations of the empirical approach to modeling OM thermal maturation are that 1) T_{max} reached in the system may be difficult to determine, 2) correction of borehole temperature data to equilibrium reservoir conditions is poorly understood, and 3) vitrinite reflectance (R_m) data may be subject to significant error. The model is a simple approximation of a complex chemical system; however, as shown by numerous studies (Waples, 1984) and by examples presented in this chapter, OM thermal maturation is successfully modeled by simple empirical functions.

A Kinetic Model of Thermal Maturation

First-order reactions are symbolically expressed:

$$C \rightarrow \text{products}$$

where C is the concentration of unimolecular reactant (i.e., the molecule spontaneously reacts without requiring other external components or phases); in this case, it is the amount of OM transformable to products. The number of molecules that decompose by this unimolecular process per unit time (equal to the reaction rate constant, k) is proportional to the concentration of transformable molecules present in the system ($[C]$). The rate of decrease in $[C]$ over some duration of reaction, t, is $-d[C]/dt$ and this is equal to the decomposition rate multiplied by the concentration of C (Equation 1),

$$-d[C]/dt = k[C] \tag{1}$$

Rearranging Equation 1,

$$d[C]/[C] = -kdt \tag{1a}$$

Integrating this differential equation with the condition that reaction duration is measured from the time that the initial amount of C is $[C]_0$ gives Equation 2 (Pauling, 1965),

$$\ln([C]/[C]_0) = -kt \tag{2}$$

$[C]/[C]_0$ is the fraction of reactive OM (f) remaining after some geologically functional heating duration (t). Substituting f into Equation 2 and solving for k,

$$k = -\ln(f)\cdot(1/t) \tag{3}$$

This equation is used to estimate k, from a geologically reasonable estimate of t and a measure of the reactive OM. A model for OM thermal maturation can be empirically calibrated from this equation by calculating k as a function of reaction temperature and analyzing these data using an Arrhenius diagram. This method is necessary because of problems in determining chemically meaningful reaction parameters—elapsed time, reaction extent, and temperature—in sedimentary environments. These factors, coupled with only a schematic understanding of the reaction mechanisms (Tissot and Espitalié, 1975), make it difficult to rigorously define OM thermal maturation as a chemical system.

Evaluating the fraction of reactive OM (f) is difficult because of hydrocarbon migration in the rock. It is not reasonable to assume that the reaction products in the OM are always the direct result of its thermal maturation, because closed geological environments are rare. If the present OM chemical composition (equivalent to $[C]$ in Equation 2) is used to measure reaction extent, a starting composition must be determined (to find $[C]_0$). OM is also potentially variable in chemical composition both laterally and vertically in a geological system (Kantsler et al., 1978; and others), and the carbon-rich residue remaining after burial and heating may not be comparable to less mature OM near the surface. Thus, the elemental composition as it exists now is easily measured but is not valid unless hydrocarbon migration into the rock was not significant. Furthermore, the starting composition must be assumed to use bulk chemical parameters as a measure of the fraction of reactable OM. These problems preclude using a direct measure of reaction extent, and an empirical model of f is substituted.

Tissot and Espitalié (1975) quantified a transformation ratio, r, for thermal maturation of vitrinite (Fig. 5.1). The utility of r is its empirical relation to R_m, a convenient rank measure, and this model gives reasonable r values for OM thermal maturation in sedimentary environments and laboratory studies (Tissot and Welte, 1984). The transformation ratio measures the degree of product formation relative to the total generating capacity of the OM. Thus, r is a measure of the amount of converted OM, the complement of f, which is the remaining reactive OM (i.e., $f = 1 - r$). R_m (and consequently the estimate of r) does vary between types of OM (Price and Barker, 1985), but this change seems small in comparison to errors in measuring t and T_{max}. As shown later in Figure 5.4A, there is no apparent difference between the various types of OM and the calculated kinetic parameters in systems where R_m suppression is not a major factor.

Functional Heating Duration

Burial Diagenesis

Methods of computing heating duration for OM thermal maturation are attempts to estimate reaction time during burial where temperature slowly changes, or stabilizes, over geological time. For example, Hood et al. (1975) defined the effective heating time for OM thermal maturation as the elapsed time when the system is within 15°C of T_{max}. Mackenzie and McKenzie (1983) concluded that this is a workable measure of reaction time for burial diagenesis. Estimates of heating duration can emphasize elapsed time near T_{max} because lower temperature reactions are not significant in determining final OM rank. For example, OM thermal maturation, using the classic approximation that the reaction rate doubles for each 10°C increase, will be approximately 1,000 times faster at 150°C (approximate cessation of liquid hydrocarbon generation) than at 50°C (approximate initiation of hydrocarbon generation) (Hunt, 1979). The contribution to OM rank from low-temperature reactions appears to be overwhelmed by reaction at higher temperature.

Given sufficient heating duration at T_{max}, OM thermal maturation could stabilize by reactions consuming all potentially cleavable bonds. Experiments indicate that thermal maturation of OM proceeds by parallel reactions that have a wide E_a

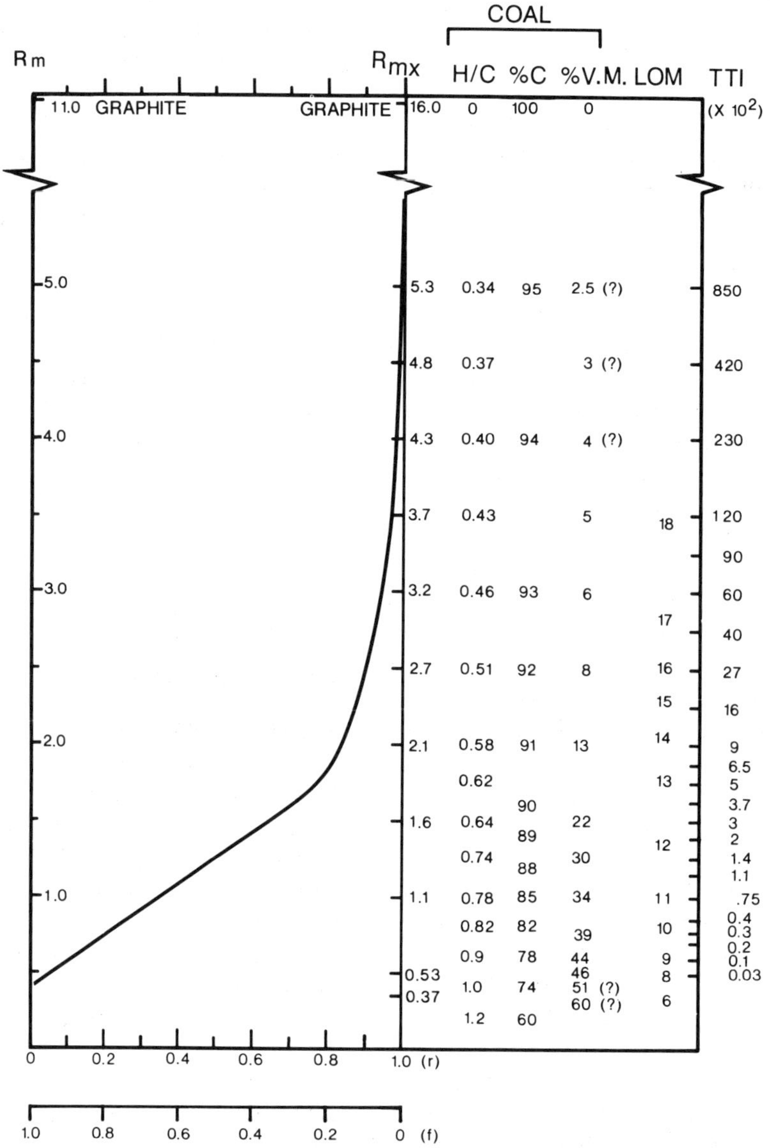

FIGURE 5.1. The fraction of reactive OM (f) and transformation ratio (r) compared to other measures of thermal maturation (modified from Tissot and Espitalié, 1975). Mean vitrinite reflectance in oil (R_m) and (r) correlation from Tissot and Espitalié (1975). Atomic hydrogen-carbon ratio (H/C), carbon content (%C), and volatile matter content (V.M.) correlation with R_m from McCartney and Teichmüller (1972). Level of organic metamorphism (LOM) correlation with R_m from Hood and Castano (1974). Lopatin time-temperature index (TTI) correlation with R_m from Waples (1980). R_m converted to mean maximum reflectance (R_{mx}) using an equation of Ting (1978).

range (Juntgen and Klein, 1975; Tissot and Espitalié, 1975). For a reasonable burial temperature, the wide E_a range over which OM thermal maturation occurs suggests that the reactions: 1) with a low E_a are complete and not generating products, 2) with a moderate E_a are generating significant products, or 3) with a high E_a will be slow and not complete reaction in geological time. Recent summaries of the chemistry of hydrocarbon generation (Johns, 1979; Saxby, 1982) indicate that, up to moderate burial temperature, the important reactions are decarboxylation of OM

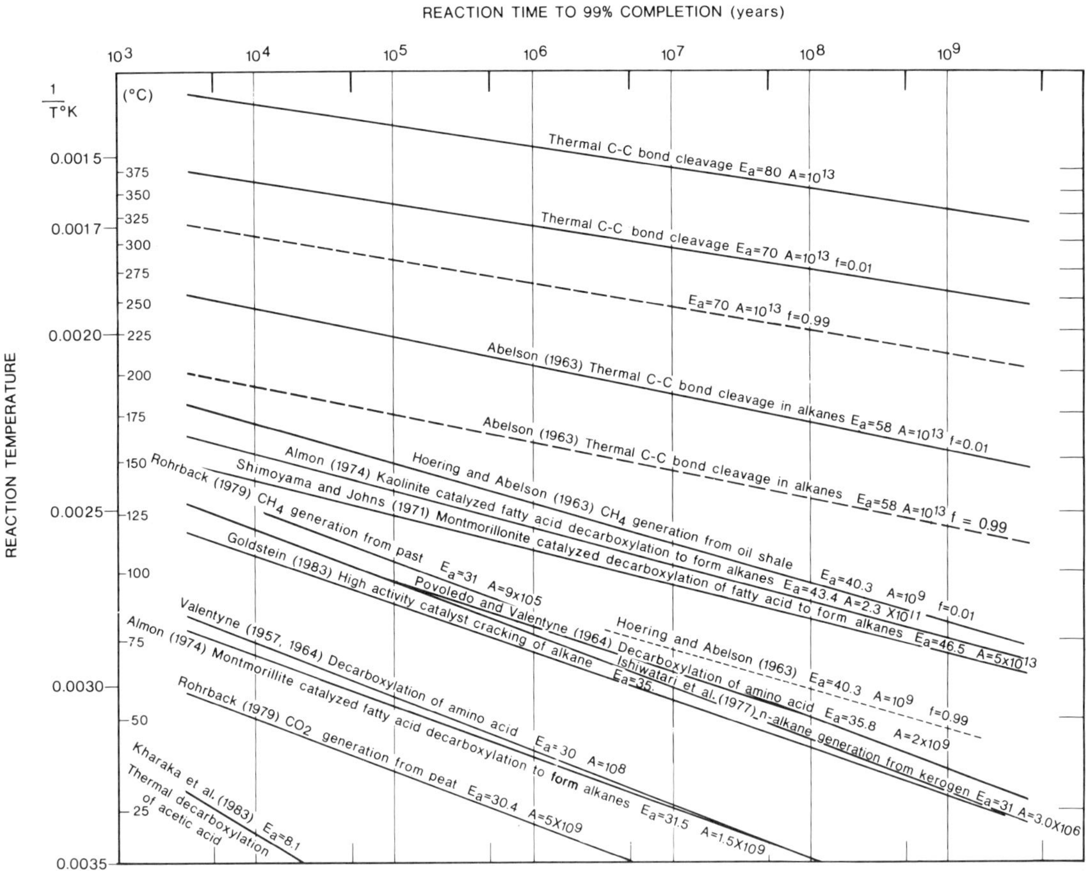

FIGURE 5.2. Arrhenius plot of experimentally determined kinetic parameters extrapolated into geological time. Pairs of curves annotated with $f = 0.99$ and $f = 0.01$ refer to the approximate initiation of reaction (1% complete) and to the end of reaction (99% complete), respectively. All other curves are reaction time to 99% completion ($f = 0.01$).

with activation energies in the range of 10 to 50 kcal/mol (42 to 210 kJ/mol) (Fig. 5.2). OM thermal maturation, at temperatures above about 125°C, proceeds by a series of carbon-carbon (C–C) bond cleavage reactions increasing from 50 to 90 kcal/mol (210 to 380 kJ/mol) (Abelson, 1963; Benson, 1965). However, reactions with activation energies of about 10 kcal/mol (42 kJ/mol) appear never to be significant in controlling OM maturation because they rapidly approach completion (reaction to the 99% level) in less than 10^5 years at near-surface temperature (Fig. 5.2). The general correlation of R_m with T_{max} (Neruchev and Parparova, 1972; and others, see Waples, 1984) suggests that at a given temperature only a limited suite of reactions control OM thermal maturation, and increased heating duration at that temperature will not make the slower (high E_a) reactions geologically significant. At a temperature of 100°C, for example, decarboxylation reactions with an E_a of 30 to 40 kcal/mol (130 to 170 kJ/mol) require on the order of 10^6 to 10^7 years to approach completion, while the lowest E_a C–C bond cleavage reactions would require about 10^9 years (Fig. 5.2). C–C bond cleavage at this temperature requires a heating duration on the order of the earth's age to approach completion, illustrating the fact that reactions with a considerably higher E_a can be neglected in modeling OM thermal maturation.

For the purpose of this empirical calibration, heating duration estimates should not exceed the elapsed time necessary for the controlling reactions to approach completion. Considering only reactions that at a given temperature can approach

TABLE 5.1. Estimates of the time required for the stabilization of kerogen thermal maturation.[a]

Reference	Stabilization time (yr)	Notes
Seyer (1933)	Short	Petroleum generation above 200°C
McNab et al. (1952)	10^6	Petroleum generation
Vallentyne (1964)	$<10^6$	Complete amino acid decarboxylation at 100°C
Tan (1965)	6×10^7	Coal maturation
Abelson (1967)	10^6	Methane generation from oil shale pyrolysis complete in 10^6 years at 115°C
Abelson (1967)	10^6	Analysis of Los Angeles Basin hydrocarbon generation data (Philippi, 1965) suggests effective duration of the total (heating) exposure is equivalent to roughly 2×10^6 years at 150°C
Brooks (1970)	Insignificant (Waples, 1984)	States that coal rank is a maximum-recording geothermometer
Neruchev and Parparova (1972)	10^6	Coal and kerogen maturation
Lopatin and Bostick (1973)	Insignificant (Waples, 1984)	State that coal rank is a maximum-recording geothermometer
Bartenstein and Teichmüller (1974)	Insignificant (Waples, 1984)	State that coal rank is a maximum-recording geothermometer
Nagornyi and Nagornyi (1974)	Insignificant	"Geological time does not limit the coalification process."
Juntgen and Klein (1975)	10^7	Coal maturation
Hacquebard (1975)	10^7 to 10^8	Coal maturation
Ammosov et al. (1975, 1977)	10^6	Heating time not important in organic maturation
Demaison (1975)	10^8	Coal maturation
Cornelius (1975)	About 10^7	Equilibrium reached in petroleum generation
Harwood (1977)	Short	Kerogen pyrolysis
McTavish (1978)	Insignificant	The time factor for petroleum generation of limited significance over geologic time and probably critical only for a short period
C.G. Barker (1979)	About 5×10^7	Influence of time is minor when compared to the effect of temperature in hydrocarbon generation from kerogen
Sajgo (1980)	10^6	Petroleum generation from kerogen
Veto (1980)	Finite	"A minimum temperature is needed to start any transformation of organic material with a particular activation energy. At such a temperature a certain length of time is necessary and sufficient to complete the reaction . . . ."
Veto (1980)	Time effect exaggerated	"The role of time is probably also exaggerated . . . for the methods of Bostick and Lopatin."
Wright (1980)	Insignificant	"Temperature remains critical: a source shale with $R_m = 0.8\%$ can remain at that rank for many millions of years and never generate a drop of oil."
Teichmüller and Teichmüller (1981)	$>3 \times 10^6$	Equilibrium in organic maturation *not* established in this heating time
Price et al. (1981)	Short	Organic metamorphism in hydrothermal bombs at 350°C
Suggate (1982)	10^6	Coal maturation
Gretener and Curtis (1982)	Insignificant	Petroleum generation above 130°C
Barker (1983)	10^4	Kerogen maturation in liquid-dominated geothermal systems
Price (1983)	10^6	Kerogen maturation

[a]Insignificant stabilization time indicates that kerogen thermal maturation was found to be temperature controlled.

completion in a geologically reasonable time frame, OM thermal maturation should stabilize after about 10^6 to 10^7 years (Fig. 5.2), and increased heating duration after this time should produce negligible changes in OM. Other evidence from experimental and geological studies also indicates that OM thermal maturation stabilizes. Table 5.1 summarizes several studies of petroleum generation and coalification that show heating duration beyond 10^7 to 10^8 years has a minimal influence on OM thermal maturation during burial diagenesis. Some evidence suggests that at least 10^6 years are required for petroleum generation and/or coalification during burial diagenesis (Teichmüller and Teichmüller, 1981; Tissot and Welte, 1984). These studies and others that have found a significant effect of heating duration suggest that 10^6 to 10^7 years are required for OM thermal maturation to stabilize. Note that several geological and experimental studies (Table 5.1) assert that heating duration has an insignificant effect on OM thermal maturation, but kinetic models of diagenesis require an estimate of t, making this assessment necessary.

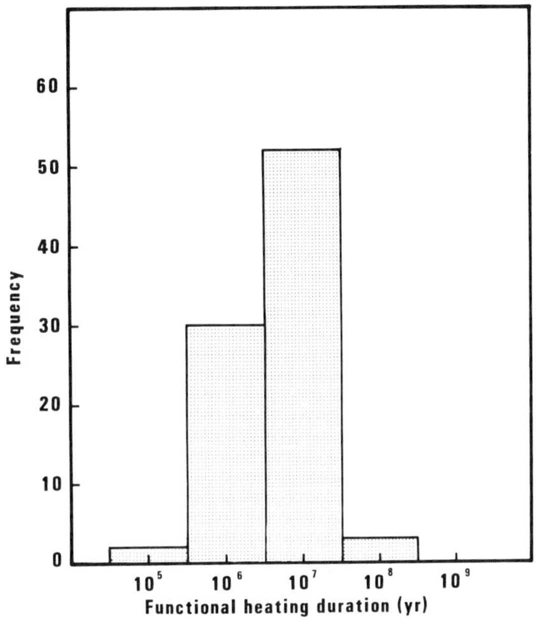

FIGURE 5.3. Histogram of functional heating duration found in sedimentary basins. Data from Table 5.2.

Burial history reconstruction of over 70 cases of OM thermal maturation (Table 5.2) during burial diagenesis shows that about 90% have been within 15°C of maximum temperature for more than 10^6 years (Fig. 5.3). Thus, in most cases of burial diagenesis, heating duration at maximum temperature has been sufficient for reaction to approach completion and for the OM to stabilize with respect to T_{max}. The stabilization of OM with respect to T_{max} does not imply that thermodynamic equilibrium is established in the system because these reactions are irreversible (Blumer, 1965; Tissot and Welte, 1984). The irreversible nature of OM thermal maturation reactions causes rank, and consequently R_m, to be set by T_{max}.

These observations indicate that the effective heating time of Hood et al. (1975) should be more limited and not consider the time elapsed during declining or stable temperature. A functional heating duration (t) for OM thermal maturation is defined as the elapsed time while temperature increases within 15°C of T_{max}. Time elapsed during temperature decline is not considered to increase R_m because OM thermal maturation is irreversible. Also, if the temperature stabilizes near T_{max} and sufficient time has elapsed for the controlling reactions to approach completion, additional time is not considered to effectively increase OM thermal maturation, and is not included in t.

The functional heating duration is still a contrived estimate of reaction time. There is no known method of determining a reaction time in sedimentary environments that is directly applicable to kinetic equations. The control of OM rank by irreversible reactions indicates that reaction time during burial diagenesis can be estimated as the elapsed time at T_{max} for the controlling reactions to approach completion. However, this type of heating duration estimate is not easily measured in sedimentary environments where temperature has increased slowly during burial (in this case, time at $T_{max} = 0$), or that have poorly known burial histories, making this an impractical definition. The definition of the functional heating duration over a 15°C range near T_{max} is a reflection of imprecise geological data. Environments that are now at T_{max}, such as geothermal systems, allow a more direct estimate of t, and elapsed time at T_{max} is used in these cases.

TABLE 5.2. Estimates of maximum reaction temperature, functional heating duration, and logarithm of the reaction rate (estimated from Equation 3) for burial diagenesis.

Reference	Reaction temperature (°C)	$\frac{1}{T°K}$	Functional heating duration (seconds)	Thermal maturity (various units)	f	ln k (per second)
Philippi (1965) Los Angeles basin, USA	115[d] 145 Dow (1978)	0.00239	3×10^{13}[c]	[a]	0.9	−33.3
Philippi (1965) Ventura basin, USA	115[d] 145 Dow (1978)	0.00239	3×10^{13}[c]	[a]	0.9	−33.3
Tissot et al. (1971) Paris basin, France	60[b]	0.00300	6×10^{14} (3×10^{14})[g]	0.6% R_m	0.89	−35.5
Myhr and Gunther (1974) Mackenzie River Delta, Canada	121	0.00254	3×10^{14}	0.6% R_m	0.89	−35.5
Laplante (1974) Pecan Lake, Gulf Coast, USA	115	0.00258	3×10^{14}[c]	82% C	0.77	−34.7
Laplante (1974) West Delta, Gulf Coast, USA	113	0.00259	6×10^{13}[c]	77% C	0.98	−35.6
Laplante (1974) West Manchester, Gulf Coast, USA	105	0.00265	1×10^{13}[c]	76% C	0.95	−35.5
Castano and Sparks (1974) Tejon, California, USA	137 Briggs et al. (1981)	0.00244	3×10^{13} Briggs et al. (1981)	0.6% R_m	0.89	−33.2
Connan (1974) Onshore Taranaki basin, New Zealand	110	0.00261	3×10^{13}[c]	[a]	0.9	−33.3
Connan (1974) Rio de Oro, Brazil	95[b]	0.00272	3×10^{14}[c]	[a]	0.9	−35.6
Connan (1974) Camargue basin, France	120[b]	0.00254	2×10^{14}[c]	[a]	0.9	−35.2
Connan (1974) Amazon basin, Brazil	75[b]	0.00287	2×10^{15}[c] (3×10^{14})[g]	[a]	0.9	−35.6
Tissot and Espitalié (1975) Douala basin, Cameroon	200[f] 128[e]	0.00247	3×10^{14}	3.3% R_m	0.04	−32.2
Ting (1975) Texas, USA	115[d]	0.00258	3×10^{14}[c]	0.5% R_m	0.95	−36.3
Cornelius (1975) Florida, USA	105[f]	0.00265	8×10^{13}[c]	0.5% R_m	0.95	−35.0
Cornelius (1975) USSR	180[f]	0.00221	6×10^{14}[c] (3×10^{14})[g]	2.5% R_m	0.10	−32.5
Cornelius (1975) SAAR	120[f]	0.00254	1×10^{15}[c] (3×10^{14})[g]	1.25% R_m	0.5	−33.7
Cornelius (1975) West Canada	135[f]	0.00245	1×10^{15}[c] (3×10^{14})[g]	2.0% R_m	0.17	−32.8
Cornelius (1975) Annet basin	190[f]	0.00216	9×10^{14}[c] (3×10^{14})[g]	2.8% R_m	0.07	−32.4
Hood et al. (1975) Shell Rumberger 5, Anadarko basin, USA	220	0.00203	6×10^{14}[c] (3×10^{14})[g]	LOM 18.5	0.02	−32.0
Hood et al. (1975) Mobil T-52-19G, Piceance basin, USA	210	0.00207	3×10^{14}	LOM 17.5	0.05	−32.2
Shibaoka and Bennett (1977) Cooper basin, Australia	110	0.00261	4×10^{14} (3×10^{14})[g]	0.6% R_m	0.89	−35.5
Scholle (1977) Cost B-2 well, Baltimore Canyon, USA	80[b]	0.00282	6×10^{14} (3×10^{14})[g]	0.5% R_m	0.95	−36.3

TABLE 5.2. (*Continued*)

Reference	Reaction temperature (°C)	$\frac{1}{T°K}$	Functional heating duration (seconds)	Thermal maturity (various units)	f	ln k (per second)
Sajgo (1980) Pannonian basin, Hungary	233[b]	0.00198	3×10^{13}	1.6% R_m	0.29	−30.8
Cassou et al. (1977) Oneida 0-25, Scotian Shelf, Canada	130[d]	0.00248	2×10^{14}	0.75% R_m	0.80	−34.4
Cassou et al. (1977) Onondaga E-84, Scotian Shelf, Canada	110[d]	0.00261	3×10^{14}	1.05% R_m	0.62	−33.9
Cassou et al. (1977) Triumph P-50, Scotian Shelf, Canada	120[d]	0.00254	8×10^{14} (3×10^{14})[g]	1.15% R_m	0.56	−34.8
Cassou et al. (1977) Dauntless D-35, Scotian Shelf, Canada	95[d]	0.00272	5×10^{14} (3×10^{14})[g]	0.68% R_m	0.84	−35.1
Cassou et al. (1977) Puffin B-90, Scotian Shelf, Canada	125[d]	0.00251	3×10^{14}	1.15% R_m	0.56	−33.9
Cassou et al. (1977) Murre G-67, Scotian Shelf, Canada	78[d]	0.00285	3×10^{14}	0.68% R_m	0.84	−35.1
Cassou et al. (1977) Adolphus K-41, Scotian Shelf, Canada	105[d]	0.00265	9×10^{14} (3×10^{14})[g]	1.0% R_m	0.65	−34.2
Cassou et al. (1977) Bonavista C-99, Scotian Shelf, Canada	75[d]	0.00287	3×10^{14}	0.72% R_m	0.81	−34.9
Cassou et al. (1977) Bjarni H-81, Scotian Shelf, Canada	65[d]	0.00296	1×10^{15} (3×10^{14})[g]	0.75% R_m	0.80	−34.8
Tissot et al. (1978) Uinta basin, USA	100[b]	0.00268	3×10^{14}	0.7% R_m	0.83	−35.0
Dow (1978); Cardoso et al. (1978) Blake–Bahama basin, Atlantic Ocean	58	0.00302	1×10^{15} (3×10^{14})[g]	0.55% R_m	0.92	−35.8
Dow (1978) Pliocene, Gulf Coast, USA	164	0.00229	6×10^{12}[c]	0.6% R_m	0.89	−31.6
Dow (1978) Upper Miocene, Gulf Coast, USA	143	0.00240	2×10^{13}[c]	0.6% R_m	0.89	−32.8
Dow (1978) Middle Miocene, Gulf Coast, USA	131	0.00248	4×10^{13}[c]	0.6% R_m	0.89	−33.5
Dow (1978) Oligocene, Gulf Coast, USA	113	0.00259	1×10^{14}[c]	0.6% R_m	0.89	−34.4
Dow (1978) Eocene, Gulf Coast, USA	97	0.00270	2×10^{14}[c]	0.6% R_m	0.89	−35.1
Dow (1978) Cretaceous, Gulf Coast, USA	84	0.00280	4×10^{14}[c] (3×10^{14})[g]	0.6% R_m	0.89	−35.8
Shibaoka et al. (1978) Barracouta-1, Gippsland basin, Australia	80	0.00283	1×10^{14}[c]	0.45% R_m	0.98	−36.1
Cornford et al. (1979) DSDP Site 397, Eastern North Atlantic	65	0.00300	2×10^{14}[c]	0.5% R_m	0.95	−35.9
Leith and Rowsell (1979) D-B1 well, Agulhas bank, South Africa	93	0.00273	2×10^{15} (3×10^{14})[g]	1.0% R_m	0.65	−34.2
Asakawa and Fujita (1979) Shimoigarashi well, Japan	95	0.00272	1×10^{13}	0.45% R_m	0.98	−33.8
Asakawa and Fujita (1979) Yoshida well, Japan	95	0.00272	2×10^{13}	0.45% R_m	0.98	−34.5

TABLE 5.2. (*Continued*)

Reference	Reaction temperature (°C)	$\frac{1}{T°K}$	Functional heating duration (seconds)	Thermal maturity (various units)	f	ln k (per second)
Thomas (1979) Bullsbrook-1, North Perth basin, Australia	103^c	0.00266	3×10^{14}	1.1% R_m	0.59	−34.0
Teichmüller and Teichmüller (1979, 1981) Landau-2, Germany	140^b	0.00242	9×10^{13}	0.8% R_m	0.77	−33.5
Teichmüller and Teichmüller (1979, 1981) Harthausen-1, Upper Rhine Graben, Germany	120^b	0.00254	9×10^{13}	0.8% R_m	0.77	−33.5
Teichmüller and Teichmüller (1979, 1981) Sandhausen-1, Upper Rhine Graben, Germany	135^b	0.00245	9×10^{13}	0.75% R_m	0.80	−33.6
Huc and Hunt (1980) Cost-1 well, Texas, USA	183^d	0.00219	5×10^{13c}	1.1% R_m	0.59	−32.2
Tissot et al. (1980) Algeria	56^b	0.00304	6×10^{14} $(3 \times 10^{14})^g$	a	0.90	−35.6
Rashid et al. (1980) Snorri well, Labrador Shelf, Canada	78	0.00285	3×10^{14c}	0.9% R_m	0.71	−34.4
Rashid et al. (1980) Karlsefni well, Labrador Shelf, Canada	48	0.00312	3×10^{14c}	0.4% R_m	0.99	−37.9
Rashid et al. (1980) Gudrid well, Labrador Shelf, Canada	55	0.00305	6×10^{14c} $(3 \times 10^{14})^g$	0.37% R_m	0.99	−37.9
Law et al. (1980) Superior No. 1, Pacific Creek, Green River basin, Wyoming, USA	155^b	0.00234	3×10^{14c}	1.1% R_m	0.59	−34.0
Connan and Cassou (1980) Sabah	105^b	0.00265	9×10^{13c}	0.59% R_m	0.90	−34.4
Connan and Cassou (1980) Sabah	105^b	0.00279	1×10^{14c}	0.36% R_m	0.99	−36.8
Connan and Cassou (1980) Kalimantan	52^b	0.00308	3×10^{14c}	0.36% R_m	0.99	−37.9
Connan and Cassou (1980) New Zealand-Oligocene	93^b	0.00273	2×10^{14c}	0.51% R_m	0.94	−35.7
Connan and Cassou (1980) New Zealand-Eocene	83^b	0.00281	3×10^{14c}	0.59% R_m	0.90	−35.6
Connan and Cassou (1980) Columbia	90^b	0.00275	2×10^{14c}	0.85% R_m	0.73	−34.1
Connan and Cassou (1980) Vietnam	112^b	0.00260	8×10^{13c}	0.53% R_m	0.93	−34.6
Connan and Cassou (1980) Australia-Permian	119^b	0.00255	9×10^{14c}	0.86% R_m	0.73	−34.5
Connan and Cassou (1980) Australia-Middle Jurassic	108^b	0.00262	8×10^{14c}	0.95% R_m	0.68	−43.3
Connan and Cassou (1980) South Africa	96^b	0.00271	6×10^{14c}	1.29% R_m	0.47	−33.6
Connan and Cassou (1980) Nigeria	63^b	0.00298	5×10^{14c}	0.65% R_m	0.86	−35.2

TABLE 5.2. (*Continued*)

Reference	Reaction temperature (°C)	$\frac{1}{T°K}$	Functional heating duration (seconds)	Thermal maturity (various units)	f	ln k (per second)
Connan and Cassou (1980) Switzerland	140[b]	0.00242	5×10^{14}[c]	1.35% R_m	0.44	−33.5
Connan and Cassou (1980) Tunisia	180[b]	0.00231	2×10^{14}[c]	1.05% R_m	0.62	−33.7
Connan and Cassou (1980) France	135[b]	0.00245	4×10^{14}[c]	2.0% R_m	0.17	−32.8
Ziegler (1980) Lower Saxony basin, Germany	105[b]	0.00265	9×10^{13}	[a]	0.90	−34.4
Ziegler (1980) Gifhorn trough, Germany	97[b]	0.00265	5×10^{14}	[a]	0.90	−35.6
Ziegler (1980) Plön trough, Germany	83[b]	0.00281	8×10^{14} (3×10^{14})[g]	[a]	0.90	−35.6
Kantsler and Cook (1979) Whicter Range-1, Perth basin, Australia	111	0.00260	6×10^{14} (3×10^{14})[g]	0.94% R_m 1.0% R_{mx}	0.68	−34.3
Price et al. (1981) Bertha Rodgers-1 well, Oklahoma, USA	252[d]	0.00190	9×10^{14}[c] (3×10^{14})[g]	4.4% R_m	0.01	−31.8
Smyth and Saxby (1981) Poolowanna-1, Pedirka-Simpson basin, Australia	87[d]	0.00278	9×10^{14}[c] (3×10^{14})[g]	0.86% R_m	0.73	−34.5
Suggate (1982) Maui-4, New Zealand	160[f]	0.00231	1×10^{14}	0.90% R_m	0.71	−33.3
Price (1982) Jacobs-1, Texas, USA	279[b]	0.00183	2×10^{14}[c]	4.4% R_m	0.01	−31.5
Barber (1982) Exmouth Plateau, Australia	141[d]	0.00242	3×10^{15} (3×10^{14})[g]	1.3% R_m	0.47	−35.6
Magoon and Claypool (1982) Inigok-1, North Slope, Alaska, USA	210[e]	0.00207	2×10^{15} (3×10^{14})[g]	5.0% R_m	0.01	−31.8
Waples (1982); Katz et al. (1982) Brunes well, Oklahoma, USA	190	0.00216	6×10^{14} (3×10^{14})[g]	2.5% R_m	0.10	−32.5
King and Claypool (1982) OCS-CAL 78-164-1 well, Offshore California, USA	118	0.00256	6×10^{13}	0.34% R_m	0.99	−36.3

[a]Unless determined in the original paper, the initiation of hydrocarbon generation is assumed to commence at $f = 0.9$ (Tissot and Welte, 1984).

[b]Present borehole temperature: It is not known if this temperature is equivalent to maximum temperature in the system thermal history.

[c]Burial history unavailable. Functional heating duration estimated by calculating an average heating rate (°C/my) and determining the time within 15°C of the maximum borehole temperature.

[d]Burial history suggests that present temperatures are the maximum reached in the borehole. However, the author(s) used uncorrected bottom-hole temperatures in their thermal history analysis.

[e]Present temperature: not the maximum in the borehole thermal history.

[f]Estimate of maximum burial temperature.

[g]Estimate of time within 15°C of maximum temperature. Functional heating duration shown in parentheses.

TABLE 5.3. Estimates of maximum reaction temperature, functional heating duration, and logarithm of the reaction rate (estimated from Equation 3) for geothermal systems.

Reference	Reaction temperature (°C)	$\frac{1}{T°K}$	Functional heating duration (seconds)	Thermal maturity (various units)	f	ln k (per second)
Barker (1983) Cerro Prieto, Mexico	350	0.00161	3×10^{11}	4.1% R_m	0.02	−25.1
Barker (1983) East Mesa, California, USA	200	0.00211	3×10^{11}	2.0% R_m	0.17	−25.9
Barker (1983) Heber, California, USA	210	0.00207	3×10^{11}	2.3% R_m	0.12	−25.7
Barker (1983) Raft River, California, USA	150	0.00236	3×10^{13}	1.5% R_m	0.35	−31.0
Barker (1983) Salton Sea, California, USA	295	0.00176	3×10^{11}	3.0% R_m	0.06	−25.4
Barker (1983) Soda Lake, Nevada, USA	170	0.00226	3×10^{11}	1.35% R_m	0.44	−26.6
Barker (1983) Wilson-1, California, USA	270	0.00184	3×10^{11}	3.0% R_m	0.06	−25.4

Geothermal Systems

The functional heating duration in active geothermal systems is simply the age of the related heat source or thermal event because the systems still exist at close to T_{max}. The age of the thermal event for the systems listed in Table 5.3 was derived by fission-track annealing studies, dating of the apparent heat source, and defining the unique features of their geological history (Barker, 1983).

Contact Metamorphism

The reaction duration and temperature in rock contact metamorphosed by an intrusive igneous sheet was estimated using a heating duration-temperature model derived from heat-flow theory (Jaeger, 1959, 1964). The intrusive sheet (at 1,000°C) is assumed to be instantly emplaced into a wet, porous sedimentary rock at a depth of about 200 m. Thus, the model was designed for the case of a hypabyssal sheet intruded into shallowly buried sedimentary rock and coal. The thermal properties of coal are significantly different from sedimentary rock, resulting in a much higher initial temperature at a coal/intrusive sheet contact and a more rapid decrease away from the contact compared to sedimentary rock. Jaeger (1959, 1964) evaluates T_{max} for a sedimentary rock, or coal, as a ratio of the distance from the contact to the thickness of the igneous sheet. This adjustment allows igneous sheets of different thickness to be compared. Contact temperature is assumed to rise instantly to maximum and subsequently decrease with time. The functional heating duration in Jaeger's model must be measured during temperature decline. The elapsed time within 15°C of T_{max} is not a useful estimate in this environment because it is essentially zero at the intrusive contact where rapid cooling occurs. Functional heating duration in this environment was measured as the elapsed time within 15% of T_{max} because this measurement is easily adapted from Jaeger's model. This larger functional heating duration is justified by laboratory experiments that indicate coal carbonization at high temperature was still reacting after 32 weeks (0.6 year) at 350°C (Goodarzi and Murchinson, 1977). This is considerably longer than the elapsed time within 15°C of T_{max} estimated by Jaeger's model.

Reaction Temperature

Most T_{max} data for systems now at peak temperature (Table 5.2) are from uncorrected bottom-hole temperature (BHT) measurements typically taken on a single logging-tool run. A minority of the remaining temperature data are from corrected BHT measurements (with the correction method sometimes unspecified). The geothermal curve for the boreholes in Table 5.2 was determined by computing a geothermal gradient from BHT and the mean annual surface temperature. The reported sample depth was used to compute the system temperature at that point by interpolation from the geothermal curve. This linear approximation can be highly inaccurate because the temperature profile can change with lithology (thermal conductivity), subsurface fluid flow, and so forth (Drury et al., 1984). However, without a range of temperature measurements in the borehole, this is the only method available. Accurate determination of present-day formation temperature also requires thermal equilibration of the borehole before measurement. The extended borehole shut-in time required to establish thermal equilibrium usually means that borehole temperature is measured soon after drilling is completed. BHT data are not generally confirmed as equilibrium reservoir temperatures by repeated measurements over a significant time interval. Correction of BHT is necessary when it is not measured at equilibrium, but experience indicates that the data necessary for calculating a correction are not included in the borehole history reports. Furthermore, attempts at correcting the BHT measurement to equilibrium conditions, although necessary, are often unsuccessful because drilling and measurement conditions vary widely and usually cannot be corrected by some uniform procedure (Drury, 1984). However, because the base of the borehole has the least exposure to mud circulation, BHT is the best estimate of the reservoir temperature in the absence of an equilibrium reservoir temperature (Bullard, 1947). In this study, the bottom-hole data were used preferentially to estimate OM reaction temperature.

In liquid-dominated geothermal systems, accurate temperature data are important and greater equilibration time is allowed. Thus, the temperature data indicate realistic reservoir conditions, as indicated by their close relationships to other temperature data such as R_m, fluid inclusion homogenization, and oxygen-isotope equilibration geothermometry (Barker and Elders, 1981; Barker, 1983). Temperature of reaction in geothermal systems was taken as the actual borehole measurement.

Temperature at the intrusive sheet contact with rock or coal was calculated from the degree of coke formation (Chandra and Bond, 1956), or R_m (Bostick, 1971). This contact temperature was used in Jaeger's (1959, 1964) model to estimate T_{max} as a function of distance from the intrusion.

Calculated Kinetic Parameters

Functional heating duration estimates for burial diagenesis range from about 10^{12} s (10^5 years) to 10^{15} s (10^8 years) and temperatures range from 40° to 280°C (Table 5.2). The calculated reaction rates show a strong linear relationship (Fig. 5.4) indicated by regression analysis (correlation coefficient, $r = -0.84$), with an E_a of 9 kcal/mol (37 kJ/mol) and an A of 7×10^{-11}/s.

Functional heating duration estimates for geothermal systems range from 10^{11} s (10^4 years) to 10^{13} s (10^6 years) and maximum reaction temperatures range from 170° to 350°C (Table 5.3). The calculated reaction rates are modeled by a linear relationship with temperature (Fig. 5.4; $r = -0.74$), with an E_a of 11 kcal/mol (46 kJ/mol) and an A of 2.5×10^{-7}/s.

Functional heating duration estimates for contact metamorphism by intrusive sheets range from 10^3 s (10^{-5} years) to 10^{10} s (10^2 years) and maximum reaction temperatures from 550° to 950°C (Table 5.4). The calculated reaction rates are modeled by a linear relationship with temperature (Fig. 5.4, $r = -0.64$) with an E_a of 12 kcal/mol (50 kJ/mol) and an A of 1.5×10^{-3}/s.

The results from this analysis confirm the low activation energy for OM thermal maturation during burial diagenesis found by others (see review by Waples, 1984). The low activation energy for OM thermal maturation in geothermal systems and contact metamorphism by intrusive sheets is the first reported for these environments. The linear correlation of ln k and $1/T$, within the resolution possible within the data scatter, indicates that OM thermal maturation is effectively modeled by a single pseudoreaction within each environment. This observation is comparable to the results from a laboratory study of OM thermal maturation by Pearson (1981, p. 220) who found "... that there is not a change in the basic reactions of the kerogen components during the heating experiments even

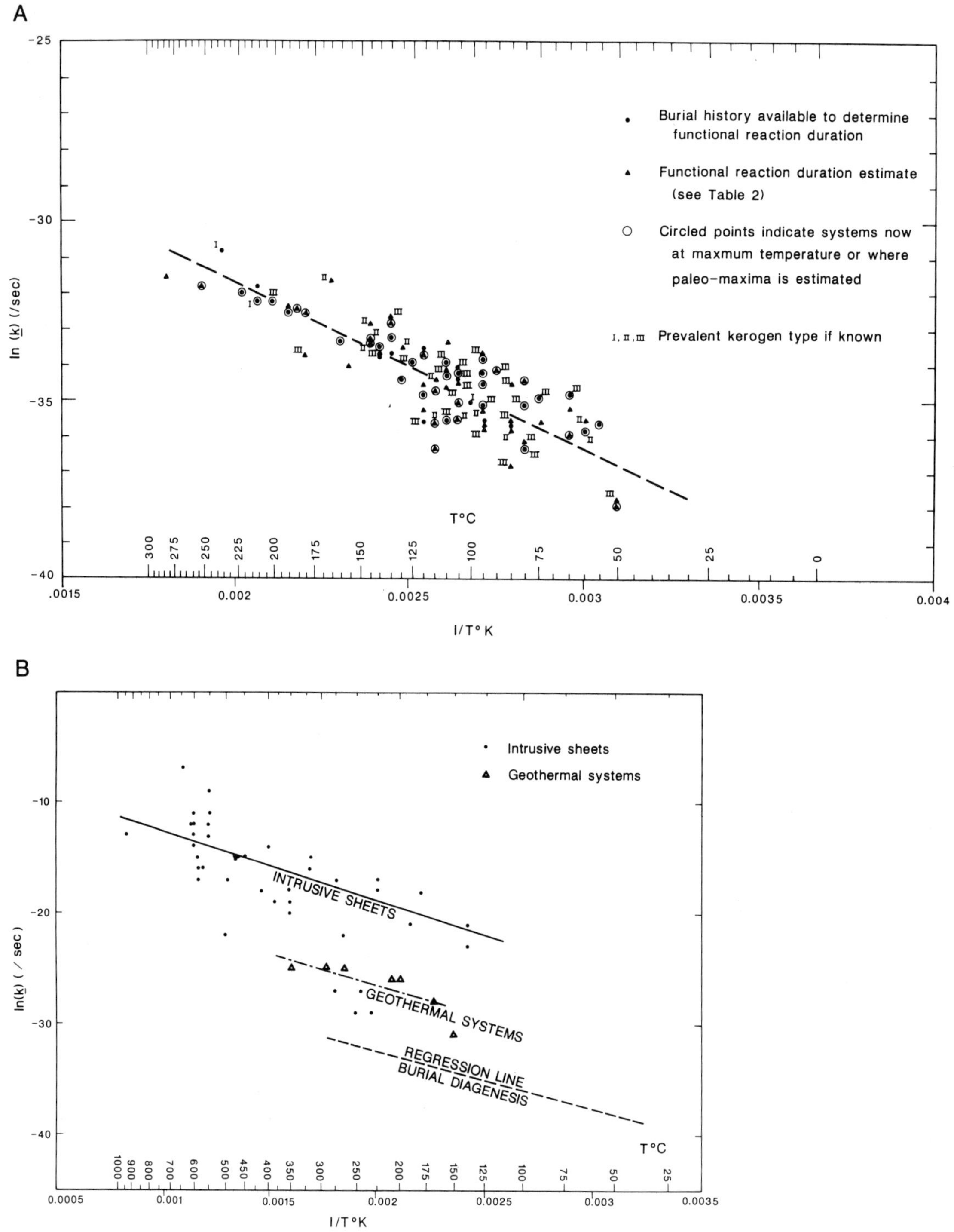

FIGURE 5.4. Arrhenius plot of reaction rates from Equation 3 and estimates of maximum reaction temperatures for burial diagenesis (A) and burial diagenesis, geothermal systems, and contact metamorphism by igneous intrusive sheets (B). The logarithm of k was rounded to the nearest whole number for plotting these data and temperature was rounded to the nearest 5°C. Systems known to be influenced by R_m suppression were not used in calculating the regression equation because of spurious k values computed from the anomalously low reaction extent indicated by R_m. For this reason, R_m and temperature data from the Los Angeles Basin were not used (Walker et al., 1983).

TABLE 5.4. Estimates of maximum reaction temperature, functional heating duration, and logarithm of the reaction rate (estimated from Equation 3) for contact metamorphism by intrusive sheets.

Reference	Reaction temperature (°C)	Functional heating duration (seconds)	Thermal maturity (various units)	(*D*) (meters)	*X/D* (no units)	ln *k* (per second)
Eby(1925) Yampa, Colorado, USA	585[a]	3×10^7	96% *C*	7.6	0.0	−15.5
Marshall (1952) Brockwell Seam, USA	585[a]	1×10^7	89.6% *C*	5.2	0.0	−16.1
Marshall (1952) Hutton Seam, USA	585[a]	1×10^8	96.4% *C*	15.5	0.0	−16.8
Johnson (1952) Colorado, USA	450[a]	1×10^7	8.3% *VM*	4.9	0.0	−15.4
Johnson (1952) Colorado, USA	410[a]	7×10^7	14.2% *VM*	12	0.0	−17.5
Clegg (1955) Harrisburg Seam, Illinois, USA	600	1×10^5	76% *C*	0.46	0.0	−13.8
Chatterjee et al. (1964) Poniati Seam, India	600	9×10^5	4.71% R_{mx}	1.4	0.0	−12.2
as above	550	9×10^5	3.43% R_{mx}	1.4	0.25	−12.6
as above	400	1×10^6	1.39% R_{mx}	1.4	0.50	−14.1
Kisch (1966) Dirty Seam, Bowan basin, Australia	950	8×10^6	12% R_{mx}	4	0.0	−13.7
Schapiro and Gray (1966) Terrance Ridge, Antarctica	500	2×10^{10}	11.5% R_{mx}	180.0	0.1	−21.8
Ghosh (1967) Dishergarh Seam, India	600	4×10^6	4.9% R_{mx}	3	0.0	−13.8
Ghosh (1967) Lower Shampur Seam, India	600	1×10^5	3.2% R_{mx}	0.46	0.0	−10.5
Ghosh (1967) Jharia-11 Seam, India	550	2×10^5	4.9% R_{mx}	0.61	0.0	−10.7
Ghosh (1967) Jharia-14 Seam, India	550	3×10^6	4.5% R_{mx}	2.3	0.0	−13.6
Ghosh (1967) Jharia-15 Seam, India	550	7×10^5	3.8% R_{mx}	1.2	0.0	−12.2
Ghosh (1967) Kalimati Seam, India	550	4×10^4	4.3% R_{mx}	0.30	0.0	−9.2
Ghosh (1967) Brindabanpur Seam, India	600	9×10^5	4.95% R_{mx}	1.4	0.0	−12.2
Ghosh (1967) Salanpur Seam, India	600	3×10^6	5.0% R_{mx}	2.6	0.0	−13.4
Ghosh (1967) Laikdih Seam, India	600	9×10^5	4.9% R_{mx}	1.4	0.0	−12.2
Ghosh (1967) Poniati Seam, India	600	4×10^5	4.7% R_{mx}	0.91	0.0	−11.5
Crelling and Dutcher (1968) Purgatoire River, USA	550	3×10^6	3.0% R_{mx}	2.44	0.75	−13.9
Bostick (1971, 1979a) Lennep, USA	475	1×10^7	2.9% R_{mx}	4.9	0.1	−15.3
as above	318	1×10^7	1.6% R_{mx}	4.9	0.25	−16.3
as above	223	2×10^7	1.1% R_{mx}	4.9	0.5	−17.7
Bostick (1971, 1979a) Highwood, USA	585	8×10^6	4.6% R_{mx}	4.0	0.25	−14.5
as above	275	1×10^7	1.3% R_{mx}	4.0	0.5	−16.5
as above	180	1×10^7	0.85% R_{mx}	4.0	1.0	−17.5

TABLE 5.4. (*Continued*)

Reference	Reaction temperature (°C)	Functional heating duration (sec)	Thermal maturity (various units)	(*D*) (meters)	*X*/*D* (no units)	ln *k* (/sec)
Bostick (1971, 1979a)						
Cascade, USA	575	4×10^7	4.5% R_{mx}	9.2	0.1	−16.0
as above	385	4×10^7	0.9% R_{mx}	9.2	0.25	−18.7
as above	270	5×10^7	0.45% R_{mx}	9.2	0.5	−22.4
Bostick (1971, 1979a)						
La Jolla, USA	475	7×10^6	2.9% R_{mx}	3.7	0.1	−14.8
as above	318	7×10^6	1.9% R_{mx}	3.7	0.25	−15.3
as above	223	9×10^6	1.3% R_{mx}	3.7	0.50	−16.7
Bostick (1971, 1979a)						
Spanish Peak, USA	475	7×10^6	3.0% R_{mx}	3.7	0.25	−14.8
as above	223	9×10^6	2.9% R_{mx}	3.7	0.50	−15.7
as above	176	9×10^6	1.15% R_{mx}	3.7	0.75	−16.7
Jones and Creaney (1977)						
Yard Seam, England	645	5×10^3	5.44% R_{mx}	0.1	0.0	−7.0
as above			0.77% R_{mx}	0.1	1.0	
Baker et al. (1977)						
DSDP Site 41-368						
Eastern Atlantic	500	1×10^8	3.3% R_m	15	0.0	−17.4
as above	355	1×10^8	0.7% R_m	15	0.25	−20.1
as above	140	1×10^8	0.6% R_m	15	0.50	−20.7
Peters et al. (1977)						
DSDP Site 41-368, Eastern Atlantic	355	1×10^8	1.85% R_m	15	0.25	−17.9
as above	188	1×10^8	0.6% R_m	15	0.50	−20.7
as above	140	1×10^8	0.4% R_m	15	0.75	−23.0
Creaney (1980)						
Whin Sill, England	275[a]	2×10^9	7.2% R_m	70.0	0.5	−26.9

Note: Only values of vitrinite reflectance that appeared to be affected (i.e., rise above background) by an intrusive sheet were used in calculating the reaction rate away from the intrusive/sediment contact.

[a]Estimated.

though certain other measurable characteristics do change such as vitrinite reflectance and elemental analysis." Pearson also found no significant change in E_a and A attributable to thermal maturation in his experiments. Karweil (1956, 1975), Lopatin (1971), Connan (1974), and others have also found that OM thermal maturation was effectively modeled by a single pseudoreaction.

Sources of Data Scatter

The errors in R_m determination and temperature estimates may account for the moderate to good fit of the regression line to the data from burial diagenesis and geothermal systems and the poor fit of the data from contact metamorphism by intrusive sheets.

Operator bias in selecting vitrinite for measurement is potentially the greatest source of error during reflectance analysis. R_m measurements by 30 microscopists on the splits of 19 different samples show that the range of R_m measurement can be up to $\pm 0.4\%$ R_m in low-rank kerogen and coal (unpublished report, International Commission on Coal Petrology, see Bostick, 1979b). This wide R_m range arises from operator bias and differences between laboratories in processing, polishing, and photometer calibration.

Other causes for imprecise R_m measurements are the variation of OM composition, rank suppression effects, and the semiquantitative nature of reflectance measurements. Studies have shown that R_m can be suppressed up to several tenths of a percent by maceral association and differences in early diagenetic history (see review by Price and Barker, 1985). Vitrinite also becomes anisotropic or bireflectant at moderate to high rank. Bireflectance in OM increases the range of the reflectance histogram and often produces polymodal distributions, making the mean value less representative. Data from studies where suppression is known to affect the R_m data are not included in the regression analysis (Fig. 5.4).

Errors in temperature determination are less of a factor in causing scatter in these data because of the low slope of the regression lines with respect to the temperature axis (Fig. 5.4). As discussed above, incorporation of uncorrected BHT data may cause some errors when used with corrected data. Temperature corrections on the order of 20° to 30°C are typical for BHT data (Hood et al., 1975), making this an important but unassessable error without more data.

These errors, while they do cause a lower correlation between the variables, do not obscure the trend in the data from burial diagenesis and geothermal systems. Contact metamorphism by intrusive sheets has a particularly poor correlation and diffuse data trend. Part of the scatter in the contact metamorphism data is due to 1) both temperature at the igneous contact and f being estimated from rank, and 2) temperature away from the intrusive contact and t being estimated from Jaeger's (1959, 1964) model. In this case, combined errors in the rank-T_{max} models for estimating temperature and Jaeger's model (Corrigan, 1982) cause additional scatter relative to data from burial diagenesis and geothermal systems where temperature, R_m, and t are directly measured or independently determined from geological data. The geochemical system near the intrusive sheet also appears important in determining R_m. System pressure, as discussed later, may also be important for OM thermal maturation near intrusive sheets because the depth of intrusion, and consequently pressure, can very widely. Also, Peters et al. (1978) found that R_m below a sill is significantly different than that found above it. This difference is attributed to the ease of product escape above the sill as compared to below it. The Peters et al. (1978) study indicates that the system closure is important in determining R_m. The range of R_m at the sill contacts would indicate significantly different temperatures in what must have been an almost symmetrical temperature distribution on either side of the sill. However, most of the data from contact metamorphism by intrusive sheets are closely clustered and the lowered correlation coefficient results from a few outlier points (Fig. 5.4). Many of these outlier points fall near those from geothermal systems, suggesting that contact metamorphism by intrusive sheets sometimes produces conditions like those found in geothermal systems. Even with these problems, the regression line still bisects the majority of the data and appears to be a reasonable model of OM thermal maturation about intrusive sheets.

Stabilization of Thermal Maturation

The t required for stabilization of OM thermal maturation is estimated by plotting a series of curves on an Arrhenius diagram generated from a T_{max}-R_m model. The T_{max}-R_m model is used to compute f at various T_{max} that, under the assumptions of the model, are the result of OM thermal maturation reactions that have approached completion and stabilized. Specifically, the T_{max}-R_m model is used to generate $1/T$ and R_m data, and t is fixed (at 10^6 and 10^7 years) for each curve, as required to estimate k from Equation 3. The time required for stabilization of thermal maturation is suggested by the curve on the Arrhenius plot that lies closest to the regression curve based on T_{max}, t, and f data from the natural system (Fig. 5.5A and B). Barker and Pawlewicz's (1985, 1986) T_{max}-R_m model is used to estimate t for burial diagenesis. Their model is based on the strong correlation of R_m with T_{max} ($r = 0.84$, n over 600) from some 35 cases of burial diagenesis. T_{max} and k data generated from Barker and Pawlewicz's model using a t of 10^6 and 10^7 years bracket and lie subparallel to the regression curve for burial diagenesis (Fig. 5.5A). This estimate confirms the t of about 10^6 to 10^7 years estimated from other studies (Table 5.1). Similarly, Barker's (1983) T_{max}-R_m model of thermal maturation based on data from six geothermal systems lies close to the regression curve from natural systems when a t of 10^4 is used (Fig. 5.5B). This t agrees with Barker's (1983) estimate of 10^4 years required for stabilization based on geological evidence. T_{max}-R_m models for thermal maturation adjacent to igneous sheets are calibrated from laboratory studies where t is directly controlled in the experiment. Bostick's (1971) and Goodarzi and Murchinson's (1977) data suggest t of about 10^{-1} to 10^0 years, but it may be longer because there is evidence that these experiments had not completed reaction.

Comparison of Thermal Maturation Models

The regression curves calculated from the data presented in this chapter (Fig. 5.4) constitute a T_{max}-t model of OM thermal maturation in each system. They are directly calibrated by a large number of examples from the environments themselves, and do not rely on assumed kinetic parameters as found in some of the other models. Because of these factors, the regression curves are used as a baseline to compare other OM thermal maturation models (Fig. 5.5A). Models that consider temperature and time were plotted on the Arrhenius diagram by assuming a t for calculating temperature, and the R_m value required for estimating f (Fig. 5.1) needed for calculating k from Equation 3.

None of the published temperature-time models of thermal maturation closely fit the data from the natural environments. Most models of OM thermal maturation for burial diagenesis fall about the regression line but show changes in slope that are not apparent in the data from natural environments. These problems may result from difficulty in determining burial history, R_m, and maximum reaction temperatures (Price, 1983). Huck and Karweil's model (as stated in Karweil, 1975) plots as a straight line subparallel to the regression line for burial diagenesis but lies well below the natural data because they assumed a lower activation energy and smaller frequency factor than others later computed for OM thermal maturation (Waples, 1984). The models developed from the classical rule of doubling the reaction rate for each 10°C temperature increase fit the burial diagenesis data fairly well. This approximation is acceptable for reactions with an activation energy of 10 to 20 kcal/mol (42 to 84 kJ/mol) (Tissot and Welte, 1984). The Lopatin (updated by Waples, 1980) model based on this rule is in fair agreement with the burial diagenesis regression line which has an E_a of about 9 kcal/mol (38 kJ/mol).

Models that consider only T_{max} in thermal maturation, when an appropriate t is used, fit the regression curves on the Arrhenius diagram about as well as the models that consider temperature and time. As discussed above, Barker and Pawlewicz's (1985, 1986) geothermometer, an example of a model that considers only T_{max} in thermal maturation, brackets the regression curve for OM thermal maturation during burial diagenesis. Barker and Pawlewicz's model would fit, with finer resolution of t, about as closely to the natural systems regression line, as the models

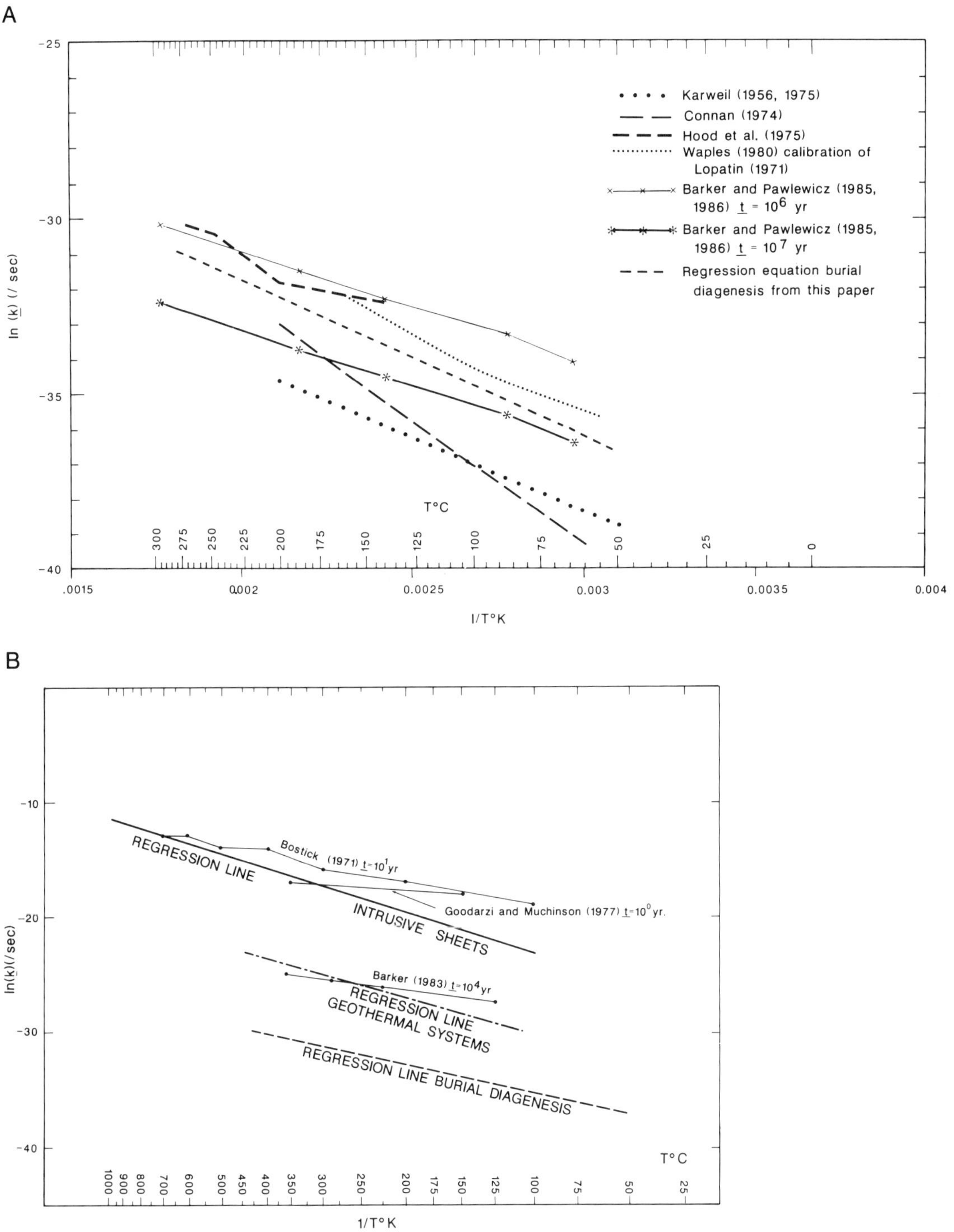

FIGURE 5.5. Comparison of linear regression curves (Fig. 5.4) to published models of kerogen thermal maturation for burial diagenesis (A) and geothermal systems and contact metamorphism by intrusive sheets (B). These models were plotted by assuming a functional heating duration of 10^6 years unless otherwise noted. Karweil's (1956, 1975) model was plotted using his estimates of E_a = 8.4 kcal/mol (35 kJ/mol) and $A = 7 \times 10^{-12}$/s for coalification.

that consider temperature and time. Barker's (1983) T_{max}-R_m model for geothermal systems is approximately linear, like the regression line, but it has a lower slope (or E_a). A T_{max}-R_m model for contact metamorphism by intrusive igneous sheets (Bostick, 1971) also suggests a lower E_a (slope) than that observed in the natural environments.

Comparison of thermal maturation models is limited by the problems in determining the rank, temperature, and reaction time in natural systems. In relatively hot environments, the strong temperature signal imparts a distinct thermal maturity and geological history record to the rocks resulting in reproducible and apparently accurate data. Conversely, in colder environments, the rank, T_{max}, and especially t, can be obscure and may only represent an interpretive fantasy. Consequently, comparison of thermal maturation tends to show that, within the scatter of the data, various models are similar in precision (Veto, 1980; Wright, 1980; and others).

The comparable fit of models that consider T_{max} and t, or T_{max} alone, to the kinetic data from natural systems could result from the emphasis that the T_{max}-t models place on reaction near T_{max}. The T_{max}-t model of Lopatin (Waples, 1980) considered that the integrated time-temperature history determines OM rank. The Lopatin time-temperature index (TTI) of maturation uses a temperature factor of 2^n for each 10°C temperature interval (n). The temperature factor is multiplied by the elapsed time spent in each temperature interval and summed over the entire burial history. TTI is heavily weighted toward reaction occurring near T_{max} because of the exponential temperature factor. In this respect it is similar to the method of Hood et al. (1975), which only considers the elapsed time within 15°C of T_{max}.

The emphasis of these models on OM thermal maturation at T_{max} is necessary because, as discussed above, temperature is recorded by irreversible chemical changes in the OM. Therefore, for rank to accurately record T_{max} it must cease reaction at this temperature. Thus, the relevant reaction time is that elapsed close to T_{max}. Models that emphasize T_{max} alone should be comparable in accuracy to those that consider T_{max} and t.

Discussion

Although the differences in reaction rate between these three environments may be due to the methods of calculating the heating duration and other factors (Snowdon, 1979), the effect of pressure is an attractive explanation. Pressure influences the diffusion of products out of the OM and, as discussed below, can change the apparent reaction rate. The effect of pressure is expressed in the kinetic parameters by the significant difference in reaction rate between environments. Because these reactions have a similar E_a (indicated by subparallel lines on an Arrhenius diagram), the reaction rate, k, is determined mainly by temperature and Arrhenius factor, A. Parallel lines on an Arrhenius plot differ only in their value of A. The Arrhenius factor is a coefficient representing the chance that molecules in the OM or coal structure will transform into separate hydrocarbons. The successively larger Arrhenius factor observed from burial diagenesis (smallest), to geothermal systems, to contact metamorphism by intrusive sheets (largest), suggests that there is a greater chance for transformation to occur in contact metamorphism relative to geothermal systems and burial diagenesis. The change in Arrhenius factor between these environments could be strongly influenced by the system pressure that existed in these systems when they were near T_{max}.

Thermodynamic considerations suggest that pressure should inhibit catagenesis by retarding devolatilization of OM (Lopatin and Bostick, 1973). Artificial coalification studies confirm that pressure inhibits devolatilization of OM apparently by inhibiting diffusion of the reaction products. Davis and Spackman (1964), using cypress wood in bomb experiments at up to 470,000 kPa (68,000 lb/in^2) and 400°C, showed that at higher temperatures, pressure is an important factor. This is partially confirmed by experimental reactions lasting a few hours in a closed system at high temperature and pressure, in which it was found that coal alteration was retarded (Berkowitz, 1967). Gorohkov et al. (1973), in experiments on artificial graphitization of marine algal detritus, determined that temperature, pressure, and experiment duration were important parameters in OM metamorphism. They found the effect of pressure in their closed-system experiments was to retard graphitization by the accumulation of gaseous products. Cecil et al. (1979) observed that high hydrostatic pressures retarded maturation of cypress peat heated for 1 month in open and closed systems. Stach et al. 1982, p. 59) concluded in their review that "Experiments show that static pressure does not promote chemical coalification change but even retards them—obviously because removal of gas is made more difficult." Huck and Patteisky (1964) have shown that at pressures up to 8,000 atm, the higher the pressure, the smaller the rank increase measured by R_m. Huck and Patteisky's results indicate that the lower the pressure, the greater the rate of catagenesis at a given temperature.

Pressure retardation of R_m is also observed in natural environments. Diessel et al. (1978) found that OM metamorphosed in rapidly subsiding basins displayed anomalously low rank levels. McTavish (1978) studied the effect of pressure on vitrinite reflectance in borehole samples from normal hydrostatic and overpressured zones in offshore northwest Europe. He found that R_m was retarded (lower) in the overpressured zones when compared to normal hydrostatic zones.

The evidence from natural environments and experimental systems suggests that the effect of pressure is to inhibit diffusion and escape of OM transformation products. Thus, in a relatively open and/or low pressure system, catagenesis is promoted and the reaction rate at a given temperature is higher. The effect of pressure on OM thermal maturation is apparently expressed by a change in the reaction rate at a given temperature on the Arrhenius plot for each environment (Fig. 5.4).

Conclusions

The data presented here support the following conclusions:

1. OM maturation can be modeled by considering temperature and heating duration, or considering T_{max} alone. Both types of model can be used to predict paleotemperature to about the same degree of precision.
2. Reaction rate and temperature data analyzed on an Arrhenius plot fall along three subparallel segments that correspond to OM thermal maturation in burial diagenesis, geothermal systems, and sediments adjacent to intrusive sheets. The similar slope of these subparallel lines indicates that the activation energy of OM maturation is about 10 kcal/mol (42 kJ/mol) and that only a single reaction mechanism is observed in each environment.
3. OM stabilizes with respect to temperature after about 10^6 to 10^7 years in burial diagenesis, 10^4 years in geothermal systems, and about 10^{-1} to 10^0 years in contact metamorphism by intrusive sheets.
4. The effect of pressure appears to determine the ease of product escape from the OM and may influence the reaction rate at a given temperature within each environment. Because of the differences in determining heating duration in each environment, the change in reaction rate may be due to other sources besides pressure.

References

Abelson, P.H. 1963. Organic geochemistry and the formation of petroleum. In: Proceedings of the Sixth World Petroleum Congress, Geophysics and Geology, section 1, Hamburg, Verien Zur Forderung, pp. 397–407.

Abelson, P.H. 1967. Conversion of biochemicals to kerogen and n-paraffins. In: Abelson, P.H. (ed.): Researches in Geochemistry: Vol. 2. New York, Wiley, pp. 63–86.

Almon, W.R. 1974. Petroleum-forming reactions: Clay catalyzed fatty acid decarboxylation. Ph.D. dissertation, University of Missouri, Columbia, 117 pp.

Ammosov, I.I., Babashkin, B.G., and Sharkova, L.S. 1975. Bituminit nizhnekembriyskikh otlozheniy Irkuts-kay neftegazonosnoy oblasti (Bituminite of Lower Cambrian deposits in the Irkutsk oil and gas region). In: Yeremin, I.V. (ed.): Paleotemperatury zon Nefteobrazovaniya. Moscow, Nauka Press, pp. 25–29.

Ammosov, I.I., Gorshkov, V.I., Greshnikov, N.P., and Kalmykov, G.S. 1977. Paleogeotermicheskiye Kriteriyi Razmeshcheniya Neftyanykh Zalezhen (Paleogeothermal Criteria of the Location of Petroleum Deposits). Leningrad, Nedra Press.

Asakawa, T., and Fujita, Y. 1979. Organic metamorphism and hydrocarbon generation in sedimentary basins of Japan. In: Generation and Maturation of Hydrocarbons in Sedimentary Basins. United National Economic and Social Commission for Asia and the Pacific, Technical Publication Series 6, pp. 142–162.

Baker, E.W., Huang, W.Y., Rankin, J.G., Castano, J.G., Guinn, J.R., and Fuex, A.N. 1977. Electron paramagnetic resonance study of thermal alteration of kerogen in deep-sea sediments by basaltic sill intrusion. In: Lancelot, Y., et al. (eds.): Initial Reports of Deep Sea Drilling Project: Vol. 41. Washington, DC, U.S. Government Printing Office, pp. 839–847.

Barber, P.M. 1982. Palaeotectonic evolution and hydrocarbon genesis of the central Exmouth Plateau. Australian Petroleum Exploration Association Journal 22:131–144.

Barker, C.E. 1983. The influence of time on metamorphism of sedimentary organic matter in selected geothermal systems, western North America. Geology 11:384–388.

Barker, C.E., and Elders, W.A. 1981. Vitrinite reflectance geothermometry and apparent heating duration in the Cerro Prieto geothermal field. Geothermics 10:207–223.

Barker, C.E., and Pawlewicz, M.J. 1985. The correlation of vitrinite reflectance with maximum temperature in humic organic matter. Abstracts, Society of Economic Paleontologists and Mineralogists Annual Midyear Meeting, Golden, Colorado, p. 8.

Barker, C.E., and Pawlewicz, M.J. 1986. The correlation of vitrinite reflectance with maximum temperature in humic organic matter. In: Buntebarth, G., and Stegena, L. (eds.): Paleogeothermics. New York, Springer-Verlag, pp. 79–93.

Barker, C.G. 1979. Organic geochemistry in petroleum exploration. American Association of Petroleum Geologists Continuing Education Course Note Series 10, 159 pp.

Bartenstein, H., and Teichmüller, R. 1974. Inkohlungsuntersuchugen, ein Schlussel zur Prospektierrung von palaozoischen Kohlenwasserstoff-Lagerstatten. Fortschritte in der Geologie von Rheinland und Westfalen 24:129–160.

Benson, S.W. 1965. Bond energies. Journal of Chemical Education 42:502–518.

Berkowitz, N. 1967. The coal-carbon transformation: Basic mechanisms. In: Symposium on the Science and Technology of Coal. Ottawa, Canada, Mines Branch, Department of Energy, Mines, and Resources, pp. 148–156.

Blumer, M. 1965. Organic pigments: Their long term fate. Science 149:722–726.

Bostick, N.H. 1971. Thermal alteration of clastic organic particles as an indicator of contact and burial metamorphism in sedimentary rocks. Geoscience and Man 3:83–92.

Bostick, N.H. 1979a. Microscopic measurement of the level of catagenesis of solid organic matter in sedimentary rocks to aid exploration for petroleum and to determine former burial temperatures: A review. In: Scholle, P.A., and Schluger, P.R. (eds.): Aspects of Diagenesis. Society of Economic Paleontologists and Mineralogists Special Publication 26, pp. 17–44.

Bostick, N.H. 1979b. Organic petrography of nineteen rocks, a split of each analyzed in thirty different laboratories. Abstracts of Papers, Ninth International Congress of Carboniferous Stratigraphy and Geology, Urbana, Illinois, p. 24.

Briggs, N.D., Naeser, C.W., and McCulloh, T.H. 1981. Thermal history by fission-track dating, Tejon oil field area, California (abst.). American Association of Petroleum Geologists Bulletin 65:906.

Brooks, J.D. 1970. The use of coals as indicators of the occurrence of oil and gas. Australian Petroleum Exploration Association Journal 10:35–40.

Bullard, E.C. 1947. The time necessary for a borehole to attain temperature equilibrium. Monthly Notices, Royal Astronomical Society 5:127–130.

Cardoso, J.N., et al. 1978. Preliminary organic geochemical analyses. Site 391, leg 44 of the deep sea drilling project. In: Benson, W.E., and Sheridan, R.E. (eds.): Initial Reports of the Deep Sea Drilling Project: Vol. 44. Washington, DC, U.S. Government Printing Office, pp. 617–624.

Cassou, A.-M., Connan, J., and Porthault, B. 1977. Relations between maturation of organic matter and geothermal effect, as exemplified in Canadian east coast offshore wells. Canadian Petroleum Geologists Bulletin 25:174–194.

Castano, J.R., and Sparks, D.M. 1974. Interpretation of vitrinite reflectance measurements in sedimentary rocks and determination of burial history using vitrinite reflectance and authigenic minerals. In: Dutcher, R.R., Hacquebard, P.A., Schopf, J.M., and Simon, J.A. (eds.): Carbonaceous Materials as Indicators of Metamorphism. Geological Society of America Special Paper 153, pp. 31–52.

Cecil, C.B., Stanton, R.W., Allshouse, S.D., and Cohen, A.D. 1979. Effects of pressure on coalification. Abstracts of Papers, Ninth International Congress Carboniferous Stratigraphy and Geology, Urbana, Illinois, p. 32.

Chandra, D., and Bond, R.L. 1956. The reflectance of carbonized coals. In: Proceedings of the International Committee for Coal Petrology: Vol. 2. pp. 47–51.

Chatterjee, N.N., Chandra, D., and Ghosh, T.K. 1964. Reflectance of Poniati seam affected by a mica-peridotite dyke. Journal of Mines, Metals, and Fuels (India) 12:346–348.

Clegg, K.E. 1955. Metamorphism of coal by peridotite dikes in southern Illinois. Illinois Geological Survey Report of Investigations 178:18 pp.

Connan, J. 1974. Time-temperature relation in oil genesis. American Association of Petroleum Geologists Bulletin 58:2516–2521.

Connan, J., and Cassou, A.M. 1980. Properties of gases and petroleum liquids derived from terrestrial kerogen at various maturation levels. Geochimica et Cosmochimica Acta 44:1–23.

Cornelius, C.D. 1975. Geothermal aspects of hydrocarbon exploration in the North Sea areas. Norges Geologiske Undersokelse 316:25–66.

Cornford, C., Rullkotter, J., and Welte, D.H. 1979. Organic geochemistry of DSDP leg 47A, site 397, eastern North Atlantic: Organic petrography and extractable hydrocarbons. In: Rad, von U., et al. (eds.): Initial Reports of the Deep Sea Drilling Project: Vol. 47, Part 1. Washington, DC, U.S. Government Printing Office, pp. 511–522.

Corrigan, G.M. 1982. Cooling rate studies of rocks from two basic dykes. Mineralogical Magazine 46:387–394.

Creaney, S. 1980. Petrographic texture and vitrinite reflectance variation on the Alston block, north-east England. Yorkshire Geological Society Proceedings 42:553–580.

Crelling, J.C., and Dutcher, R.R. 1968. A petrologic study of a thermally altered coal from the Purgatoire River Valley of Colorado. Geological Society of America Bulletin 79:1375–1386.

Davis, A., and Spackman, W. 1964. The role of the cellulosic and lignitic components of wood in artificial coalification. Fuel 43:215–224.

Demaison, G.J. 1975. Relationships of coal rank to paleotemperatures in sedimentary rocks. In: Alpern, B. (ed.): Petrographie de la Matiere Organique des Sediments, Relations avec la Paleotemperature et le Potentiel Petrolier. Paris, Centre National de la Recherche Scientifique, pp. 217–224.

Diessel, C.F.K., Brothers, R.N., and Black, P.M. 1978. Coalification and graphitization in high-pressure schists in New Caledonia. Contributions to Mineralogy and Petrology 68:63–78.

Dow, W.G. 1978. Petroleum source beds on continental slopes and rises. American Association of Petroleum Geologists Bulletin 62:1584–1606.

Drury, M.J. 1984. On a possible source of error in extracting equilibrium formation temperatures from borehole BHT data. Geothermics 13:175–180.

Drury, M.J., Jessop, A.M., and Lewis, T.J. 1984. The detection of groundwater flow by precise temperature measurements in boreholes. Geothermics 13:163–174.

Eby, J.B. 1925. Contact metamorphism of some Colorado coals by intrusives. American Institute of Mining and Metallurgical Engineers Transactions 71:246.

Ghosh, T.K. 1967. A study of temperature conditions at igneous contacts with certain Permian coals of India. Economic Geology 62:109–117.

Goldstein, T.P. 1983. Geocatalytic reactions in formation and maturation of petroleum. American Association of Petroleum Geologists Bulletin 67:152–159.

Goodarzi, F., and Murchinson, D.G. 1977. Effect of prolonged heating on the optical properties of vitrinite. Fuel 56:89–96.

Gorohkov, S.S., Petrova, N.I., and Kovalenko, V.S. 1973. Experimental study of the alteration of biogenic carbon at high temperature and pressure. Doklady Akademii Nauk SSSR 209:194–196.

Gretener, P.E., and Curtis, C.D. 1982. Role of temperature and time on organic metamorphism. American Association of Petroleum Geologists Bulletin 66: 1124–1129.

Hacquebard, P.A. 1975. Pre- and postdeformational coalification and its significance for oil and gas exploration. In: Alpern, B. (ed.): Petrographie de la Matiere Organique des Sediments, Relations avec la Paleotemperature et le Potentiel Petrolier. Paris, Centre National de la Recherche Scientifique, pp. 225–241.

Harwood, R.J. 1977. Oil and gas generation by laboratory pyrolysis of kerogen. American Association of Petroleum Geologists Bulletin 61:2082–2102.

Hoering, T.C., and Abelson, P.H. 1963. Hydrocarbons from kerogen. Carnegie Institution of Washington Yearbook 62:229–234.

Hood, A., and Castano, J.R. 1974. Organic metamorphism: Its relationship to petroleum generation and application to studies of authigenic minerals. United Nations Economic Commission for Asia and the Far East Technical Bulletin 8:85–118.

Hood, A., Gutjahr, C.C.C., and Heacock, R.L. 1975. Organic metamorphism and the generation of petroleum. American Association of Petroleum Geologists Bulletin 59:986–996.

Huc, A.Y., and Hunt, J.M. 1980. Generation and migration of hydrocarbons in offshore south Texas Gulf Coast sediments. Geochimica et Cosmochimica Acta 44:1081–1089.

Huck, Von G., and Patteisky, K. 1964. Inkohlungsreaktionen unter Druck. Fortschritte in der Geologie von Reinland und Westfalen 12:551–558.

Hunt, J.M. 1979. Petroleum Geochemistry and Geology. San Francisco, Freeman, 617 pp.

Ishiwatari, R., Ishiwatari, M., Rohrback, B.G., and Kaplan, I.R. 1977. Thermal alteration experiments on organic matter from recent marine sediments in relation to petroleum genesis. Geochimica et Cosmochimica Acta 41:814–828.

Jaeger, J.C. 1959. Temperatures outside a cooling intrusive sheet. American Journal of Science 257:44–54.

Jaeger, J.C. 1964. Thermal effects of intrusions. Reviews of Geophysics 2:443–466.

Johns, W.D. 1979. Clay mineral catalysis and petroleum

generation. Annual Reviews of Earth and Planetary Science 7:183–198.

Johnson, V.H. 1952. Thermal metamorphism and ground-water alteration of coking coal near Paonia, Colorado. Mining Engineering 4:391–395.

Jones, J.M., and Creaney, S. 1977. Optical character of thermally metamorphosed coals of northern England. Journal of Microscopy 109:105–118.

Juntgen, H., and Klein, J. 1975. Origin of natural gas from coaly sediments. Erdol und Kohle 28:65–73.

Kantsler, A.J., and Cook, A.C. 1979. Maturation patterns in the Perth Basin. Australian Petroleum Exploration Association Journal 19:94–107.

Kantsler, A.J., Smith, G.C., and Cook, A.C. 1978. Lateral and vertical rank variation: Implications for hydrocarbon exploration. Australian Petroleum Exploration Association Journal 18:143–156.

Karweil, J. 1956. Die Metamorphose der Kohlen von Stand punkt der physikalischen Chemie. Deutsche Geologische Gesellschaft Zeitschrift 107:132–139.

Karweil, J. 1975. The determination of paleotemperatures from the optical reflectance of coaly particles in sediments. In: Alpern, B. (ed.): Petrographie de la Matiere Organique des Sediments, Relations avec la Paleotemperature et le Potentiel Petrolier. Paris, Centre National de la Recherche Scientifique, pp. 195–203.

Katz, B.J., Liro, L.M., Lacey, J.E., and White, H.E. 1982. Time and temperature in petroleum formation: Application of Lopatin's method to petroleum exploration: Discussion. American Association of Petroleum Geologists Bulletin 66:1150–1151.

Kharaka, Y.K., Carothers, W.W., and Rosenbauer, R.J. 1983. Thermal decarboxylation of acetic acid: Implications for the origin of natural gas. Geochimica et Cosmochimica Acta 47:397–402.

King, J.D., and Claypool, G.E. 1982. Biological marker compounds and implications for generation and migration of petroleum in rocks of the Point Conception deep-stratigraphic test well, OCS-CAL 78-164 no. 1, offshore California. In: Issacs, C.M., and Garrison, R.E. (eds.): Petroleum Generation and Occurrence in the Miocene Monterey Formation, California. Los Angeles, Pacific Section, Society of Economic Paleontologists and Mineralogists, pp. 191–200.

Kisch, H.J. 1966. Carbonization of semi-anthracitic vitrinite by an analcime basanite sill. Economic Geology 61:1043–1063.

Laplante, R.E. 1974. Hydrocarbon generation in Gulf Coast Tertiary sediments. American Association of Petroleum Geologists Bulletin 58:1281–1289.

Law, B.E., Spencer, C.W., and Bostick, N.H. 1980. Evaluation of organic matter, subsurface temperature and pressure with regard to gas generation in low-permeability Upper Cretaceous and Lower Tertiary sandstones in Pacific Creek area, Sublette and Sweetwater Counties, Wyoming. Mountain Geologist 17:23–35.

Leith, M.J., and Rowsell, D.M. 1979. Burial history and temperature-depth conditions for hydrocarbon generation and migration on the Agulas Bank, South Africa. In: Anderson, A.M., and van Biljon, W.J. (eds.): Some Sedimentary Basins and Associated Ore Deposits of South Africa. Geological Society of South Africa Special Publication 6, pp. 205–217.

Lopatin, N.V. 1971. Temperature and geological time as factors of carbonification. Akademiya Nauk SSSR Series Geologicheskaya Izvestiya 3:95–106.

Lopatin, N.V., and Bostick, N.H. 1973. Geologicheskiye faktory katagneza ugley. In: Priroda Organischeskogo Veshchestva Sovremennykh i Isokopaemykh Osadkov. Moskow, Nauka Press, pp. 79–90.

Mackenzie, A.S., and McKenzie, D. 1983. Isomerization and aromatization of hydrocarbons in sedimentary basins formed by extension. Geological Magazine 120:417–528.

Magoon, L.B., and Claypool, G.E. 1982. NPRA Inigok no. 1: Use of Lopatin's method to reconstruct thermal maturity. Pacific Section, American Association of Petroleum Geologists Newsletter, January 1982, pp. 4–6.

Marshall, C.E. 1952. The thermal metamorphism of coal seams. Sir Douglas Mawson Anniversary Volume, Adelaide, Scotland, Adelaide University, pp. 109–142.

McCartney, J.T., and Teichmüller, M. 1972. Classification of coals according to degree of coalification by reflectance of the vitrinite component. Fuel 51:64–68.

McNab, J.G., Smith, P.V., and Betts, R.L. 1952. The evolution of petroleum. Industrial and Engineering Chemistry 44:2556–2563.

McTavish, R.A. 1978. Pressure retardation of vitrinite reflectance, offshore northwest Europe. Nature 271: 648–650.

Myhr, D.W., and Gunther, P.R. 1974. Lithostratigraphy and coal reflectance of a Lower Cretaceous deltaic succession in the Gulf-Mobil Parsons F-09 borehole, N.W.T. Geological Survey of Canada Paper 74-1b, pp. 24–28.

Nagornyi, V.N., and Nagornyi, Y.N. 1974. The question of the quantitative evaluation of the role of the time in processes of the regional metamorphism of coals. Solid Fuel Chemistry 8:30–36.

Neruchev, S.G., and Parparova, G.M. 1972. The role of geologic time in processes of metamorphism of coal and dispersed organic matter in rocks (in Russian). Otdeleniye Geologiya i Geofizika (Akademiya Nauk SSSR Sibirsk) 10:3–10.

Pauling, L. 1965. College Chemistry (3rd ed.). San Francisco, Freeman, 832 pp.

Pearson, D.B. 1981. Experimental simulation of thermal maturation in sedimentary organic matter. Ph.D. dissertation, Rice University, Houston, TX, 563 pp.

Peters, K.E., Ishiwatari, R., and Kaplan, I.R. 1977. Color of kerogen as index of organic maturity. American Association of Petroleum Geologists Bulletin 61:504–510.

Peters, K.E., Simoneit, B.R.T., Brenner, S., and Kaplan, I.R. 1978. Vitrinite reflectance-temperature determination for intruded Cretaceous black shale in the eastern Atlantic. In: Oltz, D.F. (ed.): Low Temperature Metamorphism of Kerogen and Clay Minerals. Los Angeles, Pacific Section, Society of Economic Paleontologists and Mineralogists, pp. 65–96.

Philippi, G.T. 1965. On the depth, time, and mechanism of petroleum generation. Geochimica et Cosmochimica Acta 29:1021–1049.

Povoledo, D., and Vallentyne, J.R. 1964. Thermal reaction kinetics of the glutamic acid-pyroglutamic acid system in water. Geochimica et Cosmochimica Acta 28:731–734.

Price, L.C. 1982. Organic geochemistry of core sample from an ultra deep hot well (300°C, 7 km). Chemical Geology 37:215–228.

Price, L.C. 1983. Geologic time as a parameter in organic metamorphism and vitrinite reflectance as an absolute paleogeothermometer. Journal of Petroleum Geology 6:5–38.

Price, L.C., and Barker, C.E. 1985. Suppression of vitrinite in amorphous-rich kerogen: A major unrecognized problem. Journal of Petroleum Geology 8: 59–84.

Price, L.C., Clayton, J.L., and Rumen, L.L. 1981. Organic geochemistry of the 9.6 km Bertha Rogers no. 1 well, Oklahoma. Organic Geochemistry 3:59–77.

Rashid, M.A., Purcell, L.P., and Hardy, I.A. 1980. Source rock potential for oil and gas of the east Newfoundland and Labrador Shelf areas. In: Miall, A.D. (ed.): Facts and Principles of World Petroleum Occurrence. Canadian Society of Petroleum Geologists Memoir 6, pp. 589–608.

Rohrback, B.G. 1979. Analysis of low molecular weight products generated by thermal decomposition of organic matter in recent sedimentary environments. Ph.D. dissertation, University of California, Los Angeles, 195 pp.

Sajgo, C. 1980. Hydrocarbon generation in a super-thick Neogene sequence in south-east Hungary: A study of the extractable organic matter. In: Douglas, A.G., and Maxwell, J.R. (eds.): Advances in Organic Geochemistry 1979. New York, Pergamon Press, pp. 103–113.

Saxby, J.D. 1982. A reassessment of the range of kerogen maturities in which hydrocarbons are generated. Journal of Petroleum Geology 5:117–128.

Schapiro, N., and Gray, R.J. 1966. Physical variations in highly metamorphosed Antarctic coals. In: Gould, F.E. (ed.): Coal Science. American Chemical Society Advances in Chemistry Series, vol. 55, pp. 196–217.

Scholle, P.A. 1977. Geological studies on the COST no. B-2 well, U.S. mid-Atlantic outer continental shelf area. U.S. Geological Survey Circular 750, 71 pp.

Seyer, W.F. 1933. The conversion of fatty and waxy substances into petroleum hydrocarbons. Institute of Petroleum Technologists Journal 19:773–783.

Shibaoka, M., and Bennett, A.J.R. 1977. Patterns of diagenesis in some Australian sedimentary basins. Australian Petroleum Exploration Association Journal 17:58–63.

Shibaoka, M., Saxby, J.D., and Taylor, G.H. 1978. Hydrocarbon generation in Gippsland basin, Australia: Comparison with Cooper basin, Australia. American Association of Petroleum Geologists Bulletin 62:1151–1158.

Shimoyama, A., and Johns, W.D. 1971. Catalytic conversion of fatty acids to petroleum-like paraffins. Nature Physical Science 232:140–144.

Smyth, M., and Saxby, J.D. 1981. Organic petrology and geochemistry of source rocks in the Pedirka-Simpson Basins, central Australia. Australian Petroleum Exploration Association Journal 21:187–199.

Snowdon, L.R. 1979. Errors in the extrapolation of experimental kinetic parameters to organic geochemical systems. American Association of Petroleum Geologists Bulletin 63:1128–1138.

Stach, E., Mackowsky, M.-T., Teichmüller, M., Taylor, G.H., Chandra, D., and Teichmüller, R. 1982. Stach's Textbook of Coal Petrology (3rd ed.). Berlin, Gebruder Borntraeger, 535 pp.

Suggate, R.P. 1982. Low-rank sequences and scales of organic metamorphism. Journal of Petroleum Geology 4:377–392.

Tan, L.-P. 1965. The metamorphism of Taiwan Miocene coals. Bulletin of the Geological Survey of Taiwan 16:44 pp.

Teichmüller, M., and Teichmüller, R. 1979. Zur Geothermischen Geschichte des Oberrhein-Grabens Zusammenfassung und Auswertung eines Symposiums. Fortschritte in der Geologie von Rheinland und Westfalen 27:109–120.

Teichmüller, M., and Teichmüller, R. 1981. The significance of coalification studies to geology: A review. Bulletin Centres Recherches Exploration-Production Elf-Aquitaine 5:491–534.

Thomas, B.M. 1979. Geochemical analysis of hydrocarbon occurrences in northern Perth Basin, Australia. American Association of Petroleum Geologists Bulletin 63:1092–1107.

Ting, F.T.C. 1975. Reflectivity of disseminated vitrinite in the Gulf Coast region. In: Alpern, B. (ed.): Petrographie de la Matiere Organique des Sediments Relations avec la Paleotemperature et le Potentiel Petrolier. Paris, Centre National de la Recherche Scientifique.

Ting, F.T.C. 1978. Petrographic techniques in coal analysis. In: Karr, C., Jr. (ed.): Analytical Methods for Coal and Coal Products. New York, Academic Press, pp. 3–26.

Tissot, B.P., Bard, J.F., and Espitalié, J. 1980. Principal factors controlling the timing of petroleum generation. In: Miall, A.D. (ed.): Facts and Principles of World Petroleum Occurrence. Canadian Society of Petroleum Geologists Memoir 6, pp. 143–152.

Tissot, B.P., Debyser-Califet, Y., Deroo, G., and Oudin, J.L. 1971. Origin and evolution of hydrocarbons in early Toarcian shales, Paris Basin, France. American Association of Petroleum Geologists Bulletin 55: 2177–2193.

Tissot, B.P., Deroo, G., and Hood, A. 1978. Geochemical study of the Uinta basin: Formation of petroleum from the Green River Formation. Geochimica et Cosmochimica Acta 42:1469–1485.

Tissot, B.P., and Espitalié, J. 1975. The thermal evolution of organic matter in sediments: Applications of mathematical simulation. Revue de la Institut Francais du Petrole 30:743–777.

Tissot, B.P., and Welte, D.H. 1984. Petroleum Formation and Occurrence (2nd ed.). New York, Springer-Verlag, 699 pp.

Vallentyne, J.R. 1957. Thermal degradation of amino acid. In: Annual Report of the Director of the Geophysical Laboratory, 1956–1957. Carnegie Institution of Washington, Geophysical Laboratory Paper 1277, pp. 185–186.

Vallentyne, J.R. 1964. Biogeochemistry of organic matter: II. Geochimica et Cosmochimica Acta 28:157–188.

Veto, I. 1980. An examination of the timing of catagenesis of organic matter using three published models. In: Douglas, A.G., and Maxwell, J.R. (eds.): Advances in Organic Geochemistry 1979. New York, Pergamon Press, pp. 163–167.

Walker, A.L., McCulloh, T.H., Petersen, N.F., and Stewart, R.J. 1983. Discrepancies between anomalously low reflectance of vitrinite and other maturation indicators from an Upper Miocene oil source rock, Los Angeles Basin, California (abst.): American Association of Petroleum Geologists Bulletin 67:565.

Waples, D.W. 1980. Time and temperature in petroleum exploration: Application of Lopatin's method to petroleum exploration. American Association of Petroleum Geologists Bulletin 64:916–926.

Waples, D.W. 1982. Time and temperature in petroleum formation: Application of Lopatin's method to petroleum exploration: Reply. American Association of Petroleum Geologists Bulletin 66:1152.

Waples, D.W. 1984. Thermal models for oil generation. In: Brooks, J., and Welte, D.H. (eds.): Advances in Petroleum Geochemistry: Vol. 1. London, Academic Press, pp. 8–67.

Wright, N.J.R. 1980. Time, temperature, and organic maturation: The evolution of rank within a sedimentary pile. Journal of Petroleum Geology 2: 411–425.

Ziegler, P.A. 1980. Northwest European Basin: Geology and hydrocarbon provinces. In: Miall, A.D. (ed.): Facts and Principles of World Petroleum Occurrence. Canadian Society of Petroleum Geologists Memoir 6, pp. 653–706.

6
Chemical Geothermometers and Their Application to Formation Waters from Sedimentary Basins

Yousif K. Kharaka and Robert H. Mariner

Abstract

Chemical geothermometers, based on the concentration of silica and proportions of sodium, potassium, lithium, calcium, and magnesium in water from hot springs and geothermal wells, have been used successfully to estimate the subsurface temperatures of the reservoir rocks. Modified versions of these geothermometers and a new chemical geothermometer, based on the concentrations of magnesium and lithium, are developed to estimate the subsurface temperatures (30°C to 200°C) in sedimentary basins where water salinities and hydraulic pressures are generally much higher than those in geothermal systems. The new Mg-Li geothermometer, which can be used to estimate subsurface temperatures as high as 350°C for waters from sedimentary basins and geothermal systems, is given by:

$$t = \frac{2200}{\log\left(\frac{\sqrt{Mg}}{Li}\right) + 5.47} - 273,$$

where t is temperature (°C) and Mg and Li concentrations are in mg/L.

Quartz, Mg-Li, Mg-corrected Na-K-Ca, and Na-Li geothermometers give concordant subsurface temperatures that are within 10°C of the measured values for reservoir temperatures higher than about 70°C. Mg-Li, Na-Li, and chalcedony geothermometers give the best results for reservoir temperatures from 30°C to 70°C. Subsurface temperatures calculated by chemical geothermometers are at least as reliable as those obtained by conventional methods. Chemical and conventional methods should be used together where reliable temperature data are required.

Introduction

Knowledge of subsurface temperatures is essential to exploration for geothermal energy. Subsurface temperatures are also becoming an important tool in exploration for petroleum (see Hitchon, 1985; Meyer and McGee, 1985, for recent discussions). This use of temperatures in the petroleum industry stems from the fact, known for more than 100 years, that many oil and gas fields are associated with mappable, positive geothermal anomalies at producing levels (Ovnatanov and Tamrazyan, 1970; Meyer and McGee, 1985). Also, all investigations involving water-rock interactions require accurate subsurface temperatures.

Oil and gas wells are the primary sources of temperature data in sedimentary basins. Unfortunately, the accuracy of temperatures obtained from oil wells varies widely (Meyer and McGee, 1985). Many operators do not recognize the importance of temperatures in exploration and production of petroleum and make no attempt to gather accurate data. Also, most temperatures are obtained from wells during drilling when the temperature distribution in the hole is under maximum thermal disturbance. The most reliable temperatures are those obtained from static bottom-hole pressure and temperature surveys generally conducted in production wells. Temperatures obtained from drill-stem tests are of intermediate accuracy. However, the majority of temperatures are obtained from electric log headings; these give the least reliable data and are generally lower than the true subsurface values. Meyer and McGee (1985) showed that the discrepancy between log-header and static bottom-hole test temperatures from the Wattenberg field, Colorado, is large, ranging from 11°C to 61°C.

In this study we investigate the applicability of chemical geothermometers, developed for geothermal systems, to formation waters from sedimentary basins. The chemical geothermometers were applied to 88 water analyses; 54 of these were from sedimentary basins where subsurface temperatures are known, and 34 were from geothermal systems. This data base is used to develop a new chemical geothermometer based on the concentrations of magnesium and lithium that gives the most reliable estimates of temperatures in sedimentary basins (30°C to 200°C). Results further show that chemical geothermometers commonly used for geothermal systems must be modified prior to their application to oil-field waters. The equations and parameters needed for these modifications are derived and reported in this chapter.

Chemical Geothermometers for Geothermal Systems

Concentrations of dissolved constituents in geothermal fluids are a function of the temperature of the aquifer and the alteration mineral assemblage (White, 1965; Ellis, 1970; Truesdell, 1976). Any constituent whose concentration is controlled by a temperature-dependent reaction could, theoretically, be used as a geothermometer. However, to be useful as a geothermometer, additional conditions must be met. These conditions include an adequate supply of the reactants, establishment of equilibrium between water and minerals in the aquifer, and absence of additional interactions as the water (gas) flows to the sampling point (Fournier et al., 1974). Mixing of waters from aquifers with different temperatures could alter the concentrations of constituents used in geothermometers and require the application of specialized mixing models (Fournier and Truesdell, 1974; Fournier, 1981).

Chemical geothermometers and the conditions for their application to geothermal systems have been reviewed recently by Truesdell (1976), Ellis and Mahon (1977), Fournier (1981), and Henley et al. (1984). The commonly used chemical geothermometers (Table 6.1) include the silica geothermometers (Fournier and Rowe, 1966; Fournier, 1973), the Na-K-Ca geothermometer (Fournier and Truesdell, 1973), the Na-K geothermometers (White, 1965; Ellis, 1970; Truesdell, 1976; Fournier, 1979), and the Mg-corrected Na-K-Ca geothermometer (Fournier and Potter, 1979).

Silica geothermometry is based on solubility of quartz, chalcedony, cristobalite, or amorphous silica (Fournier and Rowe, 1966; Fournier, 1973). The equations for quartz geothermometry closely approximate the solubility of quartz at the vapor pressure of the solution (±2°C) in the temperature range from 0°C to 250°C; above 250°C, the equations depart from the experimentally determined solubility of quartz (Fournier, 1981). In applying the silica geothermometers, ambiguity may arise as to which silica mineral is controlling the dissolved silica concentration. Arnórsson (1975) noted that in Iceland quartz controlled dissolved silica concentrations at temperatures of more than about 180°C, and chalcedony controlled dissolved silica below about 110°C. In the 180°C to 110°C range, it was not possible to determine, a priori, which silica mineral was controlling the dissolved silica concentration. In granitic terrains, quartz may control dissolved silica concentrations down to about 80°C (Brook et al., 1979). Also, mixing may sharply reduce the dissolved silica concentration, resulting in low calculated temperatures from silica geothermometers in some geothermal springs.

It should further be noted that an assumption of silica geothermometry is that the dissolved silica is present in solution as silicic acid ($H_4SiO_4^0$). A notable exception to this general assumption may occur in low-temperature geothermal systems in granitic rocks where the pH of water may be high (Mariner et al., 1983). In waters with pH values higher than about 8.5, a significant part of dissolved silica is present as $H_3SiO_4^-$. The actual concentrations of $H_4SiO_4^0$ in these waters must be calculated for use in silica geothermometers.

All geothermometers based on cation ratios are empirical; that is, they are based on temperature-dependent changes in the cation ratios of a large number of samples where aquifer temperatures are

TABLE 6.1. Equations for the most reliable chemical and isotope geothermometers applied to geothermal waters.

Geothermometer	Equation
Quartz–no steam loss	$t°C = \dfrac{1309}{5.19\text{-log } SiO_2} - 273$
Quartz–maximum steam loss	$t°C = \dfrac{1522}{5.75\text{-log } SiO_2} - 273$
Chalcedony	$t°C = \dfrac{1032}{4.69\text{-log } SiO_2} - 273$
Na-K (Fournier)	$t°C = \dfrac{1217}{\log (Na/K) + 1.483} - 273$
Na-K (Truesdell)	$t°C = \dfrac{885.6}{\log (Na/K) + 0.8573} - 273$
Na-K-Ca	$t°C = \dfrac{1647}{\log (Na/K) + \beta[\log(\sqrt{Ca}/Na) + 2.06] + 2.47} - 273$ $\beta = 4/3$ for $t < 100°C$; $= 1/3$ for $t > 100°C$
Mg-Corrected Na-K-Ca	(See text)
$\Delta^{18}O(SO_4 - H_2O)$	$t°C = \dfrac{1700}{(1000 \ln\alpha + 4.1)^{0.5}} - 273$ $\alpha = \dfrac{1000 + \delta^{18}O(HSO_4^-)}{(1000 + \delta^{18}O(H_2O)}$ and $\delta^{18}O$ in per mil

Note: Concentrations are in mg/kg. Modified from Fournier (1981).

known. Historically, the relation between Na, K, and aquifer temperatures was noted first by White (1965). Subsequently, slightly different curves (equations) relating Na/K ratios to temperature were presented by Ellis (1970), Truesdell (1976), and Fournier (1979). The Na-K geothermometer, however, is useful only at temperatures of more than about 150°C. At lower temperatures, calcium usually makes up a significant fraction of the cations, and the Na-K geothermometer gives anomalously high temperature estimates in calcium-rich waters. This led to the development of the Na-K-Ca geothermometer (Fournier and Truesdell, 1973). In many low-temperature environments, however, magnesium concentrations are high and the Na-K-Ca geothermometer gives excessively high temperature estimates. An empirical correction for the magnesium concentration was determined and applied to the Na-K-Ca geothermometer for use in these waters (Fournier and Potter, 1979). As a result, the Na-K-Ca geothermometer, with magnesium correction where appropriate, may be applied to waters with temperatures of 0° to 350°C.

Other chemical and isotope geothermometers have been suggested but have not been widely used. Examples include the empirical Na-Li geothermometer of Fouillac and Michard (1981); solubilities of minerals, such as anhydrite (Sakai and Matsubaya, 1974); isotopic fractionation between dissolved constituents and water, such as sulfate and water (Mizutani, 1972; McKenzie and Truesdell, 1977); and various gas geothermometers (Giggenbach, 1980; D'Amore and Panichi, 1980; Nehring and D'Amore, 1984). The Na-Li geothermometer is not widely utilized, perhaps because it was developed later and has a chloride dependence. Isotopic geothermometers can give reliable temperatures, but their use is restricted because the mass spectrometer and isotope extraction lines required to prepare and determine the isotopic compositions are not widely available.

Finally, chemical geothermometers based on solubilities of specific minerals other than quartz or chalcedony have not generally been utilized, because chemical complexing and activity coefficients of the dissolved constituents must be calculated at the aquifer-temperature. These calculations require complex chemical speciation programs, such as SOLMNEQ (Kharaka and Barnes, 1973), which can calculate mineral saturation states at various temperatures.

Application of Chemical Geothermometers to Oil-Field Waters

Oil-field waters differ from geothermal waters in several significant ways that affect the application of the chemical geothermometers discussed above. These differences are a function of the generally higher pressures, lower temperatures, and higher salinities of oil-field waters. Application of chemical geothermometers to natural gas wells may be complicated because chemical analyses of waters from these wells may not represent the true chemical composition of water from the production zone. This complication arises because of dilution by condensed water vapor produced with natural gas, especially in the case of wells from geopressured geothermal systems. On the other hand, problems related to mixing of waters from different zones or to water-rock interactions prior to sampling are generally less severe than those encountered in geothermal systems.

Pressure Correction for Silica Geothermometers

Silica geothermometers are based on the equilibrium constant (K) for the dissolution of silica minerals given by the reaction:

$$SiO_{2(s)} + 2H_2O \rightleftharpoons H_4SiO_4^0 \qquad (I)$$

The equilibrium constant is given by the following equation:

$$K = \frac{a_{H_4SiO_4^0}}{a_{SiO_{2(s)}} \cdot a^2_{H_2O}} \qquad (1)$$

where a is the activity of the subscripted species. Assuming unit activity for the silica mineral reacted and replacing $a_{H_4SiO_4^0}$ with the product of its molality (m) and activity coefficient (γ), Equation 1 becomes

$$K = \frac{m_{H_4SiO_4^0} \times \gamma_{H_4SiO_4^0}}{a^2_{H_2O}} \qquad (2)$$

Chemical geothermometers based on the solubilities of silica minerals (Table 6.1) are derived assuming a hydraulic pressure equal to the vapor pressure of water at the specified temperatures. This assumption introduces only small errors in the case of high-enthalpy geothermal systems because hydraulic pressures in the reservoir rocks of these systems approximate the boiling-point curve (Ellis and Mahon, 1977), resulting in pressures that are generally less than 100 bars (1,500 psi). However, petroleum wells are generally deeper and can have much higher hydraulic pressures, especially in geopressured systems where values greater than 1,000 bars (15,000 psi) may be attained (Kharaka et al., 1985).

Silica geothermometers (Table 6.1) must be modified to account for the increased solubility resulting from higher pressures encountered in oil-field waters. Solubility values, as a function of pressure (0 to 1,000 bars) and temperatures (50°C to 350°C) were calculated using equations given by Fournier and Potter (1982). These calculations show (Fig. 6.1), that at any given temperature, the increase in solubility is approximately linear with increasing pressure. The slope of this relationship steepens at higher temperatures, showing that the increase in solubility for a given increase in pressure is higher at higher temperatures.

The increase in solubility of quartz per unit pressure normalized to its solubility at the vapor pressure of water was fitted to the following equation:

$$y = a \times e^{bt} \qquad (3)$$

where y is the relative increase in solubility per bar, t is temperature (50°C to 350°C), and a and b are regression coefficients. The values obtained are $a = 7.862 \times 10^{-5}$ and $b = 3.61 \times 10^{-3}$. An excellent correlation coefficient (r) of 0.97 was obtained when results calculated using Equation 3 were compared with those given by Fournier and Potter (1982).

To correct for pressure effects, the concentrations of silica in formation waters should be multiplied by a correction factor (pf) given by:

$$pf = (1 - 7.862 \times 10^{-5} \times e^{(3.61 \times 10^{-3} \times t)} P) \qquad (4)$$

where P is the hydraulic pressure in bars and t is the measured or calculated subsurface temperature in °C. Computed temperatures without pressure corrections are always higher than measured subsurface temperatures. The computed temperatures, for example, are higher than the true values by 5°C at 100°C and 12°C at 200°C in a reservoir at 1,000 bars.

Activity Coefficient of $H_4SiO_4{}^0$ and Activity of Water

Chemical geothermometers based on solubilities of silica minerals (Table 6.1) are derived assuming that the activity coefficient of $H_4SiO_4{}^0$ ($\gamma_{H_4SiO_4{}^0}$) and the activity of water (a_{H_2O}) in Equation 2 are equal to unity. These assumptions introduce only minor errors in the computed temperatures for geothermal systems where salinities are almost always lower than that of seawater and temperatures are higher than about 200°C. The salinities encountered in sedimentary basins, on the other hand, are generally much higher than seawater, reaching values greater than 350,000 mg/L of dissolved solids (White, 1965; Hitchon et al., 1971; Carpenter et al., 1974; Kharaka et al., 1985). Values for the activity coefficient of $H_4SiO_4{}^0$ and activity of water (Equation 2) depart significantly (>5%) from unity at salinities greater than seawater (35,000 mg/L); silica geothermometers (Table 6.1) should be corrected for these departures. No correction is necessary for the assumption that $H_4SiO_4{}^0$ is the only silica species in oil-field waters because the pH values of these waters are almost always lower than about 8.0 (Kharaka et al., 1985).

There are several methods for calculating the activity coefficients of neutral species like $H_4SiO_4^0$ that give different results. The activity coefficients of all neutral species are generally assumed equal to that of dissolved CO_2 in NaCl solutions (Helgeson, 1969). The activity coefficients for CO_2 (γ_{CO_2}) are calculated using:

$$\gamma_{CO_2}(T) = \frac{k_m}{k} \qquad (5)$$

where k and k_m are the Henry's law coefficients in pure water and sodium chloride solutions of molality (m) at temperature T in °K. Values for k_m and k are available as a function of temperature (0°C to 350°C) and molality of NaCl (0 to 6.0 molal) from Ellis and Golding (1963) and Drummond (1982).

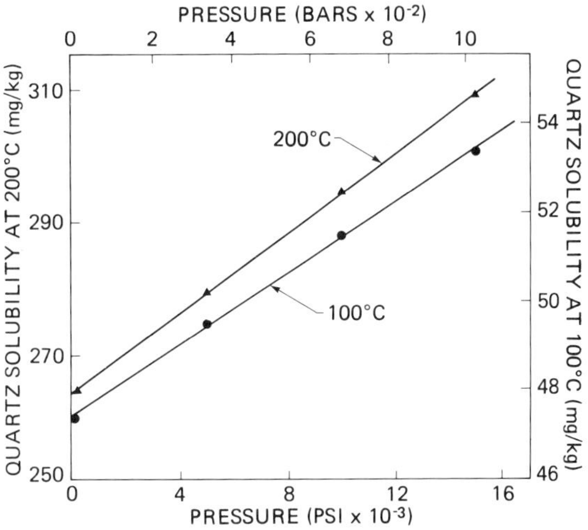

FIGURE 6.1. Solubility of quartz as a function of pressure at 100°C and 200°C.

We prefer the expression of Marshall (1980) and Chen and Marshall (1982) for calculating the activity coefficient of $H_4SiO_4{}^0$. The activity coefficient of $H_4SiO_4{}^0$ was derived from data on the solubility of amorphous silica as a function of temperature (0°C to 350°C) and salinity (0 to 6 molal). The expression is:

$$\log(\gamma_{H_4SiO_4^0}) = (0.00978 \times 10^{(280/T)}) \times m \qquad (6)$$

Sodium and chloride are by far the dominant species in most formation waters from sedimentary basins (Kharaka et al., 1985), and Equation 6 can be applied directly to these waters. However, the concentrations of other species, especially calcium, can be high, requiring modification of Equation 6. Data in Marshall (1980) and Chen and Marshall (1982) show that Equation 6 can be generalized to:

$$\log(\gamma_{H_4SiO_4{}^0}) = (0.00489 \times 10^{(280/T)}) \times \Sigma_i z_i^2 m_i \qquad (7)$$

where z_i and m_i are the charge and analytical molality of species (i). The summation covers all the species in the formation water.

An estimate of the errors involved in calculating subsurface temperatures assuming that $\gamma_{H_4SiO_4{}^0}$ is equal to unity can be made from Equation 7. This equation shows that $\gamma_{H_4SiO_4{}^0}$ increases with increasing salinity of the water, but decreases with increasing temperature. The activity coefficients of $H_4SiO_4{}^0$ at 100°C are equal to 1.00, 1.21, and

TABLE 6.2. Activity of water in NaCl and $CaCl_2$ solutions at 25°C to 200°C.

Source of a_{H_2O} and temp. (°C)	All temperatures[a]	From osmotic coefficients[b] 25	100	200
m_{NaCl}				
0	1.00	1.00	1.00	1.00
0.2	0.993	0.993	0.994	0.994
0.5	0.983	0.984	0.984	0.985
1.0	0.966	0.967	0.967	0.967
2.0	0.932	0.932	0.932	0.937
3.0	0.898	0.893	0.894	0.900
4.0	0.864	0.852	0.854	0.871
5.0	0.830	0.807	0.812	0.837
6.0	0.796	0.760	0.768	0.802
m_{CaCl_2}				
0	1.00	1.00	1.00	1.00
0.2	0.993	0.991	0.991	0.992
0.5	0.983	0.976	0.977	0.980
1.0	0.966	0.945	0.949	0.959

[a]From Equation 8.
[b]From Equation 9.

1.47 in NaCl solutions with salinities of 0, 3, and 6 molal, respectively. Assuming an activity coefficient of unity for $H_4SiO_4{}^0$ in these solutions results in calculated subsurface temperatures that are 9°C and 17°C lower than the true values in the 3- and 6-molal salinity samples. Calculations further show that these errors do not change appreciably with changes in temperatures from 50°C to 200°C.

The activity of water (Equation 2) can be calculated from the expression given by Garrels and Christ (1965) as

$$a_{H_2O} = 1 - 0.017\ \Sigma_i\ m_i \qquad (8)$$

The summation covers the molalities (m_i) of all the species in solution. A more accurate value for the activity of water can be obtained from the expression given by Helgeson et al. (1970) as

$$\log\ (a_{H_2O}) = 0.00782\ \Sigma \nu_e\ m_e\ \varphi_e \qquad (9)$$

where ν_e is the number of moles of ions in the formula for the electrolyte (e) (e.g., $\nu_{NaCl} = 2$, and $\nu_{CaCl_2} = 3$) and m_e and φ_e are the molality and osmotic coefficient of this electrolyte. Equations 8 and 9 give approximately the same value for the activity of water in NaCl solutions (0 to 6 molal) and temperatures of 25°C to 200°C (Table 6.2). The activity of water is somewhat lower in $CaCl_2$ solutions (Table 6.2). The activity of water can be approximated for the majority of oil-field waters using Equation 8. Equation 9, however, should be used for waters where the concentrations of divalent cations comprise more than about 20% of the total cations. Values for the osmotic coefficients of NaCl, $CaCl_2$, and other electrolytes as a function of temperature and molality are given in Staples and Nuttall (1977), Holmes et al. (1978, 1981), and Pitzer (1981).

An estimate of the errors involved in calculating subsurface temperatures assuming that $a_{H_2O} = 1$ can be made from Equations 8, 9, or Table 6.2. These data show that a_{H_2O} decreases with increasing salinity of water, but is essentially independent of temperature over the temperature range of 25°C to 200°C. Table 6.2 shows that a_{H_2O} ranges from 1.00 to 0.80 as salinities increase from 0 to 6 molal NaCl. The activity of water is particularly important because the solubility of quartz, as indicated by Equation 2, is inversely proportional to a_{H_2O}. Assuming $a_{H_2O} = 1$ results in calculated subsurface temperatures that are about 9°C and 19°C lower than the true values in the 3- and 6-molal NaCl solutions.

The errors in calculated temperatures resulting from assuming $a_{H_2O} = 1$ are additive to those assuming $\gamma_{H_4SiO_4{}^0} = 1$ as indicated by Equation 2. Thus, assuming $a_{H_2O} = 1$ and $\gamma_{H_4SiO_4{}^0} = 1$ results in calculated temperatures that are about 18°C and 35°C lower than the correct values in the 3- and

6-molal NaCl solutions at 100°C. It should be noted that corrections for pressure effects will somewhat reduce the magnitude of errors resulting from assuming that a_{H_2O} and $\gamma_{H_4SiO_4^0}$ are equal to unity.

The possible total errors in estimated subsurface temperatures neglecting the effects of pressure on the solubility of quartz and assuming that $\gamma_{H_4SiO_4^0}$ and a_{H_2O} are equal to unity are shown in Figure 6.2. Using these assumptions, the calculated subsurface temperatures of 26 samples from the central Mississippi Salt Dome basin (Kharaka et al., 1986; unpublished data) and coastal Louisiana and Texas (Kharaka et al., 1978, 1979) are always lower than the measured values (Fig. 6.2). The selected samples have salinities that range from 100,000 mg/L to 330,000 mg/L of dissolved solids; the hydraulic pressures range from 200 bars to 900 bars; and the corrected temperatures should plot on the lines shown in Figure 6.2. The lower calculated temperatures show that corrections for the effects of $\gamma_{H_4SiO_4^0}$ and a_{H_2O} are much greater in these waters than the pressure correction, which is in the opposite direction. The errors in calculated temperatures in samples with lower salinities, of course, will be lower than those indicated in Figure 6.2.

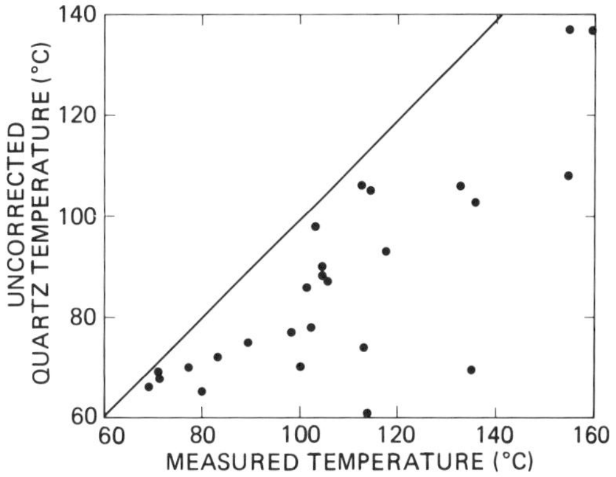

FIGURE 6.2. Temperature calculated using uncorrected quartz geothermometer versus measured subsurface temperatures for brines from sedimentary basins. Note that the calculated temperatures plot below the ideal line showing that they are always lower than the subsurface temperatures.

New Mg-Li Geothermometer

It is generally recognized that the concentrations and proportions of magnesium in subsurface waters are much lower than those in seawater and generally decrease with increasing temperatures (White, 1965; Fournier and Potter, 1979; Kharaka et al., 1985). The concentrations and proportions of lithium, on the other hand, increase with increasing temperatures (Fouillac and Michard, 1981; Kharaka et al., 1985), suggesting that the magnesium-to-lithium ratio may be a sensitive indicator of temperature. The geochemical reasons for this behavior are not fully understood (Ellis and Mahon, 1977), but it is known that Mg^{2+} and Li^+ commonly substitute for each other in amphiboles, pyroxenes, micas, and clay minerals. This substitution takes place mainly because the two cations have almost identical crystalline ionic radii.

The methodology for developing chemical geothermometers from chemical composition of waters is detailed in Fournier and Truesdell (1973). In the case of Mg-Li, an exchange reaction is written of the type:

$$Li^+ + (0.5\ Mg)\ Solid \rightleftharpoons 0.5\ Mg^{2+} + (Li)\ Solid \tag{II}$$

The equilibrium constant for this reaction at temperature T (K_T) is given by

$$K_T = \frac{(m_{Mg^{2+}})^{0.5}}{m_{Li^+}} \cdot \left[\frac{\gamma^{0.5}_{Mg^{2+}} \cdot a_{(Li)\ Solid}}{\gamma_{Li^+} \cdot a_{(0.5\ Mg)\ Solid}}\right] \tag{10}$$

An assumption of chemical geothermometers is that the terms within the brackets is equal to unity, resulting in:

$$K_T \simeq \frac{(m_{Mg})^{0.5}}{m_{Li}} \tag{11}$$

The chemical geothermometers are based on the van't Hoff equation given by

$$\frac{\delta \ln K}{\delta T} = \frac{\Delta H^0_r}{RT} \tag{12}$$

which can be integrated (assuming ΔH^0_r is constant) and simplified to

$$\log K_T = \log K_{298} - \frac{\Delta H^0_r}{4.576}\left(\frac{1}{T} - \frac{1}{298}\right) \tag{13}$$

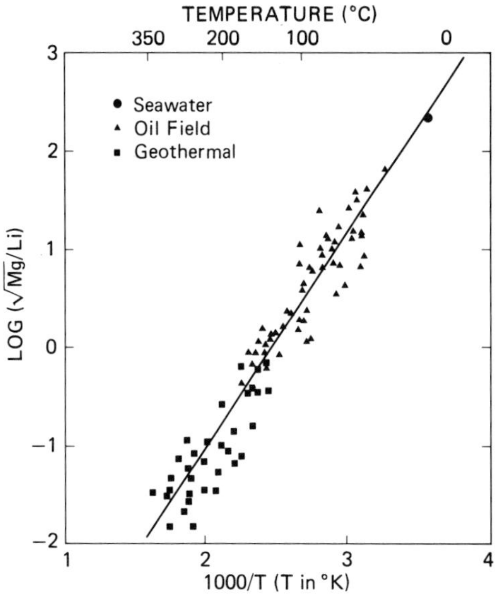

FIGURE 6.3. Magnesium/lithium ratios as a function of subsurface temperatures for oil field, geothermal, and seawater.

where T is temperature in °K, $\Delta H^0{}_r$ is the standard enthalpy of Reaction II in cal/mole. Equation 13 can be rearranged to give

$$\log K_T = \log K_{298} + \frac{\Delta H^0{}_r}{1{,}364} - \frac{\Delta H^0{}_r}{4.576}\left(\frac{1}{T}\right) \quad (14)$$

Substituting for the value of $\log K_T$ from Equation 11 yields:

$$\log\left(\frac{(m_{\mathrm{Mg}})^{0.5}}{m_{\mathrm{Li}}}\right) = \log K_{298} + \frac{\Delta H^0{}_r}{1{,}364} - \frac{\Delta H^0{}_r}{4.567}\left(\frac{1}{T}\right) \quad (15)$$

Equation 15 can be simplified into a linear equation by assuming that $\Delta H^0{}_r$ is a constant. The equation with constants A and B is

$$\log\left(\frac{(m_{\mathrm{Mg}})^{0.5}}{m_{\mathrm{Li}}}\right) = A + B\left(\frac{1}{T}\right) \quad (16)$$

where A is equal to $\log K_{298} + \frac{\Delta H^0{}_r}{1{,}364}$ and B is equal to $-\frac{\Delta H^0{}_r}{4.576}$.

The Mg-Li geothermometer is developed from a plot of

$$\log\left(\frac{(m_{\mathrm{Mg}})^{0.5}}{m_{\mathrm{Li}}}\right) \text{ versus } \frac{1}{T}$$

that yields a line with a slope equal to B and an intercept equal to A. This geothermometer is obtained by rearranging Equation 16 to yield:

$$t_{\mathrm{Mg\text{-}Li}} = \frac{B}{\log\left(\frac{(m_{\mathrm{Mg}})^{0.5}}{m_{\mathrm{Li}}}\right) - A} - 273 \quad (17)$$

where t is temperature in °C.

The data base for the development of the Mg-Li and other geothermometers is shown in Table 6.3. Reliable subsurface temperatures are generally not reported with chemical analyses of oil-field waters. However, we have temperature and chemical data from more than 250 formation waters from about 30 oil and gas fields located in Texas, Louisiana, Mississippi, California, and Alaska (Kharaka et al., 1985). Fifty-four samples were selected mainly from our own files based on: 1) reliability of reported subsurface temperatures, 2) absence of dilution by condensed water in the case of samples from gas wells, and 3) relatively uniform temperature distribution over the range of temperatures of oil-field waters (33°C to 170°C). For data from geothermal systems, only samples from wells were selected. In the case of samples from wells where boiling had occurred, the reported in situ chemical compositions of waters were used.

The reliability of the Mg-Li geothermometer is indicated by examination of Figure 6.3. An excellent correlation coefficient (r) of 0.96 is obtained for all the data in Figure 6.3. A very good correlation coefficient of 0.90 is obtained when the regression is carried out using only samples from oil-field waters. The correlation coefficients obtained (see "Discussion and Recommendations") using other cation geothermometers are lower than those for the Mg-Li geothermometer.

The least-squares line drawn through Figure 6.3 gives a slope of 2.20 and an intercept of −5.47. The Mg-Li geothermometer for all the data in this study is given by:

$$t_{\mathrm{Mg\text{-}Li}} = \frac{2{,}200}{\log\left(\frac{(C_{\mathrm{Mg}})^{0.5}}{C_{\mathrm{Li}}}\right) + 5.47} - 273 \quad (18)$$

TABLE 6.3. Cation concentrations (mg/L) from selected oil-field and geothermal waters with known subsurface temperatures.

Sample no.	Location	Temperature (°C)	Mg	Ca	Li	Na	K	Reference
Seawater	Average composition	3	1,350	400	0.17	10,500	380	Goldberg (1963)
83-TX-1	High Island, TX	55	869	1,370	2.0	28,400	172	Kharaka et al. (1985)
83-TX-6	do	53	499	679	.6	15,400	110	Do
83-TX-7	do	52	349	850	.6	13,300	103	Do
83-TX-9	do	49	1,010	2,250	2.2	31,800	167	Do
78-AK-54	Prudhoe Bay, AK	94	20	182	4.0	7,600	86	Do
78-AK-55	do	90	16	151	3.3	5,720	68	Do
79-GG-201	Pleasant Bayou #2, TX	170	210	6,500	34	32,100	1,900	Kharaka et al. (1979)
79-GG-204	do	154	660	9,100	39	38,000	840	Do
81-GG-51	Crown Zellerback #2, LA	147	39	460	5.5	11,000	97	Kraemer and Kharaka (1986)
79-GG-50	F.F. Sutter #1, LA	132	670	7,670	19.0	48,300	990	Do
80-GG-1	W. Girouard #1, LA	134	17	115	3.5	8,700	43	Do
79-GG-251	B. Simon #2, LA	141	270	3,090	19.0	32,200	510	Do
77-GG-58	E. Delcambre #1, LA	112	335	2,150	8.2	47,600	305	Do
77-GG-55	do	114	270	1,850	7.2	40,800	260	Do
80-GG-301	L. Koelemay #1, LA	127	6.2	27	1.8	7,280	32	Do
77-GG-117	La Blanca, TX	148	3.3	150	1.2	2,680	46	Kharaka et al. (1980)
77-GG-1	Erath, LA	77	800	2,570	2.1	47,000	200	Kharaka et al. (1978)
77-GG-15	Bayou Sale, LA	159	920	14,800	35	51,700	467	Do
77-GG-21	Weeks Island, LA	101	750	3,560	4.0	57,400	390	Do
WR-2YK74	Wheeler Ridge, CA	56	158	375	1.0	8,550	160	Kharaka (unpublished data)
WR-6YK74	do	50	106	370	1.6	4,200	81	Do
WR-8YK74	do	80	158	2,370	2.0	12,050	112	Do
WR-10YK74	do	89	102	6,300	1.7	11,550	170	Do
81-NSV-2	Black Butte, CA	33	143	392	.19	2,220	12.1	Do
81-NSV-3	do	44	141	319	.30	7,380	28.8	Do
81-NSV-19	Grimes, CA	58	80	274	.36	8,100	40.3	Do
82-SSV-16	Suisun Bay, CA	65	76	130	1.30	5,800	36.5	Do
82-SSV-18	Rio Vista, CA	70	26	121	.44	5,250	43.5	Do
82-SSV-22	River Island, CA	49	50	215	.52	2,450	35	Do

TABLE 6.3. *Continued.*

Sample no.	Location	Temperature (°C)	Mg	Ca	Li	Na	K	Reference
82-SSV-26	do	83	80	163	.37	2,650	65	Do
82-SSV-23	Lindsey Slough, CA	92	35	322	.94	7,400	83	Do
82-SSV-28	Union Island, CA	99	40	165	1.69	6,000	14.8	Do
82-SSV-3	do	66	49	100	.42	4,000	29	Do
82-SSV-4	do	71	17.8	76	.59	4,300	37.3	Do
82-SSV-5	Winters, CA	48	48.5	120	.32	4,550	25.5	Do
82-SSV-6	Lindsay Slough, CA	76	60	279	.62	4,600	76.5	Do
912-72	Kettleman North Dome, CA	81	206	797	1.43	13,200	88.4	Kharaka and Berry (1974)
912-97	do	98	109	1,650	2.41	12,200	174	Do
32-32J	do	101	169	2,100	1.19	9,300	183	Do
912-45	do	122	2.43	88	3.26	3,800	66.5	Kharaka and Berry (1976)
912-2	do	136	3.42	30.7	3.05	3,760	94.2	Do
76-GG-2	Chocolate Bayou, TX	94	60	280	3.3	15,250	120	Do
76-GG-3	do	98	40	180	3.5	15,750	110	Do
76-GG-5	do	103	30	130	3.7	16,500	120	Do
76-GG-1	do	102	35	160	3.2	14,000	90	Do
76-GG-24	Halls Bayou, TX	150	170	1,800	15	20,500	180	Do
76-GG-25	do	138	185	1,600	13	17,750	240	Do
76-GG-29	Alta Loma, TX	119	90	700	6.0	19,500	190	Kharaka et al. (1977)
76-GG-51	White Point East, TX	61	445	2,500	5.1	29,500	215	Do
76-GG-52	do	46	475	3,040	2.6	22,750	77	Do
76-GG-54	do	68	340	2,880	5.4	25,250	125	Do
PM2a	Paris Basin, France	80	141	581	1.39	4,090	74.3	Bouléque (1978)
CH 101	do	72	119	373	1.11	3,100	54.7	Do
CGEH (9)	Coso, CA	205	.57	51	14.0	1,480	132	Fournier et al. (1980)
Well #1	El Tatio, Chile	211	1.1	270	30.2	4,480	420	Cusicanqui et al. (1975)
Well #3	do	253	2.08	268	31.1	3,512	168	Do
Well #4	do	229	3.7	228	28.0	4,537	193	Do
Well #6	do	180	1.3	99	17.1	1,900	111	Do
Well #5 (666)	Cerro Prieto, Mexico	295	.50	505	22.9	8,000	1,900	Mañon et al. (1977)

TABLE 6.3. *Continued.*

Sample no.	Location	Temperature (°C)	Mg	Ca	Li	Na	K	Reference
M-25	do	256	.21	260	14.2	4,750	1,090	Fausto et al. (1979); Truesdell et al. (1979)
M-29	do	260	.85	314	15.5	4,830	850	Do
M-19a	do	304	.24	301	15.7	5,240	1,170	Do
Well #2	Broadlands, New Zealand	260	.1	2.2	11.7	1,050	210	Mahon and Finlayson (1972)
Well #3	do	281	.8	3.0	12.2	1,045	213	Do
Well #9	do	294	.35	2.0	12.7	930	203	Do
Well #9	Otake, Japan	248	.19	12.3	5.2	936	131	Koga (1970)
Well #7	do	230	.025	9.9	4.5	846	105	Do
Well #1	Hatchobaru, Japan	300	.158	9.9	11.1	1,396	289	Do
Well #1	Salton Sea, CA	340	54	28,000	215	50,400	17,500	Muffler and White (1969)
Well #2	do	300	10	28,800	210	53,000	16,500	White (1968)
Utah State 14-2	Roosevelt, UT	268	.28	11	25	2,070	384	Capuano and Cole (1982)
Schmitt Well	Raft River, ID	147	.24	50	1.4	535	22	Nathenson et al. (1982)
7-G-009-81	Calistoga, CA	135	.5	2.0	1.95	206	9	Murray et al. (1985)
272	Klamath Falls, OR	148	.05	35	.38	230	7.4	Janik et al. (1985)
GTO3RM79	Brady #7, NV	154	.1	41	2.0	900	67	Mariner (unpublished data)
GTO6RM79	Brady #8, NV	154	.32	45	1.5	850	36	Do
Y-8	Yellowstone, WY	169	.04	1.2	2.6	360	15	Fournier (1981)
East Mesa 6-1	Imperial Valley, CA	190	15.2	1,020	45	9,130	1,180	Howard et al. (1978)
East Mesa 6-2	do	182	.3	13	4.0	1,400	125	Do
East Mesa 8-11	do	170	1.6	41	2.0	723	42	Do
Reynolds #1	Stillwater, NV	136	1.7	110	1.9	1,500	42	Mariner et al. (1974, 1975)
Chevron 1-29	Carson Desert, NV	199	.36	58	2.3	1,200	130	Olmsted et al. (1984)
Well #8	Kawerau, New Zealand	262	.39	1.1	5.5	740	130	Ellis and Mahon (1977)
Well #137	Rotorua, New Zealand	160	.22	1.0	1.4	565	31	Do
Well #1	Ngauha, New Zealand	225	1.4	29	10.7	900	78	Do
Well #24	Wairakei, New Zealand	250	.04	12	13.2	1,250	210	Do
Well #1A	Kizildere, Turkey	200	.2	2.5	4.5	1,280	135	Do

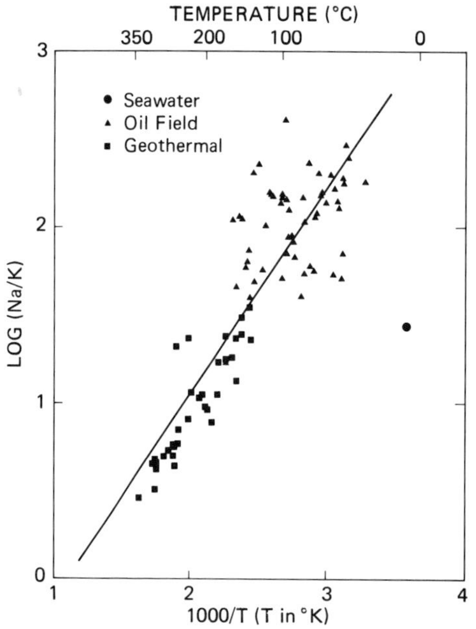

FIGURE 6.4. Sodium/potassium ratios as a function of subsurface temperatures for oil field, geothermal, and seawater. Note the scatter in data for waters from low-temperature environments.

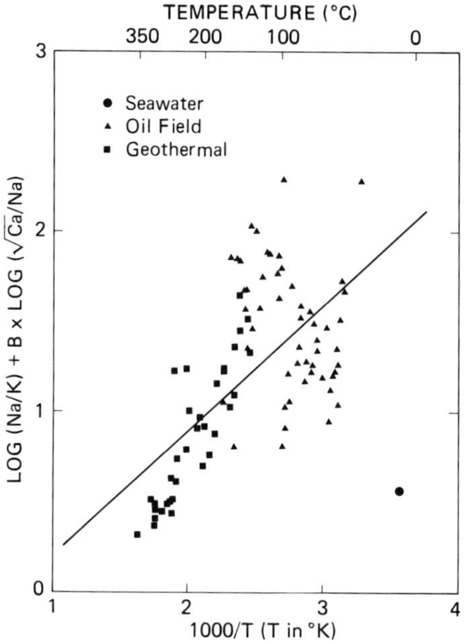

FIGURE 6.5. Sodium/potassium/calcium ratios as a function of subsurface temperatures for oil field, geothermal, and seawater. Note the large scatter in the data for low-temperature environments.

where C is the concentration in mg/L of the subscripted cation. A least-squares line was also drawn through samples from oil-field waters alone; the equation is:

$$t_{\text{Mg-Li}} = \frac{1{,}910}{\log \dfrac{(C_{\text{Mg}})^{0.5}}{C_{\text{Li}}} + 4.63} - 273 \quad (19)$$

Other Cation Geothermometers

The data in Table 6.3 were used further to test the applicability of other chemical geothermometers to oil-field waters. Results are shown in Figures 6.4 to 6.8 and in Table 6.4. These results show that the Na-K geothermometer (Fig. 6.4) cannot be used to calculate reliable subsurface temperatures for oil-field waters. This geothermometer is much more reliable when applied to geothermal systems, especially at temperatures higher than about 150°C.

The Na-K-Ca geothermometer gives totally random results ($r = 0.00$) when applied to oil-field waters. Much better results are obtained ($r = 0.88$) when this geothermometer is applied to geothermal waters alone (Fig. 6.5 and Table 6.4). The Na-K-Ca geothermometer with a Mg correction (Fournier and Potter, 1979) gives excellent results for all waters ($r = 0.95$) (Table 6.4). The calculations of the corrections are somewhat cumbersome, but are critical in the case of low-temperature waters with high concentrations of Mg.

The Mg-corrected Na-K-Ca geothermometer is based on the temperature estimated from the Na-K-Ca geothermometer and a variable R given by:

$$R = \frac{C_{\text{Mg}}}{C_{\text{Mg}} + 0.61 C_{\text{Ca}} + 0.31 C_{\text{K}}} \times 100 \quad (20)$$

where concentrations are expressed in mg/L. For R values from 5 to 50 the correction is given by

$$\begin{aligned} \Delta t_{\text{Mg}} = {} & 10.66 - 4.7415\,R + 325.87\,(\log R)^2 \\ & - 1.032 \times 10^5\,(\log R)^2/T - 1.968 \\ & \times 10^7\,(\log R)^2/T^2 + 1.605 \\ & \times 10^7\,(\log R)^3/T^2 \end{aligned} \quad (21)$$

and for R from 0.5 to 5

TABLE 6.4. Chemical geothermometers recommended for use in formation waters from sedimentary basins.

Geothermometer	Equation (concentrations in mg/L)	Correlation coefficients (r)		Recommendations
		Oil-field waters	All waters	
1. Quartz	$t = \frac{1309}{0.41 - \log (K \cdot pf)} - 273$			
	K and pf from Equations 2 and 4 (see text)			70°C to 250°C
2. Chalcedony	$t = \frac{1032}{-0.09 - \log (K \cdot pf)} - 273$			30°C to 70°C
3. Mg-Li	$t = \frac{2200}{\log(\sqrt{Mg}/Li) + 5.47} - 273$	0.90	0.96	0°C to 350°C
4. Na-K	$t = \frac{1180}{\log(Na/K) + 1.31} - 273$	0.40	0.87	Do not use in oil-field waters
5. Na-K-Ca	$t = \frac{699}{\log(Na/K) + \beta\,[\log(\sqrt{Ca}/Na) + 2.06] + 0.489} - 273$	0.00	0.63	Do not use in oil-field waters
6. Mg-corrected Na-K-Ca	Same as Na-K-Ca (Table 6.1) with Mg-corrections (see text)	0.83	0.95	0°C to 350°C
7. ($\beta = 1/3$)	$t = \frac{1120}{\log(Na/K) + 1/3\,[\log(\sqrt{Ca}/Na) + 2.06] + 1.32} - 273$	0.58	0.86	Use only where no Mg data
8. Na-Li	$t = \frac{1590}{\log(Na/Li) + 0.779} - 273$	0.80	0.91	0°C to 350°C

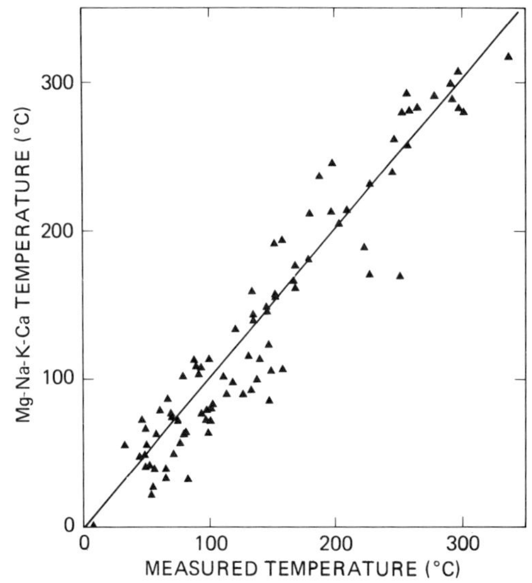

FIGURE 6.6. Temperature calculated using magnesium-corrected Na-K-Ca geothermometer versus measured temperatures for all samples in the study.

$$\Delta t_{Mg} = 1.03 + 59.971 \log R + 145.05 (\log R)^2 - 36711 (\log R)^2/T - 1.67 \times 10^7 \log R/T^2 \quad (22)$$

where Δt_{Mg} is the temperature correction in °C that should be subtracted from the Na-K-Ca temperature and T is the temperature in °K calculated from Na-K-Ca geothermometer. The Mg-corrected Na-K-Ca temperature (t_{Mg}) is given by

$$t_{Mg} = t_{Na\text{-}K\text{-}Ca} - \Delta t_{Mg} \quad (23)$$

The Mg correction is negligible where R values are less than 0.5. No correction is available for waters with R values greater than 50, and the Na-K-Ca geothermometer should not be applied to these waters. Rarely, Δt_{Mg} may be negative; in this case, use the temperature calculated from the Na-K-Ca geothermometer, ignoring the magnesium correction.

Temperatures calculated using Na-K-Ca geothermometer with Mg correction are plotted in Figure 6.6 against the subsurface temperatures (Table 6.3). The least-squares line through all the samples gives an excellent correlation coefficient of 0.95. A greatly improved (compared to no Mg correction) correlation coefficient of 0.83 is obtained when the regression is carried out using oil-field waters alone.

An improved correlation (Fig. 6.7) is obtained with the Na-K-Ca geothermometer when all the samples are plotted using $\beta = 0.333$ instead of $\beta = 1.333$ for the samples with calculated temperatures less than 100°C. A good correlation coefficient of 0.86 (Table 6.4) is obtained for all the waters, but the correlation coefficient obtained for oil-field waters alone is still a poor 0.58.

The Na-Li geothermometers developed independently by Fouillac and Michard (1981) and Kharaka et al. (1982) give better results (Fig. 6.8; Table 6.4) with oil-field waters than with waters from geothermal systems. A good correlation coefficient of 0.80 is obtained with oil-field waters alone using Kharaka et al.'s equation (Table 6.4).

Discussion and Recommendations

Several chemical geothermometers can be used to estimate the subsurface temperatures of formation waters from sedimentary basins. The silica geothermometers are based on rigorous thermodynamic principles and should give the best estimates of subsurface temperatures. Silica geothermometers, however, may be ambiguous because the water may be in equilibrium with quartz or one of its polymorphs. The quartz geothermometer is generally applicable to waters from sedimentary basins with temperatures higher than about 70°C (Kharaka et al., 1977). At temperatures lower than 70°C the chalcedony geothermometer gives the best results for oil-field waters. A pressure correction, given by Equation 4 and corrections for the activity coefficient of $H_4SiO_4{}^0$ (Equation 7) and activity of water (Equation 8 or 9), should be applied to silica geothermometers where hydraulic pressures are higher than 200 bars and salinities are higher than 10,000 mg/L of dissolved solids.

Chemical geothermometers based on the concentrations of cations are all empirical and not related to the solubility of any single mineral. The Mg-Li geothermometer can be applied successfully to low-temperature systems, including the ocean, indicating that relatively fast exchange reactions involving clay minerals may control these ratios. For this chapter, however, we have made no attempt to identify the specific clay mineral(s) responsible for the Mg-Li geothermometer.

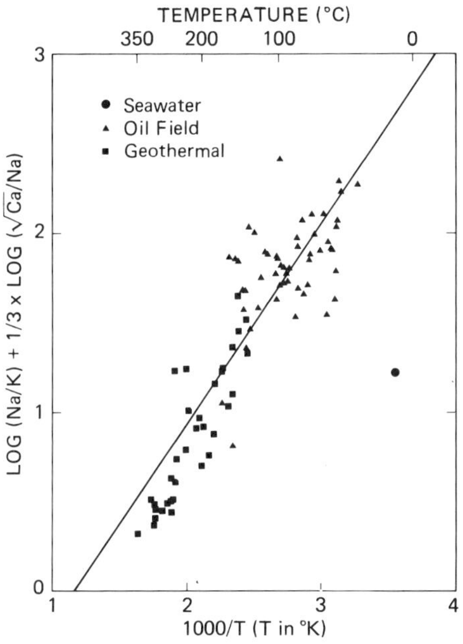

FIGURE 6.7. Sodium/potassium/calcium ratios as a function of subsurface temperature for oil field, geothermal, and seawater. Note the improved correlation obtained for oil-field waters where $\beta = 0.333$ is used as compared to that shown in Figure 6.5.

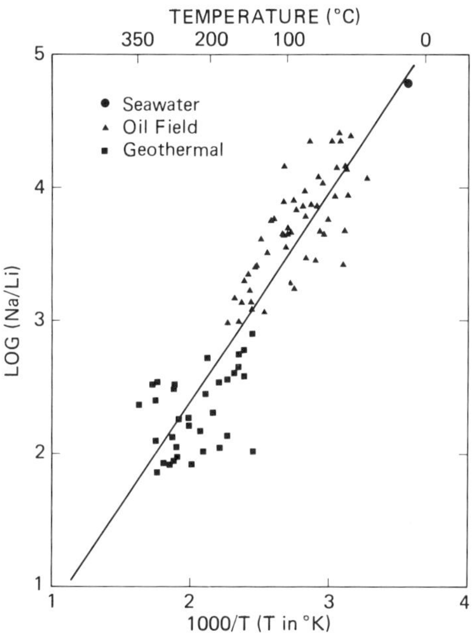

FIGURE 6.8. Sodium/lithium ratios as a function of subsurface temperatures for oil field, geothermal, and seawater.

The Mg-Li geothermometer gives improved results over the Mg-corrected Na-K-Ca geothermometer for both geothermal and oil-field waters; the improvement is dramatic for oil-field waters. The improved results obtained with the Mg-Li geothermometer are for samples obtained from wells. Additional testing is needed to evaluate the application of this geothermometer to samples from hot springs (Table 6.4). Temperatures calculated for oil-field samples (Table 6.3) using Equations 18 and 19 are slightly different. However, exactly the same correlation coefficient ($r = 0.90$) was obtained when these temperatures were regressed against the subsurface temperatures. We suggest that Equation 18 be used for all subsurface waters because Equation 19 cannot be used reliably for geothermal waters with temperatures higher than about 200°C.

The uncorrected Na-K-Ca geothermometer cannot be used to calculate subsurface temperatures lower than about 200°C in sedimentary basins. Improved, but still poor, results are obtained with the Na-K-Ca geothermometer if $\beta = 0.333$ is always used with the coefficients derived in this study (Table 6.4). A regression on temperatures calculated by this procedure and those measured gives a correlation coefficient of only 0.58. However, the use of this geothermometer may be necessary in oil-field waters where Mg and Li concentrations are not available.

The Na-K geothermometer gives poor results for oil-field waters. The Na-Li geothermometer (Table 6.4), on the other hand, gives good results that are comparable to the Mg-corrected Na-K-Ca geothermometer. The Na-Li geothermometer is recommended for use in both oil-field and geothermal waters, but gives slightly better results for oil-field waters (Table 6.4).

The ratios of cations (Table 6.3) were combined in equations other than those shown in Table 6.4 to investigate other possible geothermometers. The best results were obtained by combining the magnesium/lithium ratios with the magnesium/sodium ratios in the relationship

$$\log\left(\frac{(C_{Mg})^{0.5}}{C_{Li}}\right) + 0.25 \log\left(\frac{(C_{Mg})^{0.5}}{C_{Na}}\right)$$

The correlation coefficient obtained for the oil-field samples using the above relationship, however, was lower ($r = 0.85$ versus $r = 0.90$) than that obtained using the magnesium/lithium ratios alone.

Chemical geothermometers should not be applied to chemical analyses of waters from gas wells as these waters are often diluted by condensed water vapor produced with natural gas. Kharaka and associates (1977, 1985) have shown that chemical analysis from many gas wells, especially those from geopressured geothermal systems, are diluted and do not represent the true chemical composition of formation water from that production zone. The criteria to distinguish such samples are given by these authors, but temperatures calculated from these samples are erratic and generally lower than the measured temperatures.

All the chemical geothermometers recommended in this study are incorporated in a modified version of the geochemical model SOLMNEQ (Kharaka and Barnes, 1973). The program computes the activity of H_2O and $H_4SiO_4^0$ and makes the required pressure correction in the case of silica geothermometers. Models such as this can be used to accurately determine subsurface temperatures from the chemical composition of formation water and solubilities of individual minerals known to be present in the aquifer.

Acknowledgments. We would like to thank D.J. Specht who worked with the senior author on an earlier version of the Mg-Li geothermometer. We would also like to thank our colleagues R.O. Fournier, J.D. Hem, N.D. Naeser, and D.K. Nordstrom for a thorough review of this manuscript.

References

Arnórsson, S. 1975. Application of the silica geothermometer in low temperature hydrothermal areas in Iceland. American Journal of Science 275:763–784.

Boulégue, J. 1978. Metastable sulfur species and trace metals (Mn, Fe, Cu, Zn, Cd, Pb) in hot brines from the French Dogger. American Journal of Science 278:1394–1411.

Brook, C.A., Mariner, R.H., Mabey, D.R., Swanson, J.R., Guffanti, M., and Muffler, L.J.P. 1979. Hydrothermal convection systems with reservoir temperatures ≥90°C. In: Muffler, L.J.P. (ed.): Assessment of Geothermal Resources of the United States– 1978. U.S. Geological Survey Circular 790, pp. 18–85.

Capuano, R.M., and Cole, D.R. 1982. Fluid-mineral equilibria in a hydrothermal system, Roosevelt Hot Springs, Utah. Geochimica et Cosmochimica Acta 46:1353–1364.

Carpenter, A.B., Trout, M.L., and Pickett, E.E. 1974. Preliminary report on the origin and chemical evolution of lead- and zinc-rich brines in central Mississippi. Economic Geology 69:1191–1206.

Chen, C.A., and Marshall, W.L. 1982. Amorphous silica solubilities: IV. Behavior in pure water and aqueous sodium chloride, sodium sulfate, magnesium chloride, and magnesium sulfate solutions up to 350°C. Geochimica et Cosmochimica Acta 46:279–287.

Cusicanqui, H., Mahon, W.A.J., and Ellis, A.J. 1975. The geochemistry of the El Tatio geothermal field, Northern Chile. Proceedings of the Second United Nations Symposium on the Development and Use of Geothermal Resources, San Francisco, vol. 1, pp. 703–711.

D'Amore, F., and Panichi, C. 1980. Evaluation of deep temperature of hydrothermal systems by a new gas geothermometer. Geochimica et Cosmochimica Acta 44:549–556.

Drummond, S.E., Jr. 1982. Boiling and mixing of hydrothermal fluids: Chemical effects on mineral precipitation. Ph.D. thesis, Pennsylvania State University, University Park, PA, 380 pp.

Ellis, A.J. 1970. Quantitative interpretation of chemical characteristics of geothermal systems. Geothermics 2:516–528.

Ellis, A.J., and Golding, R.M. 1963. The solubility of carbon dioxide above 100°C in water and in sodium chloride solutions. American Journal of Science 261: 47–60.

Ellis, A.J., and Mahon, W.A.J. 1977. Chemistry and Geothermal Systems. New York, Academic Press, 392 pp.

Fausto, J.J., Sanchez, A.A., Jimenez, M.E.S., Esquer, I.P., and Ulloa, F.H. 1979. Hydrothermal geochemistry of the Cerro Preito geothermal field. Second Symposium on the Cerro Prieto Geothermal Field, Baja California, Mexico, pp. 199–223.

Fouillac, C., and Michard, G. 1981. Sodium/lithium ratio in water applied to geothermometry of geothermal reservoirs. Geothermics 10:55–70.

Fournier, R.O. 1973. Silica in thermal water: Laboratory and field investigations. In: Proceedings of the International Symposium on Hydrogeochemistry and Biogeochemistry, Japan, 1970: Vol. 1. Washington, DC, The Clark Company, pp. 122–139.

Fournier, R.O. 1979. A revised equation for the Na/K geothermometer. Geothermal Resources Council Transactions 3:221–224.

Fournier, R.O. 1981. Application of water chemistry to geothermal exploration and reservoir engineering. In:

Rybach, L., and Muffler, L.J.P. (eds.): Geothermal Systems: Principles and Case Histories. New York, Wiley, pp. 109–143.

Fournier, R.O., and Potter, R.W., II. 1979. A magnesium correction for the Na-K-Ca geothermometer. Geochimica et Cosmochimica Acta 43:1543–1550.

Fournier, R.O., and Potter, R.W., II. 1982. An equation correlating the solubility of quartz in water from 25° to 900°C at pressures up to 10,000 bars. Geochemica et Cosmochimica Acta 46:1969–1973.

Fournier, R.O., and Rowe, J.J. 1966. Estimation of underground temperatures from the silica content of water from hot springs and wet-steam wells. American Journal of Science 264:685–697.

Fournier, R.O., Thompson, J.M., and Austin, C.F. 1980. Interpretation of chemical analyses of waters collected from two geothermal wells at Coso, California. Journal of Geophysical Research 85:2405–2410.

Fournier, R.O., and Truesdell, A.H. 1973. An empirical Na-K-Ca chemical geothermometer for natural waters. Geochemica et Cosmochimica Acta 37:1255–1275.

Fournier, R.O., and Truesdell, A.H. 1974. Geochemical indicators of subsurface temperatures: Part 2. Estimation of temperature and fraction of hot water mixed with cold water. Journal of Research of the U.S. Geological Survey 2:263–270.

Fournier, R.O., White, D.E., and Truesdell, A.H. 1974. Geochemical indicators of subsurface temperatures: Part 1. Basic assumptions. Journal of Research of the U.S. Geological Survey 2:259–262.

Garrels, R.M., and Christ, C.L., 1965. Solutions, Minerals, and Equilibria. New York, Harper & Row, 450 pp.

Giggenbach, W.F. 1980. Geothermal gas equilibria. Geochimica et Cosmochimica Acta 44:2021–2032.

Goldberg, E.D. 1963. Chemistry—the oceans as a chemical system. In: Hill, M.N. (ed.): The Sea: Vol. 2. Composition of Seawater, Comparative and Descriptive Oceanography. New York, Interscience, pp. 3–25.

Helgeson, H.C. 1969. Thermodynamics of hydrothermal systems at elevated temperatures and pressures. American Journal of Science 267:729–804.

Helgeson, H.C., Brown, T.H., Nigrini, A., and Jones, T.A. 1970. Calculation of mass transfer in geochemical processes involving aqueous solutions. Geochimica et Cosmochimica Acta 34:569–592.

Henley, R.W., Truesdell, A.H., and Barton, P.B., Jr. 1984. Fluid-mineral equilibria in hydrothermal systems. Reviews in Economic Geology: Vol. 1. Chelsea, MI, Society of Economic Geologists, 267 pp.

Hitchon, B. 1985. Geothermal gradients, hydrodynamics, and hydrocarbon occurrences, Alberta, Canada. American Association of Petroleum Geologists 68: 713–743.

Hitchon, B., Billings, G.K., and Klovan, J.E. 1971. Geochemistry and origin of formation waters in the western Canada sedimentary basin: III. Factors controlling chemical composition. Geochimica et Cosmochimica Acta 35:567–598.

Holmes, H.F., Baes, C.F., Jr., and Mesmer, R.E. 1978. Isopiestic studies of aqueous solutions at elevated temperatures: I. KCl, $CaCl_2$, and $MgCl_2$. Journal of Chemical Thermodynamics 10:983–996.

Holmes, H.F., Baes, C.F., Jr., and Mesmer, R.E., 1981. Isopiestic studies of aqueous solutions at elevated temperatures: III. $(1\text{-}\gamma)$ NaCl + $\gamma CaCl_2$. Journal of Chemical Thermodynamics 13:101–113.

Howard, J.H., Apps, J.A., Benson, S.M., Goldsten, N.E., Graf, A.N., Haney, J.P., Jackson, D.D., Juprasert, S., Majer, E.L., McEdward, D.G., McEvilly, T.V., Narasimhan, T.N., Schechter, B., Schroeder, R.C., Taylor, P.C., van de Kamp, P.C., and Wolery, T.J. 1978. Geothermal resource and reservoir investigations of U.S. Bureau of Reclamation Leasehold at East Mesa, Imperial County, California. Berkeley, CA, Lawrence Berkeley Laboratory, University of California, Report LBL-7094, 305 pp.

Janik, C.J., Truesdell, A.H., Sammel, E.A., and White, A.F. 1985. Chemistry of low-temperature geothermal waters at Klamath Falls, Oregon. Geothermal Resources Council Transactions 9(1):325–331.

Kharaka, Y.K., and Barnes, I. 1973. SOLMNEQ: Solution-mineral equilibrium computations. Springfield, VA, U.S. Department of Commerce, NTIS Report PB 215-899, 81 pp.

Kharaka, Y.K., and Berry, F.A.F. 1974. The influence of geological membranes on the geochemistry of subsurface waters from Miocene sediments at Kettleman North Dome, California. Water Resources Research 10:313–327.

Kharaka, Y.K., and Berry, F.A.F. 1976. The influence of geological membranes on the geochemistry of subsurface waters from Eocene sediments at Kettleman North Dome, California: An example of effluent-type waters. In: Cadek, J., and Paces, T. (eds.): Proceedings of the International Symposium on Water-Rock Interaction, Prague, 1974. Prague, Czechoslovakia, The Geological Survey, pp. 268–277.

Kharaka, Y.K., Brown, P.M., and Carothers, W.W. 1978. Chemistry of waters in the geopressured zone from coastal Louisiana: Implications for the geothermal development. Geothermal Resources Council Transactions 2:371–374.

Kharaka, Y.K., Callender, E., and Carothers, W.W. 1977. Geochemistry of geopressured geothermal waters from the Texas Gulf Coast. Proceedings, Third Geopressured-Geothermal Energy Conference, University of Southwestern Louisiana, Lafayette, Louisiana: Vol. 1, pp. G1121–G1165.

Kharaka, Y.K., Hull, R.W., and Carothers, W.W. 1985. Water-rock interactions in sedimentary basins. In: Relationship of Organic Matter and Mineral Diagenesis. Society of Economic Paleontologists and Mineralogists Short Course 17. Center for Energy Studies, University of Southwestern Louisiana, pp. 79–176.

Kharaka, Y.K., Lico, M.S., and Carothers, W.W. 1980. Predicted corrosion and scale-formation properties of geopressured-geothermal waters from the northern Gulf of Mexico Basin. Journal of Petroleum Technology 32:319–324.

Kharaka, Y.K., Lico, M.S., and Law, L.M. 1982. Chemical geothermometers applied to formation waters, Gulf of Mexico and California basins (abst.). American Association of Petroleum Geologists Bulletin 66:588.

Kharaka, Y.K., Lico, M.S., Wright, V.A., and Carothers, W.W. 1979. Geochemistry of formation waters from Pleasant Bayou No. 2 well and adjacent areas in coastal Texas. In: Dorfman, N.H., and Fisher, W.L. (eds.): Proceedings, Fourth United States Gulf Coast Geopressured-Geothermal Energy Conference: Research and Development. Austin, TX, University of Texas at Austin, pp. 178–193.

Kharaka, Y.K., Maest, A.S., Fries, T.L., Law, L.M., and Carothers, W.W. 1986. Geochemistry of lead and zinc in oil field brines: Central Mississippi Salt Dome basin revisited. Proceedings of Conference on the Genesis of Stratiform Sediment-Hosted Pb-Zn Deposits, Stanford University, California, pp. 50–54.

Koga, A. 1970. Geochemistry of the waters discharged from drillholes in the Otake and Hatchobaru Areas. Geothermics (Special Issue 2), 2(2):1422–1425.

Kraemer, T.F., and Kharaka, Y.K. 1986. Uranium geochemistry in U.S. Gulf Coast geopressured-geothermal systems. Geochimica et Cosmochimica Acta 50:1440–1455.

Mahon, W.A.J., and Finlayson, J.B. 1972. The chemistry of the Broadlands geothermal area, New Zealand. American Journal of Science 232:48–68.

Mañon, A.M., Mazor, E., Jimenez, M.E.S., Sanchez, A.A., Faustu, J.J., and Zenizo, C. 1977. Extensive geochemical studies in the geothermal field of Cerro Prieto, Mexico. Berkeley, CA, Lawrence Berkeley Laboratory, University of California, Report LBL-7019, 113 pp.

Mariner, R.H., Brook, C.A., Reed, M.J., Bliss, J.D., Rapport, A.L., and Lieb, R.J. 1983. Low-temperature geothermal resources in the western United States. In: Reed, M.J. (ed.): Assessment of Low-Temperature Geothermal Resources of the United States—1982. U.S. Geological Survey Circular 892, pp. 31–50.

Mariner, R.H., Presser, T.S., Rapp, J.B., and Willey, L.M. 1975. The minor and trace elements, gas and isotope compositions of the principal hot springs of Nevada and Oregon. U.S. Geological Survey Open-File Report, 27 pp.

Mariner, R.H., Rapp, J.B., Willey, L.M., and Presser, T.S. 1974. Chemical composition and estimated minimum thermal reservoir temperatures of the principal hot springs of northern and central Nevada. U.S. Geological Survey Open-File Report, 32 pp.

Marshall, W.L. 1980. Amorphous silica solubilities: III. Activity coefficient relations and predictions of solubility behavior in salt solutions, 0–350°C. Geochemica et Cosmochimica Acta 44:925–931.

McKenzie, W.F., and Truesdell, A.H. 1977. Geothermal reservoir temperatures estimated from the oxygen isotope compositions of dissolved sulfate and water from hot springs and shallow drill holes. Geothermics 5:51–62.

Meyer, H.J., and McGee, H.W. 1985. Oil and gas fields accompanied by geothermal anomalies in Rocky Mountain region. American Association of Petroleum Geologists Bulletin 69:933–945.

Mizutani, Y. 1972. Isotope composition and underground temperature of the Otake geothermal water, Kyushu, Japan. Geochemical Journal 6:67–73.

Muffler, L.J.P., and White, D.E. 1969. Active metamorphism of upper Cenozoic sediments in the Salton Sea geothermal field and the Salton Trough, southeastern California. Geological Society of America Bulletin 80:157–182.

Murray, K.S., Jonas, M.L., and Lopez, C.A. 1985. Geochemical exploration of the Calistoga geothermal resource area, Napa Valley, California. Geothermal Resources Council Transactions 9(1):339–344.

Nathenson, M., Nehring, N.L., Crosthwaitie, E.G., Harmon, R.S., Janik, C.J., and Borthwick, J. 1982. Chemical and light-stable isotope characteristics of waters from the Raft River Geothermal Area and environs, Cassia County, Idaho; Box Elder County, Utah. Geothermics 11(4):215–237.

Nehring, N.L., and D'Amore, F. 1984. Gas chemistry and thermometry of the Cerro Prieto, Mexico, geothermal field. Geothermics 13:75–89.

Olmstead, F.H., Welch, A.H., Van Denburgh, A.S., and Ingebritsen, S.E. 1984. Geohydrology, aqueous geochemistry, and thermal regime of the Soda Lakes and Upsal Highback geothermal systems, Churchill County, Nevada. U.S. Geological Survey Water Resources Investigations Report 84-4054, 166 pp.

Ovnatanov, S.T., and Tamrazyan, G.P. 1970. Thermal studies in subsurface structural investigations, Apseron Peninsula, Azerbaijan, USSR. American Association of Petroleum Geologists Bulletin 54:1677–1685.

Pitzer, K.S. 1981. Characteristics of very concentrated aqueous solutions. In: Rickard, D.T., and Wickman, F.E. (eds.): Chemistry and Geochemistry of Solutions at High Temperatures and Pressures. Physics and

Chemistry of the Earth: Vol. 13-14. New York, Pergamon Press, pp. 249–272.

Sakai, H., and Matsubaya, O. 1974. Isotope geochemistry of the thermal waters of Japan and its bearing on the Kuroko ore solutions. Economic Geology 69: 974–991.

Staples, B.R., and Nuttall, R.L. 1977. The activity and osmotic coefficients of aqueous calcium chloride at 298.15°K. Journal of Physical Chemistry Reference Data 6:385–407.

Truesdell, A.H. 1976. Geochemical techniques in exploration, summary of section III. Proceedings of the Second United Nations Symposium on the Development and Use of Geothermal Resources, San Francisco: Vol. 1. Berkeley, CA, University of California, pp. 53–78.

Truesdell, A.H., Thompson, J.M., Coplen, T.B., Nehring, N.L., and Janik, C.J. 1979. The origin of Cerro Prieto geothermal brine. Second Symposium on the Cerro Prieto Geothermal Field, Baja California, Mexico, pp. 224–240.

White, D.E. 1965. Saline waters of sedimentary rocks. In: Young, A., and Galley, G.E. (eds.): Fluids in Subsurface Environments. American Association of Petroleum Geologists Memoir 4, pp. 342–366.

White, D.E. 1968. Environments of generation of some base-metal ore deposits. Economic Geology 63: 301–335.

7
Paleotemperatures from Fluid Inclusions: Advances in Theory and Technique

Robert C. Burruss

Abstract

Fluid inclusions in diagenetic minerals can be used to determine paleotemperatures. Three sets of observations are necessary to make accurate interpretations: 1) detailed petrography to establish the relative time of formation of the inclusions, 2) careful analysis of the burial and tectonic history of the host rocks to tie diagenesis to the geologic history of the basin, and, finally, 3) analysis of the phase behavior and chemical composition of individual inclusions to define the pressure-volume-temperature (PVT) properties of the trapped fluids.

Once these observations are complete, two major limitations remain in the interpretation of paleotemperature. First is the assumption that the inclusions have not altered in composition or volume since entrapment. Recently published work shows that inclusions can re-equilibrate, but the extent to which that process affects most observations in sediments is unknown. Second, an independent measure of "paleopressure" during inclusion formation is necessary to distinguish whether the pressure was hydrostatic or approached lithostatic. Data from coexisting hydrocarbon and aqueous fluid inclusions in core samples from the Mission Canyon Limestone (Mississippian), Williston basin, North Dakota, illustrate a method for independently determining *both* paleotemperature and paleopressure from a single set of fluid inclusion measurements. The technique requires petrographic evidence for simultaneous trapping of two immiscible fluids. The PVT properties of such coexisting fluids require that the isochores for the two different fluids intersect at the temperature *and* pressure of entrapment of the inclusions. Calculation of the PVT properties of each fluid is based on detailed chemical analyses of both fluids. Recent results from new analytical techniques—especially capillary column gas chromatography to analyze hydrocarbon inclusions and laser Raman spectroscopy to analyze gases in aqueous inclusions—demonstrate that paleotemperature studies can be widely applicable in sedimentary environments.

Introduction

The measurement of temperatures using fluid inclusions is straightforward. Accurate interpretation of these measurements to yield paleotemperatures is not always easy. This chapter describes the basic theory for interpreting inclusion temperatures in sedimentary rocks. The topics include discussions of the two most important limitations on the technique, methods overcoming these effects, and a discussion of two new analytical techniques for obtaining the composition of inclusions. The basic procedures and instrumentation to make temperature measurements on inclusions are reviewed by Hollister and Crawford (1981), Roedder (1984), and Shepherd et al. (1985) and are not discussed here. The goal of this chapter is to demonstrate that study of fluid inclusions in sedimentary rocks is a technically and theoretically well-founded method for determining paleotemperatures. Furthermore, advances in understanding the properties of subsurface fluids and new analytical techniques will allow this method to be widely applied in studies of the thermal history of sedimentary basins.

Accurate interpretation of temperature measurements on fluid inclusions requires three types of data. First, detailed petrography of the inclusions, including UV (ultraviolet, 365 nm) fluorescence observations, is necessary to establish the relative time of trapping of the inclusions in the diagenetic history of the sediments (Burruss, 1981b). Second, the burial and tectonic history of the rocks containing the inclusions must be known so they can be understood in light of the geological history of the sediments (Burruss et al., 1983, 1985). Third, the phase behavior, composition, and pressure-volume-temperature (PVT) properties of the fluids trapped in the inclusions must be understood. Fluids in sedimentary rocks are complex mixtures of organic, inorganic, and aqueous components, making this last piece of information difficult to obtain in many cases. Some of the problems involved are discussed by Hanor (1980), Burruss (1981a, 1981b), and Roedder (1984).

The first two types of data simply reflect good practice in sedimentary petrology and geology. This background information is necessary before application of any geothermometer, not just fluid inclusion studies. The fluid composition and PVT data are specifically needed for fluid inclusion techniques. These properties control not only the basic interpretation of observations on inclusions, but also whether the inclusion techniques can be applied. There are distinct sets of subsurface pressure and temperature conditions during basin evolution that cannot be analyzed by inclusion techniques.

The composition and PVT properties of fluids also control the two basic limitations on the determination of paleotemperatures. The first limitation is that there must be an independent measure of pressure at the time of inclusion trapping. The second is the extent to which inclusions have physically or chemically reequilibrated to temperature conditions different from those at trapping.

These aspects of inclusion measurements are most easily discussed by reviewing the five assumptions basic to inclusion studies and considering how these assumptions apply to a simple aqueous fluid. The assumptions (Roedder and Bodnar, 1980) are:

1. The inclusion traps a one-phase fluid.
2. The volume of the inclusion does not change after trapping.
3. The composition of the trapped fluid does not change nor are any substances gained or lost after trapping.
4. The effects of pressure are known.
5. The time and mechanism of trapping are known.

Assumption 1 allows the properties of the inclusion to be representative of a subsurface fluid. It also allows determination of a *minimum* trapping temperature. Most inclusions contain two or more phases at room temperature. If the inclusion trapped a one-phase fluid, the temperature measured in the laboratory at which the last additional phase disappears, causing the inclusion to become single phase (the homogenization temperature, T_h), is the minimum trapping temperature. There are examples of two-phase trapping that violate this assumption (Roedder, 1984), but in general it is valid.

When Assumption 2 is valid, any temperature measurement on an inclusion is meaningful, and the homogenization temperature can be used to calculate a paleotemperature. Natural subsurface processes that change the volume of an inclusion will cause the calculations of paleotemperature to yield temperatures different from conditions at trapping. These processes are discussed below.

The third assumption, that the composition of the inclusion is constant and nothing is gained or lost after trapping, overlaps with Assumption 2. If the composition (total number of all different molecules in the inclusion) is constant, and the volume of the inclusion is constant, then the fluid trapped in the inclusion has a fixed molar volume (volume per Avogadro's number of molecules) or constant density (mass per unit volume).

The fourth assumption, that the effects of pressure are known, allows calculation of the trapping temperature from the homogenization temperature. This calculation is commonly called the pressure correction. It is critical information for inclusions in sedimentary rocks. At depths below 1,600 to 3,200 m (5,000 to 10,000 ft), the pressure can vary from hydrostatic (in a few cases, less than hydrostatic, "subnormal") to, more commonly, greater than hydrostatic, even as much as the lithostatic load. This variation can cause significant differences in the interpretation of paleotemperatures. In other words, knowledge of the paleopressure at the time of inclusion formation is needed to determine a paleotemperature.

The final assumption points out the need for thorough petrographic study of the inclusions before any paleotemperature measurements are attempted. This can be restated as the fact that, without knowledge of the time during basin evolution when an inclusion was trapped, interpretation of paleotemperature is difficult. The question of mechanism of trapping is principally related to whether the inclusions are primary (trapped during crystal growth) or secondary (trapped after crystal growth along healed microfractures). Roedder (1984, Chapters 1 and 2) discusses these problems thoroughly. Examples of petrographic evidence for the time of formation of inclusions in sedimentary rocks are shown by Burruss (1981b) and Burruss et al. (1983, 1985).

The physical basis for the first three assumptions can be illustrated by the PVT properties of pure water as shown in Figure 7.1. The specific volume of water for coexisting liquid and vapor as a function of temperature is shown in Figure 7.1A. Also shown is a line of constant volume corresponding to a hypothetical fluid inclusion for which all the previous assumptions are true. At room temperature (25°C), the inclusion consists of liquid and a small vapor bubble in proportions indicated by the tie line shown. As the inclusion is heated, the relative amounts of liquid and vapor change. The volume of the vapor bubble decreases until, at 200°C, the bubble vanishes. This temperature is the homogenization temperature.

The consequences of violating the first three assumptions are best illustrated by an enlarged view (Fig. 7.1B) of the liquid side of the diagram in Figure 7.1A. Figure 7.1B shows the behavior of a hypothetical inclusion with a specific volume given by Line B. With all assumptions valid, this inclusion will homogenize to liquid at 135°C. However, if by some process the volume of this inclusion increases by 5%, homogenization will occur along Line C at 179°C. If material is lost from the inclusion but the volume remains the same, this would be equivalent to the case of volume expansion and the homogenization temperatures would increase. On the other hand, if the volume of the original inclusion B decreased by 5%, then homogenization would occur at a lower temperature, 75°C, along Line A. This would be equivalent to *adding* material to the original inclusion while holding its volume constant.

Of these processes, the one with greatest geological significance is the increase in volume. By con-

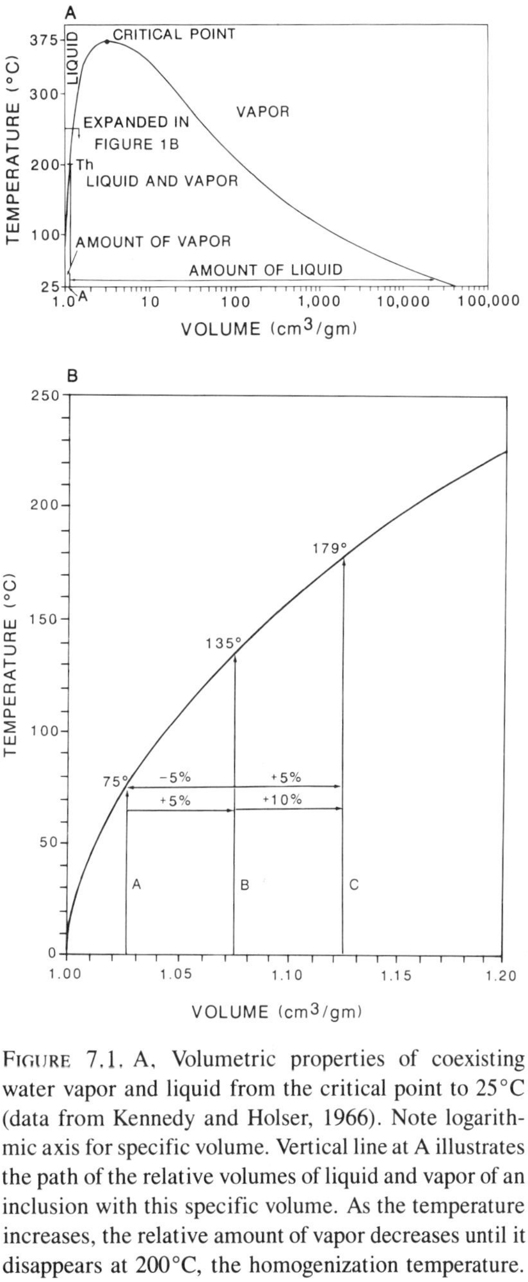

FIGURE 7.1. A, Volumetric properties of coexisting water vapor and liquid from the critical point to 25°C (data from Kennedy and Holser, 1966). Note logarithmic axis for specific volume. Vertical line at A illustrates the path of the relative volumes of liquid and vapor of an inclusion with this specific volume. As the temperature increases, the relative amount of vapor decreases until it disappears at 200°C, the homogenization temperature. The liquid side of the two-phase field is expanded in B. B, enlarged view of the volumetric properties of liquid water coexisting with vapor. The volumetric behavior of three hypothetical inclusions (A, B, and C) is shown by vertical lines of constant volume. Horizontal lines illustrate effects of changing inclusion volume after entrapment as discussed in the text.

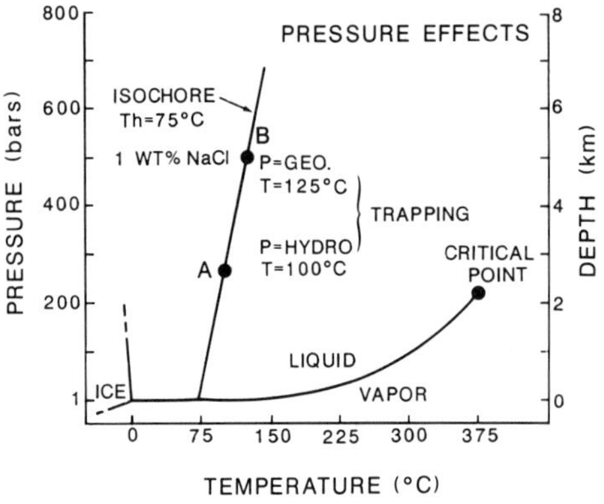

FIGURE 7.2. The effect of pressure on the determination of paleotemperatures from fluid inclusions. An inclusion containing water with 1 weight % NaCl that homogenizes at 75°C will have a trapping temperature (paleotemperature) of 100°C (shown at A), if the pressure gradient was hydrostatic. However, if trapping occurred under a geostatic pressure gradient, the paleotemperature will be 125°C (shown at B).

sidering a hypothetical inclusion corresponding to Line A in Figure 7.1B, the consequences for inclusion studies of sedimentary rocks will be obvious. This inclusion homogenizes at 75°C. If, for some reason, the volume of the inclusion increases, (i.e., the inclusion "stretches" by 5%), then T_h increases to 135°C. If the volume increase is 10%, then T_h goes up to 179°C, more than 100°C higher than the original T_h! Clearly, if some process is operating that can "stretch" inclusions, accurate interpretation of paleotemperatures from T_h measurements will be difficult.

Paleotemperature Determinations

Using this knowledge of the homogenization temperature measurement and the effects of changing the molar volume of the trapped fluid, the trapping temperature (T_t) can be determined from the homogenization temperature. Figure 7.2 is a pressure-temperature (PT) projection of the PVT properties of an aqueous solution of 1 weight % NaCl (data from Potter and Brown, 1977). An inclusion with this composition, homogenizing to liquid at 75°C, will have the isochore shown and could have formed at any set of P and T conditions along this line. If there is independent evidence that this inclusion formed at 2.7 km (8,230 ft) depth under a normal hydrostatic gradient (≈ 100 bars/km or ≈ 0.5 psi/ft), then the trapping temperature of the inclusion is defined by Point A, 100°C. However, even if there is evidence of trapping at this depth (e.g., at maximum depth of burial), there remains the problem of knowing the *pressure* at this depth. In the absence of evidence of a hydrostatic pressure gradient, the pressure during inclusion formation could approach geostatic (e.g., 500 bars; 7,350 psi) at this depth, causing the temperature of trapping to increase to 125°C. Clearly, accurate estimates of paleotemperatures (trapping temperatures) require an independent method of estimating pressure at the time of trapping of the inclusion. One method for dealing with this problem is discussed in a later section. The generic problems of making pressure estimates from inclusions are discussed by Roedder and Bodnar (1980).

Reequilibration

A serious problem with inclusion measurements is the possibility of reequilibration to new PT conditions. The most common type of reequilibration is known as "stretching" and is discussed in detail by Bodnar and Bethke (1984). This problem is especially significant in studies of sediments because it may affect inclusions formed early in burial history that are then subjected to higher temperature conditions during continued subsidence. The physical principles involved in stretching are shown in Figure 7.3 using the same hypothetical inclusion illustrated in Figure 7.2. In this example, the original inclusion homogenizes to liquid at 75°C in a diagenetic cement that formed at 2.7 km (8,230 ft) depth, 270 bars (3,970 psi), and 100°C. As the rock containing this inclusion continues to subside along a constant geothermal gradient appropriate for the initial conditions (30°C/km, 100 bars/km), the pressure inside the inclusion increases dramatically over the external pore-fluid pressure on the cementing phase. By the time the rock has subsided an additional 2 km (as shown in Fig. 7.3), the internal pressure is about 400 bars (5,880 psi) greater than the external pressure. Depending on the mechanical strength of the host mineral at these pressure and temperature conditions, the host will either plastically deform ("stretch") or brittle fracture. If the host stretches to the extent that the internal pressure equals the external pressure, then the inclusion will be "reset" to the new pressure and temperature conditions defined by Point B. The inclusion will no longer homogenize at the temperature (75° C) corresponding to its density at original trapping but will homogenize at a higher temperature (at least 100°C, and possibly as high as 179°C for the example in Fig. 7.1). Thus, despite petrographic evidence that the cement containing this inclusion is a relatively "early" cement, it will only yield temperatures corresponding to some later stage of deeper burial. The difference between internal and external pressures also may cause the inclusion to decrepitate, fracturing the host mineral. If this fracture extends to the grain boundary, material can be gained or lost by the inclusion. Not only will the fluid density be reset to a new pressure and temperature conditions, but the composition will be altered as well, possibly to the extent that the original fluid is completely replaced by a new, later fluid.

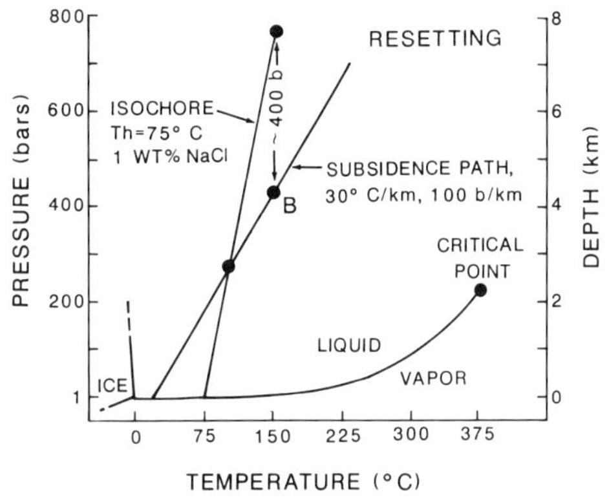

FIGURE 7.3. The process of natural resetting of fluid inclusion temperatures shown with the same hypothetical inclusion discussed in Figure 7.2. This inclusion will develop an internal pressure greater than a hydrostatic pore fluid pressure if the host rock continues to subside along the PT path shown. After burial by about 1.7 km of additional sediment, the excess internal pressure will be about 400 bars. This differential stress may cause the inclusion to either "stretch" or decrepitate.

Thus, there are two major types and consequences of reequilibration. First is the case of simple stretching, where the volume changes but the composition of the fluid does not. This is essentially a physical reequilibration causing a change in density of the inclusion. Second is decrepitation, where the chemical composition of the fluid changes as well as the density. There are variations on these themes involving gain or loss of components by diffusion through the host mineral (Kelly and Turneaure, 1970; Burruss, 1977) or possibly by chemical reactions within the inclusion resulting in changes in the amounts of the original components (e.g., cracking of heavy hydrocarbons to methane) (Hanor, 1980). Although these effects may be important in some cases, they are second-order effects compared with the two basic reequilibration mechanisms discussed above.

Although no single quantitative model predicts the occurrence and extent of reequilibration, there are some reasonably well-founded experimental and empirical observations to keep in mind. Stretching requires a significant differential between internal and external pressure. During basin subsidence, it will only occur when there is a significant difference between the slopes of the isochore for an inclusion and the slope of the pressure-temperature path of the external pore fluid. Thus, as the slope of the isochore approaches that of the PT path of the pore fluid, the change in the differential pressure approaches zero regardless of any increase in temperature with continued subsidence.

In general, aqueous brines at low temperature (<150°C) have steep isochores compared with typical sedimentary PT paths (Barker, 1972). Inclusions trapping such fluids will tend to stretch, whereas many hydrocarbon-rich fluids are more compressible and have isochores with lesser slopes, closer to sedimentary PT paths. This difference in behavior between aqueous and hydrocarbon inclusions will be made clearer in the discussion of hydrocarbon inclusions later in this chapter.

Other important variables in the reequilibration process are discussed by Bodnar and Bethke (1984). The most important are the size of the inclusion and Moh's scale hardness of the host mineral. The empirical observations are that the smaller the inclusion, the more resistant it is to stretching (Roedder, 1984, pp. 70–95). Also, evidence suggests that the harder the mineral, the more resistant it is to stretching. Soft minerals like halite, gypsum, anhydrite, and barite are most likely to stretch, and quartz is least likely (Bodnar and Bethke, 1984). Leroy (1979) presented evidence that small inclusions, 2 to 5 μm in largest dimension, in quartz can maintain differential pressures of as much as 2 kbars (29,000 psi) without stretching or fracturing the host. This relation is not well quantified and should only be used as a guide because the strength of any host mineral

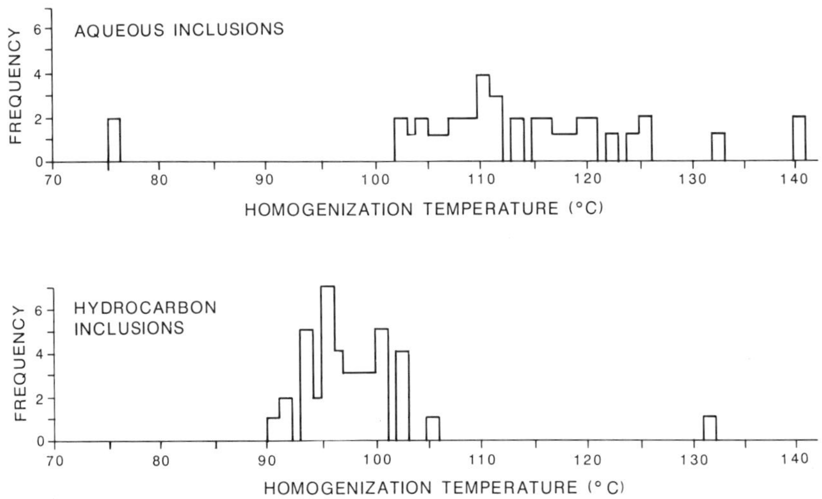

FIGURE 7.4. Homogenization temperatures of aqueous and hydrocarbon fluid inclusions in fracture-filling cements in the Mission Canyon Limestone, Little Knife Field, North Dakota. Note that hydrocarbon inclusions, although trapped at the same time as the aqueous inclusions, homogenize at consistently lower temperatures than the aqueous inclusions. (From Narr and Burruss, 1984; reprinted by permission of the American Association of Petroleum Geologists.)

is a function of temperature, pressure, density of dislocations, strain rate, and many other factors.

Underlying this discussion are the fundamental physical properties of the trapped fluid: the PVT relationships of the fluid that are controlled by the composition of the fluid. Unlike fluids in igneous and metamorphic rocks and in hydrothermal ore deposits, fluids in sedimentary rocks are complex mixtures of aqueous electrolytes, gases, and organic compounds. The complexity of possible compositions makes it difficult, if not impossible, to estimate the composition and density of inclusions through interpretation of phase equilibria using techniques described by Crawford (1981) and Burruss (1981a, 1981c).

To routinely use fluid inclusion observations to estimate paleotemperature of sediments, there must be ways of overcoming these limitations. If inclusions have stretched, this may be detected by studying the self-consistency of the observations. If apparently primary inclusions in petrographically defined, early cements have the same T_h as inclusions formed at maximum burial, this agreement suggests that stretching has occurred. If there is a strong correlation of increasing size of inclusion with increasing homogenization temperature within a single generation of inclusions, Bodnar and Bethke's (1984) observations strongly suggest that these inclusions have reequilibrated. If stretching does appear to be significant in a set of observations, it does not rule out the importance of these observations. If the inclusions stretched and reequilibrated to the maximum paleotemperature, then the inclusions behaved like a maximum-recording thermometer. They may define a maximum paleogeothermal gradient as described by Burruss and Hollister (1979) for a geothermal well site. Such results may not be the original goal of a research project, but they may be useful observations for understanding the thermal evolution of a basin.

Paleopressure and Paleotemperature

Once the question of stretching is resolved, we are still faced with estimating pressure at the time of inclusion formation. As pointed out by Roedder and Bodnar (1980), one convenient solution is to use the inclusions themselves to define paleopressure as well as paleotemperature. This is possible if, at the time of inclusion formation, two immiscible fluids coexist in the pore space and *both* are trapped as inclusions. Because both fluids (e.g.,

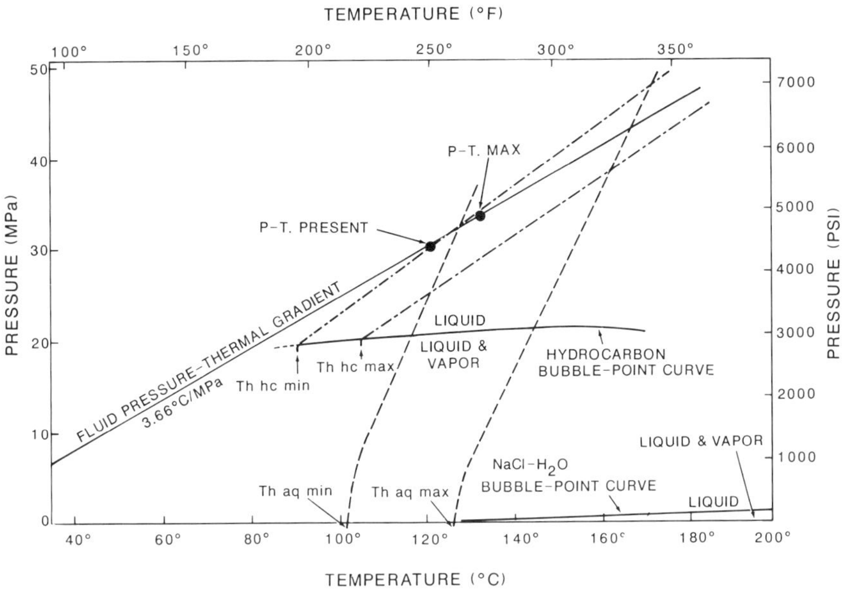

FIGURE 7.5. Pressure-temperature projection of the PVT properties of the petroleum and aqueous fluid inclusions trapped in fracture-filling cements in the Mission Canyon Limestone, Little Knife Field, North Dakota. Isochores are shown for the maximum and minimum homogenization temperatures of the two inclusion types in Figure 7.4. Also shown is the temperature and pressure gradient in fluids in the reservoir rocks. (Modified from Narr and Burruss, 1984; reprinted by permission of American Association of Petroleum Geologists.) *Symbols*: T_h aq min and max, temperature of homogenization of aqueous inclusions, minimum and maximum; T_h hc min and max, corresponding values for hydrocarbon inclusions; MPa, megapascals.

formation water and oil) coexisted in the rock at the same pressure and temperature, then the isochores for the inclusions of these fluids must intersect at that pressure and temperature.

As part of a study of the origin of fracture porosity and permeability in a carbonate reservoir, Narr and Burruss (1984) applied this concept in an attempt to estimate both the paleotemperature and paleopressure of fracture formation. Their fluid inclusion observations and interpretations are summarized here to illustrate the technique. Partially mineralized, thin, vertical fractures cross-cut all diagenetic features in the Mississippian, Mission Canyon Limestone at the Little Knife Field, North Dakota, as shown in Figures 5 and 6 in the original paper by Narr and Burruss (1984). Fluid inclusions occur in bridging minerals in these fractures as either primary, isolated inclusions or along secondary healed microfractures as shown in Narr and Burruss (1984, Fig. 10, A and B). The inclusions contain either vapor and aqueous brine of about 25 weight % NaCl equivalent salinity or vapor and fluorescent petroleum, or in rare cases, vapor, fluorescent petroleum, and an aqueous brine (see Narr and Burruss, 1984, Fig. 17). This last observation is critical because it establishes that the brine and petroleum were coexisting in the fracture porosity during inclusion formation.

Homogenization temperatures measured on both the aqueous inclusions and the hydrocarbon inclusions are shown in Figure 7.4. With a few exceptions, the hydrocarbon inclusions homogenize in the range of 90° to 106°C and the aqueous inclusions homogenize in the range of 102° to 126°C. If the T_h measurement directly indicated the paleotemperature, then those ranges should be the same for simultaneously trapped inclusions—clearly not the case here. The paleotemperature is determined by considering the PVT properties of both the hydrocarbon fluid and the aqueous fluid as shown in Figure 7.5. Homogenization of the hydrocarbon inclusions occurs at relatively high pressure on the bubble-point curve for this fluid (see Neumann et al., 1981, for a review of the PVT

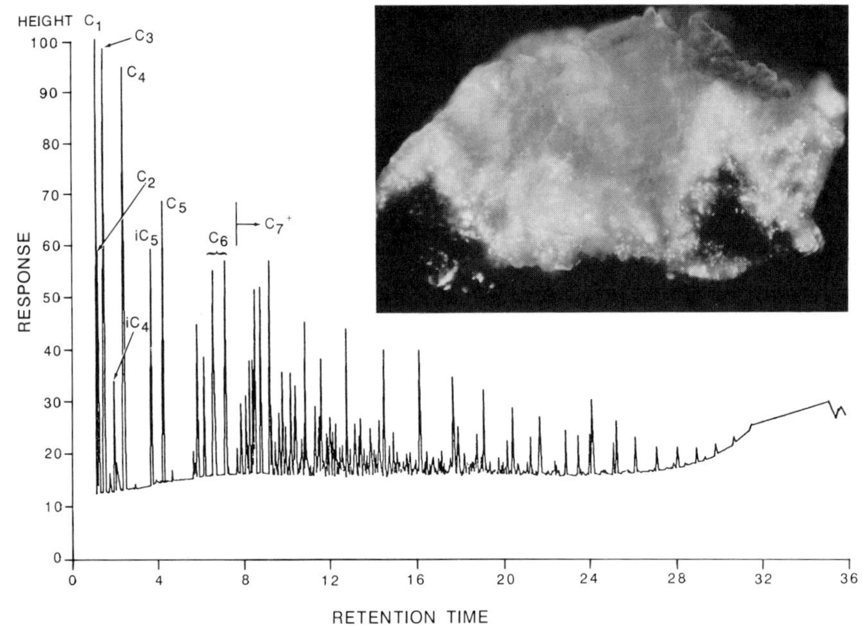

FIGURE 7.6. Gas chromatogram of hydrocarbon fluid inclusions in a chip of iron-rich calcite weighing 2.5 mg, shown under ultraviolet excited fluorescence in the inset photomicrograph. Bright spots are individual fluid inclusions containing fluorescent petroleum. The chromatogram shows response versus retention time. The individual peaks at the left are identified as individual hydrocarbon compounds that correspond with the hydrocarbon components listed in Table 7.1. Note that this sample is not from the Little Knife Field.

properties and phase behavior of petroleum). Homogenization of the aqueous inclusions occurs on the bubble-point curve ("boiling curve") for 25 weight % NaCl brines at relatively low pressure. Because the inclusions were trapped in latest diagenetic features in the hydrocarbon reservoir, Narr and Burruss (1984) assumed that the composition and PVT properties of the hydrocarbon inclusions were the same as the reservoir fluid described by Nettle et al. (1981). Isochores for these fluids were calculated with a standard Peng-Robinson equation of state (Peng and Robinson, 1976). Isochores for the aqueous inclusions were interpolated from the tables of Potter and Brown (1977).

Because these are coexisting fluids, the isochores must intersect. The minimum and maximum temperature isochores for both fluids intersect to define a rather large area of pressure-temperature space shown in Figure 7.5. The most important relationships are defined by the intersection of the subsurface PT gradient with this area in PT space. The present-day (original) reservoir PT conditions are coincident with the intersection of the PT gradient with the minimum temperature hydrocarbon isochore, but outside the PT space defined by the isochores for the hydrocarbon and aqueous inclusions. Because present-day PT conditions fall outside the range of latest diagenetic PT conditions defined by the inclusions, either PT conditions have changed with time since trapping the inclusions or some error exists in interpreting the position of the isochores.

Reservoir PT conditions at Little Knife may have changed because approximately 300 m of overburden were removed by late Tertiary erosion. By assuming the reservoir PT gradient was constant through the later Tertiary to the present, then by adding back this missing part of the total section, the present-day PT conditions shift to the point labeled "P-T. MAX" on Figure 7.5, well within the conditions defined by the isochores. This gives some confidence that the general technique works.

Effects of Fluid Composition

Two problems stand out when considering possible errors in the position of isochores for diagenetic fluid inclusions. First, there must be a routine way to calculate the PVT properties of hydrocarbon fluids trapped in inclusions. Equations of state are available to do this (e.g., Peng and Robinson, 1976), but applying them requires knowledge of the composition of the fluid. Reservoir engineers usually obtain the composition of a reservoir fluid by gas chromatography, resulting in data similar to that in Table 7.1, an analysis of the Little Knife reservoir fluid from Nettle et al. (1981). Gas chromatography can be applied to inclusions in small samples of diagenetic phases (Andrawes et al., 1984; Horsfield and McLimans, 1984) and compositions can be obtained. Figure 7.6 is a gas chromatogram of a 2.5-mg chip of an iron-rich calcite cement filling a septarian vein that contains hydrocarbon inclusions described by Burruss and Goldstein (1980) and Burruss et al. (1980). By using capillary column techniques, all the hydrocarbon components necessary for input to an equation of state can be resolved. Although the techniques described by Andrawes et al. (1984) allow resolution of additional, nonhydrocarbon components, such as N_2, CO_2, and H_2S, their approach has not yet been applied to analysis of these components and hydrocarbons in the same inclusions. Gas chromatographic techniques currently under development at the U.S. Geological Survey should allow both the nonhydrocarbon and hydrocarbon components to be resolved. These techniques will ultimately allow calculation of isochores for any hydrocarbon fluid trapped as inclusions.

The second problem with the position of the isochores in Figure 7.5 is the effect of hydrocarbon and nonhydrocarbon gases in the aqueous inclusions. As pointed out by Hanor (1980), the presence of dissolved methane in aqueous fluids shifts the isochores to higher pressures, an effect not considered in the interpretation of the Little Knife data. The importance of this problem can be seen by reconsidering that example. As shown in Table 7.1, the Little Knife reservoir fluid contains about 35 mole % CH_4. Because this fluid coexists with an aqueous fluid, some of the CH_4 and other gases will be dissolved (partitioned) in the aqueous phase. Although the amount of CH_4 in solution is relatively low, it shifts the bubble point of the 25 weight % NaCl equivalent brine to higher pressures and consequently shifts the position of the isochores. Figure 7.7 schematically shows the direction of the effect. In the Little Knife example, this effect makes the interpretation "better" from the standpoint that the area of intersecting isochores in PT space more closely corresponds both to present-day reservoir PT conditions and maximum, preerosion PT conditions.

TABLE 7.1. Reservoir-fluid composition in the Little Knife Field, North Dakota.

Component	Mole %
H_2S	7.99
N_2	0.98
CO_2	1.54
C_1	35.36
C_2	9.93
C_3	6.39
iC_4	1.27
nC_4	3.60
iC_5	1.48
nC_5	2.05
C_6	1.80
C_7^+	27.61
GOR(SCF/B)	1,100

Note: Data from Nettle et al. (1981).

An analytical technique capable of determining the gas composition of aqueous inclusions would help solve this problem. However, such an analysis is a challenge quite different from that posed by the analysis of hydrocarbon inclusions. Because aqueous and hydrocarbon inclusions occur in the same sample, commonly in the same crystal, any technique that opens many inclusions (e.g., decrepitation or crushing) will open many aqueous and hydrocarbon inclusions. Thus, the analysis will not represent the gas content of only the aqueous inclusions. A technique is necessary that allows analysis of single inclusions. This is not trivial because, as Roedder (1967) points out, a single 10-μm diameter inclusion contains only about 10^{-9} g of material and the gas content of that fluid may only be about 1,000 ppm.

Qualitative and semiquantitative estimates of gas contents are possible with the crushing stage microscope (Roedder, 1970). However, this technique is destructive and is capable of detecting only one or two components in each analysis. Another approach, technologically more complex than the crushing stage, but versatile and nondestructive, is microfocused, laser Raman spectroscopy. This technique was initially applied to fluid inclusions by Rosasco et al. (1975) and Delhaye and Dhamelincourt (1975).

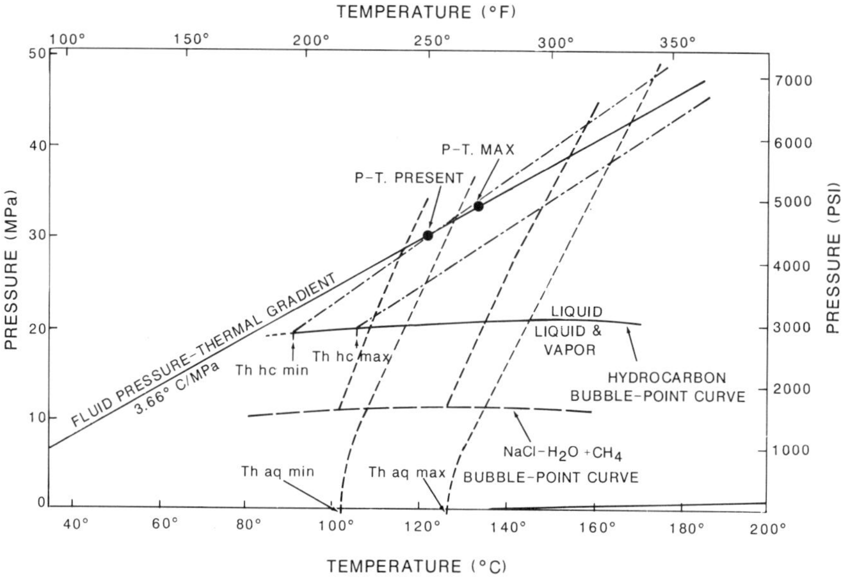

FIGURE 7.7. Modified version of Figure 7.5 showing the probable effect of dissolved gas in the aqueous inclusions on the position of the isochores for these inclusions. Heavy dashed lines show the shift of the aqueous bubble-point curve and isochores to higher pressure. These shifts improve the correspondence of the observations of Narr and Burruss (1984) to modern reservoir conditions. All symbols are defined in Figure 7.5.

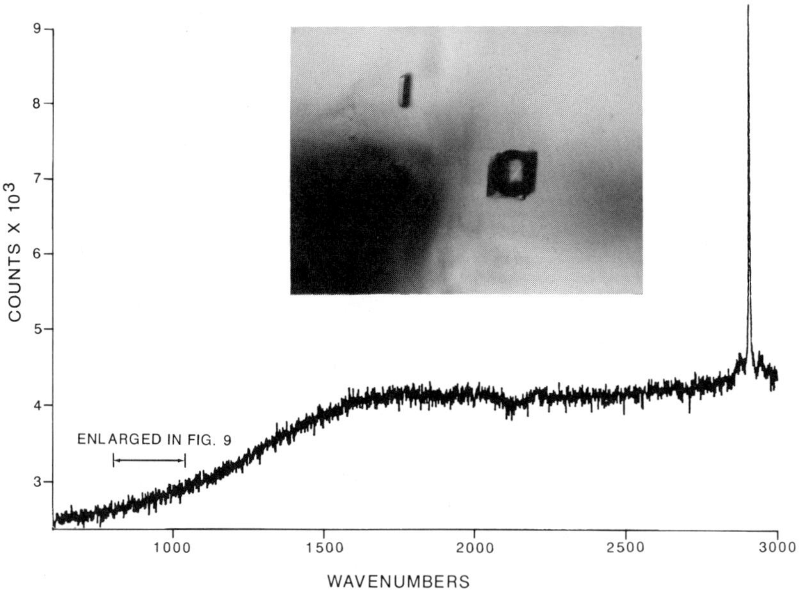

FIGURE 7.8. Raman spectrum of vapor bubble of aqueous inclusion in halite shown in inset photomicrograph. Sharp peak at right at 2,917 cm^{-1} corresponds to methane. This spectrum illustrates one method of detecting gaseous components of individual aqueous inclusions. *Analytical conditions*: Ramanor U-1000 spectrometer laser power 100 mW at 514.5 nm, slits 500 μm, 1 scan at 0.5 s per wavenumber, analyzed 5/4/84.

The basic technique of Raman spectroscopy is described in a number of textbooks (e.g., Long, 1977). Reviews of the microfocused technique applied to fluid inclusions are available in Guilhaumou (1982) and Roedder (1984). Raman spectroscopy measures the frequency of vibrations of the chemical bond between atoms in a molecule, the same property measured by infrared spectroscopy. Raman spectroscopy has an advantage over infrared techniques in that the frequency is measured as a shift in the frequency of incident light scattered from the molecule. The incident light must be monochromatic but it can be in the visible wavelength range (typically the 514.5-nm line of an Ar ion laser), thus allowing use of standard microscope optics for both sample viewing and spectral analysis. The major disadvantage of Raman spectroscopy is that Raman scattering is an extremely inefficient process. This requires the instrumentation for Raman studies to be extremely sensitive and typically rather expensive.

The Raman technique is easily performed on fluid inclusions and readily allows qualitative and semiquantitative analyses of molecular species within the inclusions. For the present problem of determining the gas content of aqueous inclusions, the technique allows identification of the gases present and their relative abundance. Figure 7.8 is the Raman spectrum of the vapor bubble in a large aqueous inclusion in halite. The complete spectrum shows only the presence of methane as a single sharp peak at 2,917 cm^{-1}. However, longer counting time in the region of the second strongest vibrational frequencies of ethane and propane, shown in Figure 7.9, reveals their presence in very limited amounts. The steep slope of the baseline for these peaks and the "plateau" under the methane peak in Figure 7.8 are due to fluorescence of higher molecular weight organic compounds.

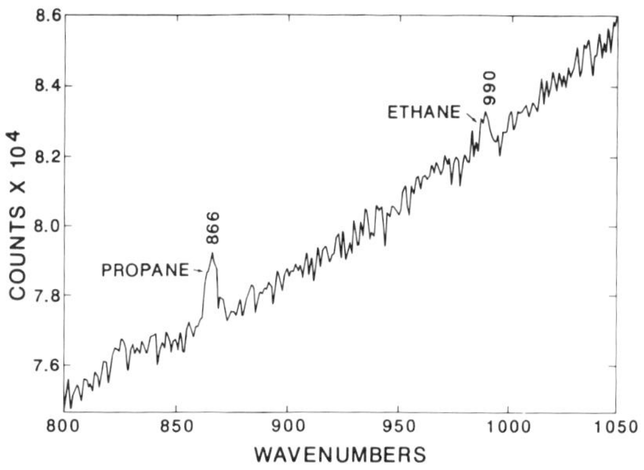

FIGURE 7.9. High-resolution spectrum of part of the spectral region shown in Figure 7.8, measured on the same fluid inclusion. Longer counting time reveals small peaks corresponding to ethane and propane. Analytical conditions as in Figure 7.8 except: 3 scans at 5 s per wavenumber summed together.

Conclusions

This chapter demonstrates that fluid inclusions are a useful tool for determining paleotemperatures during sedimentary diagenesis. The basic points are as follows:

1. The thermodynamic basis for interpreting phase transitions in fluid inclusions is well known. The basic principles of paleotemperature determination are readily illustrated with simple aqueous fluids.
2. The basic limitations on the paleotemperature are the need for an estimate of paleopressure and the possibility of natural reequilibration of early inclusions by later, higher temperature, burial conditions.
3. Possible methods for overcoming these limitations are available. One approach is the analysis of aqueous and petroleum inclusions that were trapped at the same time.
4. Wide application of fluid inclusion measurements will require chemical analysis of hydrocarbon and aqueous inclusions by such techniques as gas chromatography and Raman spectroscopy. The analytical data can then be used in equations of state for fluid mixtures to calculate isochores for complex sedimentary diagenetic fluids. These isochores will allow estimates of paleotemperatures at the time of trapping of the fluid inclusions.

Acknowledgments. I thank several people for help and collaboration in my fluid inclusion studies during the last several years. Two former colleagues at Gulf Research and Development Company, Wayne Narr, now at Princeton University, and Dick

Rupert, now retired, made major contributions. Jill Pasteris, Washington University, St. Louis, Missouri, made her Raman laboratory available to me. Jill Duncan Burruss provided remarkable patience and support during our two years of commuting from Pittsburgh to Houston and Denver.

References

Andrawes, F., Holzer, G., Roedder, E., and Gibson, E.K., Jr. 1984. Gas chromatographic analysis of volatiles in fluid and gas inclusions. Journal of Chromatography 304:181–193.

Barker, C. 1972. Aqua-thermal pressuring: Role of temperature in development of abnormal-pressure zones. American Association of Petroleum Geologists Bulletin 56:2068–2071.

Bodnar, R.J., and Bethke, P.M. 1984. Systematics of stretching of fluid inclusions: I. Fluorite and sphalerite at 1 atmosphere confining pressure. Economic Geology 79:141–161.

Burruss, R.C. 1977. Analysis of fluid inclusions in graphitic metamorphic rocks from Bryant Pond, Maine, and Khtada Lake, British Columbia: Thermodynamic basis and geologic interpretation of observed fluid compositions and molar volumes. Ph.D. dissertation, Princeton University, Princeton, NJ, 174 pp.

Burruss, R.C. 1981a. Analysis of phase equilibria in C-O-H-S fluid inclusions. In: Hollister, L.S., and Crawford, M.L. (eds.): Fluid Inclusions: Applications to Petrology. Mineralogical Association of Canada Short Course Handbook, vol. 6, pp. 39–74.

Burruss, R.C. 1981b. Hydrocarbon fluid inclusions in studies of sedimentary diagenesis. In: Hollister, L.S., and Crawford, M.L. (eds.): Fluid Inclusions: Applications to Petrology. Mineralogical Association of Canada Short Course Handbook, vol. 6, pp. 138–156.

Burruss, R.C. 1981c. Analysis of fluid inclusions: Phase equilibria at constant volume. American Journal of Science 281:1104–1126.

Burruss, R.C., Cercone, K.R., and Harris, P.M. 1983. Fluid inclusion petrography and tectonic-burial history of the Al Ali No. 2 well: Evidence for the timing of diagenesis and migration, northern Oman foredeep. Geology 11:567–570.

Burruss, R.C., Cercone, K.R., and Harris, P.M. 1985. Timing of hydrocarbon migration: Evidence from fluid inclusions in calcite cements, tectonics and burial history. In: Schneidermann, N.M., and Harris, P.M. (eds.): Carbonate Cements. Society of Economic Paleontologists and Mineralogists Special Publication 36, pp. 277–289.

Burruss, R.C., and Goldstein, R.H. 1980. Time and temperature of hydrocarbon migration: Fluid inclusion evidence from the Fayetteville Formation, NW Arkansas. Geological Society of America Abstracts with Programs 12:396.

Burruss, R.C., and Hollister, L.S. 1979. Evidence from fluid inclusions for a paleo-geothermal gradient at the geothermal test well sites, Los Alamos, New Mexico. Journal of Volcanology and Geothermal Research 5:163–177.

Burruss, R.C., Toth, D.J., and Goldstein, R.H. 1980. Fluorescence microscopy of hydrocarbon fluid inclusions: Relative timing of hydrocarbon migration events in the Arkoma basin, NW Arkansas (abst.). EOS, Transactions of the American Geophysical Union 61:400.

Crawford, M.L. 1981. Phase equilibria in aqueous fluid inclusions. In: Hollister, L.S., and Crawford, M.L. (eds.): Fluid Inclusions: Applications to Petrology. Mineralogical Association of Canada Short Course Handbook, vol. 6, pp. 75–100.

Delhaye, M., and Dhamelincourt, P. 1975. Raman microprobe and microscope with laser excitation. Journal of Raman Spectroscopy 3:33–43.

Guilhaumou, N. 1982. Analyse ponctuelle des inclusions fluides par microsonde moléculaire a laser (MOLE) et microthermometry. Paris, Presses de l'Ecole Normale Superieure, Travaux du Laboratoire de Geologie, no. 14, 68 pp.

Hanor, J.S. 1980. Dissolved methane in sedimentary brines: Potential effect on the PVT properties of fluid inclusions. Economic Geology 75:603–609.

Hollister, L.S., and Crawford, M.L. (eds.). 1981. Fluid inclusions: Applications to petrology. Mineralogical Association of Canada Short Course Handbook, vol. 6, 304 pp.

Horsfield, B., and McLimans, R.L. 1984. Geothermometry and geochemistry of aqueous and oil-bearing fluid inclusions from Fateh Field, Dubai. In: Schenck, P.A., deLeeuw, J.W., and Lijmbach, G.S.M. (eds.): Advances in Organic Geochemistry 1983. New York, Pergamon Press, pp. 733–740.

Kelly, W.C., and Turneaure, F.S. 1970. Mineralogy, paragenesis and geothermometry of the tin and tungsten deposits of the eastern Andes, Bolivia. Economic Geology 65:609–680.

Kennedy, G.C., and Holser, W.T. 1966. Pressure-volume-temperature and phase relations of water and carbon dioxide. In: Clark, S.P., Jr. (ed.): Handbook of Physical Constants. Geological Society of America Memoir 97, pp. 321–384.

Leroy, J. 1979. Contribution a l'etalonnage de la pression interne des inclusions fluides lors de leur decrepitation. Bulletin de Mineralogie 102:584–593.

Long, D.A. 1977. Raman Spectroscopy. New York, McGraw-Hill, 453 pp.

Narr, W.M., and Burruss, R.C. 1984. Origin of reservoir fractures in Little Knife Field, North Dakota. American Association of Petroleum Geologists Bulletin 68:1087–1100.

Nettle, R.L., Lindsay, R.F., and Desch, J.B. 1981. Well test report and CO_2 injection plan for the Little Knife field CO_2 minitest, Billings County, North Dakota. First Annual Report, Department of Energy, Minitest Contract No. 08383-26, 87 pp.

Neumann, H.-J., Paczynska-Lahme, B., and Swerin, D. 1981. Composition and Properties of Petroleum. New York, Holsted Press, 137 pp.

Peng, D.Y., and Robinson, D.B. 1976. A new two-constant equation of state. Industrial and Engineering Chemistry Fundamentals 15:59–64.

Potter, R.W., II, and Brown, D.L. 1977. The volumetric properties of aqueous sodium chloride solutions from 0° to 500°C and pressures up to 2000 bars based on a regression of available data in the literature. U.S. Geological Survey Bulletin 1421-C, 36 pp.

Roedder, E. 1967. Fluid inclusions as samples of ore fluids. In: Barnes, H.E. (ed.): Geochemistry of Hydrothermal Ore Deposits. New York, Holt, Rinehart & Winston, pp. 515–574.

Roedder, E. 1970. Application of an improved crushing microscope stage to studies of the gases in fluid inclusions. Schweizerische Mineralogische and Petrographische Mitteilungen 50(1):41–58.

Roedder, E. 1984. Fluid Inclusions. Reviews in Mineralogy: Vol. 12. Washington, DC, Mineralogical Society of America, 644 pp.

Roedder, E., and Bodnar, R.J. 1980. Geologic pressure determinations from fluid inclusion studies. Annual Reviews of Earth and Planetary Science 8: 263–301.

Rosasco, G.J., Roedder, E., and Simmons, J.H. 1975. Laser-excited Raman spectroscopy for nondestructive partial analysis of individual phases in fluid inclusions in minerals. Science 190:557–560.

Shepherd, T.J., Rankin, A.H., and Alderton, D.H.M. 1985. A Practical Guide to Fluid Inclusion Studies. London, Blackie, 239 pp.

8
The Thermal Transformation of Smectite to Illite

A.M. Pytte and R.C. Reynolds

Abstract

Mixed-layered illite/smectite minerals composed of 80% illite layers have been identified from different argillaceous rocks that have been subjected to estimated peak temperatures that ranged from 250°C to 70°C, for durations of approximately 10 yr to 300 my. These observations strongly suggest that the reaction progress or extent is controlled by kinetic factors rather than by equilibrium factors. A sixth-order kinetic expression (first-order with respect to the pore-fluid activity ratio K/Na, and fifth-order with respect to the mole fraction of smectite) was successfully applied to the progressive illitization of smectite in the contact metamorphic zone adjacent to an 8.5-m-thick basalt dike that penetrates the upper Pierre Shale near Walsenberg, Colorado. The kinetic expression, together with its preexponential constant and activation energy (33 kcal/mol), provides a fair to good transformation model for a young geothermal sequence, and for burial diagenetic profiles that range in stratigraphic age from approximately a few million to 300 Ma.

The sixth-order model is an empirical device that explains the field evidence, but probably has little or no fundamental physical-chemical significance. The correct kinetic law is likely to be a chain of low-order reactions, each of which has kinetic constants that differ from the others.

Introduction

The transformation of smectite to illite through an intermediate mixed-layered series is probably the most volumetrically important clay mineral reaction in sedimentary rocks. The reaction progress correlates with hydrocarbon maturation (Perry and Hower, 1972); the transformation may cause zones of overpressure in shales which drive the expulsion of hydrocarbons and water to more permeable rocks (Burst, 1969; Bruce, 1984) and it may release ions that are responsible for cementation and other diagenetic reactions (Boles and Franks, 1979). Indeed, it appears that measurable mineralogical changes, correlated with temperature, occur over the continuum between room temperature (soils) and perhaps 300°C, which marks the beginning of greenschist facies metamorphism (Środoń and Eberl, 1984). In addition, the various representatives of this mineral reaction series are ubiquitous in sedimentary rocks and are major constituents of shales.

Evidence for the thermal transformation of smectite to illite comes from four distinct types of studies. The most convincing demonstration of smectite-illite diagenesis was provided by Perry and Hower (1970), who demonstrated a continuous increase, with respect to depth, in the proportion of illite layers in mixed-layered illite/smectite (hereafter designated as I/S) in shales from a Gulf Coast Tertiary well. Since that time, the Perry and Hower measurements have been repeated many times by numerous investigators. Few geologists these days doubt the essential validity of their conclusions.

Most bentonites are composed of smectite (usually montmorillonite) in young, shallow sedimentary sequences. Weaver (1953) noted, however, that clay layers whose stratigraphic characteristics certainly identify them as bentonites tend to be composed of I/S in older rocks. Indeed, the I/S in such beds often consists of 80% or more illite in rocks older than middle Paleozoic. Originally, these so-called potash bentonites were thought to represent a chemical response to long residence times on the sea floor (Weaver, 1953), but the work of Hoffman and Hower (1979), in the Cretaceous of the disturbed belt of Montana, showed that burial temperatures were responsible for the high illite contents of the meta- or potash bentonites. They demonstrated that bentonites on the Sweetgrass arch are nearly pure smectite, whereas the same beds in the nearby disturbed belt are highly illitic due to increased temperatures associated with burial under accumulated thrust sheets. Nadeau and Reynolds (1981) studied Cretaceous bentonites in the southern Rocky Mountains and reported a wide range in illite content that correlated with regions of deep burial or proximity to igneous intrusives. One such bed contained I/S (75% illite) near the contact with an Oligocene intrusive, and yielded a K-Ar age that is nearly identical to the age of the intrusion (Aronson and Lee, 1986), yet an exposure approximately 20 km from the intrusive consists of essentially pure smectite for which the zircon fission-track age is 100 Ma (Kramer, 1981).

Contact metamorphosed Cretaceous shales, in proximity to igneous dikes, have been analyzed from southcentral Colorado (Reynolds, 1981; Pytte, 1982). The concentration of illite in I/S increases monotonically toward the dikes, demonstrating the compositional response of the mixed-layered phase to temperature. The reaction environment here consists of very high temperatures for short time durations, compared to the other geological settings described above.

Finally, I/S representing almost the entire compositional range has been prepared in the laboratory by means of hydrothermal synthesis (Eberl and Hower, 1976; Eberl, 1978a).

The major questions concern the reaction controls. What chemical and physical factors control the reaction progress or extent? Certainly temperature is important, and this fact makes the smectite-illite reaction interesting to geologists. The reaction extent ideally constitutes a diagenetic grade indicator or paleothermometer and could serve as a useful tool for tracing the thermal and burial history of sedimentary basins. Its utility, however, depends on the extent to which the effects of temperature can be isolated from those of time, pore-fluid composition, and the composition of the mineral reactants. Nevertheless, the reaction extent correlates quite well with temperature and time for rock suites from different geological situations (Środoń and Eberl, 1984), suggesting that either many of the possible variables do not exert powerful influences on the reaction extent, or, on the other hand, that the ranges of such variables are fixed or buffered by geochemical systems that are widely extant in time and space.

Major questions concerning the smectite-illite reaction are as follows:

1. Is the reaction controlled by chemical equilibrium or kinetic principles, or both?
2. What are the relative molar quantities of illite and smectite that make up the reaction stoichiometry?
3. What are the effects of pore-fluid and mineral composition (smectite) on the reaction extent?
4. Does the reaction proceed as layer-by-layer transformation or by a dissolution reprecipitation mechanism?

This chapter presents evidence and arguments that support the role of kinetic factors in controlling reaction extent. The treatment is meant to apply to shales and perhaps carbonate rocks where solution chemistry is apt to be rock dominated. Clay diagenetic reactions in sandstones may be quite different due to the effects of chemically exotic fluids and temperatures that may or may not be controlled by the local geothermal gradient.

Effects of Solution Composition on Reaction Progress

Pure illite, or its solid solution components celadonite or muscovite, may be stable in the presence of seawater and at surface temperatures. If so, montmorillonite and I/S are metastable and the composition of the mixed-layered material represents the reaction extent or progress caused by some combination of time and temperature effects, probably conditioned by pore-water chemistry (see Lippmann, 1982, for a discussion of the thermodynamic status of clays). Eberl and Hower (1976), Eberl (1978b), and Roberson and Lahann (1981)

TABLE 8.1. Approximate times at temperatures exceeding 90% of peak values for argillaceous rocks containing I/S with 80% illite.

Approximate time	Estimated peak temperature (°C)	Geological conditions	Reference
10 yr	250	Contact metamorphism	Reynolds (1981)
10,000 yr	150	Hydrothermal well	Jennings and Thompson (1986)
1 my	127	Burial diagenesis	Perry and Hower (1972)
10 my	100	Burial diagenesis	Perry and Hower (1972)
300 my	70	Burial diagenesis	Środoń and Eberl (1984)
450 my	70	K-Bentonite	Huff and Turkmenoglu (1981)

have studied the I/S reaction series by means of hydrothermal laboratory synthesis. Experimental results indicate that high potassium activity increases the reaction rate, and that magnesium, calcium, and sodium activities impede the reaction to diminishing degrees in the order given. Altaner et al. (1984) describe a 2.5-m-thick bentonite bed from Montana that is chemically and mineralogically zoned. The upper and lower contacts contain more illitic I/S than the core. Potassium diffusion from the enclosing shale is cited as the cause of the zonation. The role of potassium as either rate limiting or equilibrium controlling is demonstrated, whether or not kinetics are crucial here.

Time-Temperature Associations

Field occurrences of I/S in rocks of different ages and subjected to different temperature regimes are best explained as reaction limits imposed by time-temperature conditions. Table 8.1 shows some of these data. They refer to the composition 80% illite because this is a common one in old or deeply buried rocks, and problems of potassium availability may become crucial for more illitic compositions. Thermal histories in any location may be complex, and time-at-temperature is difficult to estimate accurately, so Table 8.1 shows estimated maximum temperatures and rough approximations of the times that temperatures exceeded 90% of peak values. The data make the point that I/S has apparently formed over a wide range of temperatures, and that an inverse correlation exists between temperature and time for a fixed composition. The overwhelming importance of temperature is noteworthy, though potassium availability may have been a factor in controlling the composition of the Ordovician K-bentonite. For the fixed composition of 80% illite, the range in temperatures is only about 4-fold, whereas the time range is approximately a factor of 10^7. The I/S reaction extent is thus a geothermometer for all practical purposes. Other variables may be involved that, for some reason or the other, were not operative during the reactions that produced the compositions shown here. But the most important point illustrated by the data in Table 8.1 is the likely role of kinetics in conditioning the reaction progress.

Illite/Smectite in the Contact Metamorphic Environment

The transformation of smectite to illite at high temperatures provides strong evidence for kinetic control. Studies were made of the I/S reaction profile in the upper Pierre Shale where it is penetrated by a 8.5-m-thick basalt dike near Walsenburg, Colorado (Reynolds, 1981; Pytte, 1982). The discussion below is given in detail because the work has been published as a master's thesis and is not easily available. Samples from a single bed were collected at 0.3-m intervals starting at the contact and continuing to a distance of 12 m. The <0.5-μm fractions (equivalent spherical diameter) were separated, pipetted onto glass slides, and dried at 90°C. They were solvated by exposure to ethylene glycol vapor at 60°C for a minimum of 12 hr, and analyzed by means of a Siemens D-500 diffractometer equipped with a copper tube and a graphite monochrometer. Ordering in I/S was determined by peak positions in the low angle region, and composition was estimated from the angular position of the I/S diffraction peak near 17° 2θ (Reynolds and Hower,

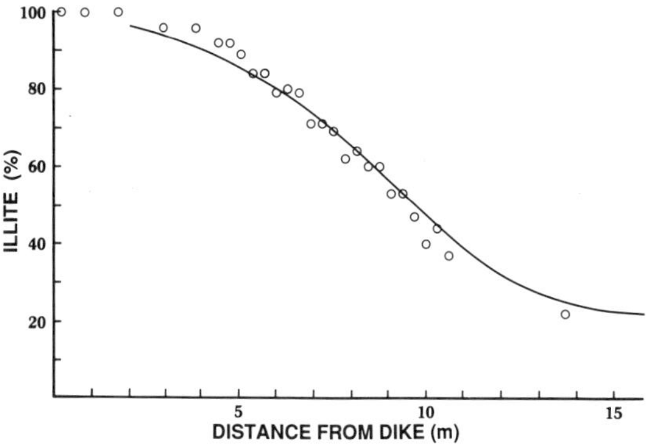

FIGURE 8.1. Composition of I/S in the Pierre Shale adjacent to a basalt dike near Walsenberg, Colorado. Curve is calculated from the kinetic model.

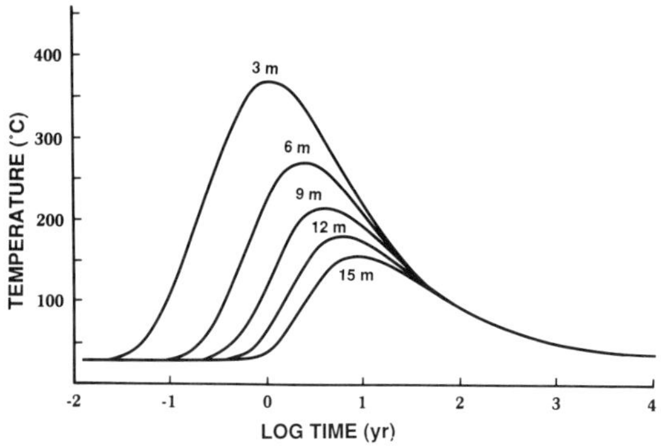

FIGURE 8.2. Temperature versus time curves in shale adjacent to an 8.5-m-thick dike, calculated for different distances from the contact.

1970; Reynolds, 1980). The results are shown on Figure 8.1, which also illustrates (solid curve) a fit of the data to a kinetic scheme described below.

The reaction profile shows several interesting features. Unlike Gulf Coast profiles, the reaction progress continues smoothly through and beyond the composition of 80% illite to the end product, which is essentially pure illite. Here, K-feldspar diminishes in concentration toward the dike (Lynch and Reynolds, 1984), but is not eliminated. These data are consistent with the conclusion that the termination of the reaction in Gulf Coast shales is due to exhaustion of K-feldspar at that point (Hower et al., 1976). The transformation from random to ordered interstratification takes place at about 60% illite, just as it does in Gulf Coast and other burial sequences. The reason for ordering at this composition, instead of at 50% illite, is unknown, but its occurrence in both the burial and contact metamorphosed sequences suggests that the same reaction principles were in effect despite the very different time-temperature settings.

The time-temperature relation is very different for contact as opposed to burial metamorphism. The contact metamorphosed shales were affected by thermal pulses whose peak temperatures and durations increased with proximity to the dike. We assumed a simple conductive heat transfer process and computed the time-temperature functions by means of a model published by Jaeger (1964). A temperature-constant thermal diffusivity coefficient of 0.0064 cm^2/s was computed from data given by Tyler et al. (1978) and the effect of latent heat was incorporated by assuming a temperature of 1200°C for the intrusive. A low wall-rock temperature of 30°C was assumed quite arbitrarily because of the stratigraphically high position of the sequence and the late (Pliocene) date of emplacement of the dike. Figure 8.2 shows several time-temperature curves calculated for different distances from the contact. The curves are not expressible by a closed-form equation, but are easily utilized in computer calculations.

A kinetic expression was developed from chemical principles and the constants in it were optimized by adjustment so that the results conformed to those for the dike and for burial diagenetic reaction profiles. Let S be the mole fraction of smectite in I/S. Then

$$-dS/dt = k\,S^{a} \quad (1)$$

where t is the time, a is a constant that describes the order of the reaction, and k is the rate constant given by the Arrhenius equation

$$k = A \exp(-U/RT) \quad (2)$$

The constant A is the frequency factor, U is the activation energy for the reaction, R is the gas constant, and T is the temperature in °K. A solution term is necessary to account for the effect of the activity of potassium on the reaction rate. We assumed equilibrium between albite and K-feldspar, and, using data from Robie et al. (1978), calculated the ratio of the activities of K/Na which increases with temperature. For these rocks, at least, this procedure seems justified because other work (Lynch and Reynolds, 1984) showed that the molar concentration of albite in the rocks increases toward the dike and is antipodal to the molar concentration of K-feldspar. A general kinetic expression may be written

$$-dS/dt = S^a (\mathrm{K/Na})^b A \exp(-U/RT) \quad (3)$$

in which the form of K/Na has been arbitrarily selected. The superscripts a and b represent integers whose sum is the reaction order. Experimentation with Equation 3 revealed that many different sets of variables yielded excellent agreement with the data of Figure 8.1, but that only high-reaction orders allowed extrapolation to long duration, low-temperature conditions such as those that apply to burial diagenesis. The best compromise gave the parameters $a = 5$, $b = 1$, $A = 5.2 \times 10^7\ \mathrm{s}^{-1}$, and $U = 33$ kcal/mol. These results were used to construct the curve shown on Figure 8.1, and they are retained for the treatment below.

The quantity K/Na is given by the van't Hoff equation, which, after rearrangement and collection of constants, yields

$$\mathrm{K/Na} = 74.2 \exp(-2490/T) \quad (4)$$

Equation 4 is substituted into Equation 3 for which $a = 5$ and $b = 1$, the differential equation is integrated, the integration constant is inserted, and the result is

$$S^4 = \frac{S_0^4}{1 + 4 \times 74.2\, t\, S_0^4 A \exp(-2490/T - U/RT)} \quad (5)$$

where S_0 is the mole fraction of smectite in the initial mixed-layered clay and t is the time in seconds. Note that the van't Hoff equation, when expressed in this way (Equation 4), contributes to the exponential term as additional activation energy. For our study, the K/Na ratio could have been eliminated and the empirical analysis would have achieved identical results (so long as the K/Na ratio contributes a first-order effect) by the use of a higher activation energy.

Kinetic Model and Time-Temperature Associations

The validity of the adjusted values for A and U was tested using the data of Table 8.1, though the fixed time and temperature approximations for these rocks represent thermal histories that have been simplified and are, at best, only approximations. We assumed that $S_0 = 1$ and substituted $S = 0.2$ into Equation 5 because all of the compositions are 80% illite. Equation 5 was rearranged in the logarithmic form to give

$$\ln(t) = (1/T)(2490 + U/R) - \ln(A) + 0.743 \quad (6)$$

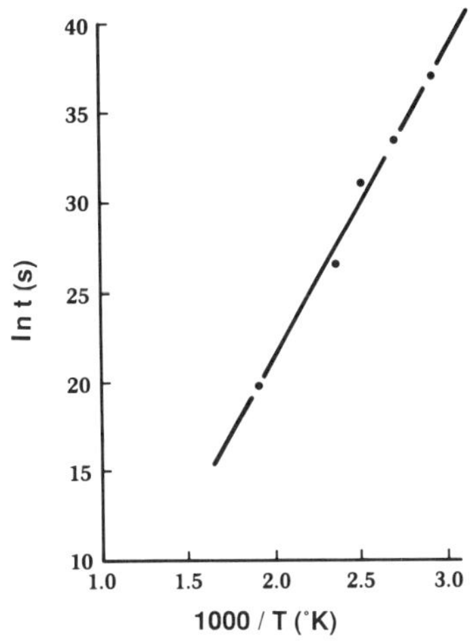

FIGURE 8.3. Plot of the time-temperature data of Table 8.1. The line was calculated by linear regression.

Figure 8.3 shows a plot of Equation 6 for which the values of time and temperature have been taken from the data of Table 8.1. The regression line has a slope of 2,490 + U/R, an intercept of $-\ln(A) + 0.743$, with the correlation coefficient = 0.99. Solving for the kinetic parameters yields $A = 5.6 \times 10^7\ \mathrm{s}^{-1}$ and $U = 33.2$ kcal/mol, in satisfactory agreement with the values deduced by trial-and-error methods from a more careful consideration of the time-temperature relations for the contact metamorphic reaction profile studied. Admittedly $\ln(t)$ is imperfectly known, but it is noteworthy that an error in t by a factor of 2 would cause the displacement of a given data point by only the amounts shown by the two points of Figure 8.3 that do not lie on the regression line.

Kinetics of Burial Metamorphism

The integrated form of Equation 3 cannot be used for modeling contact metamorphism and is cumbersome to use for burial diagenesis. The difficulty arises because temperature is a function of time,

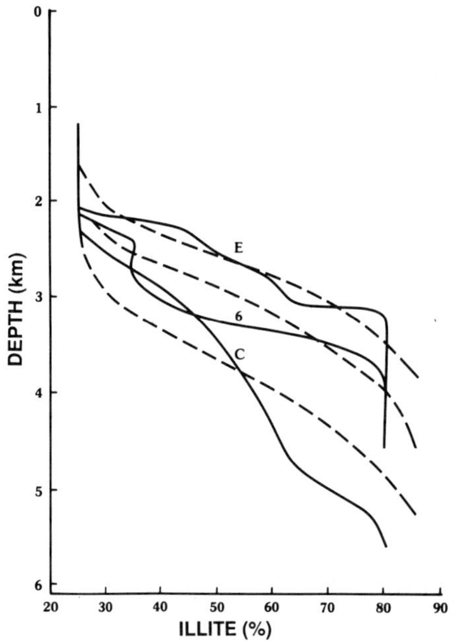

FIGURE 8.4. Burial diagenesis profiles for Gulf Coast Tertiary wells. Solid curves are smoothed observed data for Wells E and C (Perry and Hower, 1972) and Well 6 (Aronson and Hower, 1976). Dashed curves are calculated from the kinetic model.

and that function must be entered for T in Equation 3. The integration is easily accomplished by means of a digital computer and the replacement of the product $t \exp(-2490/T - U/RT)$ in Equation 5 by Q where

$$Q = \exp(-2490/T(t) - U/RT(t))\, dt \qquad (7)$$

and the quantity $T(t)$ is the temperature in °K at time t. For contact metamorphism, $T(t)$ is given, for example, by the curves of Figure 8.2, the lower limit of integration is zero and the upper limit is the time at which the temperature has fallen to the preintrusive value. For burial diagenesis, the upper limit is the stratigraphic age, and $T(t)$ is equal to the surface temperature plus the product of the sedimentation rate, the geothermal gradient, and time. Proper attention must be given to the units to assure their consistency, and concentrations must be expressed as mole fractions.

Figure 8.4 shows calculated reaction profiles and measured I/S compositions with respect to depth for three Gulf Coast wells. Published values for I/S have been smoothed to give reasonably simple curves. Well E is based on a final stratigraphic age of 40 Ma and a geothermal gradient of 0.0308 °C/m and Well C on 10 Ma and 0.0243 °C/m (Perry and Hower, 1972); both calculated profiles are based on continuous and constant sedimentation rates. Calculated data for Well 6 utilized the more complicated burial history summarized by Aronson and Hower (1976). A surface temperature of 10°C is assumed for all calculations, and kinetic constants are identical to those used to fit the dike reaction profile (Fig. 8.1).

The fit for Well E is good and Well 6 is adequately simulated by the model except for the strange reversal in the observed profile. The agreement for Well C is poor and suggests either inadequacies in the model or an imperfectly known burial history for this section.

The model is very sensitive to the uniformity and magnitude of the sedimentation rate, the geothermal gradient, and the surface temperature. So much so, in fact, that detailed agreement with burial diagenetic profiles may always be poor because these factors probably will never be known with the accuracy required for an exhaustive and convincing test. But the kinetic model proposed has generally proved extrapolatable over a wide range of time and temperature conditions. If the extrapolation is warranted, it demonstrates the dominating effect of temperature and the finite but small sensitivity of the reaction progress to time. Bruce (1984) has come to similar conclusions based on his study of Tertiary Gulf Coast reaction profiles of different ages.

Comments on the Kinetic Model

The model proposed here is incomplete. It takes into account only one aspect of solution chemistry, the K/Na ratio, and no provision is included to provide for different reactivities of compositionally different smectites (Boles and Franks, 1979; Bruce, 1984). The fact that the model works as well as it does suggests that large variations in these quantities have not been operative for many of the reaction sequences studied to date. But there are exceptions. We cannot account for the composition of 90% illite in an Ordovician K-bentonite in New York State. Johnsson (1984) has shown that partial resetting of the zircon fission-track age in this bentonite indicates a minimum temperature of approximately 175 °C, and any sensible burial and uplift history for this region predicts a composition of 96+% illite. The difference between 90 and 96% may seem small, but the sharply reduced reaction rate at high illite contents means that the two compositions require very different temperatures. Potassium availability may have limited the reaction extent, but the illite in adjacent limestones is less than 5% expandable, suggesting that potassium deficiency may not account for the anomalous composition of the K-bentonite. Another difficulty exists with the Carboniferous reaction profile described by Środoń and Eberl (1984). Agreement is quite good for the latter portions of the reaction, but the model predicts significantly less illitization for the lower temperature samples.

A reaction order of five (with respect to smectite) is an empirical formulation that is difficult to reconcile with physical-chemical principles. Probably a very different kinetic law is operative that is approximated by a simple fifth-order equation. C. M. Bethke and S. P. Altaner (personal communication, 1984) have proposed a scheme consisting of three sequential first-order steps, each of which has a different activation energy, and their calculated results correlate well with several published Gulf Coast Tertiary reaction profiles.

Conclusions

1. The correlation of I/S composition with temperature and time strongly suggests that the reaction progress is kinetically controlled.
2. A sixth-order kinetic expression, fifth-order with respect to smectite content and first-order with respect to the ratio of K/Na, describes the I/S reaction extent for conditions that range from contact metamorphism to low-temperature, long-term burial diagenesis.
3. The high reaction-order causes the reaction rate to slow drastically, for a given temperature, as the composition approaches that of pure illite.
4. Temperature is the dominant control on reaction progress; time is much less important.
5. Refinement of the proposed model requires that formulations be developed to account for the effects on reaction rate of the activities of magnesium and calcium in pore fluids, and the composition of prediagenetic smectite.

Acknowledgments. The authors gratefully acknowledge the National Science Foundation for supporting this research by means of grants EAR 79-03984 and EAR 84-07783.

References

Altaner, S.P., Hower, J., Whitney, G., and Aronson, J.L. 1984. Model for K-bentonite formation: Evidence from zoned K-bentonites in the disturbed belt, Montana. Geology 12:412–415.

Aronson, J.L., and Hower, J. 1976. Mechanism of burial metamorphism of argillaceous sediment: 2. Radiogenic argon evidence. Geological Society of America Bulletin 87:738–744.

Aronson, J.L., and Lee, M. 1986. K/Ar systematics of bentonite and shale in a contact metamorphic zone, Cerrillos, New Mexico. Clays and Clay Minerals 34:483–487.

Boles, J.R., and Franks, S.G. 1979. Clay diagenesis in Wilcox sandstones of southwest Texas: Implications of smectite diagenesis on sandstone cementation. Journal of Sedimentary Petrology 49:55–70.

Bruce, C.H. 1984 Smectite dehydration—its relation to structural development and hydrocarbon accumulation in northern Gulf of Mexico Basin. American Association of Petroleum Geologists Bulletin 68:673–683.

Burst, J.F., Jr. 1969. Diagenesis of Gulf Coast clayey sediments and its possible relation to petroleum migration. American Association of Petroleum Geologists Bulletin 53:73–93.

Eberl, D.D. 1978a. Reaction series for dioctahedral smectites. Clays and Clay Minerals 26:327–340.

Eberl, D.D. 1978b. The reaction of montmorillonite to mixed-layered clay: The effect of interlayer alkali and alkaline earth cations. Geochimica et Cosmochimica Acta 42:1–7.

Eberl, D.D., and Hower, J. 1976. Kinetics of illite formation. Geological Society of America Bulletin 87: 1326–1330.

Hoffman, J., and Hower, J. 1979. Clay mineral assemblages as low grade metamorphic geothermometers: Application to the thrust faulted disturbed belt of Montana, U.S.A. In: Scholle, P.A., and Schluger, P.R. (eds.): Aspects of Diagenesis. Society of Economic Paleontologists and Mineralogists Special Publication 26, pp. 55–79.

Hower, J., Eslinger, E., Hower, M.E., and Perry, E.A. 1976. Mechanism of burial metamorphism of argillaceous sediment: 1. Mineralogical and chemical evidence. Geological Society of America Bulletin 87: 725–737.

Huff, W.D., and Turkmenoglu, A.G. 1981. Chemical characteristics and origin of Ordovician K-Bentonites along the Cincinnati Arch. Clays and Clay Minerals 29:113–123.

Jaeger, J.C. 1964. Thermal effects of intrusions. Reviews of Geophysics 2:443–466.

Jennings, S., and Thompson, G.R. 1986. Diagenesis of Plio-Pleistocene sediments of the Colorado River delta, southern California. Journal of Sedimentary Petrology 56:89–98.

Johnsson, M.J. 1984. The thermal and burial history of south-central New York: Evidence from vitrinite reflectance, clay mineral diagenesis, and fission track dating of apatite and zircon. M.A. thesis, Dartmouth College, Hanover, NH, 155 pp.

Kramer, M.S. 1981 Contact metamorphism of the Mancos Shale associated with the intrusion at Cerrillos, New Mexico. M.A. thesis, Dartmouth College, Hanover, NH, 102 pp.

Lippmann, F. 1982. The thermodynamic status of clay minerals. In: van Olphen, H., and Veniale, F. (eds.): International Clay Conference 1981. Amsterdam, Elsevier Scientific Publishing Co., pp. 475–485.

Lynch, L., and Reynolds, R.C. 1984. The stoichiometry of the illite-smectite reaction (abst.). Twenty-First Annual Meeting of the Clay Minerals Society, Baton Rouge, LA, p. 84.

Nadeau, P.H., and Reynolds, R.C. 1981. Burial and contact metamorphism in the Mancos shale. Clays and Clay Minerals 29:249–259.

Perry, E.A., and Hower, J. 1970. Burial diagenesis in Gulf Coast peletic sediments. Clays and Clay Minerals 18:165–177.

Perry, E.A., and Hower, J. 1972. Late-stage dehydration in deeply buried peletic sediments. American Association of Petroleum Geologists Bulletin 56:2013–2021.

Pytte, A.M. 1982. The kinetics of the smectite to illite reaction in contact metamorphic shales. M.A. thesis, Dartmouth College, Hanover, NH, 78 pp.

Reynolds, R.C. 1980. Interstratified clay minerals. In: Brindley, G.W., and Brown, G., (eds.): Crystal Structures of the Clay Minerals and Their X-Ray Identification. London, Mineralogical Society, 495 pp.

Reynolds, R.C. 1981. Mixed-layered illite-smectite in a contact metamorphic environment (abst.). Eighteenth Annual Meeting of the Clay Minerals Society, University of Illinois at Urbana-Champaign, Urbana, IL, p. 5.

Reynolds, R.C., and Hower, J. 1970. The nature of interlayering in mixed layer illite/montmorillonite. Clays and Clay Minerals 18:25–36.

Roberson, H.E., and Lahann, R.W. 1981. Smectite to illite conversion rates: Effect of solution chemistry. Clays and Clay Minerals 29:129–135.

Robie, R.A., Hemingway, B.S., and Fisher, J.R. 1978. Thermodynamic properties of minerals and related substances at 298.15 K and 1 bar (10^5 Pascals) pressure and at higher temperatures. U.S. Geological Survey Bulletin 1452, 456 pp.

Środoń, J., and Eberl, D.D. 1984. Illite. In: Bailey, S.W. (ed.): Micas: Reviews in Mineralogy: Vol. 13. Reno, NV, Mineralogical Society of America, 584 pp.

Tyler, L.D., Cuderman, J.F., Kruhansl, J.L., and Lappin, A.R. 1978. Near-surface heater experiments in argillaceous rocks. In: Seminar on In Situ Heating Experiments in Geologic Formations, Ludvika, Stripa, Sweden. Brussels, Belgium, Organization of Economic Cooperation and Development, pp. 31–43.

Weaver, C.E. 1953. Mineralogy and petrology of some Ordovician K-bentonites and related limestones. Geological Society of America Bulletin 64:921–964.

9

$^{40}Ar/^{39}Ar$ Thermochronology of Sedimentary Basins Using Detrital Feldspars: Examples from the San Joaquin Valley, California, Rio Grande Rift, New Mexico, and North Sea

T. Mark Harrison and Kevin Burke

Abstract

$^{40}Ar/^{39}Ar$ age spectrum analysis of low temperature K-feldspars generally reveals age gradients indicative of the time the sample cooled between 200°C to 100°C. Reheating the feldspar above ≈120°C for geologically significant intervals sets up a radiogenic ^{40}Ar ($^{40}Ar^*$) concentration gradient characteristic of diffusion loss that can yield information related to the original cooling age and the timing and thermal intensity of the later event. Detrital microcline (and other K-feldspar) in sedimentary rocks can be used in this way to monitor the thermal evolution of a basin because accumulated $^{40}Ar^*$ is lost in response to rising temperature.

Applications of this approach to deep drill-hole samples from the San Joaquin Valley, California, Rio Grande Rift, New Mexico, and North Sea basin demonstrate the utility of this method in providing thermal and temporal constraints for geodynamic models. Improvements in analytical methods promise to reduce present limitations on interpretation attributable to sample heterogeneity.

Introduction

The challenge of obtaining reliable low-temperature geothermometric data in sedimentary rocks has taxed the ingenuity of workers from fields as diverse as nuclear magnetic resonance and paleontology. This has resulted in a bewildering array of techniques, each with a unique thermally activated mechanism deemed capable of recording ancient heating episodes. Our new approach to this venerable problem utilizes the thermal instability in retention of the daughter product of the ^{40}K-^{40}Ar decay system in K-feldspars to investigate the temperature-time history of the rock. Harrison and Bé (1983) introduced the practice of analyzing detrital microcline from sedimentary rocks by the $^{40}Ar/^{39}Ar$ age spectrum technique to obtain the internal distribution of radiogenic ^{40}Ar ($^{40}Ar^*$) within the crystals. This distribution is a sensitive function of the temperature history of the sample and can be used to reveal the geological age of thermal disturbances.

Microcline has suffered from a reputation of being a poor K-Ar geochronometer that is capable of losing $^{40}Ar^*$ by diffusion at room temperature (e.g., Faure, 1977). In fact, most perthitic K-feldspars have closure temperatures (see Dodson, 1973) between about 130° and 200°C, with a relatively narrow temperature transition between behavior as open and closed systems. This means that at temperatures above about 120°C (depending on the time scale of heating), accumulated $^{40}Ar^*$ produced by the in situ decay of ^{40}K will be lost from the K-feldspar crystal lattice by volume diffusion causing a zero age at the crystal margins during heating and an internal concentration profile characteristic of the *amount* of $^{40}Ar^*$ loss. Since the mobility of Ar in the K-feldspar lattice is a sensitive function of temperature, the integrated flux of $^{40}Ar^*$ out of the crystal is a measure of the thermal intensity of the heating episode. The principal advantages of this method applied to basin evolution studies are that it can simultaneously provide information about the age of basin cooling, the temperature (and perhaps the duration) of a heating episode, and the age of the sediment source. The main limitation of the method is that only rocks containing stable detrital microcline can be used.

$^{40}Ar/^{39}Ar$ Method

The $^{40}Ar/^{39}Ar$ dating technique was first suggested by Merrihue and Turner (1966) and involves irradiating geological materials in a fast neutron flux so that a portion of the ^{39}K atoms are transmuted to ^{39}Ar by a neutron-proton (n,p) reaction. Because the $^{40}K/^{39}K$ ratio is essentially constant in geological materials, the amount of ^{39}Ar produced during an irradiation is proportional to the ^{40}K concentration, making the relatively easily determined $^{40}Ar^*/^{39}Ar$ ratio a measure of the $^{40}Ar^*/^{40}K$ ratio, which can be used to determine the age of the sample. The Ar isotopes are liberated and analyzed using conventional high-vacuum fusion, purification, and isotopic analysis techniques. After correction for interfering isotopes produced during irradiation and for atmospheric argon, the $^{40}Ar^*/^{39}Ar$ ratio allows calculation of an age if the fast neutron dose is known. Because of uncertainties in accurately determining the threshold energy of neutrons capable of the specific (n,p) reaction, and thus measuring the fast neutron dose, Merrihue and Turner (1966) suggested placing a mineral of known K-Ar age together with the unknown to monitor the dose. By solving the age equations for both minerals simultaneously, a constant is obtained that defines the integrated fast neutron flux.

The K-Ar age equation is

$$t = \frac{1}{\lambda} \cdot \ln \left(1 + \frac{\lambda}{\lambda_e + \lambda_{e'}} \cdot \frac{^{40}Ar^*}{^{40}K}\right) \quad (1)$$

where λ = ^{40}K total decay constant, $\lambda_e + \lambda_{e'}$ = partial decay constant for the ^{40}K to ^{40}Ar branch, and $^{40}Ar^*/^{40}K$ = ratio of $^{40}Ar^*$ to parent ^{40}K at present day. Equation 1 can be rewritten as:

$$^{40}Ar^* = {^{40}K} \cdot \frac{\lambda_e + \lambda_{e'}}{\lambda} (e^{\lambda t} - 1) \quad (2)$$

The amount of ^{39}Ar produced during irradiation has been given by Mitchell (1968) as

$$^{39}Ar = {^{39}K}\, \Delta T\, \varphi(\varepsilon)\, \sigma(\varepsilon) d\varepsilon \quad (3)$$

where ^{39}Ar = number of ^{39}Ar atoms produced during irradiation, ^{39}K = original number of ^{39}K atoms in sample, ΔT = irradiation duration, $\varphi(\varepsilon)$ neutron flux at energy ε, and $\sigma(\varepsilon)$ = neutron capture cross section of $^{39}K(n,p)^{39}Ar$ at energy ε. The equation is then integrated over all incident neutron energies. From Equations 2 and 3 it follows that for an irradiated mineral of age, t:

$$\frac{^{40}Ar^*}{^{39}Ar} = \frac{^{40}K}{^{39}K} \cdot \frac{\lambda_e + \lambda_{e'}}{\lambda} \frac{1}{\Delta T} \frac{e^{\lambda t}-1}{\int \varphi(\varepsilon)\sigma(\varepsilon) d\varepsilon} \quad (4)$$

Mitchell (1968) defined a dimensionless irradiation parameter, J, derived from the monitor mineral, which is a measure of the neutron absorption by ^{39}K, an identical quantity for both monitor and unknown. If,

$$J = \frac{^{39}K}{^{40}K} \frac{\lambda}{\lambda_e + \lambda_{e'}} \Delta T \int \varphi(\varepsilon)\sigma(\varepsilon) d\varepsilon \quad (5)$$

then Equation 5 can be substituted into Equation 4 giving

$$\frac{^{40}Ar^*}{^{39}K} = \frac{(e^{\lambda t} - 1)}{J} \quad (6)$$

By measuring the $^{40}Ar^*/^{39}Ar$ ratio of the monitor mineral of known age, t, then

$$J = \frac{(e^{\lambda t} - 1)}{^{40}Ar^*/^{39}Ar} \quad (7)$$

As this quantity is the same for the unknown, then the age of the unknown sample is given by

$$t = \frac{1}{\lambda} \ln \left(1 + J \frac{^{40}Ar^*}{^{39}Ar}\right) \quad (8)$$

where $^{40}Ar^*/^{39}Ar$ is the isotopic ratio of the unknown sample.

An implication of Equation 8 is that because only isotopic ratios need be known to calculate an age, a sample containing a uniform distribution of $^{40}Ar^*$ need not be completely outgassed in order to obtain a valid age, in contrast with conventional K-Ar dating. The spatial distribution of $^{40}Ar^*$ can be inferred if the gas is extracted over many temperature increments, provided the sample releases the $^{40}Ar^*$ and ^{39}Ar by a diffusion mechanism. This property affords the greatest potential of the technique – a capability of resolving the concentration profile of $^{40}Ar^*$ within crystals. Typically this is done by plotting the apparent age obtained in any heating step against the percent of ^{39}Ar released. The derived plot is called an age spectrum or release pattern. If ^{39}Ar is homogeneously distributed throughout the sample, this plot gives a graphic indication of the internal ^{40}Ar distribution. Two other advantages in the $^{40}Ar/^{39}Ar$ technique are the ability to analyze both potassium and argon on the same sample aliquant, circumventing uncertainty due to sampling, and the ability to analyze samples with very low potassium concentrations.

With the application of the $^{40}Ar/^{39}Ar$ age spectrum technique to a variety of meteorites, Merrihue and Turner (1966) and Turner et al. (1966) demonstrated that both the age of crystallization and of reheating could be obtained from the data. A theoretical model proposed by Turner (1968) showed that these results could be interpreted as resulting from the diffusion loss of $^{40}Ar^*$. When lunar material became available, many excellent examples were provided supporting a model of $^{40}Ar^*$ gradients due to partial outgassing (Turner, 1970). Later work confirmed this behavior and it was further observed that the plateau segment of $^{40}Ar/^{39}Ar$ age spectra often corresponded to ages obtained from internal Rb-Sr isochrons.

Concurrent with the lunar studies, the suitability of the $^{40}Ar/^{39}Ar$ dating technique for the analysis of terrestrial materials was investigated. Total fusion ages produced by this method were found to be in good general agreement with the conventional K-Ar ages (Mitchell, 1968; Dalrymple and Lanphere, 1971; McDougall and Roksandic, 1974). However, investigations of the behavior of age spectra were ambiguous. Results of several studies of samples that had undergone a postcrystallization heating (Lanphere and Dalrymple, 1971; Berger, 1975; Hanson et al., 1975) did not reflect in any obvious way the ages of crystallization or reheating. However, experience gained through this work pointed to several experimental artifacts as the possible cause of these disappointing results (Harrison, 1983). These sources of confusion included the effects of impure mineral separates, thermal gradients across the sample aggregate during gas extraction, analysis of material crushed below its effective diffusion radius, structural changes in the mineral during argon extraction in the vacuum system, and large temperature increments. The possible incorporation of ambient argon during a reheating episode may further complicate an age spectrum (Harrison and McDougall, 1981). Of the several minerals later demonstrated to yield meaningful natural age gradients, the most easily interpreted is low-temperature K-feldspar because of its simple structure and lack of structural water.

From the time of the early application of the K-Ar dating method to geological material, it was apparent that low-temperature K-feldspars do not quantitatively retain radiogenic $^{40}Ar(^{40}Ar^*)$ at moderately elevated temperatures. This effect was variously ascribed to the degree of perthitization (Sardarov, 1957), differing unmixing rates of potassium and sodium (Brandt and Bartnitskiy, 1964), and deformation (Kuz'min, 1961; Albarede et al., 1978). In a review of argon diffusion literature, Musset (1969) noted that a number of studies using K-feldspars provided diffusion coefficients that varied by as much as six orders of magnitude at any one temperature. On the basis of a diffusion study using a homogeneous orthoclase, Foland (1974) concluded that the low retentivity of argon and variability in diffusion parameters in perthites are a result of the thickness of exsolution lamellae behaving as the effective diffusion radius of the mineral. Foland (1974) calculated that substantial $^{40}Ar^*$ loss could occur from these regions at relatively low temperatures ($\approx 150°C$) over geological time. In a study of the cooling history of a forcefully emplaced granitoid, Harrison et al. (1979) interpolated an effective retention or closure temperature (Dodson, 1973) for a perthitic K-feldspar of about 150°C. Harrison and McDougall (1982) reported age spectrum analyses of microclines from the Separation Point Batholith, northwest Nelson, New Zealand, which cooled slowly ($\approx 5°C/my$) through the temperature zone of partial $^{40}Ar^*$ retention (Fig. 9.1). These release patterns are characterized by a smooth age gradient rising from an initial age corresponding to cooling to a temperature of about 110°C to terminal ages indicating closure in the central portion of the crystals had occurred at $\approx 160°C$. Diffusion coefficients were calculated from the measured ^{39}Ar release during heating and were subsequently plotted against the reciprocal absolute temperature of the heating step to yield the Arrhenius parameters, activation energy (E) and frequency factor (D_0/ℓ^2) for argon transport in that sample. Using this custom diffusion law, a closure temperature of $\approx 130°C$ was calculated, which is consistent with

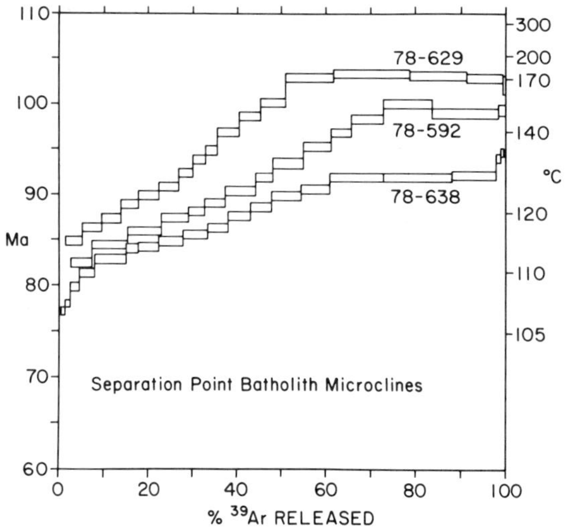

FIGURE 9.1. $^{40}Ar/^{39}Ar$ age spectra of slowly cooled microclines from the Separation Point Batholith, New Zealand. The cooling history for 78-592 is shown on the right vertical axis. The low-temperature ages of between 77 and 85 Ma correspond to the time the samples cooled below 120°C. The differences in age spectra for these three samples reflect relatively small temperature differences within the batholith during cooling.

the independently derived regional cooling history (Harrison and McDougall, 1980). With the establishment of appropriate Arrhenius parameters and closure temperature, the stage was set for using detrital microcline as an indicator of basin thermal history.

Case Study: San Joaquin Basin

Geological Setting

The San Joaquin Valley of central California has yielded in excess of 4×10^9 barrels of oil this century (Ziegler and Spotts, 1976) and as a result has been extensively studied by geologists (Hoots et al., 1954; Simonson, 1958; Callaway, 1971). The Tejon and Basin Blocks of the southern San Joaquin Valley have been the focus of a variety of geological and geochemical investigations including detailed stratigraphy (Bandy and Arnal, 1969; MacPherson, 1978; Webb, 1981), organic geochemistry (Castano and Sparks, 1974), zeolite thermometry (Briggs et al., 1981), and fission-track paleothermometry (Briggs et al., 1981; N.D. Naeser, C.W. Naeser, and T.H. McCulloh, personal communication, 1985). On the basis of laumontite crystallization and fission-track dating of detrital apatites, Briggs et al. (1981) suggested that although both the Basin and Tejon Blocks had experienced a thermal event since the late Miocene, the Basin Block is presently at peak thermal conditions, whereas the Tejon Block had cooled somewhat from a previously elevated temperature distribution. Because of their well-documented nature, Harrison and Bé (1983) chose to perform $^{40}Ar/^{39}Ar$ analysis on detrital microcline grains from the drill cores intersecting the Eocene to Miocene strata used in the Briggs et al. (1981) study. Figure 9.2 shows schematically the stratigraphy of Basin and Tejon Blocks in the sampled area and the inset indicates the map location of the various deep drill holes from which the cores were obtained.

The sedimentary history of the area has been well documented in several studies (Bandy and Arnal, 1969; MacPherson, 1978; Webb, 1981). Eocene to Pliocene sandstones deposited in a marine environment were buried by a significant pile of Pleistocene to Holocene terrigenous sediments. Movement of the Tejon Block along the White Wolf and Pleito faults has obscured this record and hydrothermal activity has apparently cooled the Cenozoic rocks. In contrast, the Basin Block contains several kilometers of young, probably post-Pliocene sediment. The Basin Block has been the prime depocenter since the late Miocene. Bounded on the south by the Tejon Block platform, the northern limit of the Maricopa subbasin is the east-west trending Bakersfield arch. Clastic sediments in the basin have been derived from two major sources: the Sierra Nevada Range, and more recently, the Tejon Block (MacPherson, 1978). Essentially all post-Pliocene sediment in the Tejon Block originated from the Tehachapi Range.

$^{40}Ar/^{39}Ar$ Results

Age spectra of samples taken from the Tejon Block and shallow (<4.4 km) depths in the Basin Block are characterized by an age gradient between about 60 Ma in the initial gas fractions and terminal ages of 85 Ma. Sample 14 (Fig. 9.3) typifies these results. This profile reflects slow cooling, between 200°C and ≈ 100°C in this 25-million-year span,

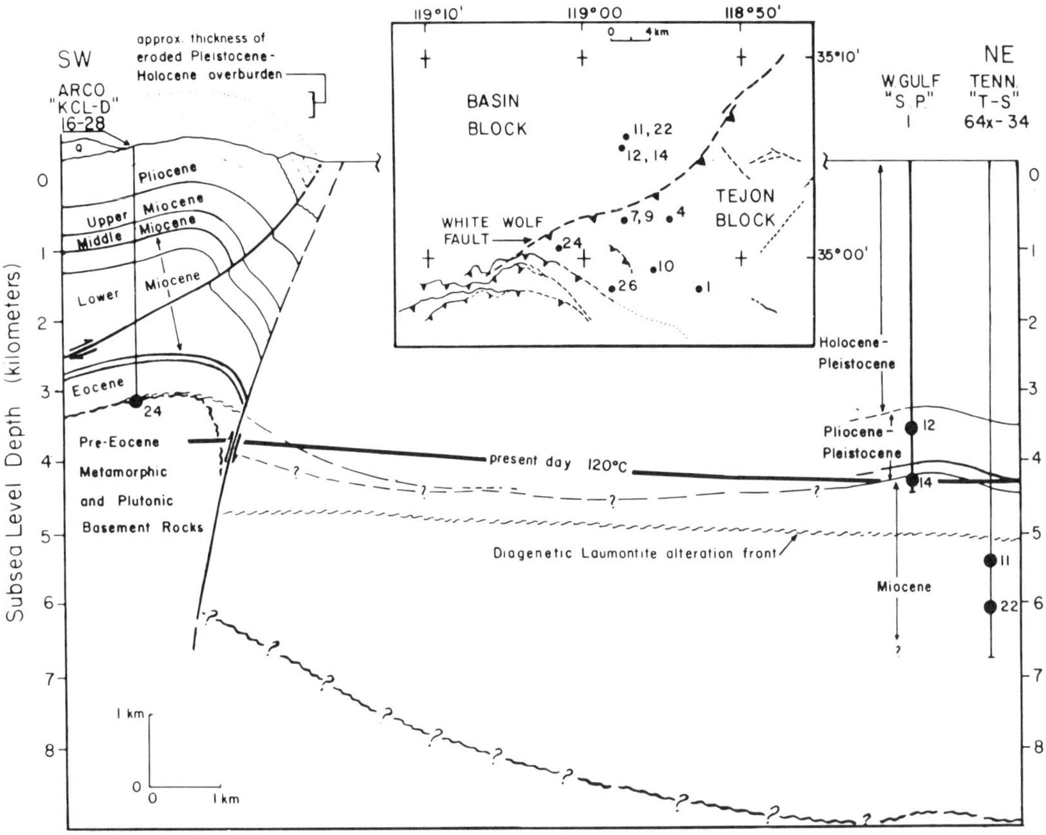

FIGURE 9.2. Sample location map (inset) and structural profile of the southern San Joaquin basin (from N.D. Naeser, C.W. Naeser, and T.H. McCulloh, personal communication, 1985) showing the uplifted Tejon Block on the left and the flat lying Basin Block sediments on the right.

of the source rocks in the uplifting Sierra Nevada Range. Subsequent to erosion from the Sierra Nevada and deposition, the K-feldspars in the deeper parts of the basin experienced burial temperatures sufficient to partially outgas some of the in situ produced $^{40}Ar^*$. This process produced a concentration profile typical of diffusion loss exemplified by the age spectrum for Sample 22 taken from 6.2 km, shown in Figure 9.4 together with a shallower sample (Sample 12) for contrast. The age spectrum rises from an extrapolated zero age to intersect the slow cooling gradient at about 30% ^{39}Ar release. This $^{40}Ar^*$ deficit amounts to about 18% of the gas originally present in the crystal, and can be used to estimate the duration of heating at the present maximum temperature of 157°C (Harrison and Bé, 1983; N.D. Naeser, C.W. Naeser, and T.H. McCulloh, personal communication, 1985). On the basis of Arrhenius law parameters obtained with an extraction furnace with poor temperature control (±50°C), Harrison and Bé (1983) estimated an equivalent "square-pulse" heating of 200,000-yr duration; that is, a sudden rise to the present temperature at about 200 ka.

A more plausible thermal history involving slow heating over the past million years could be accommodated by these data. The enhanced resolution of our new extraction furnace, which is capable of a several degree temperature accuracy (±0.3°C precision), indicates that these estimates are too low. Using $E = 33.0$ kcal/mol and $D_0/\ell^2 = 100$/s the square pulse heating duration is 500,000 yr and the linear heating scenario begins about 2 Ma ago. These results are consistent with both the fission-

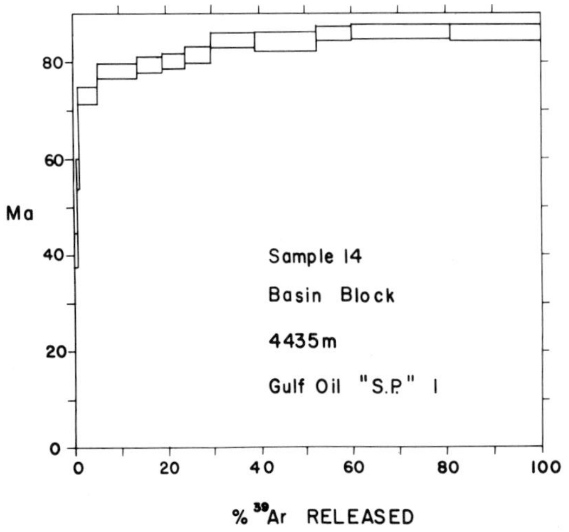

FIGURE 9.3. $^{40}Ar/^{39}Ar$ age spectrum for Sample 14 microcline separated from core recovered at 4.44-km depth in the Basin Block.

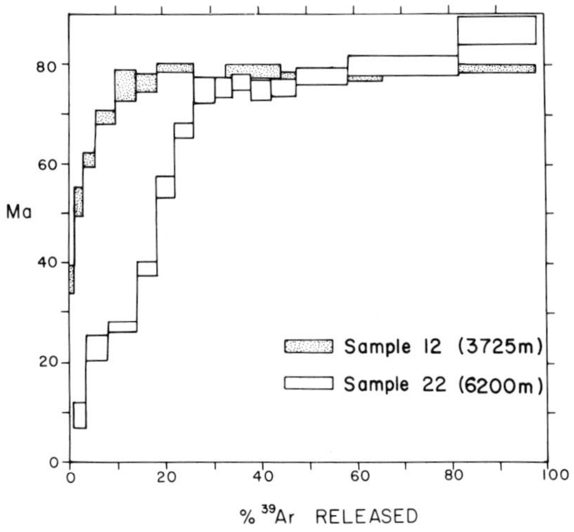

FIGURE 9.4. $^{40}Ar/^{39}Ar$ age spectra of K-feldspars from Basin Block Samples 12 and 22. Sample 12, from a depth of 3.7 km, has experienced virtually no $^{40}Ar^*$ loss in the course of erosion from the adjacent Sierra Nevada Range and subsequent burial in the San Joaquin basin. Sample 22, obtained at a depth of 6.2 km, exhibits 18% $^{40}Ar^*$ loss due to a recent thermal event.

track annealing data and the burial history independently determined by N.D. Naeser, C.W. Naeser, and T.H. McCulloh (personal communication, 1985).

These age spectra constrain a thermal model that indicates that the measured thermal gradient of ≈23°C/km does not reflect the equilibrium state, but is a transient effect reflecting the slow thermal relaxation following the dumping of cold sediment into the basin. Harrison and Bé (1983) calculated a value of 30°C/km as more truly representing the presubsidence gradient.

Case Study: Albuquerque Basin

Geological Setting

Extensional tectonics have played a progressively important role in the Cenozoic geology of the Cordillera that can be related, in a very general way, to post-Eocene plate interactions along the western seaboard of North America (e.g., Burchfiel et al., 1982). The Rio Grande Rift forms a distinctive extensional element within the Cordillera that has been the subject of comprehensive stratigraphic and structural studies, reflection and refraction seismic studies, heat flow, magnetic, gravity, and magnetotelluric studies (see, for example, papers in Riecker, 1979).

The tectonic and burial history of the Albuquerque basin section of the rift (Fig. 9.5) has been well documented (e.g., Kelley, 1977). It presents an opportunity to assess the significance of $^{40}Ar/^{39}Ar$ age spectrum results on detrital microcline from deep drill holes in an active rift with a coherently modeled thermal history (e.g., Morgan and Golombek, 1984). The Albuquerque basin is one of the larger north-south trending basins that make up the ≈1000-km-long Rio Grande Rift structure. Prebasin Precambrian, Paleozoic, and Mesozoic outcrops in the region are restricted to the margins of the basin, while the basin-fill deposits related to subsidence in response to extension are Miocene and younger. Mesozoic and Paleozoic strata (typically 3 km thick) are probably present in the basement of the rift valley and under the Neogene basins (Kelley, 1977). About 50 oil and gas wells have been drilled in the Albuquerque basin (Black, 1982) penetrating the Miocene Santa Fe Group, which represents the bulk of the basin-fill. The bulk of detritus shed into the basin was of basement origin until about the mid-Miocene, but volcanic debris become significant in younger rocks.

We obtained cores and cuttings from the two deepest wells in the basin, Isleta #2 and Federal

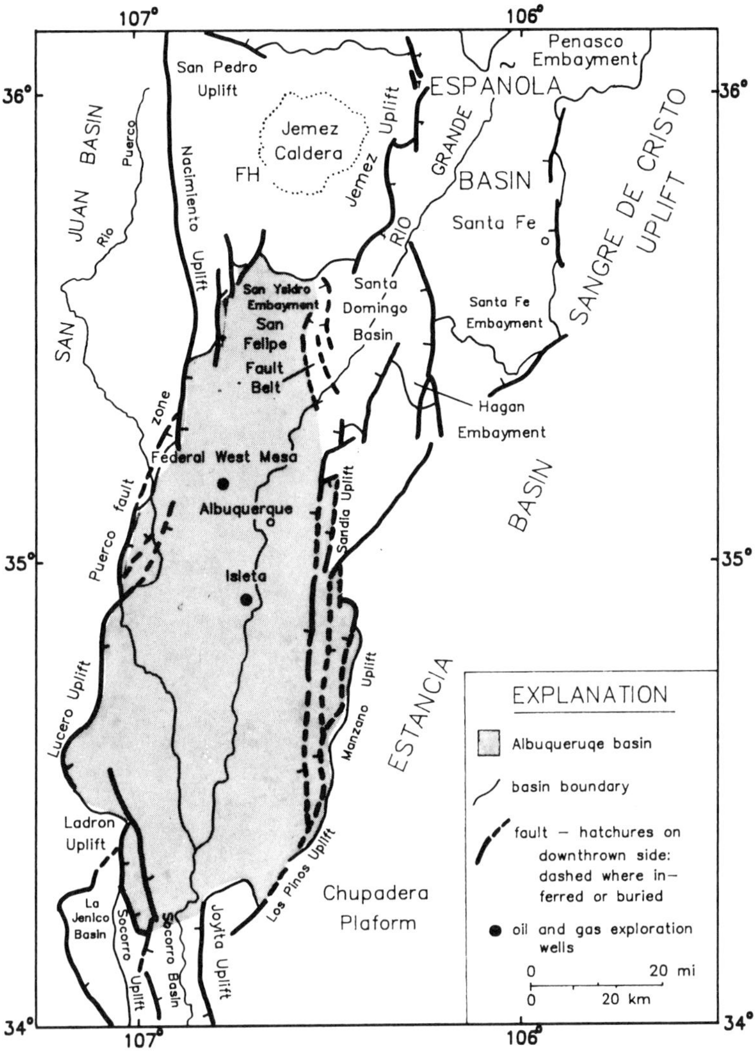

FIGURE 9.5. Map of Albuquerque basin and adjacent basement uplifts showing the locations of the Isleta #2 and Federal West Mesa #1 drill wells as well as the Fenton Hill site (FH) (From Lozinsky, 1984.)

West Mesa #1, which reached 6.5-km and 6.6-km depths, respectively. The Federal West Mesa #1 hole left Tertiary at 5.8 km and the bottom of the Isleta #2 well yielded Eocene pollen (Scott Amos, personal communication, 1984). Core material from various intervals in the depth range 4.8 km to 6.5 km of the Isleta #2 well was made available to us by Shell Oil Company along with temperature and stratigraphic documentation. Cuttings from 0.16 km to 5.1 km were also provided. For the most part these rocks are extremely well-preserved coarse sandstones and, with the exception of the deepest cores, yielded abundant microcline separates. The rocks within the 6.1-km to 6.5-km interval have been extensively hydrothermally altered and produced only small, albite-rich feldspar concentrates. Although less attractive as a source of samples, mineral separates were also prepared from the cuttings. Cuttings from the depth interval 4.5 km to 5.12 km from the Federal West Mesa #1 well, also supplied by Shell Oil Company, provided good K-feldspar concentrates.

In order to assess the age variations induced into the Santa Fe detrital feldspars as a result of the recent heating, associated with burial and an increased thermal gradient, we have also analyzed K-feldspar by the $^{40}Ar/^{39}Ar$ spectrum method from a variety of prebasin source rocks. These include the Precambrian Sandia Granite, the Permian Abo Sandstone, Precambrian gneisses from Fenton Hill

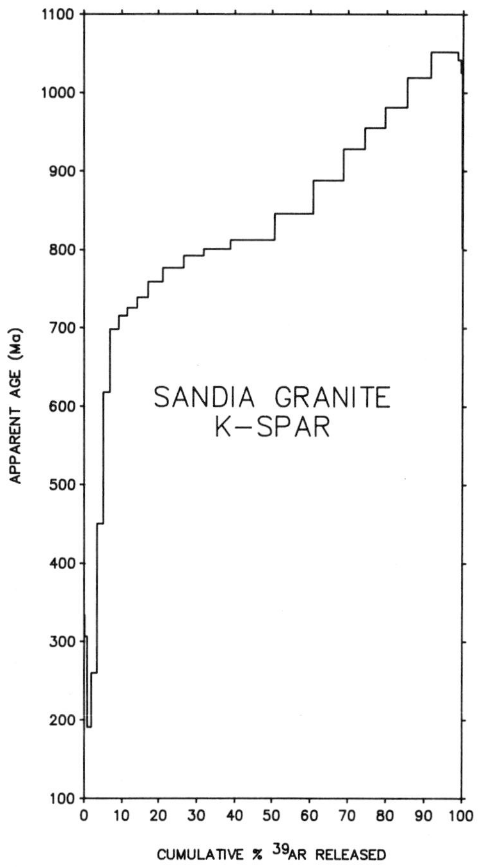

FIGURE 9.6. $^{40}Ar/^{39}Ar$ age spectrum of microcline from the Sandia Granite showing $^{40}Ar^*$ loss superimposed on a Precambrian slow cooling gradient.

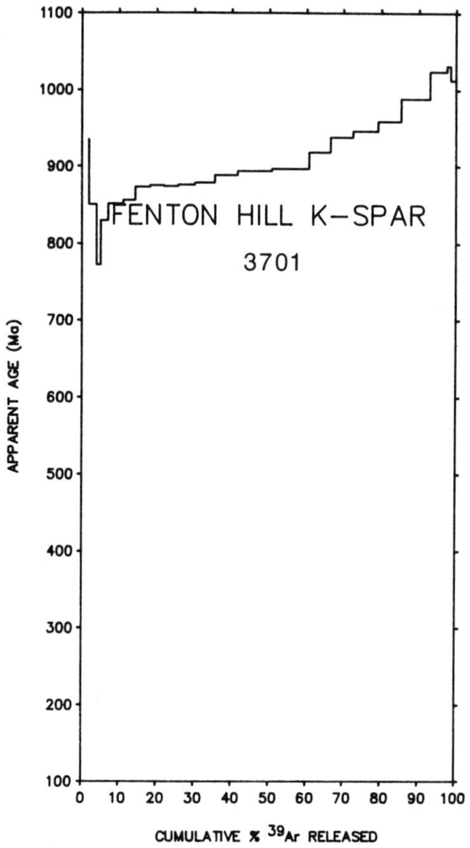

FIGURE 9.7. $^{40}Ar/^{39}Ar$ age spectrum of K-feldspar from Sample 3701 from 1.13-km (3,701-ft) depth in the Fenton Hill drill hole GT-2. The spectrum shows a Precambrian slow cooling gradient.

and north of the Espanola basin, which give coverage around the northern half of the bordering rim of the Albuquerque basin (Fig. 9.5).

Analytical techniques are described in detail in Harrison and Bé (1983) and Harrison et al. (1986). Results of these age spectrum experiments are shown in Figures 9.6–9.16.

A detailed stratigraphic and sedimentologic study of the Isleta #2 cores and cuttings is in progress at the New Mexico Institute of Mining and Technology (Lozinsky, 1984). Results of this project should yield a high-resolution burial curve with which we can constrain detailed heat-flow calculations. In this chapter, we describe our data in general terms but postpone further thermal analysis for a future paper when the complete burial history becomes available.

Prebasin K-Feldspars

Sandia Granite

The 1,440 $\pm$ 40 Ma Sandia Granite (Majumdar et al., 1984) is exposed along the fault-bounded eastern edge of the Albuquerque basin (Fig. 9.5). This Precambrian outcrop has experienced up to 6.4 km of vertical displacement relative to the bottom of the basin, at least a third of this uplift occurring during the Tertiary (Kelley, 1977; Kelley et al., 1982). Following emplacement, the granite cooled slowly, with Rb/Sr biotite systems closing about 1,330 Ma ago. $^{40}Ar/^{39}Ar$ analysis of a K-feldspar separated from this body gives an age spectrum (Fig. 9.6) suggestive of slow cooling between $\approx$250°C and 150°C in the interval 1,050 to $\approx$700 Ma. Additional, minor, $^{40}Ar^*$ loss has occurred

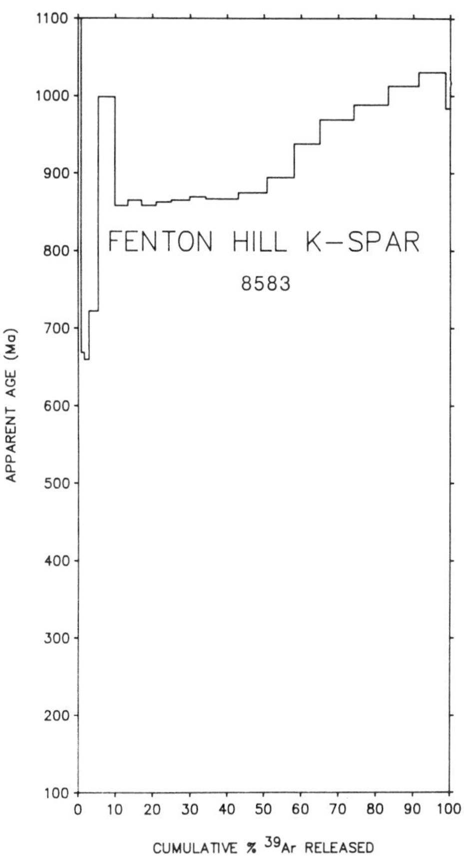

FIGURE 9.8. $^{40}Ar/^{39}Ar$ age spectrum of K-feldspar from Sample 8583 from 2.62-km (8,583-ft) depth in the Fenton Hill drill hole GT-2. The spectrum shows a Precambrian slow cooling gradient and possible, but minor, later $^{40}Ar^*$ loss.

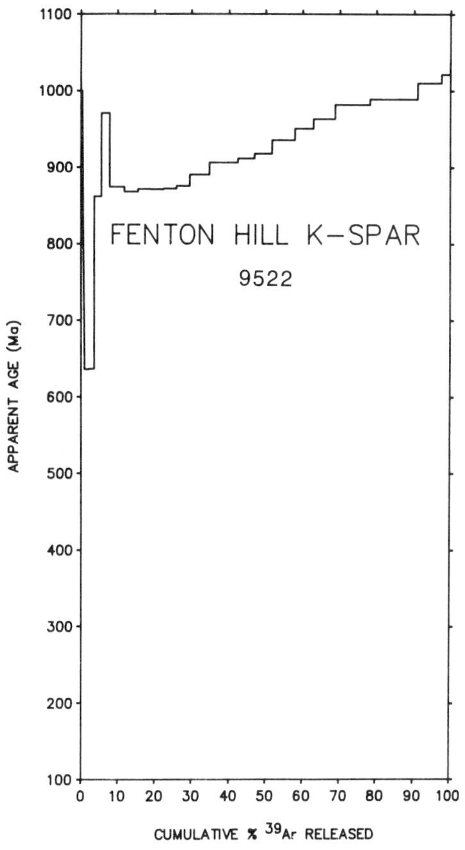

FIGURE 9.9. $^{40}Ar/^{39}Ar$ age spectrum of K-feldspar from Sample 9522 from 2.90-km (9,522-ft) depth in the Fenton Hill drill hole GT-2. The spectrum indicates slow cooling during the late Precambrian.

until recently, consistent with the observation of Miocene apatite fission-track ages from rocks at this structural level which would require temperatures as high as about 100°C until ≈ 15 Ma ago (Kelley et al., 1982; Kelley and Duncan, 1984). The upper portion of the Sandia Granite is a possible source for detritus in the Santa Fe Group sandstones.

Precambrian Gneiss

K-feldspar from an exposure of Precambrian gneiss adjacent to the Espanola basin (Fig. 9.5) and from intersections of similar rocks in the deep drill holes at Fenton Hill (Harrison et al., 1986) (Fig. 9.5) yielded $^{40}Ar/^{39}Ar$ age spectra similar in some respects to the Sandia Granite sample. The three Fenton Hill samples, from depths of 1.13 km, 2.62 km, and 2.90 km (Figs. 9.7–9.9), yield an early plateau of about 870 Ma and rise monotonically to an age of ≈ 1,030 Ma. The outcrop sample yields ages between ≈ 830 and 900 Ma over 70% of the release pattern with $^{40}Ar^*$ loss continuing to as late as 500 Ma ago (Fig. 9.10). Clearly these four samples have not experienced the same amount of recent uplift as the Sandia Granite (Fig. 9.6) and may represent typical K-feldspars available for deposition from the Precambrian basement during the mid-Tertiary.

Abo Sandstone

A K-feldspar separate was made from a sample of the Permian Abo Sandstone exposed on the north-

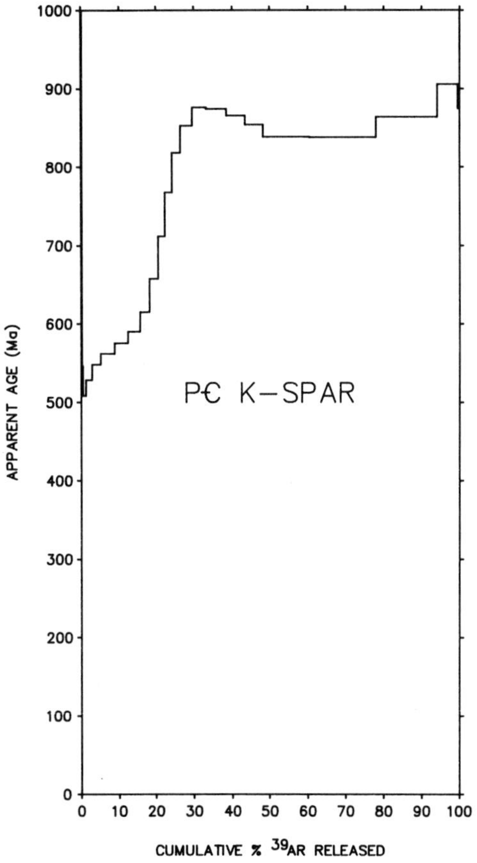

FIGURE 9.10. $^{40}Ar/^{39}Ar$ age spectrum of microcline from outcrop of Precambrian gneiss adjacent to the Espanola Basin.

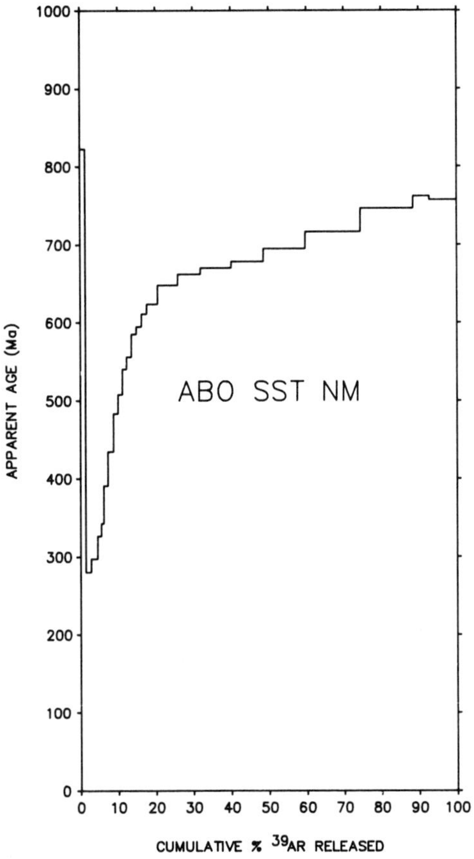

FIGURE 9.11. $^{40}Ar/^{39}Ar$ age spectrum of detrital K-feldspar separated from the Permian Abo Sandstone, which indicates ^{40}Ar loss as recently as 200 Ma ago.

ern rim of the basin. As much as 3 km of Paleozoic and Mesozoic overburden at this site has been estimated by Kelley (1977), which could cause burial temperatures high enough to cause late-Mesozoic $^{40}Ar^*$ loss from the detrital Precambrian feldspars in the Abo Sandstone. Such a history would be consistent with the release pattern shown in Figure 9.11 of an age gradient rising from an extrapolated age of about 100 Ma to ages as old as $\approx$750 Ma. This second cycle feldspar is also a possible candidate for Tertiary redeposition.

Santa-Fe Group – Basin K-Feldspars

The thermal gradient recorded in both the Isleta #2 and Federal West Mesa #1 wells averages about 31 °C/km (Scot Amos, personal communication, 1983), placing our samples from 4.2 km to 5.2 km at present temperatures between 145 °C and 175 °C. Age spectra for the two samples from the Federal West Mesa #1 well, 14900 (4.54 km to 4.57 km) and 16070 (4.88 km to 4.90 km), reveal considerable ^{40}Ar loss (Figs. 9.12 and 9.13) with respect to microcline from any of the five source candidates (Figs. 9.6 to 9.11) representing the initial distribution. Federal West Mesa #1 14900 may have lost as much as 35% of the preexisting $^{40}Ar^*$ during post-Oligocene basin heating while Sample 16070 has been outgassed between 50% and $\approx$65%, depending on the starting distribution. The age intercepts at 0% ^{39}Ar release do not reveal the age of the responsible heating event because of contamination on the periphery of the grains by excess ^{40}Ar (for an explanation of this phenome-

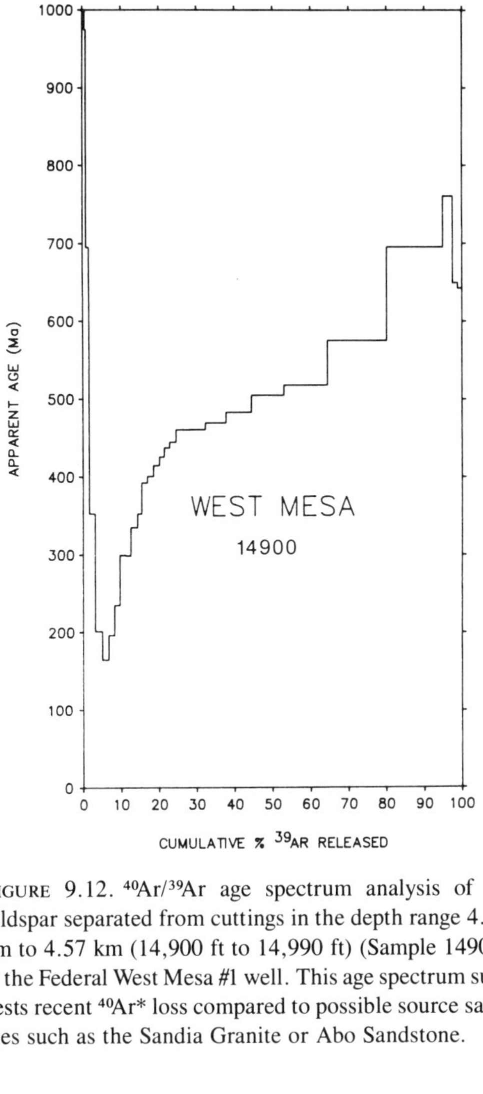

FIGURE 9.12. $^{40}Ar/^{39}Ar$ age spectrum analysis of K-feldspar separated from cuttings in the depth range 4.54 km to 4.57 km (14,900 ft to 14,990 ft) (Sample 14900) at the Federal West Mesa #1 well. This age spectrum suggests recent $^{40}Ar^*$ loss compared to possible source samples such as the Sandia Granite or Abo Sandstone.

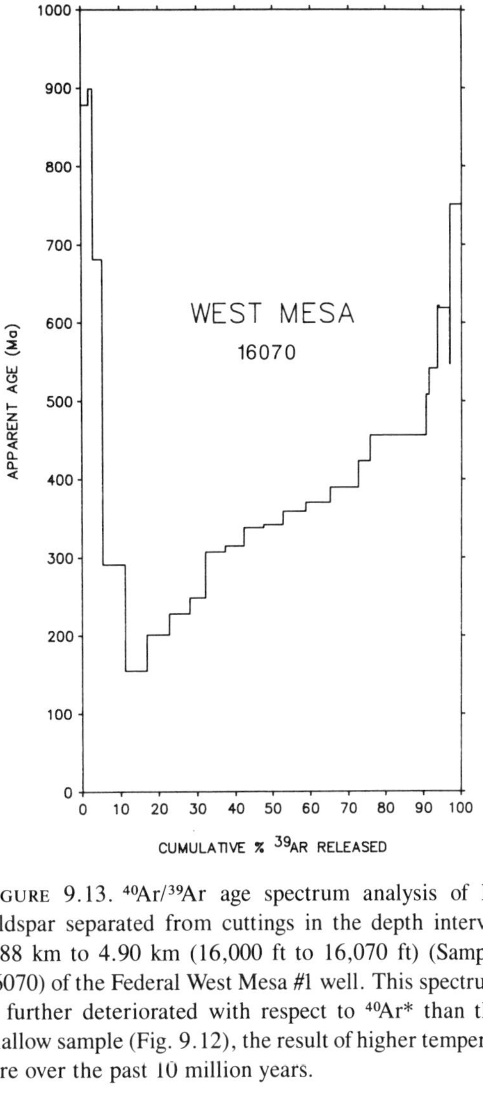

FIGURE 9.13. $^{40}Ar/^{39}Ar$ age spectrum analysis of K-feldspar separated from cuttings in the depth interval 4.88 km to 4.90 km (16,000 ft to 16,070 ft) (Sample 16070) of the Federal West Mesa #1 well. This spectrum is further deteriorated with respect to $^{40}Ar^*$ than the shallow sample (Fig. 9.12), the result of higher temperature over the past 10 million years.

non, see Harrison and McDougall, 1981; Harrison, 1983; Harrison et al., 1986). Although this contamination obscures the age of outgassing, extrapolation of the crystal interior age gradients suggests a near-zero age for this event, suggesting that the bottom of the Albuquerque basin is now as hot as it has been at any time during the current rifting episode.

The three Isleta #2 samples yield steep age gradients that indicate that outgassing has occurred during an event within the past 80 million years. Extrapolation of these curves and appreciation of the geological environment suggest that this event is the modern basin heating. The two K-feldspar separates (Fig. 9.14) 16224 (4.95 km) and 17161 (5.23 km), yield generally coherent spectra that indicate $^{40}Ar^*$ loss in recent time of $\approx 45\%$ and 35%, respectively. Again, the resolution of the time of gas loss is obscured by the existence of excess $^{40}Ar^*$ at the periphery of the grains, but a curve fit to the 17161 data reveals a virtually zero age for this event. However, the temperatures presently measured at 4.95 km and 5.23 km are approximately 165°C and 175°C, respectively.

An apparent discrepancy of these results is that the sample at the *lower* temperature (16224) has experienced the *greater* $^{40}Ar^*$ loss. The explanation for this paradox underscores the value of our technique. Diffusion coefficients calculated from

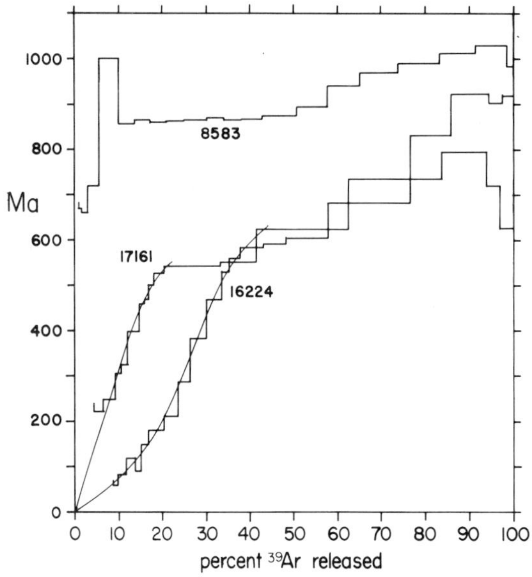

FIGURE 9.14. $^{40}Ar/^{39}Ar$ age spectra of microcline extracted from core samples in the Isleta #2 well at 5.23 km (17,161 ft) and 4.95 km (16,224 ft). Also shown is an age spectrum from one of the Fenton Hill samples (8583) (Fig. 9.8). Paradoxically, the sample at the shallower depth in Isleta #2 (16224) has lost more $^{40}Ar^*$ than the deeper sample. This is a result of the differing $^{40}Ar^*$ retentivities of the samples.

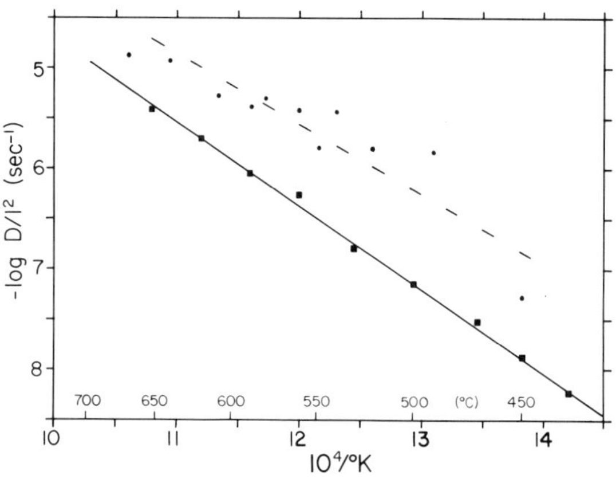

FIGURE 9.15. Arrhenius diagram of diffusion coefficients calculated from measured ^{39}Ar loss from Isleta #2 Samples 17161 and 16224. Data from 17161 (solid squares) show a good alignment and a steeper slope than the scattered data from 16224 (solid circles). The implication of this result is that Sample 16224 contains a spectrum of grain sizes with an overall $^{40}Ar^*$ retentivity less than sample 17161.

the measured ^{39}Ar release during the gas extraction are plotted in Figure 9.15 against the reciprocal absolute temperature of the extraction step. The results from Sample 17161 reveal an extremely well-correlated ($r = 0.997$) array corresponding to an activation energy (E) of 36.6 kcal/mol, and a frequency factor (D_0/ℓ^2) of 2,300/s. In contrast, the results from Sample 16224 plot in a scattered fashion well above the 17161 line. A likely explanation for the poor correlation is that a mixture of different feldspar structural states and variable diffusion dimensions have obscured the coherence that would have been yielded by a sample within a single K-feldspar structure. Evidence for the latter mechanism is inferred from the age spectrum itself, which shows a convex upwards gradient that is characteristic of a complex distribution of grain sizes (e.g., Turner, 1968). Despite the scatter, it is clear that argon is lost more easily from Sample 16224 (and with a lower temperature dependence) than 17161.

The burial history of the Santa Fe Group in this area is at present sufficiently well known (Kelley, 1977; Lee Russell, personal communication, 1983; John Hawley, personal communication, 1984) to allow us to estimate the temperature increase for these samples since the late Miocene to be about 8°C/my. Assuming linear heating and using the kinetic parameters determined for 17161, the calculated cumulative ^{40}Ar loss for this sample is 42%, close to the value of 35% actually observed. This agreement is probably unrealistically good in view of the proximity of the Isleta #2 well to recent volcanic activity, but demonstrates the value of a method of paleothermometry that supplies specific Arrhenius parameters for the sample as a by-product of analysis.

Sample 13800 (Fig. 9.16) from 4.18-km to 4.21-km depth in the Isleta well shows only slightly less ^{40}Ar loss than Sample 17161 from 5.23 km. This may again be due to a lower activation energy for ^{40}Ar transport or may reflect our uncertainty in the actual starting age distribution in the detrital K-feldspars. The latter explanation seems unlikely in that Sample 13800 yields lower ages over the first $\approx 70\%$ ^{39}Ar release than any of the five non-Tertiary samples analyzed. In summary, K-feldspars from the Albuquerque basin yield age spectra consistent with active heating of rift fill sediments over the last several million years, but stratigraphic studies in progress may allow refinement of this simple picture.

Case Study: North Sea Basin

A dozen analyses of K-feldspar separated from Paleozoic horizons in the North Sea basin reveal small $^{40}Ar^*$ loss gradients in response to Tertiary to present subsidence (T.M. Harrison, unpublished data; Phelps and Harrison, 1985, 1986). Analyses of potassium feldspars from sandstones from the southern Viking Graben and southern North Sea area yield plateau ages that are consistent with the age of surrounding basement rocks (Phelps and Harrison, 1986). On the basis of the amount of recent argon loss in the Viking Graben samples, a temperature history was constructed that is consistent with that derived independently by standard backstripping techniques.

Five samples were analyzed from the southern Viking Graben: four Mesozoic deltaic sandstones and one Paleocene turbiditic sandstone. The potassium feldspar separates from the deltaic sandstones yielded plateau ages that cluster around 400 Ma, suggesting derivation from Caledonian aged crystalline rocks, which are the dominant basement rocks in the Viking Graben area. Potassium feldspars from the Paleocene turbidite yielded a minimum age of Late Precambrian. Facies relationships suggest that the Paleocene turbidite was derived from the Shetland platform, an area underlain by Precambrian crystalline rocks.

Three samples of the Permian Rotliegendes Sandstone from the southern North Sea area were also analyzed. Potassium feldspars from two samples yielded complex spectra indicating components from both Caledonian and Precambrian aged crystalline rocks. One sample yielded a well-defined plateau age of 398 Ma, suggesting derivation from Caledonian aged crystalline rocks.

Only the release spectra from the southern Viking Graben samples were of sufficient quality for thermal history analysis. The present-day temperature of these samples ranges from 68°C to 115°C. Recent $^{40}Ar^*$ loss (2 to 3%) was detected in the hottest sample. Following the methods of Harrison and McDougall (1982) and Harrison and Bé

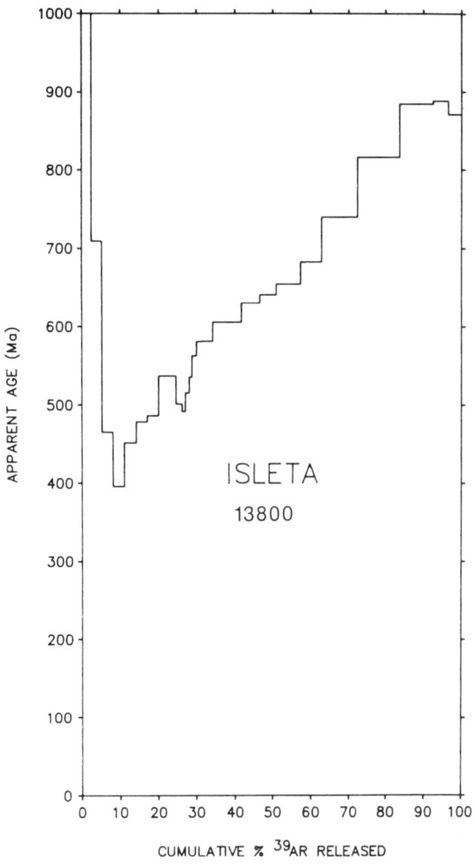

FIGURE 9.16. $^{40}Ar/^{39}Ar$ age spectrum of K-feldspar extracted from Isleta #2 cuttings in the depth interval 4.18 km to 4.21 km (13,700 ft to 13,800 ft) (Sample 13800).

(1983), an activation energy of 36 kcal/mol was calculated. With this activation energy, the observed amount of argon loss should have occurred in only 2 or 3 million years at the present-day temperature of 115°C. A more geologically reasonable temperature history requires that present-day temperatures have been maintained for less than 1 million years, and that the temperatures rose above 90°C within the past 3 million years.

A temperature history for this sample was also constructed by applying standard backstripping techniques. The resulting subsidence curves are compatible with approximately 40% crustal stretching ($\beta = 0.4$) with a period of thermally controlled subsidence beginning at the end of the Late Cretaceous. A heat-flow history was constructed using 40% crustal attenuation, and the heat-flow equation was solved assuming one-dimensional, conductive heat flow. The temperature history calculated by this method predicts a rapid rise in temperature in the last 5 million years, and is compatible with that determined independently from the $^{40}Ar/^{39}Ar$ results.

Conclusion

Microcline $^{40}Ar/^{39}Ar$ thermochronology has, in the several cases presented here, shed light on aspects of the timing and thermal intensity of basin heating as well as provided provenance-related information. A strength of the approach is that laboratory degassing data can reveal sample-intensive kinetic data that may be used for the thermal calculations. Experience has shown that, depending on the somewhat variable argon transport rate in the specimen, temperatures at least as high as $\approx 110°C$ for geological periods are required before measurable $^{40}Ar^*$ loss occurs.

The most serious restriction on interpretation stems from our present need to analyze relatively large samples (thousands of grains) to obtain a sufficient signal for analysis. If this population has differing diffusion characteristics and/or source ages, a straightforward interpretation is not possible. We are currently developing the ability to analyze single (<1 mm sized) crystals with high precision and $\pm 1°C$ temperature control, removing this limitation.

Acknowledgments. Support for this research was provided through DOE grant DE-AC02-82ER13013. We also thank the Research Corporation and the Petroleum Research Fund/ACI for support for the equipment used in this study. The manuscript was improved through critical readings by Paul Morgan and Marvin Lanphere.

References

Albarede, F., Feraud, G., Kaneoka, I., and Allegre, C.J. 1978. $^{40}Ar/^{39}Ar$ dating: The importance of K-feldspars on multimineral data of polyorogenic areas. Journal of Geology 89:581–598.

Bandy, O.L., and Arnal, R.E. 1969. Middle Tertiary basin development, San Joaquin Valley, California. Geological Society of America Bulletin 80:783–819.

Berger, G.W. 1975. $^{40}Ar/^{39}Ar$ step heating of thermally overprinted biotites, hornblendes, and potassium feldspars from Eldora, Colorado. Earth and Planetary Science Letters 26:387–408.

Black, B.A. 1982. Oil and gas exploration in the Albuquerque basin. New Mexico Geological Society Thirty-Third Field Conference Guidebook, pp. 313–324.

Brandt, S.B., and Bartnitskiy, E.N. 1964. Losses of radiogenic argon in potassium-sodium feldspars on heat activation. International Geologic Review 6:1483.

Briggs N.D., Naeser, C.W., and McCulloh, T.H. 1981. Thermal history of sedimentary basins by fission-track dating. Nuclear Tracks 5:235–237.

Burchfiel, B.C., Eaton, G.P., Lipman, P.W., and Smith, R.B. 1982. The Cordilleran Orogen: Conterminous U.S. sector. In: Perspectives in Regional Geological Synthesis: Planning for the Geology of North America. Geological Society of America, Decade of North American Geology Special Publication, pp. 91–98.

Callaway, D.C. 1971. Petroleum potential of San Joaquin Basin, California. In: Cram, I.H. (ed.): Future Petroleum Provinces of the United States. American Association of Petroleum Geologists Memoir 15, pp. 239–253.

Castano, J.R., and Sparks, D.M. 1974. Interpretation of vitrinite reflectance measurements in sedimentary rocks and determination of burial history using vitrinite reflectance and authigenic minerals. Geological Society of America Special Paper 153, pp. 31–52.

Dalrymple, G.B., and Lanphere, M.A. 1971. $^{40}Ar/^{39}Ar$ technique of K-Ar dating: A comparison with the conventional technique. Earth and Planetary Science Letters 12:300–308.

Dodson, M.H. 1973. Closure temperature in cooling geochronological and petrological systems. Contributions to Mineralogy and Petrology 40:259–274.

Faure, G. 1977. Principles of Isotope Geochemistry. New York, Wiley, 464 pp.

Foland, K.A. 1974. ^{40}Ar diffusion in homogeneous orthoclase and an interpretation of Ar diffusion in K-feldspar. Geochimica et Cosmochimica Acta 38:151–166.

Hanson, G.N., Simmons, K.R., and Bence, A.E. 1975. $^{40}Ar/^{39}Ar$ spectrum ages for biotite, hornblende and muscovite in a contact metamorphic zone. Geochimica et Cosmochimica Acta 39:1269–1277.

Harrison, T.M. 1983. Some observations on the interpretation of $^{40}Ar/^{39}Ar$ age spectra. Isotope Geoscience 1:319–338.

Harrison, T.M., Armstrong, R.L., Naeser, C.W., and Harakal, J.E. 1979. Geochronology and thermal history of the Coast Plutonic Complex, near Prince Rupert, British Columbia. Canadian Journal of Earth Science 16:400–410.

Harrison, T.M., and Bé, K. 1983. $^{40}Ar/^{39}Ar$ age spectrum analysis of detrital microclines from the southern

San Joaquin Basin, California: An approach to determining the thermal evolution of sedimentary basins. Earth and Planetary Science Letters 64:244–256.

Harrison, T.M., and McDougall, I. 1980. Investigations of an intrusive contact, northwest Nelson, New Zealand: I. Thermal, chronological and isotopic constraints. Geochimica et Cosmochimica Acta 44:1985–2004.

Harrison, T.M., and McDougall, I. 1981. Excess ^{40}Ar in metamorphic rocks from Broken Hill, N.S.W.: Implications for $^{40}Ar/^{39}Ar$ age spectra and the thermal history of the region. Earth and Planetary Science Letters 55:123–149.

Harrison, T.M., and McDougall, I. 1982. The thermal significance of potassium feldspar K-Ar ages inferred from $^{40}Ar/^{39}Ar$ age spectrum results. Geochimica et Cosmochimica Acta 46:1811–1820.

Harrison, T.M., Morgan, P., and Blackwell, D.D. 1986. Constraints on the age of heating at the Fenton Hill Site, Valles Caldera, New Mexico. Journal of Geophysical Research 91:1899–1908.

Hoots, H.W., Bear, T.L., and Kleinpell, W.D. 1954. Geological summary of the San Joaquin Valley, California. In: Jahns, R.H. (ed.): Geology of Southern California. California Division of Mines Bulletin 170, chap. 2, pp. 113–129.

Kelley, S.A., and Duncan, I.J. 1984. Tectonic history of the northern Rio Grande Rift from apatite fission-track geochronology. New Mexico Geological Society Thirty-Fifth Field Conference Guidebook, pp. 67–73.

Kelley, S.A., Duncan, I.J., and Blackwell, D.D. 1982. History of uplift in the Rio Grande Rift. EOS 63:1117.

Kelley, V.C. 1977. Geology of Albuquerque Basin, New Mexico. New Mexico Bureau of Mines and Mineral Resources Memoir 33, 60 pp.

Kuz'min, A.M. 1961. Retention of argon in microcline. Geochemistry 5:485–488.

Lanphere, M.A., and Dalrymple, G.B. 1971. A test of the $^{40}Ar/^{39}Ar$ age spectrum technique on some terrestrial materials. Earth and Planetary Science Letters 12:359–372.

Lozinsky, R.P. 1984. Stratigraphy and sedimentology of the Santa Fe Group in the Albuquerque Basin, north-central New Mexico. Unpublished dissertation proposal, New Mexico Institute of Mining and Technology, Socorro, 13 pp.

MacPherson, B.A. 1978. Sedimentation and trapping mechanisms in Upper Miocene Stevens and older turbidite fans of southeastern San Joaquin Valley, California. American Association of Petroleum Geologists Bulletin 62:2243–2274.

Majumdar, A., Brookins, D.G., Wilson, A.F., and Baksi, A.K. 1984. Geochronology and geochemistry of the Sandia Granite, north-central New Mexico. EOS 65:300.

McDougall, I., and Roksandic, Z. 1974. Total fusion $^{40}Ar/^{39}Ar$ ages using HIFAR reactor. Journal of Geological Society of Australia 21:81–89.

Merrihue, G., and Turner, G. 1966. Potassium-argon dating by activation with fast neutrons. Journal of Geophysical Research 71:2852–2857.

Mitchell, J.G. 1968. The $^{40}Ar/^{39}Ar$ method for K-Ar age determination. Geochimica et Cosmochimica Acta 32:781–790.

Morgan, P., and Golombek, M.P. 1984. Two phases and two styles of extension in the northern Rio Grande Rift. New Mexico Geological Society Thirty-Fifth Field Conference Guidebook, pp. 13–19.

Musset, A.E. 1969. Diffusion measurements and the potassium-argon method of dating. Geophysical Journal of the Royal Astronomical Society 18:257– 303.

Phelps, D.W., and Harrison, T.M. 1985. Thermal history of the southern Viking Graben from $^{40}Ar/^{39}Ar$ thermochronology. Abstracts, American Association of Petroleum Geologists Research Conference on Radiogenic Isotopes and Evolution of Sedimentary Basins, New Orleans, LA.

Phelps, D.W., and Harrison, T.M. 1986. Application of $^{40}Ar/^{39}Ar$ thermochronology on detrital potassium feldspar to the study of sedimentary basins. In: Burrus, J. (ed.): Thermal Modeling in Sedimentary Basins. Paris, Editions Technip, pp. 585–600.

Riecker, R.E. 1979. Rio Grande Rift: Tectonics and Magmatism. Washington, DC, American Geophysical Union, 438 pp.

Sardarov, S.S. 1957. Retention of radiogenic argon in microcline. Geochemistry 3:233–237.

Simonson, R.R. 1958. Oil in the San Joaquin Valley, California. In: Weeks, L.D. (ed.): Habitat of Oil. Tulsa, OK, American Association of Petroleum Geologists, pp. 99–112.

Turner, G. 1968. The distribution of potassium and argon in chondrites. In: Ahrens, L.H. (ed.): Origin and Distribution of the Elements. New York, Pergamon, pp. 387–398.

Turner, G. 1970. Argon 40-Argon 39 dating of lunar rock samples. Proceedings of First Lunar Sciences Conference, Houston, TX, vol. 2, pp. 1665–1684.

Turner, G., Miller, J.A., and Grasty, R.L. 1966. The thermal history of the Bruderhiem meteorite. Earth and Planetary Sciences Letters 1:155–157.

Webb, G.W. 1981. Stevens and earlier Miocene turbidite sandstones, southern San Joaquin Valley, California. American Association of Petroleum Geologists Bulletin 65:438–465.

Ziegler, D.L., and Spotts, J.H. 1976. Reservoir and source bed history in the Great Valley of California. In: Tomorrow's Oil from Today's Provinces. Pacific Section, American Association of Petroleum Geologists Miscellaneous Publication 24, pp. 19–38.

10
The Application of Fission-Track Dating to the Depositional and Thermal History of Rocks in Sedimentary Basins

Nancy D. Naeser, Charles W. Naeser, and Thane H. McCulloh

Abstract

Fission tracks are zones of intense damage formed when fission fragments travel through a solid. One of the isotopes of uranium, ^{238}U, is the only naturally occurring isotope with a fission half-life sufficiently short to produce a significant number of fission tracks through geologic time. Uranium occurs in trace amounts in many minerals, and, because the spontaneous fission of ^{238}U occurs at a known rate, it is possible to calculate the age of a mineral by determining the number of fission tracks and the amount of uranium it contains.

If a mineral containing fission tracks is heated to a high enough temperature, however, the fission tracks will fade and disappear, resulting in a low apparent fission-track age. The temperature at which this fading, or "annealing," occurs depends on the mineral and the duration of heating. The most complete data available on annealing temperatures are for the two minerals most commonly used in fission-track studies, apatite and zircon. Fission tracks in apatite are totally annealed at temperatures that range from about 105°C (221°F) (for 10^8 yr effective heating time) to 150°C (302°F) (for 10^5 yr). The annealing temperatures of fission tracks in zircon are not as well known but are probably in the range of 200° ± 40°C (392° ± 72°F).

Apatite and zircon occur as detrital constituents in many sedimentary rocks and their annealing temperatures cover a temperature range that is useful for assessing the potential for economic resources in sedimentary basins. Consequently, fission tracks provide a powerful tool for determining the overall time-temperature history of sedimentary basins and for delineating localized temperature anomalies. Fission tracks also provide information on the timing of thermal events, on the sedimentation record, and on the provenance of the sedimentary rocks in a basin.

Application of the fission-track method to basin analysis is illustrated by two case histories – in California and Wyoming. In the southern San Joaquin Valley, California, fission-track ages of detrital apatite from drill-hole samples of Eocene to Miocene sandstones suggest that rocks in the rapidly subsiding basin northwest of the active White Wolf fault have been near their present temperature for only about 10^6 yr, whereas rocks southeast of the fault have been near their present temperature for about 10^7 yr. This agrees well with other estimates of heating duration for these rocks. Zircon fission-track data suggest that from at least Eocene to late Miocene time the dominant source of sediments in the study area was rocks related to the intrusive complexes that are presently exposed east and south of the southern San Joaquin Valley.

In the northern Green River basin, Wyoming, fission-track ages of detrital apatite from Upper Cretaceous and lower Tertiary sedimentary rocks in the El Paso Natural Gas Wagon Wheel No. 1 well indicate that significant cooling of the rocks in this part of the basin has occurred. The data suggest that the latest phase of cooling began about 4 to 2 Ma ago and involved a relatively rapid temperature decrease of 20°C (36°F) or more. Vitrinite reflectance and illite/smectite transformation data from Wagon Wheel No. 1 support this magnitude of cooling. The zircon data demonstrate a significant change in provenance in the early Tertiary that was probably related to unroofing of the crystalline core of the Wind River Mountains to the northeast.

Introduction

Fission tracks constitute a powerful tool for dating rocks and for studying their thermal, burial, and uplift and erosion history. Fission tracks have been

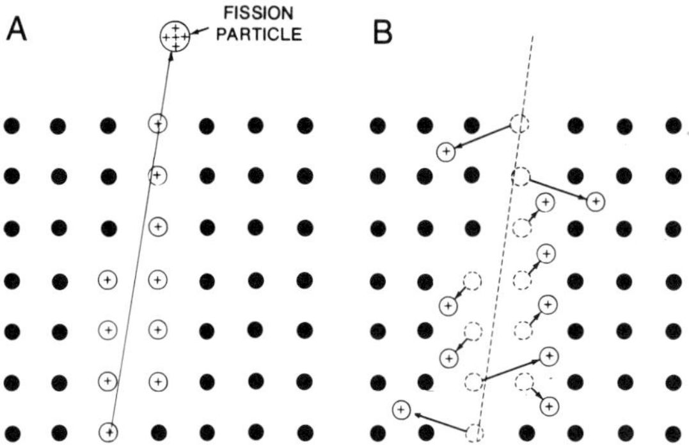

FIGURE 10.1. Schematic depiction of fission-track formation in a simple crystalline solid. A, The atoms have been ionized by the massive charged fission particle that has just passed. B, The mutual repulsion of the ions forces them to separate and be displaced from their normal crystallographic positions, forming the fission-track damage zone. (Modified from Fleischer et al., 1965; Macdougall, 1976; reprinted by permission of AAAS and Scientific American, Inc.)

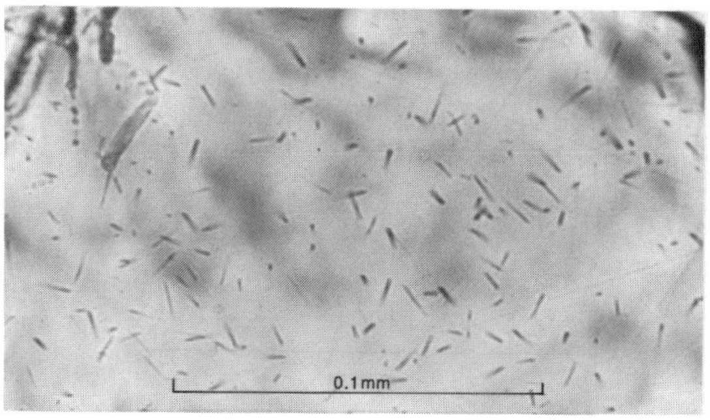

FIGURE 10.2. Etched fission tracks in apatite crystal. The tracks are straight and display the characteristic random orientation. They vary in length due to the varying angle of inclination of the fission particles and varying depth below surface of the fissioning nuclei. This apatite exhibits a rather uniform distribution of tracks, implying a uniform distribution of uranium. Many samples, however, exhibit heterogeneous track distributions.

used in a number of studies on basement rocks over the last 20 years (e.g., Naeser and Dodge, 1969; Wagner and Reimer, 1972; Wagner et al., 1977; Naeser, 1979a; Bryant and Naeser, 1980; Naeser et al., 1980; Kohn et al., 1984; numerous papers in Fleischer et al., 1984). However, only since 1979 has the fission-track method been applied to the study of sediments and to the thermal history of sedimentary basins.

Fission tracks have the potential of widespread application in basin analysis, because the two minerals most commonly used in fission-track studies, apatite and zircon, occur in many sedimentary rocks. In addition, the annealing temperatures of these two minerals span the temperature range of oil generation (Hood et al., 1975; Waples, 1980), as well as the temperatures useful for locating paleothermal anomalies associated with some mineral deposits. Fission tracks thus provide an attractive method for studying the overall thermal history and more localized temperature anomalies in sedimentary basins. Fission-track dating also has the advantage, not found in many other indicators of thermal maturity, of being able to date thermal events and marker horizons (e.g., volcanic ashes) in basins, and of providing information on the provenance of sedimentary rocks.

Naeser (1979b) first suggested the possibility of using fission-track annealing to define the thermal history of sedimentary basins. Subsequently, fission tracks have been applied to studies in a number of basins, including the southern San Joaquin Valley, California (Briggs et al., 1979, 1981a, 1981b; Naeser, 1984b; Naeser et al., in press), the northern Green River basin and southwestern Powder River basin, Wyoming (Naeser, 1984a, 1984b, 1986), and the Otway basin and Canning basin, Australia (Gleadow and Duddy, 1981; Duddy and Gleadow, 1982, 1985; Gleadow et al., 1983).

In this chapter we summarize the theory and methods of fission-track dating and the application of this method to the study of sedimentary basins. This application is illustrated by case histories in the southern San Joaquin Valley, California, and the northern Green River basin, Wyoming.

Theory of Fission-Track Dating

A fission track is the zone of intense damage formed when a fission fragment travels through a solid. Several naturally occurring isotopes fission spontaneously, but only ^{238}U has a sufficiently short fission half-life (9.9×10^{15} yr) to produce significant numbers of spontaneous tracks over a time period of geological interest.

When an atom of ^{238}U fissions, the nucleus breaks up into two lighter nuclei—one averaging about 90 atomic mass units and the other about 135 atomic mass units—with the liberation of about 200 MeV of energy. The two highly charged nuclei recoil in opposite directions and disrupt the electron balance of the atoms in the host mineral along their path. This disruption of electron charge in turn causes the mutual repulsion of positively charged ions and the displacement of the ions from their normal crystallographic positions in the host. The result is a zone of damage defining the fission track (Fleischer et al., 1975; Figs. 10.1 and 10.2). A new track is only tens of angstroms in diameter and about 10 to 20 μm in length. Tracks formed in low-density minerals are longer than those formed in dense minerals, such as zircon.

Trace amounts of uranium occur in a number of common minerals. Because ^{238}U fissions spontaneously at a constant rate, fission tracks can be used to date these minerals. An age can be calculated by determining the number of spontaneous tracks intersecting a polished surface of the mineral and the amount of uranium that produced those tracks. The techniques used in dating have been developed by physicists and geologists over the last 25 years. Fleischer et al. (1975) and Naeser (1979a) reviewed this early development of the fission-track dating method.

Annealing of Fission Tracks

Heating a mineral containing spontaneous fission tracks to a high enough temperature allows the ions displaced along the damage zone of the track to move back into normal crystallographic positions in the mineral, and thus repair the damage zone partially or totally. This results in a progressive shortening to (ultimately) a total disappearance of the fission track, leading to a reduction in the number of spontaneous tracks and an anomalously young fission-track age.

The most important factors controlling the temperature at which this "annealing" of fission tracks occurs are: 1) the mineral involved—different minerals anneal at different temperatures, and 2) the duration of heating—the longer a mineral is heated, the lower the temperature required to totally anneal its tracks. Annealing temperatures have been determined for many minerals by short-term laboratory experiments. However, temperature-time relationships that appear to be more directly applicable to most geological problems have been obtained by determining the temperatures at which minerals are totally annealed (i.e., yield a zero age) in drill holes in areas where the approximate duration of heating of the rocks is known. The most complete drill-hole data available are for apatite. Naeser's (1981) data indicate that total annealing of fission tracks in apatite occurs at temperatures that range from about 105 °C (221 °F) for relatively long-term heating (of 10^8 yr effective heating duration) to 150°C (302°F) (for 10^5 yr heating) (Fig. 10.3).

Extrapolated laboratory data predict slightly shorter heating times than the drill-hole data (Fig. 10.3). This is probably in part because the laboratory data do not take into account the number of new tracks that are formed during long-term geological annealing (see discussion in Sanford, 1981) and because the laboratory data are based on essentially square-pulse heating and thus probably underestimate heating time for most geological annealing. On the other hand, "effective" heating times are commonly difficult to estimate for drill-hole samples; it has been suggested that the times used in the drill-hole calibration have been overestimated (Harrison, 1985). If so, the effective heating times applicable to most long-term geological annealing probably lie between the times quoted for the laboratory and drill-hole data (Fig. 10.3).

The temperatures at which fission tracks in zircon are totally annealed are not as well known, but limited data show that the temperatures are higher than in apatite and suggest that they are probably in the range of 200° ± 40°C (392° ± 72°F) (e.g., see Harrison et al., 1979; Hurford, 1985; see also Zeitler, 1985).

Application to Sedimentary Basins

Fission tracks have been used in a wide range of studies in sedimentary basins. The annealing properties of fission tracks make them attractive for studying the thermal history of sediments from the time of deposition and burial through subsequent uplift and erosion. The annealing properties also give tracks the potential for defining localized temperature anomalies in a basin, such as those related to intrusions or to the passage of hot fluids through the rocks.

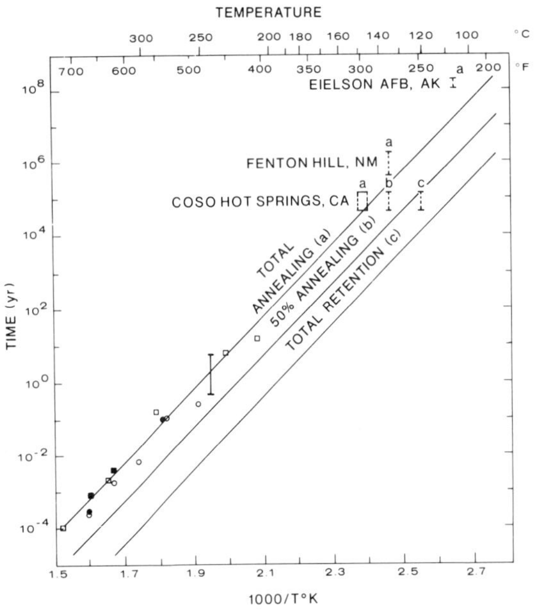

FIGURE 10.3. Temperatures required to anneal fission tracks in apatite for heating durations ranging from short-term laboratory experiments to geological heating lasting 10^8 yr (modified from Sanford, 1981, Fig. 2, reprinted by permission of the author). Temperatures for geological annealing (dotted vertical bars) (Naeser, 1981) were determined by dating apatite in drill holes in three areas where the approximate duration of heating is known. Circles and squares indicate the temperature-time conditions predicted for total annealing using the fading rates determined by the laboratory work of Naeser and Faul (1969) (open circles), Reimer (1972) (solid circles), Märk et al. (1973) (open squares), and Zimmermann and Gaines (1978) (solid squares). Extrapolation of the laboratory data (solid lines) is based on the rate constant for fission-track annealing determined from a least squares fit of the laboratory data by Zimmermann and Gaines (1978). The solid vertical bar is the 95% confidence interval for the least squares fit. a, b, and c on the solid lines and dotted vertical bars identify data for total annealing (a), 50% annealing (b), and total retention (c) of fission tracks.

Typically, most of the information about the thermal history of a sedimentary basin is provided by apatite, because many of the sedimentary rocks sampled in outcrop or drill holes have not been subjected to temperatures sufficient to anneal zircon (about 200°C; 392°F). However, even unannealed zircon can provide useful information. The lack of annealing sets some limits on maximum paleotemperatures, and the ages of the individual unannealed zircon grains are valuable in provenance studies in the basin. Zircon is also the mineral most commonly used to date volcanic ashes and their altered equivalents, bentonites and tonsteins.

Thermal History

Naeser (1979b) first suggested that fission-track annealing could be used to study the thermal history of sedimentary basins and described the expected trend of fission-track ages with depth in sedimentary rocks. This trend is typically somewhat more complicated than would be found in basement rocks because the grains that make up sediments can be of widely divergent ages. The grains may be equal in age to the stratigraphic age of the rock, as in the case of grains from contemporaneous volcanism, but much more commonly the grains will be older.

Naeser (1979b) described depth versus age trends for sedimentary sequences that 1) are at their maximum burial temperature and 2) have cooled below maximum paleotemperature. In a sedimentary sequence at maximum temperature (Fig. 10.4), fission-track ages of apatite from shallow rocks at relatively low temperatures will not have been affected by annealing during burial of the sediments ("zone of no annealing") and the ages of individual apatite grains in these rocks will reflect the age(s) of the source rocks for the detrital grains. The composite apatite age (see below) calculated for each individual sample in this zone may remain relatively constant with depth (as depicted in Fig. 10.4), but more commonly it will vary somewhat, depending on the age(s) of the detrital grains counted in a given sample (as in Fig. 10.12). In rocks subjected to progressively higher temperatures due to deeper burial, apatite will undergo partial annealing and give progressively younger ages ("zone of partial annealing"). Within this zone, the apparent apatite ages will become younger than the stratigraphic age of the rocks, and the apparent ages will finally decrease to zero at the depth where the temperature for total annealing is attained for a sufficient time ("zone of total annealing") (see Fig. 10.3).

The drill-hole annealing data and the extrapolated laboratory data (Fig. 10.3) suggest that the "zone of partial annealing" for apatite occurs over a temperature interval of only 30° to 35°C (54°

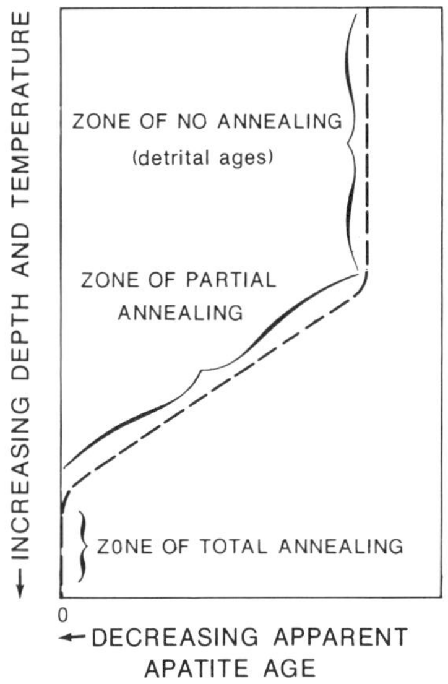

FIGURE 10.4. Expected decrease in apparent apatite age with increasing temperature and depth in a sedimentary sequence at the time of maximum burial heating (modified from Naeser, 1979b). The temperature at which the apatite age decreases to zero depends on the duration of time the rocks have been heated near their present temperature (see Fig. 10.3). Drill-hole and extrapolated laboratory data (Fig. 10.3) suggest annealing occurs over a 30° to 35°C temperature interval. See text for further explanation.

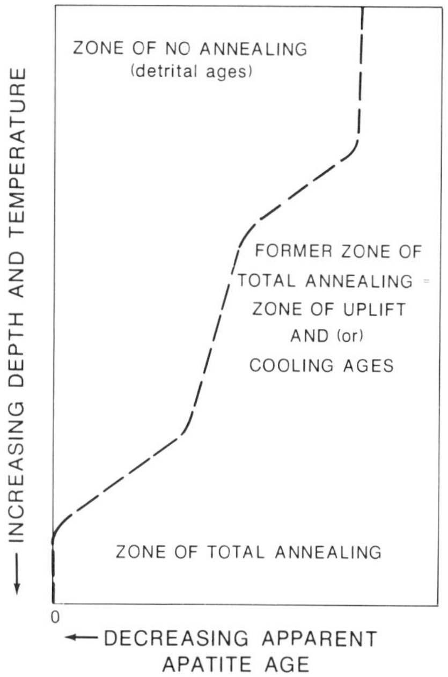

FIGURE 10.5. Expected distribution of apparent apatite ages in sedimentary rocks after uplift and/or cooling (cf. Fig. 10.4). (Modified from Naeser, 1979b.)

to 63°F) at times on the order of 10^5 to 10^8 yr. It must be emphasized that the actual temperatures covered by this zone depend, as does the temperature for total annealing, on the duration of heating. At a temperature of 105°C, for example, apatite heated for 10^8 yr will be totally annealed, but apatite subjected to short-term heating of only 10^5 yr duration should show little or no age reduction due to annealing.

When a sedimentary sequence cools below the maximum paleotemperature, in response to uplift and erosion or to a decrease in the geothermal gradient, apatite from the zone of total annealing once again begins to accumulate tracks and to record a fission-track age (Fig. 10.5). The recorded apatite age, the slope of the depth (or temperature) versus age plot, and the thickness of the zone of cooling ages give information on the time, rate, and amount of cooling, respectively.

Zircon fission-track ages should generate curves similar to those shown in Figures 10.4 and 10.5, but the annealing will occur at higher temperatures than in apatite.

Research on apatite from the Otway basin, Australia, shows that the reduction in mean track length and change in track length distribution that produce the observed age decrease with progressive annealing can in themselves offer important information about thermal history (Duddy and Gleadow, 1982; Gleadow et al., 1983; Green et al., this volume). In situations where a reasonably complete age profile cannot be obtained through a sedimentary sequence, or where the ages alone are ambiguous, length measurements often provide valuable insight into the thermal history. However, track length measurements may not be possible (or practical) in all cases, for example in young or low-uranium apatite with low spontaneous track density.

Economic Implications

The annealing of fission tracks in apatite and zircon occurs over a temperature range that makes fission-track analysis a sensitive method for assessing the potential for economic resources in sedimentary basins. For example, fission-track

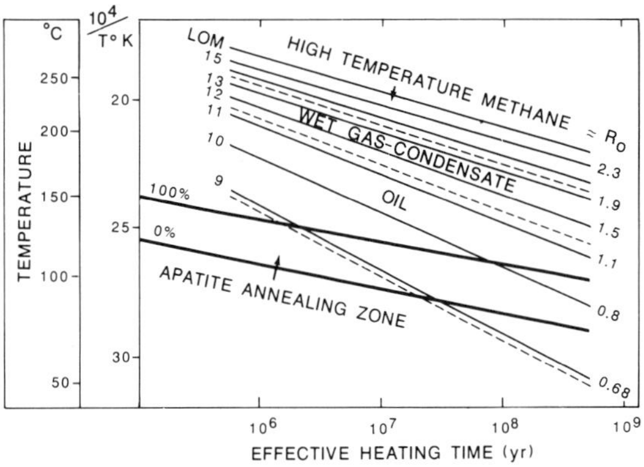

FIGURE 10.6. Correlation of the effective heating time-maximum temperature relationship of organic maturation as defined by Hood et al. (1975) to fission-track annealing in apatite. Figure modified from Hood et al. (1975, Figs. 2 and 3, reprinted by permission of American Association of Petroleum Geologist); boundaries of stages of petroleum generation based on Hood et al. (1975) and Waples (1980, Table 5); annealing data based on drill-hole data in Figure 10.3. LOM = Level of Organic Metamorphism; R_o = vitrinite reflectance.

analysis has the potential for widespread use as a relatively inexpensive exploration tool for delineating paleothermal anomalies associated with mineral deposits (Naeser et al., 1980; Cunningham and Barton, 1984; Naeser and Cunningham, 1984). Fission-track analysis also offers a powerful tool for assessing the hydrocarbon potential of sedimentary basins because the annealing temperatures of apatite and zircon span the temperature range where most oil generation is thought to occur.

A preliminary attempt to show the relationship between apatite track annealing and organic maturation is shown in Figure 10.6 (see also Gleadow et al., 1983, Fig. 6). Figure 10.6 suggests that:

1. For effective heating times greater than 1×10^6 to 2×10^6 yr (i.e., for most Tertiary and older basins), total annealing of apatite will occur within the temperature range of oil generation.
2. For progressively longer heating times, the temperature required for total annealing of apatite will correspond to a progressively greater degree of organic maturity.

However, defining the relationship between track annealing and organic maturation, as shown in Figure 10.6, is hampered at this time by several factors:

1. The uncertainties involved in defining the kinetics of vitrinite maturation and in assigning definite values of vitrinite reflectance to the stages of petroleum generation (Barker, 1983; Price, 1983, 1985; Price and Barker, 1984; McCulloh and Fan, 1985).
2. To some degree, the uncertainty in the time values used in the geological calibration of apatite annealing (Harrison, 1985).

Provenance Studies

In sedimentary rocks where zircon has been unaffected by postdepositional annealing of fission tracks, the ages of the individual zircon grains provide much information about the source rocks eroded to form the sediments. For example, within a detrital grain suite it may be possible to correlate individual age populations to probable source rocks to delineate sediment transport patterns in a basin (Hurford et al., 1984; Baldwin and Harrison, 1985; Yim et al., 1985). Changes in the zircon suite through time can be used to reconstruct changes in the sedimentation pattern, to define input from different source areas, and to reconstruct the uplift and erosion history of the source rocks (Zeitler et al., 1982; Cerveny, 1986; Cerveny et al., 1986, in press).

A general indication of the source terrane may be obtained by dating a relatively small number of grains, as illustrated in the two case histories stated here. However, detailed provenance studies may involve dating as many as 50 or more grains per sample (Zeitler et al., 1982; Hurford et al., 1984; Cerveny, 1986; Cerveny et al., 1986, in press).

Recent work (Naeser et al., 1987) has shown that such detailed studies may also require the use of a multistage etching procedure to compensate for the variable etching rate of zircons of different uranium content and age. This procedure helps ensure that all age populations present in the suite are properly etched and can be counted.

Sedimentation Record

Volcanic ash (tephra) layers and their altered equivalents, bentonites and tonsteins, form widespread marker horizons that have proved valuable for establishing chronology and correlations in sedimentary sequences. A number of studies have established fission-track dating as the most suitable method for dating tephras, particularly those older than the 40,000- to 50,000-yr limit of radiocarbon dating. A major advantage of fission-track dating, over K-Ar and conventional radiocarbon dating, is that the problem of contamination is greatly minimized (see discussions in Naeser et al., 1981, and Naeser and Naeser, 1984). This is particularly important for dating tephras, which commonly contain detrital contaminants.

Zircon phenocrysts are the preferable material used for dating tephras. Apatite occurs as phenocrysts in many tephras and may be used for dating; however, it is more subject to postdepositional annealing than zircon, and, in tephras less than about 5 Ma old, apatite usually has too low a track density to yield precise ages.

Natural glass has been dated extensively because of its abundance, particularly in Quaternary tephra. However, glasses present numerous problems, the greatest of which is the ease with which they lose spontaneous tracks by annealing. Both hydrated and nonhydrated glass can anneal at ambient surface temperatures in a million years or less (Seward, 1979; Naeser et al., 1981; Naeser and Naeser, 1984). Because of this problem, great care must be used in interpreting all glass fission-track ages; *they should always be considered minimum ages*. Unfortunately, glass is the only datable phase present in some tephras, and, in some situations, even minimum ages provide useful information.

Methods

Sample Requirements

The fission-track method can be used to date very fine sand-sized or coarser grains (coarser than about 74 μm; 200 mesh). The amount of sample required depends on the composition of the rock. In sedimentary rocks, for example, poorly sorted, arkosic sandstones yield more zircon and apatite than clean quartzose sandstones in which apatite, in particular, tends to be sparse or absent. In the two studies described here, which involved both sandstone and granitic basement rock, the amount of rock available for processing varied from about 64 to 500 g (0.14 to 1.10 lb) per sample. This yielded sufficient apatite and zircon for age determinations, but, elsewhere, larger samples may be required for a similar type of study.

When sampling material from drill holes, core samples are preferable to cuttings. Cuttings may be used, but the possibility of severe down-hole contamination must be taken into consideration. Also the cuttings must be of sufficient size that contamination from apatite or zircon grains in the drilling mud can be discounted. Such contamination is a potential problem for samples from some of the Tertiary-Cretaceous basins of the western United States where Cretaceous bentonites are commonly used as drilling mud and a large component of the detrital grains in many of the sedimentary rocks are Cretaceous in age (e.g., see Figs. 10.13 and 10.17).

In sampling drill holes it may at times be necessary to composite core or cuttings over a vertical distance to obtain sufficient material for analysis. However, this practice should be used with caution, particularly within the zone of partial annealing where significant changes in annealing can occur over relatively small temperature (and depth) intervals (e.g., see Fig. 10.16).

Laboratory Procedures

After crushing, pulverizing, and sizing bulk samples, apatite and zircon are separated by heavy liquid and magnetic separation techniques, such as those shown in Figure 10.7. The final separations should contain greater than about 80% apatite or zircon to be easily dated. In some samples careful sizing of the heavy-liquid concentrates may be needed to improve the percentage of apatite or zircon. For sedimentary rocks, the methylene iodide light fraction frequently requires treatment with a bromoform-methylene iodide mix (sp. gr. = about 3.10), and, in some samples, a further step of hand picking, to improve the segregation of apatite from contaminating grains.

Fission tracks in their natural state are too small to be seen except with an electron microscope, but a proper chemical etchant can enlarge the damage zone so that it can be observed using an optical microscope at intermediate magnifications (×200

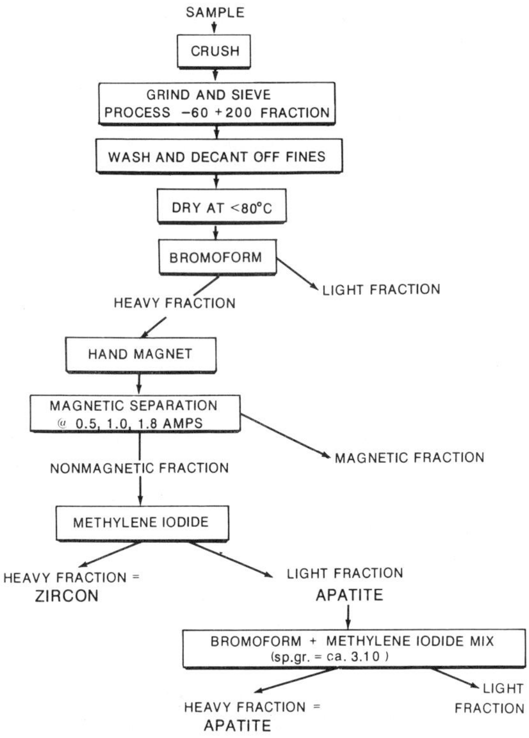

FIGURE 10.7. Steps involved in separating apatite and zircon from bulk rock samples for fission-track dating. The methylene iodide light fraction usually yields an adequate apatite separate, but in some samples further treatment with a bromoform-methylene iodide mix is necessary to remove contaminating grains.

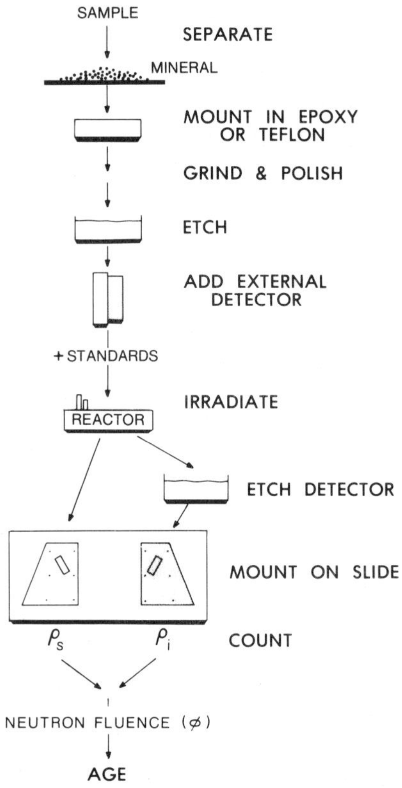

FIGURE 10.8. Steps involved in obtaining a fission-track age using the external detector method. Details of the laboratory procedures are given in Naeser (1976).

to 500) (Fig. 10.2). Common etchants used to develop tracks include nitric acid (for apatite), hydrofluoric acid (for glass and mica), concentrated basic solutions (for sphene), and basic fluxes (for zircon) (Fleischer et al., 1975, Table 2-2; Gleadow et al., 1976).

Spontaneous track density is usually determined by 1) polishing an internal surface of the glass or mineral, 2) enlarging the fission tracks intersecting this surface by etching, and 3) counting the number of tracks per unit area using an optical microscope, generally at magnifications of × 500 (glass) to × 1,500 (minerals). Because the relative abundance of ^{238}U and ^{235}U is constant in nature, the easiest and the most accurate way to determine the amount of uranium is to irradiate the sample in a nuclear reactor with thermal neutrons, which create a new set of fission tracks by inducing fission in the ^{235}U in the sample. The resulting induced track density, counted either directly in the sample or in a detector that covered the sample during irradiation, is a function of the amount of uranium in the sample and the neutron fluence the sample received in the reactor. A number of methods of determining spontaneous and induced track densities have been developed over the years. Most laboratories use a variation of one of the following two methods.

External Detector Method

The external detector method (Naeser, 1976, 1979a) (Fig. 10.8) permits induced tracks to be counted from exactly the same area of a crystal in which the spontaneous tracks are counted. It is, therefore, the preferred method for counting tracks in minerals, such as zircon and sphene, in which there typically exists a wide range in the uranium content both between and within individual grains in a single sample. *This method must also be used for all minerals from sedimentary rocks because of the potential age inhomogeneity of detrital grains derived from a number of sources.* In the external detector method the spontaneous tracks are counted in the crystal and the induced tracks are counted in a low-uranium-content (less than 10 ppb) muscovite or plastic detector that covered the crystal mount during neutron irradiation. Usually tracks in only 6 to 12 zircons are counted for an age determination of a sample, although for some applications of fission-track dating 50 or more grains are counted (e.g., see Zeitler et al., 1982; Hurford et al., 1984; Cerveny, 1986; Cerveny et al., 1986, in press).

Population Method

In the population method (Naeser, 1976, 1979a) (Fig. 10.9), spontaneous and induced track densities are determined from different splits of the sample. This method is suitable for dating glass because all the glass from a single source has a similar uranium concentration. The method is also preferred for dating apatites that contain abundant defects.

One split of the sample is mounted in epoxy, polished, and etched to reveal spontaneous tracks. The second split is irradiated, and then mounted, polished, and etched (it is standard practice to etch both mounts simultaneously). Apatite of the second split is heated prior to the irradiation to remove spontaneous tracks so that only induced tracks are revealed when the irradiated split is etched; therefore, induced track density is determined directly from the irradiated grains. Preirradiation heating of glass to remove the spontaneous tracks is not recommended because heating can alter the etching characteristics (and thus the apparent track density) of the glass. Therefore, etching glass of an irradiated split reveals both spontaneous and induced tracks. Induced track density (ρ_i) is determined by subtracting the spontaneous track density of the first split (ρ_s) from the total track density ($\rho_i + \rho_s$) in the irradiated glass.

The amount of glass that should be counted for a fission-track age determination depends on a number of factors, including the uranium content, age, and vesicularity of the glass. To date apatite by the population method, a minimum of 50 grains should be counted from each of the splits to determine spontaneous and induced track densities.

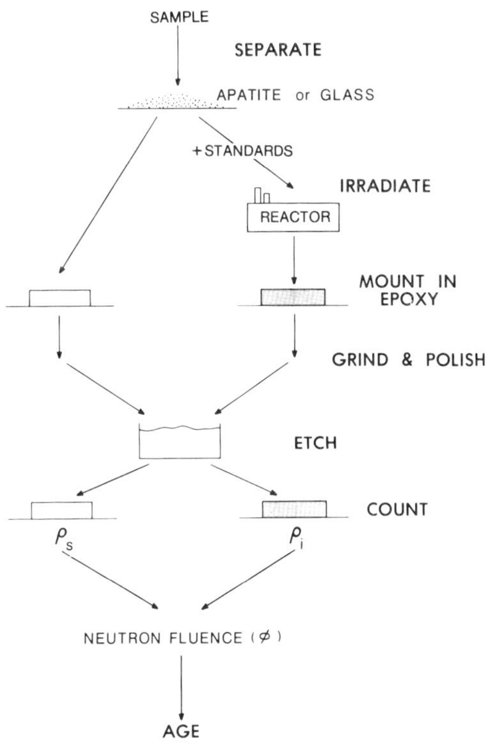

FIGURE 10.9. Steps involved in obtaining a fission-track age using the population method. Details of the laboratory procedures are given in Naeser (1976).

Age Calculation and Neutron Fluence Determination

A fission-track age is calculated as (Price and Walker, 1963; Naeser, 1967)

$$A = \frac{1}{\lambda_D} \ln \left[1 + \frac{\rho_s \lambda_D \, \sigma \, I \, \varphi}{\rho_i \, \lambda_F} \right] \quad (1)$$

where,

ρ_s = spontaneous track density from ^{238}U (tracks/cm^2)

ρ_i = neutron-induced track density from ^{235}U (tracks/cm^2)

φ = thermal neutron fluence (neutrons/cm^2)
λ_D = total decay constant for ^{238}U (1.551×10^{-10} yr^{-1})
λ_F = decay constant for spontaneous fission of ^{238}U
A number of values have been determined; most laboratories use one of the following: 6.85×10^{-17} yr^{-1} (Fleischer and Price, 1964), 7.03×10^{-17} yr^{-1} (Roberts et al., 1968), or 8.42×10^{-17} yr^{-1} (Spadavecchia and Hahn, 1967)
σ = cross-section for thermal neutron-induced fission of ^{235}U (580×10^{-24} cm^2/atom)
I = isotopic ratio ^{235}U/^{238}U (7.252×10^{-3})
A = age in years

A number of schemes have been proposed for determining the value of the neutron fluence (φ), involving the use of standard glass dosimeters of known uranium content or metal foil monitors (Carpenter and Reimer, 1974; Fleischer et al., 1975). However, studies have shown that in a given reactor run there are significant differences between the values of the neutron fluence determined from these various glasses and foils (e.g., see Hurford and Green, 1982, 1983). In practice, to circumvent the problem of determining absolute values for the neutron fluence and for the fission-track decay constant (λ_F), many laboratories have adopted an empirical calibration system by using a method of fluence determination that, when combined with one of the values of the decay constant listed above, consistently yields "correct" fission-track ages on samples of known age (that is, fission-track ages concordant with K-Ar ages of coexisting phases in the same sample) (Naeser et al., 1977).

Fleischer and Hart (1972) and Hurford and Green (1982, 1983) suggested that this practice should be simplified by determining a "zeta calibration" factor and rewriting the age equation,

$$A = \frac{1}{\lambda_D} \ln \left[1 + \frac{\rho_s \lambda_D \zeta \rho_d}{\rho_i} \right] \qquad (2)$$

where,

ρ_d = fission-track density in the detector covering the glass dosimeter during irradiation (tracks/cm^2), and
ζ = the calibration factor for a given glass dosimeter, determined by irradiating the dosimeter with samples of known age (A_{STD}), as

$$\zeta = \frac{e^{\lambda_D A_{\mathrm{STD}}} - 1}{\lambda_D \left(\frac{\rho_s}{\rho_i} \right)_{\mathrm{STD}} \rho_d} \qquad (3)$$

(modified from Hurford and Green, 1982, 1983).

The value of zeta for a given glass dosimeter will vary from one operator to another. Therefore, each individual must determine his or her own zeta values, and, in practice, the zeta adopted for a given glass dosimeter should be a mean value based on a large number of determinations using carefully chosen age standards (see discussion in Hurford and Green, 1983).

The zeta determination method, when correctly followed, probably constitutes the best method currently available for calibrating fission-track dating procedures. As a word of caution, however, it is *strongly recommended* that anyone just beginning fission-track dating should not rely solely on this method, but should also spend time working in an established fission-track laboratory to avoid developing procedural errors that may be propagated through all subsequent dating attempts.

Sedimentary Rock Samples

In the two studies described here, the composite fission-track ages of apatite and zircon from both the basement and sedimentary rock samples were calculated by the conventional method using the sums of the spontaneous and induced tracks counted in the individual grains in the sample as the values for ρ_s and ρ_i in Equation 1. The uncertainty in the ages was calculated by combining the Poisson errors on the spontaneous and induced track counts and on the track counts in the detector covering the dosimeter (Lindsey et al., 1975). These are standard methods commonly used to calculate age and uncertainty in fission-track dating, but it should be recognized that they may not be strictly applicable to sedimentary rock samples in which the individual detrital grains were most likely derived from more than one age population. The "age" calculated from the detrital grains in a sedimentary rock is in fact a composite of the individual detrital grain ages and as such has no real meaning in a stratigraphic sense.

In spite of the wide range in individual grain ages in most sedimentary rocks, our work in the southern San Joaquin Valley and northern Green River basin suggests that the age calculated from the first six grains counted in a sample will give a reasonable estimate of the composite detrital grain "age." The calculated age does not appear to change significantly with the inclusion of additional grain ages. However, this should probably be checked on an individual basin basis because it may not hold true for all detrital grain suites.

Case History: Southern San Joaquin Valley, California

Introduction

The southern San Joaquin Valley contains sedimentary rocks of Cenozoic age deposited in deep marine to nonmarine environments (Bandy and Arnal, 1969; MacPherson, 1978; Zieglar and Spotts, 1978; Webb, 1981). The White Wolf fault, an active high-angle fault at the southernmost end of the valley, divides the study area into a structurally extremely depressed, rapidly subsiding Miocene to Holocene depocenter (referred to herein as the "Basin Block") and a less depressed Eocene to Holocene depositional platform ("Tejon Block") (Fig. 10.10). From at least the early Miocene, the Tejon Block has had a relatively higher structural expression and a lower rate of deposition than the adjacent Basin Block. The Tejon Block has tended to remain high and to experience discontinuous subsidence, with marine deposition interrupted by episodes of erosion, nondeposition, and nonmarine deposition. In contrast, the Basin Block has undergone rapid subsidence and deposition that continues to dominate the area northwest of the White Wolf fault. In the Tejon Block, the total Cenozoic sedimentary section is about 4,000 m (13,120 ft) thick; in contrast, in the deepest part of the Basin Block it reaches a thickness in excess of 7,600 m (25,000 ft), including more than 3,050 m (10,000 ft) of post-Miocene, largely nonmarine, sediments (e.g., Zieglar and Spotts, 1978, p. 821, Fig. 14). Within the Tejon and Basin Blocks, most of the sedimentary rocks are at, or near, their maximum depth of burial at the present time (Hood and Castaño, 1974), although multiple unconformities in the Tejon Block attest to the lengthy history of interrupted deposition and tectonic unrest.

The distinctly different burial histories of the sedimentary rocks of the Basin and the Tejon Blocks suggest that they have experienced different temperature histories. In particular, burial history curves suggest that sediments in the rapidly subsiding and infilling part of the Basin Block have probably been near their present depth (and thus their present temperature) for a shorter time than the sediments in the Tejon Block. The fission-track study was undertaken to investigate the apparently divergent temperature history of these two blocks.

Apatite and zircon were separated from sandstones and a granitic rock sample recovered from drill holes in the Basin and Tejon Blocks and from sandstone and granitic rock samples collected from an outcrop to the south (Figs. 10.10 and 10.11). Samples TM-16, -18, -19, -27, -28, and -29 were recovered from cuttings; the remainder of the drill-hole samples are from core. Details of the localities are given in Naeser et al. (in press).

Fission-Track Data

Apatite Ages

Several lines of evidence suggest that apatite ages from the relatively shallow sedimentary rocks in both the Tejon and Basin Blocks (Fig. 10.12) have not been affected by annealing after deposition and thus belong to the "zone of no annealing" (Fig. 10.4). This includes apatite from rocks at depths shallower than about 3,000 m (about 10,000 ft) in the Tejon Block (plus possibly from TM-24) and shallower than about 4,000 m (about 13,000 ft) in the Basin Block (plus possibly TM-17). The evidence for lack of apatite annealing includes the following:

1. All of the composite apatite ages calculated for the samples from this zone (Fig. 10.12), as well as all of the individual apatite grain ages, are older than the stratigraphic age of the rocks from which they were collected.
2. No obvious relationship exists between (a) the composite apatite age calculated for each of the individual samples in this zone and (b) the temperature (or depth) of the samples.
3. The composite age calculated from all 87 of the apatite grains counted in these samples (61.5 ±

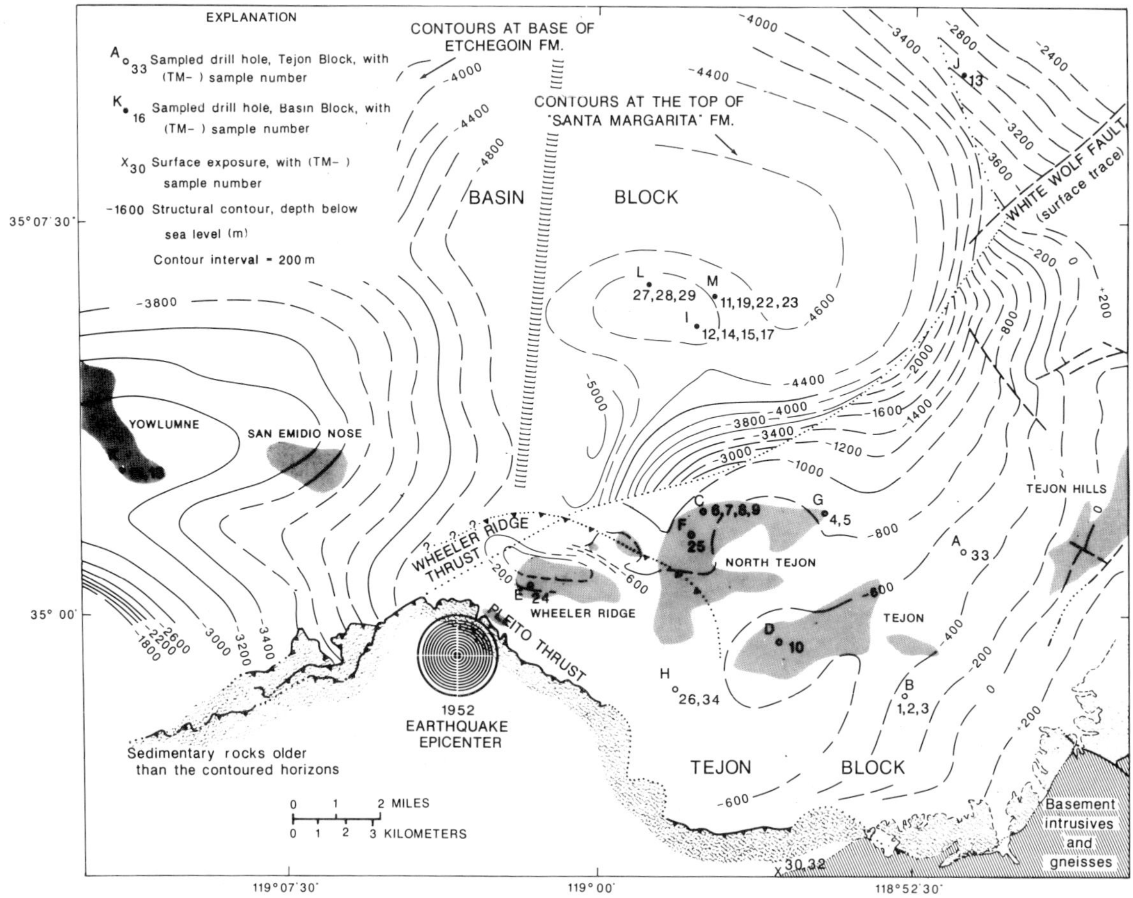

FIGURE 10.10. Map of the southern San Joaquin Valley, California, showing location of the drill holes and surface exposure that were sampled for fission-track dating. Detailed drill-hole and sample information is given in Naeser et al. (in press). Structure contours for the Upper Miocene Etchegoin and Santa Margarita Formations are modified from Bazely (1972, Fig. 4) and R.A. Ortalda, W.H. Le Roy, F.C. Porter, and W.G. Bruer (in Dibblee et al., 1965). Epicenter data from Gutenberg (1955). Shaded areas indicate the major oil fields in the study area.

3.4 Ma, ± 2 standard deviations) is statistically indistinguishable from the 60.7 ± 9.1 Ma age of apatite from the outcropping Eocene sandstone (TM-32) that has probably never been deeply buried.

4. Most of the individual apatite grain ages in the sediment samples are compatible with the fission-track ages of apatite from granitic rocks in the area that are probably similar in age to the source of at least some of the detrital grains (see below).

In contrast to the shallow drill-hole samples, apatite from the deeper samples in both the Tejon and Basin Blocks shows an increasing degree of annealing with increasing temperature (and depth). Of even greater significance from the standpoint of the thermal history of the region, the age versus present temperature trends of the two blocks are distinctly different (Fig. 10.12). In samples from the Basin Block immediately northwest of the White Wolf fault (drill holes I, L, and M in Fig. 10.10; samples bracketed by dot-dash lines in Fig. 10.12), apatite is totally annealed in rocks that are presently at a temperature of about 140°C (284°F) (a few tracks were observed in the apatite in TM-19 at 138°C [280°F], whereas, apatite in TM-11 at 144°C [291°F] is totally annealed). In contrast, in the Tejon Block, extrapolation of the annealing trend below Samples TM-34 and -26 indicates that apatite would be totally annealed in rocks that are presently at just over 115°C (239°F).

A few samples yielded apparently anomalous apatite fission-track ages compared to the trends defined by the majority of samples in the Tejon and Basin Blocks (Fig. 10.12). In the Basin Block, apatite in Well K (Samples TM-16 and -18) is totally annealed at a present temperature of just over 130°C (266°F), as compared to about 140°C (284°F) in the rocks immediately northwest of the White Wolf fault (Wells I, L, and M). Burial and temperature-time curves for these individual wells suggest that the rocks in Well K have been close to their present depth (and, assuming a constant geothermal gradient, close to their present temperature) for a somewhat longer time than the rocks in Wells I, L, and M. This would, at least in part, explain the lower annealing temperature for apatite in Well K.

In the Tejon Block, the anomalous appearance of the composite apatite age of TM-25 (Fig. 10.12) may result from the corrected temperature assigned to this sample, because correcting the measured temperatures proved to be exceptionally difficult for this drill hole.

The width of the "zone of partial annealing" (see Fig. 10.4) for apatite appears to be about 20° to 30°C (36° to 54°F), or less, in the Tejon and Basin Blocks. This range is similar to that found by Naeser (1981) and predicted by laboratory annealing experiments (Fig. 10.3).

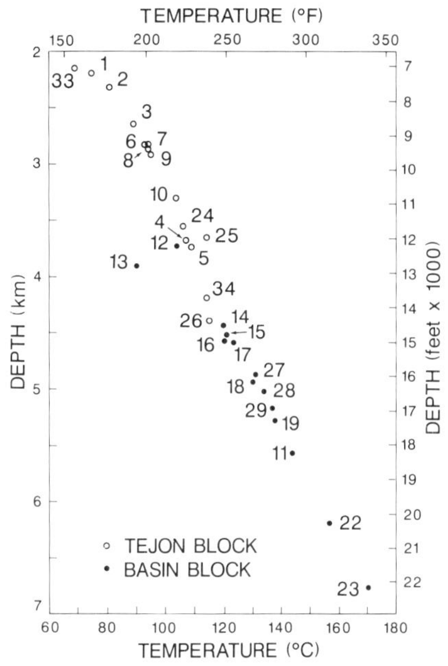

FIGURE 10.11. Depth plotted against present temperature for samples from the Tejon and Basin Blocks. Sample numbers correspond to numbers in Figure 10.10. Temperatures were determined from reservoir engineering data or well-logging measurements that have been adjusted where appropriate to compensate for the disturbing effects of drilling or hydrocarbon production, using the procedures outlined in Bostick et al. (1978, pp. 74–75) and McCulloh and Beyer (1979). For most sites the individual corrected temperatures are thought to be within about ±3°C (5.4°F) of the actual temperatures, and they match closely temperatures that would be obtained using the correction procedures of Kehle et al. (1970) and Kehle (1972). The present average geothermal gradient for the Tejon area, about 22°C/km (1.2°F/100 ft), is nearly constant beneath the entire area.

Zircon Ages

Zircons were dated from four of the drill-hole samples (TM-1, -11, -22, and -23), which are presently at temperatures of up to 170°C (338°F). These samples yielded composite zircon ages that are slightly older than the composite zircon age calculated for the outcropping Eocene sandstone (TM-32) (which has probably never been deeply buried) and that are statistically indistinguishable from the age of zircon in the outcropping Cretaceous granitic rock (TM-30) (Fig. 10.13). The range in individual zircon grain ages within each of the drill-hole samples also does not vary noticeably with depth (Fig. 10.13). Both facts suggest that even the most deeply buried samples have not been near their present (or higher) temperatures long enough for significant annealing of zircon to occur.

Thermal History of the Southern San Joaquin Valley

The significant difference in the temperatures associated with total annealing of apatite in the Basin Block versus the Tejon Block indicates that the blocks have had distinctly different thermal histories. The higher temperature associated with totally annealed apatite in the Basin Block suggests that the rocks in the basin have been near their present temperature for a relatively short time

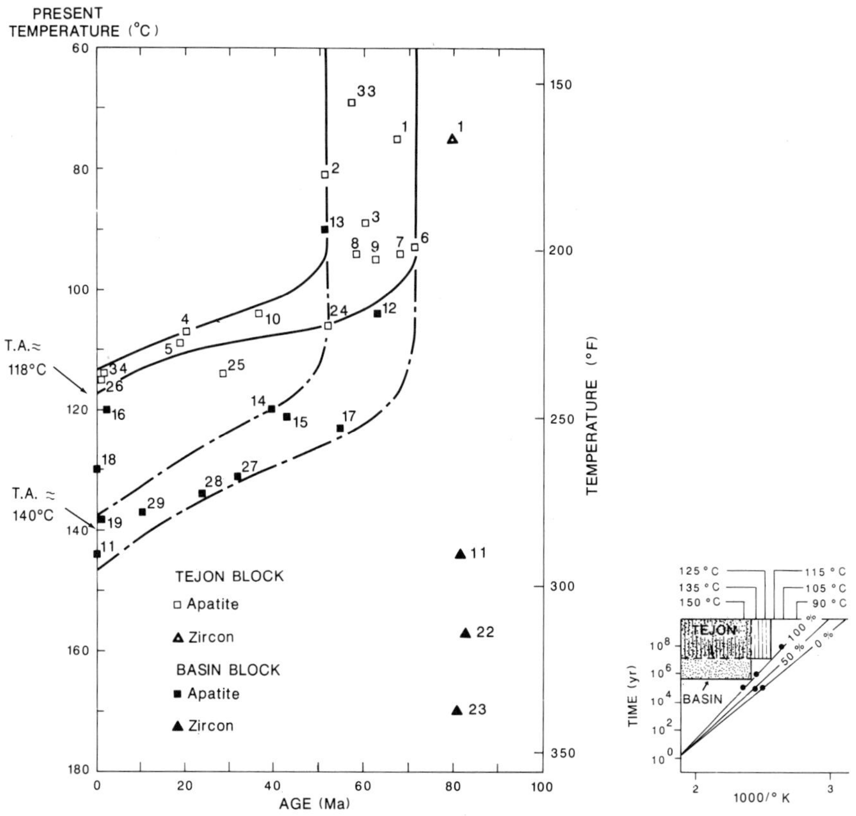

FIGURE 10.12. Fission-track age of apatite and zircon from the southern San Joaquin Valley drill-hole samples plotted against present corrected temperature. Solid lines indicate the trend of apatite ages from the Tejon Block southeast of the White Wolf fault; dot-dash lines indicate the trend in ages for samples in the Basin Block immediately northwest of the fault. T.A. = total annealing. Sample numbers correspond to numbers in Figure 10.10. Analytical data are given in Naeser et al. (in press). Note that because of the nearly constant geothermal gradient in this area (Fig. 10.11), approximately the same trend of ages would result if age were plotted against depth rather than present temperature. Small diagram compares the temperatures at which apatite is totally annealed to the drill-hole annealing data of Figure 10.3.

compared to the rocks in the Tejon Block, a fact consistent with the observed rapid subsidence and deposition in the basin compared to the intermittent deposition and lower depositional rates in the structurally high Tejon Block.

Comparison of the temperatures at which apatite is totally annealed with the calibration data (Fig. 10.12) allows estimation of the actual heating times of the rocks. In the Tejon Block apatite is totally annealed at just over 115°C (239°F), suggesting an effective heating time of about 10^7 yr. In contrast, the observed degree of annealing in the Basin Block immediately northwest of the White Wolfe fault would have been produced in just less than 10^6 yr.

These estimates of heating times for the Tejon and Basin Blocks are in good agreement with other, independent estimates based on 1) $^{40}Ar/^{39}Ar$ analysis of detrital microcline separated from a number of our samples (Harrison and Bé, 1983; Harrison and Burke, this volume) and 2) reconstructions of the burial history of the two blocks (Hood and Castaño, 1974; Zieglar and Spotts, 1978; Naeser et al., in press). The good agreement with the heating times suggested by burial history reconstructions suggests that burial of the sediments under the present geothermal gradient probably would have produced high enough temperatures to account for at least most of the observed apatite annealing.

This is significant because the presence of diagenetic laumontite in these rocks suggests that the rocks have been exposed to considerably higher temperatures than would be expected if they had simply been buried under the present gradient. In particular, some rocks on both sides of the White Wolf fault may have been heated as much as 40°C (72°F) higher than their present temperature at the time of the laumontite alteration (sometime during the last ≈5 Ma), based on the temperature-fluid pressure conditions required for laumontite formation (McCulloh and Stewart, 1979; McCulloh et al., 1981). However, the apatite fission-track data show that such alteration temperatures could have persisted for only a very brief time period (less than 10^5 yr) or else they would have produced a significantly greater degree of track annealing in apatite than is observed (N.D. Naeser, C.W. Naeser, and T.H. McCulloh, in preparation).

Provenance of the Sediments

Zircon appears to have undergone no significant annealing after deposition in the sediments under discussion, and therefore the ages obtained on the individual zircon grains give information on the rocks eroded to form these sediments. The spread and distribution of the ages of the individual grains (Fig. 10.13) indicate that they were almost certainly derived from source rocks of more than one age. The zircon ages are equal to or younger than the biotite and hornblende K-Ar ages of the Sierra Nevada intrusive complexes that were the source of most of the sediments of this area (Fig. 10.13). Zircon ages younger than the K-Ar ages may in part reflect the lower closing temperature of zircon relative to biotite and hornblende (see discussion in Naeser et al., in press). The zircon ages, and the sedimentation patterns for this area, are compatible with most of the zircon being derived from rocks related to the youngest Sierra Nevada intrusive complexes, which presently crop out to the south and east of the Tejon area (Evernden and Kistler, 1970, plates 1 and 2; MacPherson, 1978; Naeser et al., in press). Grains that appear to be related to these youngest intrusives dominate all of the zircon separates analyzed (Fig. 10.13). The detrital microcline in the sediments also yields ages compatible with the youngest intrusives. The microcline separates analyzed by Harrison and Bé

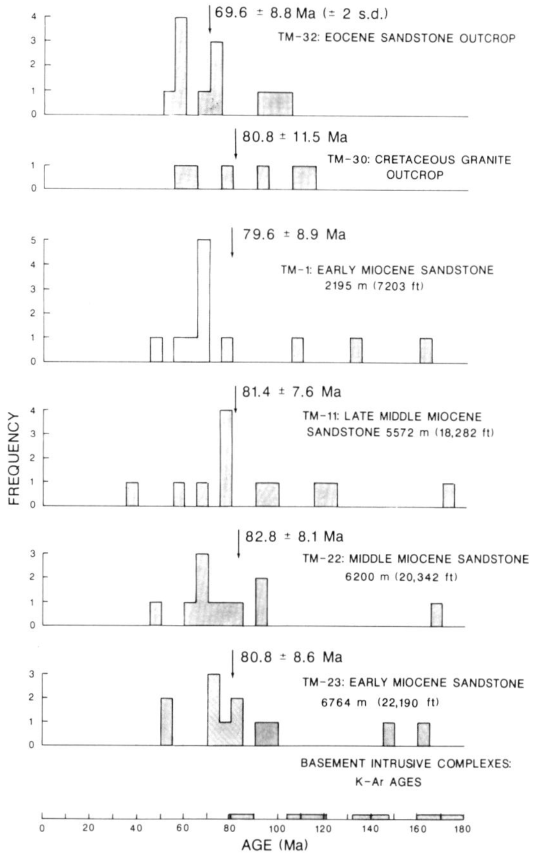

FIGURE 10.13. Histograms of the fission-track ages determined for individual zircon grains in the outcropping Eocene sandstone (TM-32), outcropping Cretaceous granitic basement rock (TM-30), and Miocene drill-hole samples (depth below surface is indicated). The composite age calculated for each sample (Fig. 10.12) is indicated by an arrow. Note that the composite ages, and the range of individual grain ages in each sample, do not vary significantly with depth or temperature. Compare with the ranges of potassium-argon ages for major granitic complexes in the Sierra Nevada (Evernden and Kistler, 1970) that were probably the source of much of the sediment in the southern San Joaquin Valley.

(1983) yielded an average K-Ar age of 78 Ma and the $^{40}Ar/^{39}Ar$ age spectra of the separates indicate slow cooling of the source rocks from 200° to ≈100°C between about 85 and 65 Ma (Harrison and Burke, this volume). The zircon and microcline data thus suggest that rocks related to the youngest Sierra Nevada intrusive complexes were a major source of the sediments in this area from at least Eocene time to late Miocene.

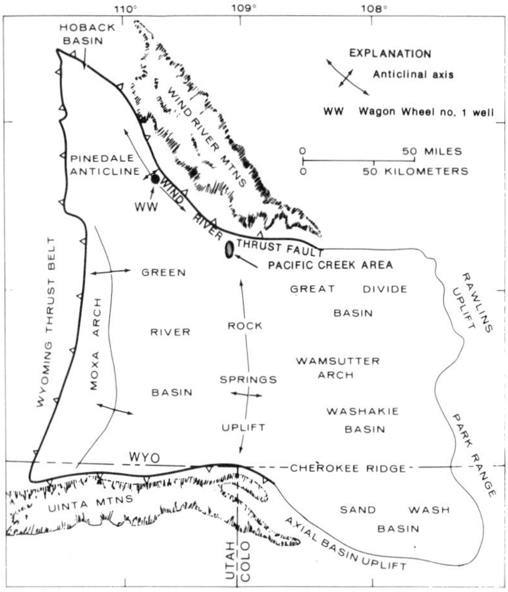

FIGURE 10.14. Map of the Green River basin, Wyoming, showing the location of the Wagon Wheel No. 1 well on the Pinedale anticline. (Modified from Law, 1984a.)

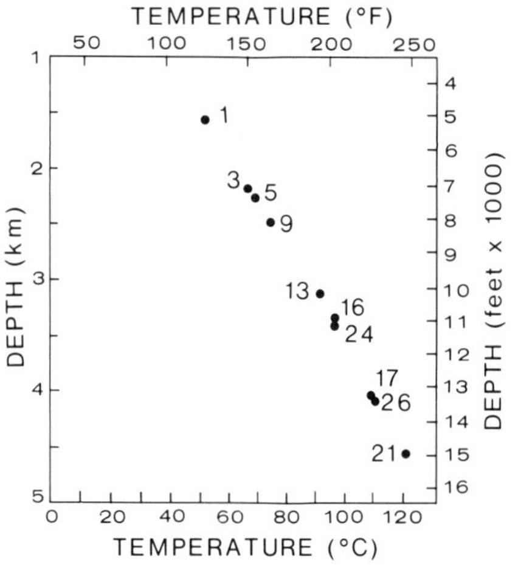

FIGURE 10.15. Depth plotted against present temperature of the Wagon Wheel No. 1 samples. Samples supplied by B.E. Law; locality data given in Naeser (1986, Table 1). Temperatures were adjusted using the procedure given in Figure 10.11. The uncertainty associated with the individual corrected temperature is probably about ±3°C (5.4°F) from the surface to 3,048 m (10,000 ft) and about ±5°C (9°F) from 3,048 m to T.D. (5,791 m; 19,000 ft). The corrected temperatures vary from 6° to 12°C (about 11° to 22°F) higher than the "measured" temperatures. They match closely temperatures that would be obtained using the correction procedures of Kehle et al. (1970) and Kehle (1972).

Case History: Northern Green River Basin, Wyoming

Introduction

In the northern Green River basin, detrital apatite and/or zircon were recovered from lower Tertiary and Upper Cretaceous sandstone core samples from the El Paso Natural Gas Wagon Wheel No. 1 well on the Pinedale anticline (Fig. 10.14). Wagon Wheel No. 1 was drilled to 5,791 m (19,000 ft) into the Upper Cretaceous Baxter Shale and is the deepest well on the Pinedale anticline. The well penetrated about 3,505 m (11,500 ft) of Upper Cretaceous rocks (from the Lance Formation to the Baxter Shale). All but the lower 853 m (2,800 ft) of the Cretaceous section, including all of the rocks sampled for this study, were deposited in nonmarine environments in a prograding wedge. Most of the Upper Cretaceous sandstones have a similar composition, dominated by a mixture of sedimentary, metamorphic, and volcanic clastic debris from source terranes to the west and northwest and can be classified as litharenites, arkosic litharenites, or arkose. One exception is the "upper part of the Ericson Sandstone" (as defined by Law, 1984b), a quartz arenite that was probably derived from Paleozoic quartzose rocks to the west. By latest Cretaceous-early Tertiary time, deposition in the Pinedale area was more complex. Structural uplift on the basin margins provided multiple source terranes for the dominantly fluvial sediments. There has been no net deposition in the northern Green River basin since the Eocene or early Oligocene (Rice and Gautier, 1983; Shuster and Steidtmann, 1983; Law, 1984a, 1984b; Pollastro and Barker, 1984b, 1986; Dickinson and Law, 1985).

The Pinedale anticline trends northwest-southeast and is about 56 km (35 mi) long and 10 km (6 mi) wide, with a structural relief of about 610 m (2,000 ft). Structural evolution of the anticline appears to be related to movement on the Wind River thrust fault, which parallels the anticline to the northeast (Fig. 10.14). Growth of the anticline may have begun during the Laramide orogeny, but the main period of structural deformation probably

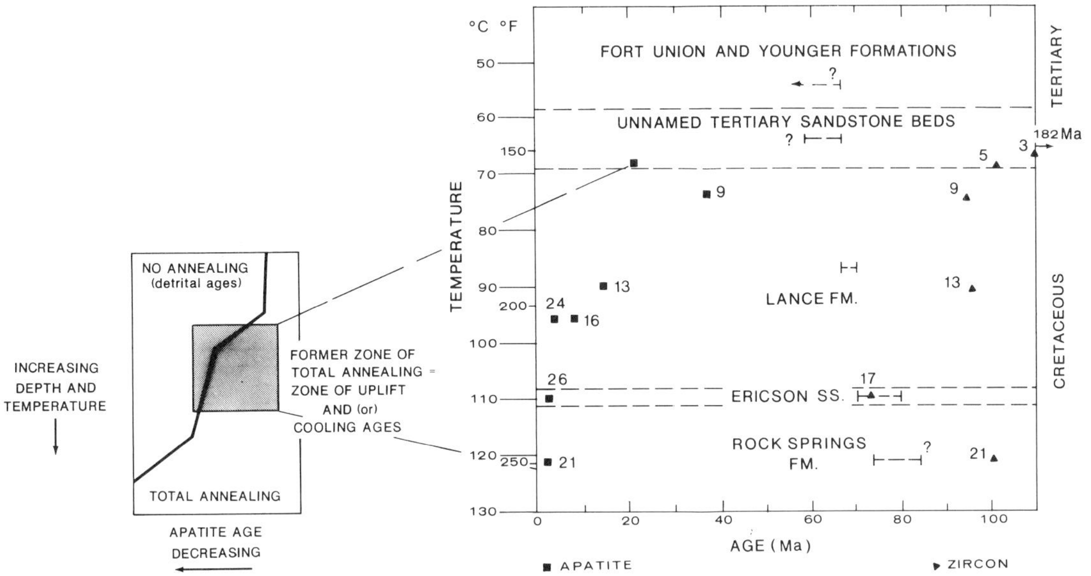

FIGURE 10.16. Fission-track ages of detrital apatite and zircon separated from sandstones in Wagon Wheel No. 1 plotted against present corrected temperature, compared to theoretical pattern of ages expected in rocks that have undergone cooling (Fig. 10.5). Bars in each stratigraphic unit indicate the approximate time of deposition of the unit (see Naeser, 1986). Stratigraphic terminology is from Law (1984b). Note that Sample WWH-17 is from the upper part of the Ericson Sandstone; WWH-26 is from the lower part. WWH-1 (not plotted) was collected in the unit labeled "Fort Union and Younger Formations" (see Fig. 10.15).

occurred in the Paleocene (B.E. Law, oral communication, 1987).

Several lines of evidence, which we shall review, suggest that the rocks in Wagon Wheel No. 1 have cooled below their maximum paleotemperatures. The fission-track study was undertaken to help clarify this cooling history. Samples for fission-track analysis were recovered over a depth interval of about 1,536 to 4,557 m (5,038 to 14,950 ft), from the lower Tertiary Fort Union Formation to the Upper Cretaceous Rock Springs Formation (Figs. 10.15 and 10.16; Naeser, 1986, Table 1).

Fission-Track Data

Apatite Ages

Apatite has been dated from seven of the Wagon Wheel No. 1 samples (Fig. 10.16; Naeser, 1986; Table 2). Sample WWH-17, from the upper part of the Ericson Sandstone (Upper Cretaceous), yielded insufficient apatite for an age determination.

All of the composite apatite ages calculated for the Wagon Wheel No. 1 samples, as well as almost all of the individual apatite grain ages, are younger than the stratigraphic age of the sedimentary units from which they were collected, indicating that significant annealing of the apatite has occurred during burial of the sediments. If the "zone of no annealing" (Fig. 10.16) is preserved in Wagon Wheel No. 1, it must be above the shallowest sample analyzed (WWH-5 at 2,237 m [7,340 ft] depth) because even that sample contains partially annealed apatite.

The apatite fission-track ages calculated for the Wagon Wheel No. 1 samples fall into two distinct trends (Fig. 10.16). In the upper part of the sampled interval, the apatite ages generally decrease with increasing temperature and approach zero at just over 96°C (205°F). At this point, the trend in ages changes abruptly and the ages remain nearly constant or decrease slightly, from about 4 to 2 Ma, over a temperature interval of at least 20°C (36°F), down through the lowest sample dated (WWH-21).

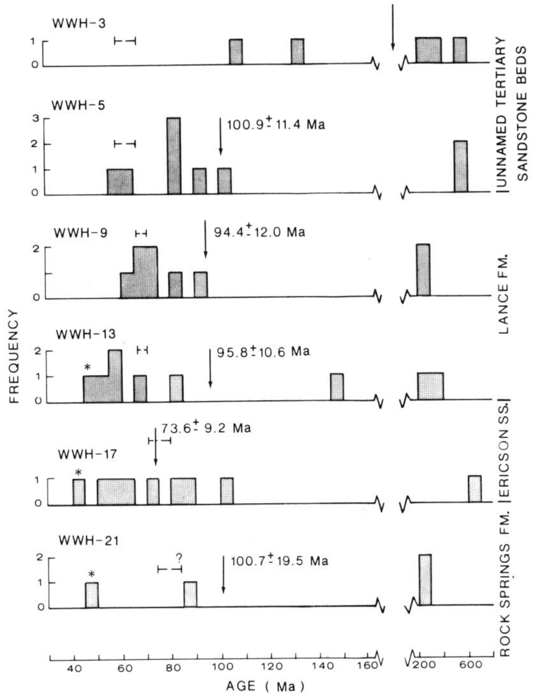

FIGURE 10.17. Histograms of the fission-track ages of individual zircon grains in six of the Wagon Wheel No. 1 samples. The composite age calculated for each sample (Fig. 10.16; Naeser, 1986, Table 2) is indicated by an arrow. Bars indicate the approximate time of deposition of the unit where the sample was collected (see Naeser, 1986). Grain ages marked with an asterisk(*) are younger than the stratigraphic age and do not overlap it at 2 standard deviations; all other ages are equal to or older than the stratigraphic age.

Zircon Ages

Zircon ages have been determined for six of the Wagon Wheel No. 1 samples (Figs. 10.16 and 10.17; Naeser, 1986, Table 2). Sample WWH-16 yielded insufficient zircon for an age determination, and WWH-1 cannot be dated because the zircon is metamict.

The composite zircon fission-track age calculated for each of the Wagon Wheel No. 1 samples as well as all but a few of the individual zircon grain ages are equal to or older than the depositional age of the unit where they were sampled (Figs. 10.16 and 10.17; Naeser, 1986). This suggests that there has been no significant annealing of the zircon in any of the dated samples since deposition.

Lack of significant zircon annealing is also suggested by the fact that the composite zircon ages calculated for most of the samples are indistinguishable from one another (Naeser, 1986, Table 2; Fig. 10.17). Some variation exists in the range of the individual zircon grain ages in the samples (Fig. 10.17), but for most samples no obvious correlation can be made between the calculated sample ages, or the range in individual grain ages, and the present temperature (or depth) of the samples. Apparent exceptions are Samples WWH-17, -3, and -1. Sample WWH-17, from the upper part of the Ericson Sandstone, yielded a significantly younger composite zircon age than many of the samples, and it apparently contains a lower percentage of old (greater than about 110 Ma) zircons (Fig. 10.17). However, as noted below, this may be due at least in part to the provenance of this stratigraphic horizon. In the upper part of the Tertiary section sampled in Wagon Wheel No. 1 (WWH-1 and -3), the detrital zircon suites appear to be older than those observed in the underlying rocks. Zircons in WWH-1 are all metamict, which implies that they are of considerable age. In WWH-3 no zircon grains younger than 100 Ma were observed, and a significantly higher percentage of the grains are metamict than in the underlying samples. Again, this probably indicates a different provenance for these units.

Thermal History in the Pinedale Anticline Area

Because of the temperature-time dependency in annealing (Fig. 10.3), the degree of annealing observed in the apatite in the upper part of Wagon Wheel No. 1 (down to the almost totally annealed apatite at about 96°C; 205°F) could have been produced either by the rocks 1) being near their present (or lower) temperature for a long time (at least 10^8 yr; see Fig. 10.3); or 2) being at a higher temperature for a shorter time, before cooling to their present temperature.

The rocks in this part of the drill hole are latest Cretaceous (Lance Formation) and younger (i.e., less than about 70 Ma) (Weimer, 1961; Obradovich and Cobban, 1975; Law, 1984b; Fig. 10.16), and it is therefore unlikely that they have been near their present temperature for a period of time even approaching 10^8 yr. We are thus left with the second alternative.

On the basis of the annealing data from the upper part of the drill hole alone, it would be difficult to speculate much further on the thermal history in this area. However, the trend of the apatite ages at temperatures greater than about 96°C (205°F) provides evidence for the nature of the cooling. These ages, which show little variation with depth down through Sample WWH-21, are thought to be recording the time of an episode of relatively rapid cooling in the rocks (cf. the "zone of uplift and (or) cooling ages," Fig. 10.6). The apatite ages thus suggest that at least the latest phase of cooling in the Pinedale anticline was initiated about 4 to 2 Ma ago and involved a relatively rapid decrease in temperature of 20°C (36°F) or more.

Pollastro and Barker (1984a, 1984b, 1986) have also studied the thermal history in Wagon Wheel No. 1, using several other paleotemperature indicators. Their depth-temperature curves determined from mean random vitrinite reflectance measured in thin coaly seams in the rocks and from illite-smectite transformation in the sandstone and shale layers are in good agreement and indicate that these rocks were subjected to temperatures 30° to 50°C (54° to 90°F) higher sometime in the past. These values are consistent with the temperature decrease of at least 20°C indicated by the fission-track analysis.

Vitrinite reflectance data suggest that the cooling recorded in the Wagon Wheel No. 1 rocks reflects uplift and erosion rather than a major decrease in the geothermal gradient (Pollastro and Barker, 1984b, 1986). This uplift and erosion may have been confined to the Pinedale anticline area or may have been more regional in extent.

Provenance of the Sediments

Through most of the section sampled in Wagon Wheel No. 1, there is little obvious variation in the age range of individual zircons in the detrital grain suites, or in the proportion of grains of various ages that are present. The available data thus indicate that the dominant source of sediments in the Pinedale anticline area remained unchanged through most of the Late Cretaceous and earliest Tertiary.

Apparent exceptions are found in Samples WWH-17, -3, and -1. Independent evidence suggests that the anomalously young age of zircon in WWH-17 is due at least in part to the provenance of this stratigraphic horizon, rather than to post-depositional annealing affects. WWH-17 was collected from the upper part of the Ericson Sandstone, which has a different composition, and has previously been interpreted as having a different provenance, than the bulk of the Cretaceous sandstone in the Western Interior (Rice and Gautier, 1983, p. 4–24; Law, 1984b; D.L. Gautier, oral communication, 1984). In Wagon Wheel No. 1, the upper part of the Ericson Sandstone is also distinguished from the remainder of the section by the fact that it yielded little or no apatite.

The anomalous zircons of WWH-1 and -3 appear to mark changes in the source terrane that are related to the structural evolution of the Wind River Mountains to the northeast. Uplift of this range began as early as part of the major Laramide activity in latest Cretaceous time. By early to middle Paleocene, this uplift and erosion had breached the crystalline core of the Wind River Mountains, producing an influx of arkosic material into the Green River basin (Prensky, 1984; B.E. Law, oral communication, 1987). A strong anomaly in the gamma-ray log for Wagon Wheel No. 1 at 2,202 m (7,225 ft), between WWH-3 and -5, has been attributed to this arkose influx (Prensky, 1984). The decrease in young zircons and increase in percentage of metamict grains in WWH-3 relative to WWH-5 are probably a direct consequence of this change in sediment source, which culminated in domination of the zircon suite in WWH-1 by metamict grains.

Conclusion

Fission tracks provide a powerful tool for determining the depositional record and thermal history of rocks in sedimentary basins. The annealing of fission tracks, particularly in apatite, provides information on the overall thermal history of the rocks from the time of deposition and burial through subsequent uplift and erosion. Annealing can also delineate more localized temperature anomalies in the basin. Fission-track ages on unannealed zircon are valuable in provenance studies and in dating tephra marker horizons within the sedimentary sequence.

In the southern San Joaquin Valley, California, apatite fission-track data clearly reflect the distinctly different thermal histories of rocks in the two structural blocks separated by the active White Wolf fault. The fission-track data suggest that rocks in the rapidly subsiding basin northwest of the fault have been near their present temperature for only about 10^6 yr, whereas rocks southeast of the fault have been near their present temperature for about 10^7 yr. These estimates of heating time agree well with estimates based on $^{40}Ar/^{39}Ar$ analysis and burial history reconstructions for these rocks. Data derived from the detrital zircon and microcline in these rocks suggest that from at least Eocene to late Miocene time the dominant source of sediments in the study area was the youngest of the Sierra Nevada granitic complexes, which presently crop out to the east and south of the southern San Joaquin Valley.

In the Pinedale anticline of the northern Green River basin, Wyoming, vitrinite reflectance and illite-smectite transformation measurements in Wagon Wheel No. 1 well suggest that present temperatures are about 30° to 50°C (54° to 90°F) cooler than the maximum paleotemperatures (Pollastro and Barker, 1984a, 1984b, 1986). The fission-track apatite ages from this drill hole support the hypothesis of higher paleotemperatures and suggest that at least the latest phase of cooling of the rocks was initiated about 4 to 2 Ma ago (in the Pliocene). This latest cooling phase has produced a relatively rapid temperature decrease of 20°C (36°F) or more. The zircon ages from Wagon Wheel No. 1 suggest that the dominant source of sediments in this area remained unchanged through most of the Late Cretaceous and earliest Tertiary, but input from different source(s) is indicated for the upper part of the Ericson Sandstone (Cretaceous). A significant change in provenance in the early Tertiary was probably related to unroofing of the crystalline core of the Wind River Mountains to the northeast.

Acknowledgments. This research was partially supported by the U.S. Geological Survey Evolution of Sedimentary Basins Program. Work in the northern Green River basin was in cooperation with B.E. Law (U.S. Geological Survey, Denver) and with the U.S. Department of Energy Morgantown Energy Technology Center. We would like to thank C.G. Cunningham, V.A. Frizzell, Jr., and D.S. Miller for their excellent reviews of this manuscript.

References

Baldwin, S.L., and Harrison, T.M. 1985. Fission track dating of detrital zircons from the Scotland Formation, Barbados, W. I. (abst.). Nuclear Tracks 10:402.

Bandy, O.L., and Arnal, R.E. 1969. Middle Tertiary basin development, San Joaquin Valley, California. Geological Society of America Bulletin 80:783–820.

Barker, C.E. 1983. Influence of time on metamorphism of sedimentary organic matter in liquid-dominated geothermal systems, western North America. Geology 11:384–388.

Bazely, W. 1972. San Emidio Nose oil field. In: Stratigraphic Oil and Gas Fields. American Association of Petroleum Geologists Memoir 16, pp. 297–316.

Bostick, N.H., Cashman, S.M., McCulloh, T.H., and Waddell, C.T. 1978. Gradients of vitrinite reflectance and present temperature in the Los Angeles and Ventura basins, California. In: Oltz, D.F. (ed.): Low Temperature Metamorphism of Kerogen and Clay Minerals. Los Angeles, Pacific Section, Society of Economic Paleontologists and Mineralogists, pp. 65–96.

Briggs, N.D., Naeser, C.W., and McCulloh, T.H. 1979. Thermal history of sedimentary basins by fission-track dating. Geological Society of America Abstracts with Programs 11:394.

Briggs, N.D., Naeser, C.W., and McCulloh, T.H. 1981a. Thermal history of sedimentary basins by fission-track dating. Nuclear Tracks 5:235–237.

Briggs, N.D., Naeser, C.W., and McCulloh, T.H. 1981b. Thermal history by fission-track dating, Tejon Oil Field area, California (abst.). American Association of Petroleum Geologists Bulletin 65:906.

Bryant, B., and Naeser, C.W. 1980. The significance of fission-track ages of apatite in relation to the tectonic history of the Front and Sawatch Ranges, Colorado. Geological Society of America Bulletin 91:156–164.

Carpenter, B.S., and Reimer, G.M. 1974. Standard reference materials: Calibrated glass standards for fission track use. National Bureau of Standards Special Publication 260-49, 16 pp.

Cerveny, P.F. 1986. Uplift and erosion of the Himalaya over the past 18 million years: Evidence from fission track dating of detrital zircons and heavy mineral analysis. M.Sc. thesis, Dartmouth College, Hanover, NH, 198 pp.

Cerveny, P.F., Naeser, N.D., Naeser, C.W., and Johnson, N.M. 1986. Uplift and erosion of the Himalayas during the past 18 million years: Evidence from fission track dating of detrital zircons in Siwalik Group sediments. Abstracts, Symposium on New Perspectives in Basin Analysis, Minneapolis, MN, 1986.

Cerveny, P.F., Naeser, N.D., Zeitler, P.K., Naeser, C.W., and Johnson, N.M. In press. History of uplift and relief of the Himalaya during the past 18 million years: Evidence from fission-track ages of detrital zircons from sandstones of the Siwalik Group. In: Kleinspehn, K.L., and Paola, C. (eds.): New Perspectives in Basin Analysis. New York, Springer-Verlag.

Cunningham, C.G., and Barton, P.B., Jr. 1984. Recognition and use of paleothermal anomalies as a new exploration tool. Geological Society of America Abstracts with Programs 16:481.

Dibblee, T.W., Jr., Bruer, W.G., Hackel, O., and Warne, A.H. 1965. Geologic map of southeastern San Joaquin Valley. In: Hackel, O. (chairman): Geology of southeastern San Joaquin Valley, California – Kern River to Grapevine Canyon. Guidebook, Pacific Section, American Association of Petroleum Geologists, scale about 1:48,000.

Dickinson, W.W., and Law, B.E. 1985. Burial history of Upper Cretaceous and Tertiary rocks interpreted from vitrinite reflectance, northern Green River basin, Wyoming. Abstracts, American Association of Petroleum Geologists Rocky Mountain Section Meeting, Denver, CO, 1985.

Duddy, I.R., and Gleadow, A.J.W. 1982. Thermal history of the Otway Basin, southeastern Australia, from geologic annealing of fission tracks in detrital volcanic apatites. Abstracts, Fission-Track Dating Workshop, Fifth International Conference on Geochronology, Cosmochronology, and Isotope Geology, Nikko National Park, Japan, pp. 13–16.

Duddy, I.R., and Gleadow, A.J.W. 1985. The application of fission track thermochronology to sedimentary basins: Two case studies (abst.). Nuclear Tracks 10:408.

Evernden, J.F., and Kistler, R.W. 1970. Chronology of emplacement of Mesozoic batholithic complexes in California and western Nevada. U.S. Geological Survey Professional Paper 623, 42 pp.

Fleischer, R.L., Harrison, T.M., and Miller, D.S. (eds.). 1984. Abstracts, Fourth International Fission Track Dating Workshop, Troy, NY, 62 pp.

Fleischer, R.L., and Hart, H.R., Jr. 1972. Fission track dating: Techniques and problems. In: Bishop, W.W., and Miller, J.A. (eds.): Calibration of Hominoid Evolution. Edinburgh, Scottish Academic Press, pp. 135–170.

Fleischer, R.L., and Price, P.B. 1964. Decay constant for spontaneous fission of ^{238}U. Physical Review 133(1B):63–64.

Fleischer, R.L., Price, P.B., and Walker, R.M. 1965. Tracks of charged particles in solids. Science 149: 383–393.

Fleischer, R.L., Price, P.B., and Walker, R.M. 1975. Nuclear Tracks in Solids: Principles and Applications. Berkeley, University of California Press, 605 pp.

Gleadow, A.J.W., and Duddy, I.R. 1981. A natural long-term track annealing experiment for apatite. Nuclear Tracks 5:169–174.

Gleadow, A.J.W., Duddy, I.R., and Lovering, J.F. 1983. Fission track analysis: A new tool for the evaluation of thermal histories and hydrocarbon potential. Australian Petroleum Exploration Association Journal 23: 93–102.

Gleadow, A.J.W., Hurford, A.J., and Quaife, R.D. 1976. Fission-track dating of zircon: Improved etching techniques. Earth and Planetary Science Letters 33: 273–276.

Gutenberg, B. 1955. Epicenter and origin time of the main shock on July 21 and travel times of major phases. In: Oakeshott, G.B. (ed.): Earthquakes in Kern County, California, During 1952. California Division of Mines Bulletin 171, pp. 157–163.

Harrison, T.M. 1985. A reassessment of fission-track annealing behavior in apatite. Nuclear Tracks 10: 329–333.

Harrison, T.M., Armstrong, R.L., Naeser, C.W., and Harakal, J.E. 1979. Geochronology and thermal history of the Coast Plutonic Complex, near Prince Rupert, British Columbia. Canadian Journal of Earth Sciences 16:400–410.

Harrison, T.M., and Bé, K. 1983. $^{40}Ar/^{39}Ar$ age spectrum analysis of detrital microclines from the southern San Joaquin Basin, California: An approach to determining the thermal evolution of sedimentary basins. Earth and Planetary Science Letters 64:244–256.

Hood, A., and Castaño, J.R. 1974. Organic metamorphism: Its relationship to petroleum generation and application to studies of authigenic minerals. United Nations Economic Commission for Asia and Far East, Committee for Coordination of Joint Prospecting for Mineral Resources in Asian Offshore Areas, Technical Bulletin 8:85–118.

Hood, A., Gutjahr, C.C.M., and Heacock, R.L. 1975. Organic metamorphism and the generation of petroleum. American Association of Petroleum Geologists Bulletin 59:986–996.

Hurford, A.J. 1985. On the closure temperature for fission tracks in zircon (abst.). Nuclear Tracks 10: 415.

Hurford, A.J., Fitch, F.J., and Clarke, A. 1984. Resolution of the age structure of the detrital zircon populations of two Lower Cretaceous sandstones from the Weald of England by fission track dating. Geological Magazine 121:269–277.

Hurford, A.J., and Green, P.F. 1982. A users' guide to fission track dating calibration. Earth and Planetary Science Letters 59:343–354.

Hurford, A.J., and Green, P.F. 1983. The zeta age

calibration of fission-track dating. Isotope Geoscience 1:285–317.

Kehle, R.O. 1972. Geothermal survey of North America 1971 annual progress report. American Association of Petroleum Geologists Research Committee (unpublished duplicated report), 31 pp.

Kehle, R.O., Schoeppel, R.J., and Deford, R.K. 1970. The AAPG geothermal survey of North America. Geothermics (Special Issue):358–367.

Kohn, B.P., Shagam, R., Banks, P.O., and Burkley, L.A. 1984. Mesozoic-Pleistocene fission-track ages on rocks of the Venezuelan Andes and their tectonic implications. In: Bonini, W.E., Hargraves, R.B., and Shagam, R. (eds.): The Caribbean-South American Plate Boundary and Regional Tectonics. Geological Society of America Memoir 162, pp. 365–384.

Law, B.E. 1984a. Introduction. In: Law, B.E. (ed.): Geological Characteristics of Low-Permeability Upper Cretaceous and Lower Tertiary Rocks in the Pinedale Anticline Area, Sublette County, Wyoming. U.S. Geological Survey Open-File Report 84-753, pp. 1–5.

Law, B.E. 1984b. Structure and stratigraphy of the Pinedale anticline, Wyoming. In: Law, B.E. (ed.): Geological Characteristics of Low-Permeability Upper Cretaceous and Lower Tertiary Rocks in the Pinedale Anticline Area, Sublette County, Wyoming. U.S. Geological Survey Open-File Report 84-753, pp. 6–15.

Lindsey, D.A., Naeser, C.W., and Shawe, D.R. 1975. Age of volcanism, intrusion, and mineralization in the Thomas Range, Keg Mountains, and Desert Mountain, western Utah. U.S. Geological Survey Journal of Research 3:597–604.

Macdougall, J.D. 1976. Fission-track dating. Scientific American 235(6):114–122.

MacPherson, B.A. 1978. Sedimentation and trapping mechanism in upper Miocene Stevens and older turbidite fans of southeastern San Joaquin Valley, California. American Association of Petroleum Geologists Bulletin 62:2243–2274.

Märk, E., Pahl, M., Purtscheller, F., and Märk, T.D. 1973. Thermische Ausheilung von Uran-Spaltspuren in Apatiten, Alterskorrekturen und Beitrage zur Geochronologie. Tschermaks Mineralogische und Petrographische Mitteilungen 20:131–154.

McCulloh, T.H., and Beyer, L.A. 1979. Geothermal gradients. In: Cook, H.E. (ed.): Geologic Studies of the Point Conception Deep Stratigraphic Test Well OCS-CAL 78-164 No. 1, Outer Continental Shelf, Southern California, United States. U.S. Geological Survey Open-File Report 79-1218, pp. 43–48.

McCulloh, T.H., and Fan, J.J. 1985. Burial thermal histories, vitrinite reflectance, and laumontite isograd (abst.). American Association of Petroleum Geologists Bulletin 69:285.

McCulloh, T.H., Frizzell, V.A., Jr., Stewart, R.J., and Barnes, I. 1981. Precipitation of laumontite with quartz, thenardite, and gypsum at Sespe Hot Springs, western Transverse Ranges, California. Clays and Clay Minerals 29:353–364.

McCulloh, T.H., and Stewart, R.J. 1979. Subsurface laumontite crystallization and porosity destruction in Neogene sedimentary basins. Geological Society of America Abstracts with Programs 11:475.

Naeser, C.W. 1967. The use of apatite and sphene for fission track age determinations. Geological Society of America Bulletin 78:1523–1526.

Naeser, C.W. 1976. Fission track dating. U.S. Geological Survey Open-File Report 76-190, 65 pp.

Naeser, C.W. 1979a. Fission-track dating and geologic annealing of fission tracks. In: Jäger, E., and Hunziker, J.C. (eds.): Lectures in Isotope Geology. New York, Springer-Verlag, pp. 154–169.

Naeser, C.W. 1979b. Thermal history of sedimentary basins: Fission-track dating of subsurface rocks. In: Scholle, P.A., and Schluger, P.R. (eds.): Aspects of Diagenesis. Society of Economic Paleontologists and Mineralogists Special Publication 26, pp. 109–112.

Naeser, C.W. 1981. The fading of fission tracks in the geologic environment: Data from deep drill holes. Nuclear Tracks 5:248–250.

Naeser, C.W., Briggs, N.D., Obradovich, J.D., and Izett, G.A. 1981. Geochronology of Quaternary tephra deposits. In: Self, S., and Sparks, R.S.J. (eds.): Tephra Studies. North Atlantic Treaty Organization Advanced Studies Institute Series C. Dordrecht, Netherlands, Reidel Publishing Co., pp. 13–47.

Naeser, C.W., and Cunningham, C.G. 1984. Age and paleothermal anomaly of the Eagle Mine ore body, Gilman district, Colorado. Geological Society of America Abstracts with Programs 16:607.

Naeser, C.W., Cunningham, C.G., Marvin, R.F., and Obradovich, J.D. 1980. Pliocene intrusive rocks and mineralization near Rico, Colorado. Economic Geology 75: 122–127.

Naeser, C.W., and Dodge, F.C.W. 1969. Fission-track ages of accessory minerals from granitic rocks of the central Sierra Nevada Batholith, California. Geological Society of America Bulletin 80:2201–2212.

Naeser, C.W., and Faul, H. 1969. Fission track annealing in apatite and sphene. Journal of Geophysical Research 74:705–710.

Naeser, C.W., Hurford, A.J., and Gleadow, A.J.W. 1977. Fission-track dating of pumice from the KBS Tuff, East Rudolf, Kenya. Nature 267:649.

Naeser, N.D. 1984a. Fission-track ages from the Wagon Wheel No. 1 well, northern Green River basin, Wyoming: Evidence for recent cooling. In: Law, B.E. (ed.): Geological Characteristics of Low-Permeability Upper Cretaceous and Lower Tertiary Rocks in the Pinedale Anticline Area, Sublette County, Wyoming.

U.S. Geological Survey Open-File Report 84-753, pp. 66–77.

Naeser, N.D. 1984b. Thermal history determined by fission-track dating for three sedimentary basins in California and Wyoming. Geological Society of America Abstracts with Programs 16:607.

Naeser, N.D. 1986. Neogene thermal history of the northern Green River basin, Wyoming. Evidence from fission-track dating. In: Gautier, D.L. (ed.): Roles of Organic Matter in Sediment Diagenesis. Society of Economic Paleontologists and Mineralogists Special Publication 38, pp. 65–72.

Naeser, N.D., and Naeser, C.W. 1984. Fission-track dating. In: Mahaney, W.C. (ed.): Quaternary Dating Methods. Amsterdam, Elsevier, pp. 87–100.

Naeser, N.D., Naeser, C.W., and McCulloh, T.H. In press. Thermal history of rocks in the southern San Joaquin Valley, California. American Association of Petroleum Geologists Bulletin.

Naeser, N.D., Zeitler, P.K., Naeser, C.W., and Cerveny, P.F. 1987. Provenence studies by fission-track dating—etching and counting procedures. Nuclear Tracks 13: 121–126.

Obradovich, J.D., and Cobban, W.A. 1975. A time-scale for the Late Cretaceous of the Western Interior of North America. Geological Association of Canada Special Paper 13, pp. 31–54.

Pollastro, R.M., and Barker, C.E. 1984a. Comparative measures of paleotemperature: An example from clay-mineral, vitrinite reflectance, and fluid inclusion studies, Pinedale anticline, northern Green River basin, Wyoming. Abstracts, Society of Economic Paleontologists and Mineralogists Midyear Meeting, San Jose, California, pp. 65–66.

Pollastro, R.M., and Barker, C.E. 1984b. Geothermometry from clay minerals, vitrinite reflectance, and fluid inclusions: Applications to the thermal and burial history of rocks cored from the Wagon Wheel No. 1 well, Green River basin, Wyoming. In: Law, B.E. (ed.): Geological Characteristics of Low-Permeability Upper Cretaceous and Lower Tertiary Rocks in the Pinedale Anticline Area, Sublette County, Wyoming. U.S. Geological Survey Open-File Report 84-753, pp. 78–94.

Pollastro, R.M., and Barker, C.E. 1986. Application of clay-mineral, vitrinite reflectance, and fluid inclusion studies to the thermal and burial history of the Pinedale anticline, Green River basin, Wyoming. In: Gautier, D.L. (ed.): Roles of Organic Matter in Sediment Diagenesis. Society of Economic Paleontologists and Mineralogists Special Publication 38, pp. 73–83.

Prensky, S.E. 1984. A gamma-ray log anomaly associated with the Cretaceous-Tertiary boundary in the northern Green River Basin, Wyoming. In: Law, B.E. (ed.): Geological Characteristics of Low-Permeability Upper Cretaceous and Lower Tertiary Rocks in the Pinedale Anticline Area, Sublette County, Wyoming. U.S. Geological Survey Open-File Report 84-753, pp. 22–35.

Price, L.C. 1983. Geologic time as a parameter in organic metamorphism and vitrinite reflectance as an absolute paleogeothermometer. Journal of Petroleum Geology 6:5–38.

Price, L.C. 1985. Examples, causes, and consequences of vitrinite reflectance suppression in hydrogen-rich organic matter: A major unrecognized problem (abst.). American Association of Petroleum Geologists Bulletin 69:298.

Price, L.C., and Barker, C.E. 1984. Suppression of vitrinite reflectance in amorphous rich kerogen: A major unrecognized problem. Journal of Petroleum Geology 8:59–84.

Price, L.C., and Walker, X.X. 1963.

Reimer, G.M. 1972. Fission track geochronology: Method for tectonic interpretation of apatite studies with examples from the central and southern Alps. Ph.D. dissertation, University of Pennsylvania, Philadelphia, 85 pp.

Rice, D.D., and Gautier, D.L. 1983. Patterns of sedimentation, diagenesis, and hydrocarbon accumulation in Cretaceous rocks of the Rocky Mountains. Society of Economic Paleontologists and Mineralogists Short Course 11, 339 pp.

Roberts, J.A., Gold, R., and Armani, R.J. 1968. Spontaneous-fission decay constant of ^{238}U. Physical Review 174:1482–1484.

Sanford, S.J. 1981. Dating thermal events by fission track annealing, Cerro Prieto geothermal field, Baja California, Mexico. M.Sc. thesis, University of California, Riverside, 105 pp.

Seward, D. 1979. Comparison of zircon and glass fission-track ages from tephra horizons. Geology 7:479–482.

Shuster, M.W., and Steidtmann, J.R. 1983. Origin and development of northern Green River basin: A stratigraphic and flexural study (abst.). American Association of Petroleum Geologists Bulletin 67:1356.

Spadavecchia, A., and Hahn, B. 1967. Die Rotationskammer und einige Anwendungen. Helvetica Physica Acta 40:1063–1079.

Wagner, G.A., and Reimer, G.M. 1972. Fission-track tectonics: The tectonic interpretation of fission-track apatite ages. Earth and Planetary Science Letters 14:263–268.

Wagner, G.A., Reimer, G.M., and Jäger, E. 1977. Cooling ages derived by apatite fission-track, mica Rb-Sr and K-Ar dating: The uplift and cooling history of the

Central Alps. Memorie degli Instituti di Geologia e Mineralogia dell'Universita di Padova XXX:1–28.

Waples, D.W. 1980. Time and temperature in petroleum formation: Application of Lopatin's method to petroleum exploration. American Association of Petroleum Geologists Bulletin 64:916–926.

Webb, G.W. 1981. Stevens and earlier Miocene turbidite sandstones, southern San Joaquin Valley, California. American Association of Petroleum Geologists Bulletin 65:438–465.

Weimer, R.J. 1961. Uppermost Cretaceous rocks in central and southern Wyoming, and northwest Colorado. Wyoming Geological Association Sixteenth Annual Field Conference Guidebook, pp. 17–28.

Yim, W.W.-S., Gleadow, A.J.W., and Van Moort, J.C. 1985. Fission track dating of alluvial zircons and heavy mineral provenance in Northeast Tasmania. Journal of the Geological Society of London 142:351–356.

Zeitler, P.K. 1985. Closure temperature implications of concordant $^{40}Ar/^{39}Ar$ potassium feldspar and zircon fission-track ages from high-grade terranes (abst.). Nuclear Tracks 10:441–442.

Zeitler, P.K., Johnson, N.M., Briggs, N.D., and Naeser, C.W. 1982. Uplift history of the NW Himalaya as recorded by fission-track ages on detrital Siwalik zircons. Abstracts, Symposium on Mesozoic and Cenozoic Geology (60th Anniversary Symposium, Geological Society of China), Beidaihe, China.

Zieglar, D.L., and Spotts, J.H. 1978. Reservoir and source-bed history of Great Valley, California. American Association of Petroleum Geologists Bulletin 62: 813–826.

Zimmermann, R.A., and Gaines, A.M. 1978. A new approach to the study of fission-track fading. In: Zartman, R.E. (ed.): Short Papers of the Fourth International Conference, Geochronology, Cosmochronology, Isotope Geology. U.S. Geological Survey Open-File Report 78-701, pp. 467–468.

11

Apatite Fission-Track Analysis as a Paleotemperature Indicator for Hydrocarbon Exploration

Paul F. Green, Ian R. Duddy, Andrew J.W. Gleadow, and John F. Lovering

Abstract

Apatite Fission-Track Analysis (AFTA) is emerging as an important new tool for thermal history analysis in sedimentary basins. At temperatures between approximately 20°C and 150°C over times of the order of 1 to 100 my, fission tracks in apatite are annealed. This is due to a rearrangement of the damage present in unetched tracks, with the result that less of a track is etchable than in fresh, newly created tracks. Because of this, the length of an etched fission track reduces with increasing annealing, and in turn, the track density (and hence the fission-track age) is also decreased. In selected boreholes in the Otway basin, southeastern Australia, apatites from the Otway Group show reduction in confined fission-track length and apparent fission-track age, in a fashion characteristic of a simple thermal history in which samples are at or near their maximum temperatures at the present day. Track lengths show a steady decrease from lengths of approximately 15 μm in outcrop or near surface samples, to zero at about 125°C. Fission-track ages, however, show little or no decrease in age until temperatures exceed about 70°C. Above this temperature, ages rapidly reduce to zero at about 125°C.

Fission-track data from the Otway basin contain more information than the simple decrease of age and length. The distributions of single grain ages show characteristic patterns, particularly above 90°C. The distribution of track lengths is also diagnostic of temperature. In particular, in samples at present temperatures between 102°C and 110°C, the distribution of lengths is almost flat, with tracks of all lengths from approximately 1 μm to 16 μm.

The temperature-sensitive fission-track parameters observed in the Otway basin may be applied in other basins to elucidate paleotemperature details. In cases of mixed provenance, individual grain ages may be identified using the external detector method. Fission-track lengths in apatites containing a significant track record at the time of deposition are generally characterized by one of two types of distributions, greatly simplifying interpretation of distributions of track lengths in samples showing significant down-hole annealing.

Presence of an inherited track component, or conversely of a total loss of tracks at some time since deposition, can be identified by a comparison of the stratigraphic age with the length-corrected fission-track age. Investigation of five fission-track parameters then allows semi-quantitative constraints to be placed on thermal history. Experiments are in progress to place this procedure on a more rigorous, quantitative basis.

AFTA offers numerous advantages over the other thermal history analysis techniques, including the ability to provide a chronology of events. The method is now established in hydrocarbon exploration as a quantitative maturation indicator and should find common application.

Introduction

The coincidence between the temperature range in which fission tracks in apatite are annealed, over times of the order of 1 to 100 my, and that in which liquid hydrocarbons are generated (Gleadow et al., 1983) has led to the emergence of Apatite Fission-Track Analysis (AFTA) as a tool of unique ability in the study of thermal histories in sedimentary basins. Earliest applications of this technique to thermal history analysis (Naeser, 1979a, 1981; Briggs et al., 1981) relied exclusively on the evaluation of fission-track ages in apatites and zircons separated from sandstone samples at various depths in boreholes. However, a detailed investigation has revealed that fission tracks in apatite contain significantly more information than simply the apparent fission-track age. For instance, the apatite age is reduced during annealing largely as a result of a reduction in the etchable range of fission tracks and, since this is revealed in the length of *confined* fission tracks, this parameter offers a more fundamental source of paleotemperature information (Gleadow et al., 1983). In addition, other fission-track parameters, such as the distribution of single crystal ages, also yield sensitive indications of temperature.

Experience in a number of sedimentary basins has led to a five-parameter approach to AFTA. This approach is described below, with reference to the Otway basin of southeastern Australia. This basin has been of particular value in developing the concepts outlined below due to its relatively simple depositional history and the presence of a thick (>3.5 km), monotonous sequence of volcanogenic sediments—the Early Cretaceous Otway Group. The sandstones in this sequence contain abundant, euhedral volcanogenic apatite, as well as zircon and sphene. As shown below, the volcanogenic origin of the Otway Group greatly simplifies the interpretation of the observed fission-track parameters.

The basis of the fission-track dating method has been described by Naeser (1979b), and the practical aspects of calibration and routine application were explained in detail by Hurford and Green (1982). In sedimentary basin studies, where a single rock sample may contain apatites from a variety of sources, use of the external detector method (Gleadow, 1981) is essential, as this allows ages to be obtained for single apatite grains. This single grain capability has further advantages as shown below.

Fission-track ages are obtained from the areal density of fission tracks revealed by chemically etching where they intersect polished surfaces of uranium-bearing minerals. However, tracks may also be revealed that do not cross the surface but are totally confined within the crystal. Such tracks, first reported by Lal et al. (1969), are etched because they intersect either fractures or other etched fission tracks that do cross the surface and thus provide a channel-way for the etchant. Because they are confined within the body of the crystal, the length of these tracks is the total distance along the track that is etchable. This "etchable range" determines the observed track density (see, for example, Fleischer et al., 1975, pp. 163–165), and therefore the mean confined track length represents a more fundamental parameter than fission-track density or age. In laboratory annealing experiments (Gleadow et al., 1983; Green et al., 1985), the mean confined track length decreases as the temperature of annealing is increased, and as a consequence the track density also falls (Laslett et al., 1984).

The reduction in mean confined track length during annealing results from a gradual rearrangement of the damaged crystal lattice due to atomic diffusion processes, with the result that less of the track remains etchable. Because this diffusion is a thermally activated process, fission-track annealing is more rapid at higher temperatures. In the laboratory, tracks are totally removed after 1 h at 370°C, whereas under geological conditions over times of the order of 1 to 100 my, total track loss occurs at a temperature of the order of 100°C. Of course, the precise conditions of temperature and time are both important in determining the degree of annealing but over geological time scales, an order of magnitude increase in time reduces the temperature responsible for a given annealing effect only by about 10°C (see below). Temperature has been shown to be the only external factor to have a significant effect on annealing of fission tracks (Fleischer et al., 1975).

The Otway Basin

The fluviatile Otway Group was deposited in a series of continental rift grabens that extended some 1,000 km across southern Australia during the Early Cretaceous. The three main areas of deposition in the Otway (Fig. 11.1), Gippsland, and Bass basins contain an estimated 100,000 km^3 of quartz-poor sediments up to 3 or 4 km thick. Most of the sediments were derived from dacitic pyroclastic volcanism, although a range of detritus representing basalts to rhyolites is present (Duddy, 1983).

Paleontological evidence indicates that while thin units at the base of the sedimentary pile may be latest Jurassic in age, the bulk of the section was deposited during the Early Cretaceous (Cookson and Dettmann, 1958; Douglas, 1969, 1975).

Using fission-track dating of sphene and zircon, Gleadow and Duddy (1981b) have shown that these minerals, and by inference the majority of the volcanogenic detritus, were derived from contemporaneous volcanism. Furthermore, Gleadow and Duddy (1981b) showed that in the outcrop section of the Otway Group, fission-track ages of sphene, zircon, and apatite, three minerals with vastly different closure temperatures, were identical, and thus concluded that the outcrop section had never been heated above approximately 70°C. Grains of nonvolcanogenic apatites are rare in the Otway Group but where they do occur, as in some samples from the Banyula-1 well, they can easily be distinguished by their petrographic character (e.g., rounded shape), generally higher U content, and, in samples that have not been significantly heated since deposition, by their much older apatite fission-track ages. Data from these "basement" grains have not been included in the analysis to follow.

Where the Otway Group occurs beneath Late Cretaceous and Tertiary quartzose sediments in basins to the west of the Otway Ranges, the burial histories are typical of those from extensional basins of passive continental margins (Hegarty, 1985). The temperature of any particular Otway Group sample has increased progressively with time, and then remained stable for approximately the last 20 my. Thermal modeling in wells from regions with thick late Cretaceous and Tertiary sedimentation is consistent with samples in these wells being at their maximum temperatures at the present time. Four such wells (Flaxman's-1, Eumeralla-1, Port Campbell-4, and Banyula-1) were selected on this basis, excluding one sample from Eumeralla-1 thought to have been slightly hotter in the past (Gleadow et al., 1983), and data from these wells are presented in subsequent sections. Maximum corrected well temperatures are approximately 125°C and present geothermal gradients are very uniform at approximately 30°C/km. Vitrinite reflectance values (R_o max) range from about 0.3 to 1.3 at depths ranging between surface and near 4 km.

Recently Storzer and Selo (1984) have attempted a reinterpretation of our Otway basin data (Gleadow and Duddy, 1981a; Duddy and Gleadow, 1982), on the basis of a number of corrections that we believe to be of highly questionable validity. In any event, their treatment leads to a revised sequence of events on the southern Australian margin that is completely incompatible with the known geology. For these reasons, their interpretation is not considered further.

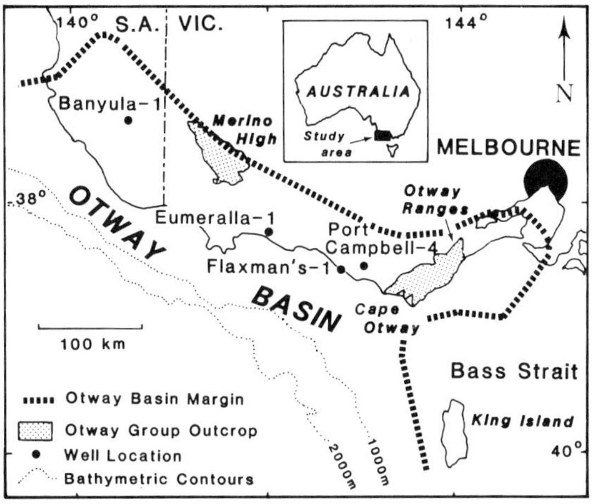

FIGURE 11.1. Location map for the Otway basin showing the outcrop regions and position of the four selected wells intersecting the Otway Group below the Late Cretaceous and Tertiary cover sequences.

Fission-Track Analysis in the Otway Basin: The Five Fission-Track Parameters

Fission-Track Ages

Apatite fission-track ages in samples from the four wells selected from the Otway basin are plotted against corrected down-hole temperature in Figure 11.2. At shallow levels (low temperature) the ages are all close to the mean age of 120 Ma observed in outcrop samples (Gleadow and Duddy, 1981b). At temperatures above 70°C the ages begin to be reduced below 120 Ma, until at temperatures above 125°C, no tracks remain and the apatites give apparent ages of zero. It is clear from Figure 11.2 that the reduction in fission-track ages is diagnostic of down-hole temperature. This illustrates the first and most obvious of our fission-track parameters, and that most widely used by other workers.

The decrease in apparent age between 70°C and 125°C in Figure 11.2 defines the "Apatite Annealing Zone" which results from long-term annealing over 120 my due to progressive temperature increase as a result of burial. Gleadow and Duddy

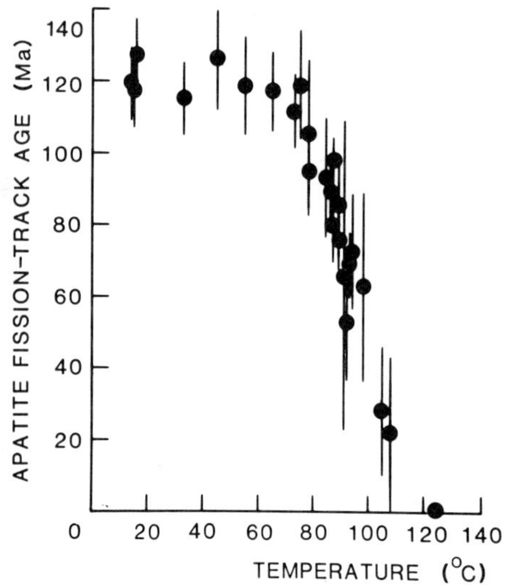

FIGURE 11.2. Variation of fission-track age of Otway Group samples with corrected present-day down-hole temperature in four Otway basin wells. At the present time these samples are thought to be at their maximum temperatures. Error bars are shown at 2 sigma.

(1981a) suggested that an equivalent time to produce the observed annealing would be of the order of 10 to 40 my.

All the apatite samples represented in Figure 11.2 come from the Otway Group and thus share a common volcanic source. In the general case, to obtain temperature information from the degree of age reduction, it is necessary to have an estimate of the original age prior to down-hole annealing. Even if the original age is unknown, a lower limit is given by the stratigraphic age. If the observed fission-track age is less than the stratigraphic age of the sample, then this gives clear evidence that significant down-hole annealing has occurred, corresponding to temperatures above 70°C in Figure 11.2.

Variation of Apparent Fission-Track Age with Depth

The *form* of the fission-track age profile with depth in Figure 11.2 is characteristic of the simple thermal history of these samples. Ages decrease steadily to zero from a plateau region of ages showing little or no annealing effect. More complex thermal histories may result in more complex profiles (Naeser, 1979a; Gleadow et al., 1983). Where temperatures have declined, the form of the age profile with depth will show evidence of the previous higher temperatures.

In the simple case, the form of the age profile (Fig. 11.2) is similar to that observed in laboratory annealing studies of track densities (Gleadow et al., 1983) and simply reflects the influence of long-term annealing during burial. Thus, the reduced ages in the annealing zone result from the partial loss of track density (due to a reduction in etchable range—see below) and are referred to as partial or apparent ages. As such, they have no meaning in their own right, other than indicating the degree of annealing that has taken place.

Distribution of Single Grain Ages

As explained above, the external detector method involves dating individual crystals, allowing inspection of the distribution of single grain ages in each apatite population. Figure 11.3 shows such distributions from outcropping Otway Group and from the four selected wells at increasing present-day down-hole temperatures, as indicated. These distributions are in every way typical of all results from the Otway basin. Also shown superimposed on each distribution in Figure 11.3 is the smoothed probability function obtained by the addition of Gaussian distributions representing each single grain age with its corresponding error, as suggested by Hurford et al. (1984). Although not rigorous, this procedure allows some useful constraints to be placed on the interpretation of the single grain age distributions.

Figure 11.3A shows that in outcrop samples, all single grain ages agree with the pooled age of approximately 120 Ma, and the smoothed probability function is narrow, symmetrical, and peaked at approximately 120 Ma. The distributions in Figure 11.3,B–D, show that as temperature increases, the single grain ages still fall within a restricted range and the smoothed function maintains its narrow peaked character. However, as shown by Figure 11.3E, above 90°C both the distribution of single grain ages and the smoothed probability function show a marked change. The distribution shows all ages from zero up to approximately the original deposition age of 120 Ma, with roughly equal probability. The smoothed function is char-

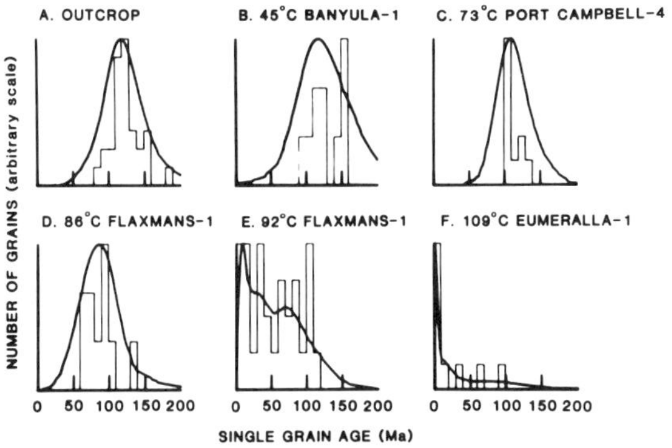

FIGURE 11.3. Distributions of single grain apatite ages from six Otway Group samples. The curve is the smoothed probability function obtained by the addition of Gaussian distributions representing each single grain age and its corresponding error. A, outcrop (composite); B, 45°C, Banyula-1; C, 73°C, Port Campbell-4; D, 86°C, Flaxman's-1; E, 92°C, Flaxman's-1; F, 109°C, Eumeralla-1. The distributions in E and F are characteristic of samples presently between 90° and 95°C and 105° and 110°C, respectively.

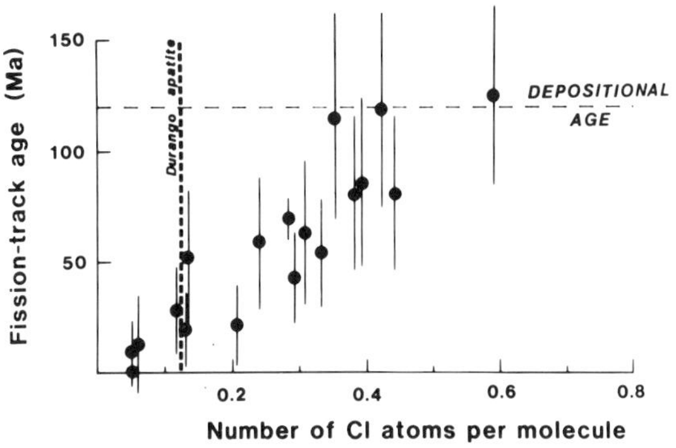

FIGURE 11.4. Relationship between apparent fission-track age and composition, for a sample of Otway Group sandstone from a depth of 2,585 m in the Flaxman's-1 well (92°C). This relationship suggests that fission tracks in chlorapatite are more resistant to annealing than those in fluorapatite. The composition of Durango apatite is also shown for comparison. From Green et al. (1985, reprinted by permission of Pergamon Journals Ltd.).

acteristically peaked at young ages showing a progressive fall toward higher ages. Single grain age data of this type are found in all samples presently in the temperature range 90°C to 95°C. Figure 11.3F shows results from a sample presently at 109°C in which most of the grains have apparent ages of zero, due to total track loss in these grains. However, a few grains have ages between zero and the original age. The smoothed curve shows a very sharp drop from a maximum near zero and persists only at a much reduced level to higher ages. Again, these data are typical of all samples presently in the temperature range 105°C to 110°C.

As explained in more detail by Green et al. (1985), the increase in the spread of single grain ages at higher temperatures is due, in large measure, to small differences in the annealing properties of individual apatites with different compositions. Figure 11.4 shows the relationship between apparent apatite fission-track age and chlorine content of apatite for the sample shown in Figure 11.3E. In this sample, from a depth at which the temperature is now 92°C, fluorapatites, being most easily annealed, have lost all tracks, while those grains richer in the chlorapatite component still retain tracks. The grains richest in chlorapatite are not totally annealed until temperatures of approximately 125°C are reached. This influence of chemical composition on apatite annealing is therefore of great advantage in paleotemperature evaluation since it is responsible for the features discussed in this section.

The degree to which individual grain ages in a sample belong to a population with a single mean age may be assessed by a χ^2 statistic devised by Galbraith (1981). From the numbers of spontaneous and induced tracks counted in individual grains, a probability can be calculated, $P(\chi^2)$, which essentially gives a measure of the probability that all grains are consistent with a single age. Values of $P(\chi^2) < 5\%$ can be taken as an indication of a real spread in single grain ages. Figure 11.5 shows values of $P(\chi^2)$ obtained from Otway Group samples at various present-day temperatures in the four selected wells. It must be borne in mind that because of the definition of the χ^2 statistic, $P(\chi^2)$ does not take a single value. Samples at temperatures below 90°C all have values of $P(\chi^2) > 5\%$, showing that all these analyses are consistent with a single age (the actual value of the age depending on the temperature as shown in Fig. 11.2). On the other hand, data from samples above 90°C show a drastic reduction in $P(\chi^2)$, to values $<0.001\%$, due to the spread in ages introduced by different

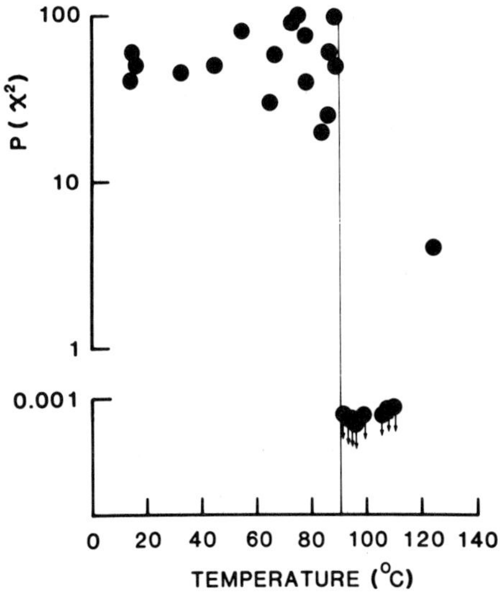

FIGURE 11.5. Variation of $P(\chi^2)$ (see text) with corrected present down-hole temperature in four Otway basin wells. Below 90°C (which is indicated by the vertical line) all single grain ages agree with a single pooled age. Above this temperature the difference in annealing properties of individual apatite grains introduces a significant spread in ages, reflected in the extremely low values of $P(\chi^2)$, which are all less than 0.01%. At temperatures greater than 120°C, $P(\chi^2)$ rises again as an increasing proportion of grains are reduced to zero age.

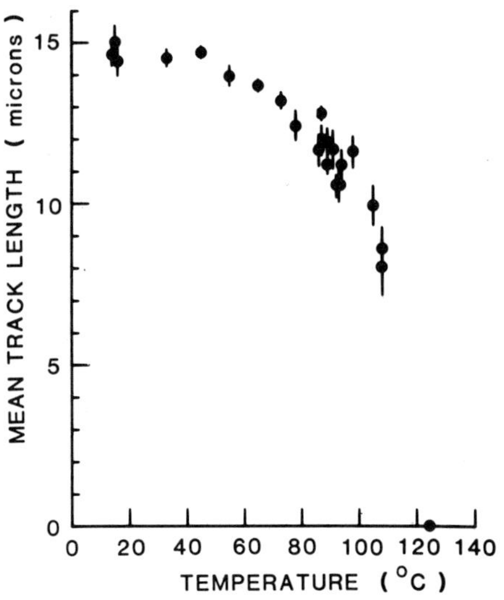

FIGURE 11.6. Variation of mean confined track length in Otway Group apatites with corrected present-day down-hole temperature. Error bars are shown at 2 sigma.

Between these latter temperature ranges, transitional types are seen. Above 120°C the vast majority of grains have zero apparent ages, with only those grains most resistant to annealing having finite ages.

annealing characteristics. The sample at 124°C shows an increase in $P(\chi^2)$ to a value of approximately 4% due to the reduction of most grains to zero age, with only one grain of 10 analyzed retaining tracks. Hence in this case the $P(\chi^2)$ value is almost indicative of a single age, which is close to zero.

Thus we have identified three features of the single grain age distribution that are sensitive to temperature. For the Otway Group, where all apatites were originally of the same age, a value of $P(\chi^2) \ll 5\%$ indicates temperatures in the region 90°C to 120°C. A flat single grain age distribution from zero up to the original age, with a smoothed probability curve as in Figure 11.3E, is indicative of 90°C to 95°C. A predominance of apparent zero ages with a few grains having ages between zero and the original age, and a smoothed probability function curve falling sharply at very young ages as in Figure 11.3F, is diagnostic of temperatures in the range 105°C to 110°C.

Variation of Mean Confined Track Length with Depth

Figure 11.6 shows the observed reduction in mean confined length with temperature in the four selected wells. The mean length shows a progressive reduction with increasing temperature from samples at present-day temperatures below 50°C, with lengths greater than 14 μm, to zero at approximately 125°C. The final length decrease is so rapid that reduction in mean length from about 8 μm to total erasure occurs in the last 15°C, between approximately 110°C and 125°C.

The reduction in track length is the result of annealing throughout burial and these data essentially represent an identical situation to laboratory annealing experiments, but carried out over a much longer time scale, and with tracks produced continuously up to the present. Thus the mean confined track length is a very sensitive indicator of temperature.

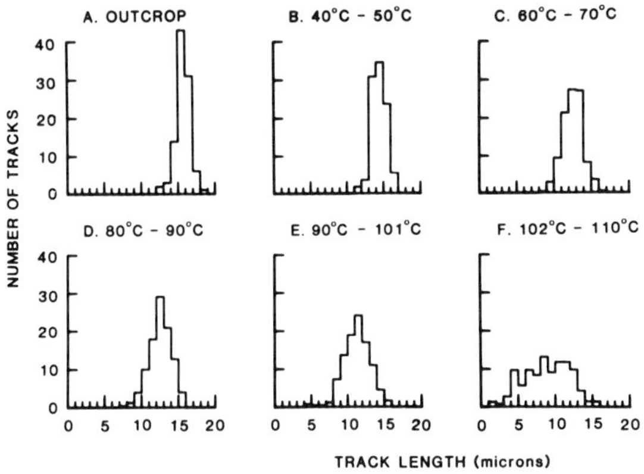

FIGURE 11.7. Distributions of confined fission-track lengths in Otway Group apatites. These are composite distributions (normalized to a total of 100 tracks) produced by summing results from samples in all four wells within the specified temperature limits. Within these limits all samples have essentially similar distributions. A, outcrop; B, 40° to 50°C; C, 60° to 70°C; D, 80° to 90°C; E, 90° to 101°C; F, 102° to 110°C. Both the mean track length and the form of the track length distribution are highly sensitive indicators of temperature. The flat distribution from lengths of 1 μm up to 16 μm shown in F is particularly diagnostic of temperatures in the range 102°C to 110°C.

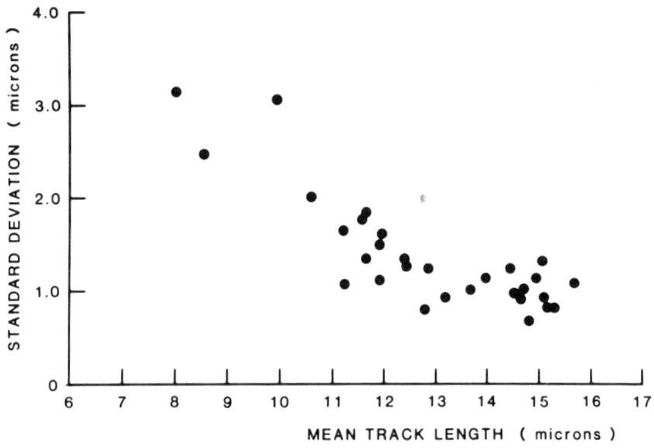

FIGURE 11.8. Relationship between standard deviation of the confined track length distribution and mean confined track length in Otway Group apatites from the four selected wells. The increase in standard deviation as mean length decreases emphasizes the increasing spread of lengths in more annealed samples as shown in Figure 11.7.

Distribution of Confined Fission-Track Lengths

Figure 11.7 shows distributions of confined fission-track lengths in Otway Group apatites in outcrop samples and from the four selected Otway basin wells. Note that Laslett et al. (1984) have shown that geometrical length bias is extremely important in the interpretation of confined track length distributions. However, the data presented here and below are not corrected for this effect as we are concerned, at this stage, only with observable parameters. In these distributions, data from samples within approximately 10°C intervals have been pooled. In each interval, the distributions obtained in individual samples were essentially identical. These distributions, together with Figure 11.6, show that as the mean track length falls the distribution maintains the narrow, symmetric form at first, but as the mean length is reduced below about 13 μm, the distributions become increasingly broad, until at temperatures between 102°C and 110°C tracks of lengths from about 1 or 2 μm up to 16 μm are present, in a very broad, almost flat, distribution. The trend of increasing width of the distribution with decreasing mean length is emphasized in Figure 11.8, which shows the relationship between the standard deviation of the length distribution and mean track length in the four selected wells. These data define the Otway basin "reference trend."

The increase in the spread of track lengths with increasing temperature arises from three dominant causes. The first source of broadening of the track length distribution is the anisotropy of the annealing process in apatite as observed in laboratory annealing (Green and Durrani, 1977; Laslett et al., 1984). A second source comes from the variation in annealing properties of individual apatite grains of different composition noted above (Fig. 11.4). Thus, shorter tracks are contributed from grains more susceptible to annealing, while in more resistant grains, tracks are longer. The third cause of the increasing spread in lengths arises because of the continuous production of tracks with time. Some tracks will have been in existence throughout the whole thermal history, whereas some will have formed only recently. Predictions of the behavior of individual fission tracks during annealing, based on a new mathematical description of annealing given by Laslett et al. (1987), suggest that for thermal histories in which temperatures are constant, or slowly rising with time, the lengths of individual tracks produced at different times converge to approximately the same length. However, tracks may take approximately 10 my to reach this length, and thus some tracks, formed within the last few million years, will be longer than those formed previously. Note, by comparison of Figures 11.3 and 11.7, that the spread in lengths lags behind the spread in ages. This arises due to the significant geometrical and observational bias involved in the detection of such short tracks, as discussed in part by Laslett et al. (1984). Only when a large proportion of the grains show significant annealing (above 100°C) do very short tracks (< 8 μm) begin to be detectable in large numbers.

In summary, both the mean track length and the form of the track length distribution are highly sensitive indicators of the temperature. For a simple burial history such as that of the Otway basin, mean track length shows a simple convex decrease with increasing temperature, and a given mean confined length is immediately indicative of temperature (Fig. 11.6).

Apatite Fission-Track Analysis in Sedimentary Basins

General

Figure 11.9 summarizes the five temperature-sensitive fission-track parameters identified in the four selected Otway basin wells. The consistency of the temperature dependence of these parameters shows that they can be taken as diagnostic of the present-day corrected down-hole temperatures. Furthermore, because the samples from these reference wells have been at their maximum temperature for about 20 my, the fission-track parameters diagnostic of a certain temperature in these wells can be considered to be diagnostic of these temperatures *in general*.

Since the processes involved in fission-track annealing are not particular to the Otway basin, but specific only to apatite, the relationship between various fission-track parameters (mean length, track length distribution, age reduction, single crystal age distribution, etc.) shown by the above data is also completely general. Therefore the data from the reference wells can be used as a basis for interpreting paleotemperatures in other basins.

In basins with thermal histories similar to that of the Otway basin, the parameters identified in the Otway basin can be applied directly to elucidate paleotemperature information. In basins with more complex thermal histories, the features identified in the Otway basin data can be used to identify signatures characteristic of certain temperature ranges, as discussed in more detail below.

In cases of longer or shorter effective heating times, as a consequence of the Arrhenius relationship, the characteristic temperatures producing a given degree of annealing will be lower or higher, respectively. Rigorous quantitative treatment of this "trade-off" between temperature and time must await a thorough understanding of the annealing kinetics of fission tracks in apatite of various compositions. Previous laboratory annealing studies of fission tracks in apatite are of little usefulness in understanding natural annealing, as they greatly overestimate the range of temperatures over which natural annealing in boreholes should be observed (Naeser, 1981; Gleadow et al., 1983). This can be attributed to the nonrecognition, in these early studies, of the influence of composition on annealing kinetics (Green et al., 1985). Naeser (1981) and Gleadow et al. (1983) ascribed natural annealing in boreholes to certain time-temperature combinations estimated on geological grounds. These were then linked to laboratory annealing data in an attempt to construct a more representative description of annealing kinetics. These studies suggest an order of magnitude increase in heating time reduces temperatures necessary for a given degree of annealing by roughly 15°C.

However, this procedure also ignores the role of composition and can therefore only be an approximation to the true description. In addition, the presentation of the observed annealing effect in terms of a single temperature and time ignores the detailed response of tracks to variable temperature annealing. Laslett et al. (1987) have developed an improved description of the annealing kinetics of fission tracks in a large, single crystal of fluorapatite, which can be extended to treat thermal histories in which temperatures vary with time, as discussed above. The predictions of this treatment clearly show the dominant role of temperature in producing fission-track annealing, with time being of less importance. For this reason we consider that the effect of an order of magnitude change in time

FIGURE 11.9. Temperature-sensitive apatite fission-track parameters observed in Otway basin wells for samples presently at their maximum temperatures since burial. Note that track length data refers only to confined tracks.

TEMP (°C)	FISSION-TRACK AGES	VARIATION OF AGE WITH DEPTH	SINGLE GRAIN AGE DISTRIBUTION	MEAN TRACK LENGTH (microns)	TRACK LENGTH DISTRIBUTION
30 40 50	little or no difference from depositional age	flat	narrow, unimodal, peaked at depositional age	14.5 μm	narrow, symmetric, unimodal, with mean decreasing with increasing temperature s.d. between 0.8 and 1.5 μm
60				14.5 – 13μm	
70 80	less than depositional age	steeply falling	as above, peaked at reduced age	13 – 11.5μm	as above, with increasing spread, s.d. between 1 and 3 um
90 100	falling rapidly with increasing temperature		spread from zero to depositional age	11.5 – 8μm	great spread from 1 to 16 μm s.d.~3μm
110			peaked at zero, few grains with intermediate ages		
120				few tracks	few tracks
130	zero	zero	all grains zero age	no tracks	no tracks

will be to change temperatures necessary to produce a given degree of annealing by about 10°C at most.

In the following discussion, we quote estimates of temperature based on the time scales of heating appropriate to the Otway basin. The change of temperature over different time scales discussed above should be borne in mind, in considering this discussion.

Use of the Otway data to interpret samples from other basins strictly depends on these samples containing a similar spread of apatite compositions. Experience to date in a variety of circumstances suggests that this assumption should be reasonable in most cases. Samples with only a limited range of apatite composition may be recognized by the narrowness of their length distributions, as in such samples the dominant source of broadening of the distribution (Figs. 11.7 and 11.8) will have been removed.

Up to this point we have considered only the relatively simple Otway-type thermal history. The application of the five temperature-sensitive fission-track parameters in elucidating details of more complex situations will now be dealt with.

Source Characterization

The understanding of fission-track parameters observed in apatites from the Otway Group is greatly simplified by the fact that most of the apatites were derived from contemporaneous volcanism and come from essentially a single (very thick) depositional unit. The majority of grains therefore had no previous track record at the time of deposition (any "basement" grains have much greater ages than the volcanogenic type and can easily be excluded).

In other basins, the situation may not always be so simple. For example, the common source of apatite in the fill of sedimentary basins will often be basement terrains rather than the contemporaneous volcanism characteristic of the Otway basin. This would seem, at first sight, to present a bewildering array of possibilities for the original fission-track parameters of detrital apatites transported into the basin. Fortunately, the single grain dating capability of the external detector method allows apatites of different age to be identified. In samples that have never been deeply buried these age groupings, and those of coexisting zircons and sphenes, can give unique provenance information (see, e.g., Duddy et al., 1984). In samples showing significant down-hole annealing, the situation is more complicated. If a source can be characterized from shallower samples of the same horizon that have remained cooler, then this information may allow interpretation of the annealed age pattern. Alternatively, investigation of the relationship between mean confined length and apparent fission-track age allows some source characterization as explained in more detail by Gleadow and Duddy (1984).

The lengths of fission tracks in apatites deposited in sedimentary basins derived from various possible source areas are, fortunately, also well characterized. Gleadow et al. (1986) have compiled a large body of data on confined track lengths from a wide variety of geological environments. They show that apatites in samples that cooled quickly and have never subsequently been heated above approximately 50°C have a narrow, symmetric length distribution with a mean of approximately 14.5 μm and a standard deviation of approximately 1.0 μm. The distributions of confined track lengths seen in outcrop and shallow boreholes in the Otway basin are typical of this type of distribution. Such distributions are termed "undisturbed volcanic" and are diagnostic of a rapidly cooled source such as volcanic rocks, or basement terrains that were cooled so rapidly during uplift that the great majority of tracks in apatites in, or derived from such rocks will have formed at low temperatures (< 50°C).

The compilation of Gleadow et al. (1986) also shows that typical basement regions have only a restricted range of track length distributions, characterized by what they term an "undisturbed basement" distribution. This distribution is negatively skewed, with a mean of approximately 12.8 μm and a standard deviation of approximately 1.6 μm. Some variation is found in different terrains, no doubt due in part to the intricacies of the specific thermal history from sample to sample. Nevertheless, the great majority of samples have mean lengths between 12 and 14 μm and standard deviations between 1 and 2 μm. In general, any old basement terrain is well characterized by the type "undisturbed basement" distribution.

Finally this compilation also shows that induced tracks in the majority of apatites have mean confined lengths within a very narrow range from 15.8 to 16.6 μm, with a standard deviation of 0.8 to 1.0 μm. This greatly simplifies interpretation since it negates the need to determine the induced distribution for each individual sample. Instead, distributions of confined spontaneous fission-track lengths can be directly compared from sample to sample.

The characteristic "induced," "undisturbed volcanic," and "undisturbed basement" confined track length distributions from Gleadow et al. (1986) are shown in Figure 11.10. Apatites containing a pre-existing track record when deposited in a sedimentary basin will, in the great majority of cases, have length distributions of either the "undisturbed volcanic" or "undisturbed basement" types.

Complex Thermal Histories

In the scheme outlined above, the primary fission-track variable is the etchable range, as manifested in the length of confined tracks. It is the response of this parameter to temperature that brings about the observed variation in the five fission-track parameters discussed above. Therefore understanding the effect of complex thermal histories must proceed from an understanding of the effect of temperature on confined track length.

Recent advances in the understanding of the annealing kinetics of fission tracks in apatite have shown that the final length of each track is dominated by the highest temperature experienced by

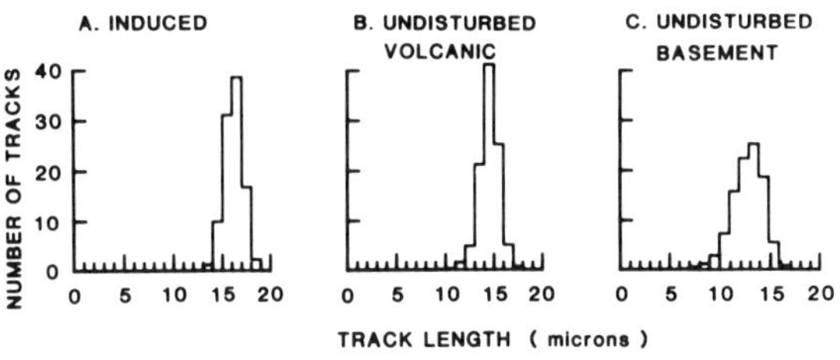

FIGURE 11.10. Representative distributions of confined fission-track lengths from Gleadow et al. (1986). A, "induced" distribution typical of induced tracks in all apatites; B, "undisturbed volcanic" distribution typical of tracks in rapidly cooled apatites; C, "undisturbed basement" distribution typical of tracks in apatites with a more protracted (but monotonic) cooling history (all normalized to 100 tracks).

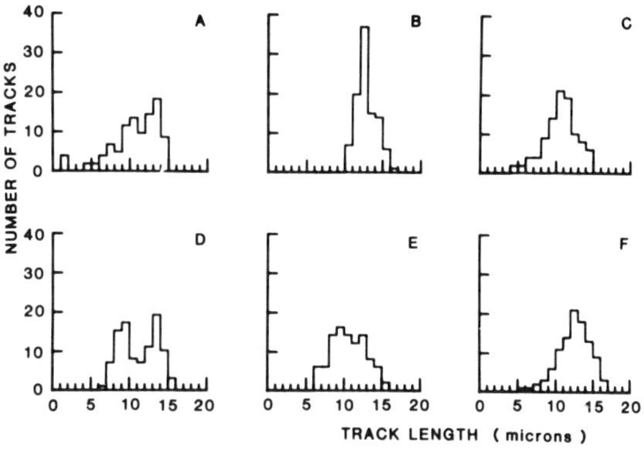

FIGURE 11.11. Six typical confined track length distributions (all normalized to 100 tracks) from samples with complex thermal histories (see text).

that track (Duddy et al., in press). As temperature is increased, tracks become progressively shorter. If temperature drops after some time, then the tracks present at this time are essentially frozen at the lengths pertaining at the thermal maximum. Tracks formed after this cooling will be longer than those formed prior to and during the heating phase. Because of the continuous production of tracks throughout time, the confined track length distribution contains, in principle, the full detail of the thermal history below approximately 150°C. This detail is perhaps most easily seen in bimodal length distributions (Laslett et al., 1982; Gleadow et al., 1986), where the two generations of tracks formed prior to and after a thermal event are clearly resolved.

In most cases the situation is not so clearly defined. Nevertheless, investigation of a complex length distribution allows broad constraints to be placed on the thermal history, using Otway basin data as a guide. Figure 11.11 shows a few cases of length distributions from samples having more complex thermal histories than that in the Otway basin. Figure 11.11A shows a very broad distribution with a broad peak between 10 μm and 13 μm but with tracks extending from 1 μm to 15 μm. By comparison with Figure 11.7, this distribution can be represented as the composite of Figure 11.7, F–H. This leads to the interpretation that the distribution contains components produced at approximately 105°C, 90°C, and 80°C, suggesting a protracted cooling through this temperature range.

The distribution in Figure 11.11B shows an unusual positive skew. This is interpreted in terms of a rather mild, recent thermal event reaching perhaps 70°C to 80°C, which reduced preexisting tracks to a mean of approximately 12 μm. Tracks formed since the postevent cooling make a minor contribution around 14 μm to 16 μm which cannot be resolved. The distribution in Figure 11.11C shows a rather more complex distribution that may be interpretable in terms of a protracted cooling between approximately 100°C and 90°C followed by more rapid cooling, although other more complex interpretations may be possible. Figure 11.11D shows the typical well-resolved bimodal distribution, in this case from Brogo, New South Wales, as reported by Laslett et al. (1982). Figure 11.11, E and F, shows other examples of natural complex distributions taken from Gleadow et al. (1986). Note that these temperature estimates assume heating times of the order of 10 my.

In contrast to the confined track length distributions, fission-track ages reveal very little of any complexity in thermal history. The ages of individual samples such as those represented in Figure 11.11 mean nothing in themselves. They are reduced from the original age (or the age that would have been obtained if no annealing had taken place) by an amount dependent on the detail of the length distribution. Without the length distribution it can often be extremely difficult to interpret such ages. If a full borehole sequence is available then some complexity may be resolved in the pattern of

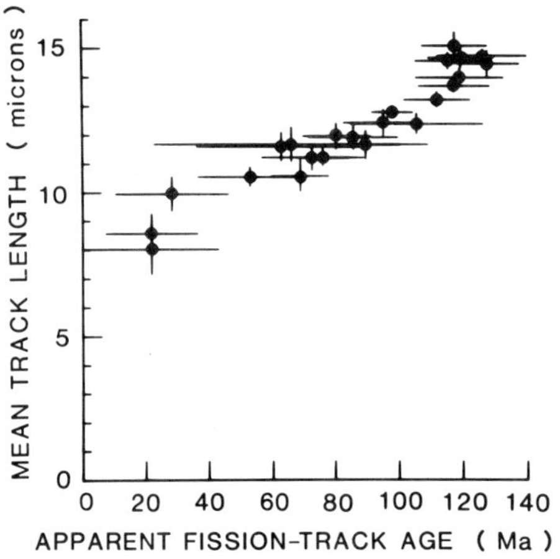

FIGURE 11.12. Relationship between mean track length and apparent apatite fission-track age in the four selected Otway basin wells (from the data in Figs. 11.1 and 11.4). This relationship allows the "correction" of apparent ages for the effects of down-hole annealing.

age variation with increasing depth (temperature). For example, in cases of rapid uplift (or downward movement of isotherms) followed by long-term stability, the variation of age with depth shows the superposition of two curves such as that shown in Figure 11.2 (as illustrated by Naeser, 1979a). However, such simple cases are rare and the situation is generally much more complex.

In most cases, therefore, age data alone allow only part of the information to be recovered. Investigation of the five fission-track parameters outlined above allows a much tighter constraint to be placed on viable thermal histories. One example of the usefulness of the combined length and age approach is given by Gleadow and Duddy (1984).

Correction of Ages in Annealed Apatites

As explained above, the reduction in age due to down-hole annealing is dominantly due to a reduction in the etchable range of fission tracks, of which the mean confined track length gives a measure. The relationship between apparent age and mean length in samples showing significant down-hole annealing affords a method of "correcting" the apparent fission-track age for the effects of this effect. Using Figure 11.12, a given track length can be related to the resulting fractional age reduction. Thus, a measured apparent age can then be divided by this factor, to obtain a corrected age, t_l. This gives an estimate of the age that would be observed if the sample had not been heated since it last began to accumulate tracks. At present, the Otway basin relationship between mean confined track length and apparent age, shown in Figure 11.12, is our only good source of information on the relationship between length and age reduction. Application of this relationship to data from another basin is only strictly valid if this basin shares a thermal history similar to that of the Otway basin. In basins with more complex histories, the length distributions, for a given mean length, may show significant differences with those in the Otway basin on which the curve in Figure 11.12 is based. In this case, the relationship between length and density will be different to the simple Otway curve, and a correction based on the latter will be invalid. In the near future we hope to model the relationship between age reduction and the distribution of confined track lengths, in which case a more rigorously defined correction may be applied for any complex case. In the meantime, however, the curve shown in Figure 11.12 allows a rudimentary form of correction, which may not be fully correct but does allow a useful insight into the thermal history of the sample.

The estimated corrected age, t_l, may be compared with the stratigraphic age, t_s, to give a parameter Δt from:

$$\Delta t = t_l - t_s$$

If Δt is close to zero, then this suggests that apatite was derived from either contemporaneous volcanism or a very rapidly uplifted source terrain. In either case, very few tracks have been inherited from the source, and thus most tracks have formed after deposition. A negative value of Δt, indicating a corrected age less that the stratigraphic age, suggests that at some time after deposition all preexisting tracks have been erased. This shows that such samples have been heated above the temperature necessary for total track loss and subsequently cooled below this value at around the time indicated by the corrected age. This of course implies that such samples have been completely through the oil window and thus would not be highly prospective for liquid hydrocarbons. Again it must be emphasized that this interpretation is dependent on a sample having an Otway-type length distribution.

A positive Δt, implying a corrected age greater than the stratigraphic age, suggests that a proportion of the observed tracks were formed prior to deposition. The behavior of this inherited component must be borne in mind in interpreting results from such samples. However, the new tracks formed since deposition can be expected to behave in a manner typical of those observed in the Otway basin.

If reliable present-day temperature information is available in a suite of borehole samples then another useful parameter that can be calculated is:

$$\Delta T = T_l - T$$

where T is the present-day temperature and T_l is the temperature obtained from Figure 11.6, corresponding to the mean length observed in a particular sample. If ΔT is close to zero then this suggests that the sample is at or near its maximum temperature at the present day and has never been significantly hotter. A negative ΔT suggests that temperatures have risen relatively recently (over the last few million years), as the effect of increased temperature is to further shorten all existing tracks to a value determined by the present temperature. A positive value of ΔT indicates that the sample is cooler now than it has been in the past, so that tracks formed during earlier times that have experienced higher temperatures are shorter than those formed in the existing thermal regime.

Care must be taken in the interpretation of both Δt and ΔT values. Neither are rigorously evaluated, both being subject to uncertain (and large) errors. However, both have proved extremely useful in the elucidation of thermal history information in the past. One example of their use is given by Gleadow and Duddy (1984) for the Canning basin of northwestern Australia. Here corrected ages are particularly useful in separating the effect of Permian intrusions from basement trends.

AFTA as a Routine Exploration Tool

In the context of paleothermal history analysis of exploration wells, the full potential of AFTA is best realized with a sequence of equally spaced samples in the depth range down to the base of the Apatite Annealing Zone. Particularly in complex burial/thermal histories, previous periods of elevated temperatures may be more easily identified at shallow levels (low present-day temperatures), since deeper in the well such effects may be masked by the present-day annealing zone.

Thus the combination of deeper and shallower samples can lead to much tighter constraints than deeper samples alone. The precise sample spacing is governed to some extent by the specific problems being tackled and the presence of suitable lithologies. In an average well, 400-m intervals would be typical, giving a total of 6 to 10 samples per well.

In addition, it is highly desirable that the present-day thermal gradient should be well controlled with a number of reliable present-day temperature measurements. The knowledge of the present-day thermal gradient is important in the application of any technique that strives to determine paleotemperature conditions, as all techniques rely to some extent on assessing perturbations from the present situation.

Apatite is a widespread minor constituent of most granitic and metamorphic rocks and is present in a wide range of volcanic rocks. As such, apatite of suitable grain size for fission-track analysis is also a common constituent of most sandstones and coarse siltstones. Higher yields usually come from the more mineralogically immature sandstones, but experience in a wide range of sedimentary basins, including examples in Australia, United States, and Europe, has shown that over 80% of sandstones contain sufficient apatite.

The size of sample required is perhaps the major limitation on using AFTA, given the generally small amount of material stored from preexisting wells. Although samples of immature sandstone weighing less than a few hundred grams have been successfully analyzed, it is generally necessary to obtain samples of 1 to 2 kg. While samples of this size may be available from conventional core, stored cuttings are usually inadequate and we have found that it is best to collect large cuttings samples during drilling operations specifically for AFTA.

Zircon and, to a lesser extent, sphene are also common detrital constituents of sandstones, and they can also be used in fission-track analysis. The track annealing temperatures in these minerals are considerably higher than in apatite, however, requiring temperatures of between about 200°C and 300°C (e.g., Naeser and Faul, 1969; Harrison and McDougall, 1980; Hurford, 1984) for total erasure. Because such temperatures are generally not reached in hydrocarbon prospective basins, fission-track analysis of these minerals can be of great benefit in providing a normalization age for volcanogenic detritus and in provenance studies.

Future Developments

At present the simple correction procedure outlined above is not very meaningful in samples with more complex length distributions, other than giving some broad limits to possible thermal histories. In principle it should be possible to "dissect" a particular length distribution and deduce the contribution of each portion to the original age. Indeed, a similar procedure should also allow the full detail of the thermal history to be ascertained. However, Laslett (1984) has shown that in most cases the confined length data are not sufficiently resolved to enable rigorous recovery of the thermal history. This problem is particularly acute for the early thermal history, which becomes obscured by cumulative errors in the estimation process.

Instead we are working toward the development of a predictive mathematical model of the behavior of fission tracks in geological systems. This model is being constructed around a quantitative analysis of annealing kinetics in apatites with a wide variety of chemical compositions, allowing the synthesis of model length distributions for any thermal history. From a thorough knowledge of the relationship between track length and track density, the model length distribution can be converted into a fission-track age. In this way, for a sequence of well samples, the five fission-track parameters can be modeled for any postulated thermal history, and tests of modeled parameters against observed values can constrain a range of viable geological possibilities. Experiments to gather the data for these models are well under way, and early results are extremely encouraging. In the interim, until these rigorous methods are developed, the semiquantitative analysis outlined above will continue to place useful constraints on thermal history.

Advantages of AFTA

This new approach to the study of thermal evolution and maturation of hydrocarbons in sedimentary basins offers several advantages over existing techniques for hydrocarbon resource evaluation. Besides offering information on maximum paleotemperatures, AFTA gives direct information on timing of thermal events and can give semiquantitative estimates of the variation of paleotemperature through time. In the near future this will be placed on a rigorous, quantitative basis. The continuous production of fission tracks with time gives the AFTA system the additional advantage of recording the entire thermal history in a single sample.

AFTA can also be applied to sequences of all ages including early Paleozoic and even Precambrian rocks. It is also one of the few thermal analysis techniques useful in highly oxidized environments such as red-bed sequences, impoverished in organic components. Finally, besides the thermal information, AFTA provides a powerful tool for identifying sedimentary provenance.

Acknowledgments. The authors would like to acknowledge ESSO Australia for substantial support and encouragement in developing this work to its present level. Continuing support for the project comes from a NERDDC grant, the University of Melbourne, and Geotrack International. In addition, thanks are due to the Victorian Department of Minerals and Energy for access to cores from Otway basin wells and to Alliance Petroleum and partners for samples from Banyula-1.

References

Briggs, N.D., Naeser, C.W., and McCulloh, T.H. 1981. Thermal history of sedimentary basins by fission-track dating (abst.). Nuclear Tracks 5:235–237.

Cookson, I.C., and Dettmann, M.E. 1958. Cretaceous "magaspores" and a closely associated megaspore from the Australian region. Micropaleontology 4:39–49.

Douglas, J.G. 1969. The Mesozoic floras of Victoria: Parts 1 and 2. Memoir of the Geological Survey of Victoria 28:310 pp.

Douglas, J.G. 1975. The Mesozoic floras of Victoria: Part 3. Memoir of the Geological Survey of Victoria 29:185 pp.

Duddy, I.R. 1983. The geology, petrology and geochemistry of the Otway Formation volcanogenic sediments. Ph.D. thesis, University of Melbourne, Australia, 426 pp.

Duddy, I.R., and Gleadow, A.J.W. 1982. Thermal history of the Otway Basin, southeastern Australia, from geologic annealing of fission tracks in detrital volcanic apatites (extended abst.). Workshop on Fission Track Dating, Fifth International Conference on Geochronology, Cosmochronology, and Isotope Geology, Nikko, Japan, pp. 13–16.

Duddy, I.R, Gleadow, A.J.W., and Keene, J.B. 1984. Fission track dating of apatite and sphene from Paleogene sediments of deep sea drilling project leg 81, site 555. In: Roberts, D.G., Schnitker, D., et al. (eds.): Initial Reports of the Deep Sea Drilling Project 81. Washington, DC, U.S. Government Printing Office, pp. 725-729.

Duddy, I.R., Green, P.F., and Laslett, G.M. In press. Thermal annealing of fission tracks in apatite: 3. Variable temperature behaviour. Chemical Geology (Isotope Geoscience Section).

Fleischer, R.L., Price, P.B., and Walker, R.M. 1975. Nuclear Tracks in Solids. Berkeley, University of California Press, 605 pp.

Galbraith, R. 1981. On statistical models for fission track counts. Mathematical Geology 13:471–488.

Gleadow, A.J.W. 1981. Fission-track dating methods: What are the real alternatives? Nuclear Tracks 5:3–14.

Gleadow, A.J.W., and Duddy, I.R. 1981a. A natural long-term annealing experiment for apatite. Nuclear Tracks 5:169–174.

Gleadow, A.J.W., and Duddy, I.R. 1981b. Early Cretaceous volcanism and the early breakup history of southeastern Australia: Evidence from fission track dating of volcanogenic sediments. In: Cresswell, M.M., and Vella, P. (eds.): Gondwana V. Rotterdam, A.A. Balkema, pp. 295–300.

Gleadow, A.J.W., and Duddy, I.R. 1984. Fission track dating and thermal history analysis of apatites from wells in the northwestern Canning Basin. In: Purcell, P.G. (ed.): The Canning Basin. Perth, Geological Society of Australia and the Petroleum Exploration Society of Australia, pp. 377–387.

Gleadow, A.J.W., Duddy, I.R., Green, P.F., and Lovering, J.F. 1986. Confined fission track lengths in apatite: A diagnostic tool for thermal history analysis. Contributions to Mineralogy and Petrology 94:405–415.

Gleadow, A.J.W., Duddy, I.R., and Lovering, J.F. 1983. Fission track analysis: A new tool for the evaluation of thermal histories and hydrocarbon potential. Australian Petroleum Exploration Association Journal 23: 93–102.

Green, P.F., Duddy, I.R., Gleadow, A.J.W., and Tingate, P.R. 1985. Fission track annealing in apatite: Track length measurements and the form of the Arrhenius plot. Nuclear Tracks 10:323–328.

Green, P.F., and Durrani, S.A. 1977. Annealing studies of tracks in crystals. Nuclear Tracks 1:33–39.

Harrison, T.M., and McDougall, I. 1980. Investigations of an intrusive contact, northwest Nelson, New Zealand: I. Thermal, chronological, and isotopic constraints. Geochimica et Cosmochimica Acta 44:1985–2004.

Hegarty, K.A. 1985. Origin and evolution of selected plate boundaries. Ph.D. thesis, Columbia University, New York, 254 pp.

Hurford, A.J. 1984. On the closure temperature for fission tracks in zircon. Abstracts, Fourth International Fission Track Dating Workshop, Troy, NY, p. 22.

Hurford, A.J., Fitch, F.J., and Clarke, A. 1984. Resolution of the age structure of the detrital zircon populations of two Lower Cretaceous sandstones from the Weald of England by fission track dating. Geological Magazine 121:269–277.

Hurford, A.J., and Green, P.F. 1982. A users' guide to fission track dating calibration. Earth and Planetary Science Letters 59:343–354.

Lal, D., Rajan, R.S., and Tamhane, A.S. 1969. Chemical composition of nuclei of $Z > 22$ in cosmic rays using meteoritic minerals as detectors. Nature 221: 33–37.

Laslett, G.M. 1984. Length-reduction correction procedures in fission track dating. Abstracts, Fourth International Fission Track Dating Workshop, Troy, NY, p. 30.

Laslett, G.M., Gleadow, A.J.W., and Duddy, I.R. 1984. The relationship between fission track length and density in apatite. Nuclear Tracks 9:29–38.

Laslett, G.M., Green, P.F., Duddy, I.R., and Gleadow, A.J.W. 1987. Thermal annealing of fission tracks in apatite: 2. A quantitative analysis. Chemical Geology (Isotope Geoscience Section) 65:1–13.

Laslett, G.M., Kendall, W.S., Gleadow, A.J.W., and Duddy, I.R. 1982. Bias in measurement of fission-track length distributions. Nuclear Tracks 6:79–85.

Naeser, C.W. 1979a. Thermal history of sedimentary basins by fission-track dating of subsurface rocks. In: Scholle, P.A., and Schluger, P.R. (eds.): Aspects of Diagenesis. Society of Economic Paleontologists and Mineralogists Special Publication 26, pp. 109–112.

Naeser, C.W. 1979b. Fission track dating and geologic annealing of fission tracks. In: Jäger, E., and Hunziker, J.C. (eds.): Lectures in Isotope Geology. Berlin, Springer-Verlag, pp. 154–169.

Naeser, C.W. 1981. The fading of fission tracks in the geologic environment: Data from deep drill holes (abst.). Nuclear Tracks 5:248–250.

Naeser, C.W., and Faul, H. 1969. Fission track annealing in apatite and sphene. Journal of Geophysical Research 74:705–710.

Storzer, D., and Selo, M. 1984. Toward a new tool in hydrocarbon resource evaluation: The potential of the fission track chrono-thermometer. In: Durand, B. (ed.): Thermal Phenomena in Sedimentary Basins. Collection Colloques et Seminaires 41, Paris, Editions Technip, pp. 89–110.

12

Significance of Combined Vitrinite Reflectance and Fission-Track Studies in Evaluating Thermal History of Sedimentary Basins: An Example from Southern Israel

Shimon Feinstein, Barry P. Kohn, and Moshe Eyal

Abstract

Vitrinite reflectance and fission-track age determinations (apatite, zircon, and sphene) from three deep boreholes in the Har HaNegev area, southern Israel, have been integrated for reconstruction of the thermal history. Drill holes, Ramon 1, Makhtesh Qatan 2, and Kurnub 1, each sited in the core of a separate breached anticline ("makhtesh"), penetrated a discontinuous Early Permian to Mesozoic sedimentary succession unconformably overlying a Precambrian arkosic and igneous complex. Post-Early Permian burial history was reconstructed from borehole data and regional stratigraphy.

The relationship between isoreflectance contours and bedding indicates that coalification predated the regional basal Cretaceous unconformity. Vitrinite reflectance profiles together with reconstructed burial curves have been used to calculate time-temperature models. These models reveal that the coalification profiles measured in the deeper part of the sedimentary section reflect a relatively short, intense Jurassic thermal event: for Ramon 1 and Makhtesh Qatan 2, 45° to 55°C/km over a period of approximately 10 to 25 my, and for Kurnub 1, 65° to 75°C/km over a period of <5 my. "Freezing" of the coalification process since at least Early Cretaceous time implies decrease of postcoalification temperatures. This cooling prevailed despite additional burial and reflects decay of the thermal gradient since Early Cretaceous time.

Fission-track dates of apatites are younger than their host strata and are interpreted as thermally reset ages. In Ramon 1, maximum paleothermal gradients derived from the apatite ages considerably constrain maximum possible postcoalification thermal gradients derived from vitrinite reflectance alone. In Makhtesh Qatan 2 and Kurnub 1, however, maximum paleothermal gradients derived by coal rank and fission-track measurements are in good agreement. In all three boreholes, apatite data independently reflect regional decay of the thermal gradient, at least through Cenozoic time. Furthermore, the ages suggest that the present-day relatively low thermal gradient (approximately 20°C/km) has prevailed in the study area since at least Early Oligocene time. Those ages obtained from samples within 150 m of total depth of the drill holes indicate cooling from the zone of total track annealing under the Early Oligocene depth-thermal gradient regime and thus provide fixed points in time on the regional cooling curve.

Zircon fission-track ages probably record cooling resulting from major regional Early Carboniferous uplift prior to Permian deposition. The fact that these ages have not been significantly reset limits the maximum possible Permian to present-day paleotemperatures to approximately 200°C ± 50°C in the section investigated. Sphene ages also record the effects of pre-Permian thermal history but of higher temperature.

The integration of vitrinite reflection and fission-track measurements provides a powerful means of thermal history analysis. The dual approach, using two parameters with different reaction kinetics, on varied lithologies in different parts of the section, yields complementary data and considerably improves constraints in developing paleothermal models.

Introduction

Thermal history is an essential component in the evaluation of hydrocarbon potential and the tectonic evolution of sedimentary basins. It may be defined as the integral of subsurface temperatures and duration of heating of a rock through geological time. Subsurface temperatures are a function of depth and thermal gradient and although the burial history may be reconstructed from stratigraphic analysis, neither paleothermal gradients nor thermal history can be directly measured.

Nevertheless, a number of thermally dependent rock properties, such as degree of coalification and fission-track annealing, may be analyzed in order to measure the effects of thermal history. Maturity (coalification rank) of sedimentary organic material is one of the most sensitive natural recorders of the relatively low temperatures frequently encountered in sedimentary basins. Because coalification reactions are irreversible and are driven by temperature and duration of heating, measured organic maturity records the integrated effect of time and temperature on the rock. The method most commonly used for measuring organic maturity is vitrinite reflectance (e.g., Stach et al., 1975; Dow, 1977; Tissot and Welte, 1978; Bostick, 1979; Robert, 1980).

Fission-track analysis of detrital mineral grains in sedimentary rocks has been shown to be useful for evaluating thermal history of sedimentary basins (Naeser, 1979, 1981; Briggs et al., 1981; Gleadow et al., 1983, 1984; Naeser, 1984; Naeser et al., this volume). The key to investigation of thermal histories by fission-track dating is based on the resetting of clocks due to the tendency for tracks to anneal upon heating. The temperature required to anneal tracks depends in part on the duration of heating and is also characteristic of the particular mineral under study.

Minerals most commonly used for fission-track dating are apatite, zircon, and sphene. Temperature conditions for annealing are best known for apatite (e.g., Gleadow and Duddy, 1981; Naeser, 1981). For heating times commonly encountered in sedimentary basins, the annealing zone for apatite occurs between about 80° to 135°C (10^6 yr) and 60° to 105°C (10^8 yr). This temperature range overlaps with that required for the generation of liquid hydrocarbons from organic source material in such sedimentary basins. Thus, fission tracks in detrital apatites may reveal whether sediments have been subjected to sufficient heating to generate liquid hydrocarbons. Other measurable parameters may also provide thermal history information, but the unique advantage of the fission-track method lies in the fact that it can also supply information on maximum paleotemperatures and their variation through time (Gleadow et al., 1983).

Fission tracks in zircon and sphene anneal at higher temperatures than in apatite. Estimated values reported are approximately 200°C $\pm$ 50°C for zircon and 250°C $\pm$ 50°C for sphene (Gleadow and Brooks, 1979; Harrison et al., 1979; Hurford, 1986; Naeser, 1979; Zaun and Wagner, 1985; Zeitler, 1984). Since these temperature ranges are not commonly attained in sedimentary basins, studies of zircon and sphene usually enable a ceiling to be placed on maximum temperatures. They may also provide information on sedimentary provenance (Gleadow et al., 1983) or thermal history during prebasin time (Kohn, Eyal, et al., 1984).

The retention of fission tracks in various minerals at different temperatures and the fact that track accumulation may be reversed under appropriate thermal conditions, as opposed to the irreversibility of organic metamorphism reactions, are special characteristics. These properties also provide the potential for identifying multiple thermal events and tectonic evolution.

In the present study, we have integrated the results obtained by vitrinite reflectance and fission-track dating for thermal history analysis of the Precambrian to Tertiary stratigraphic section in three deep boreholes in the Har HaNegev region of southern Israel (Fig. 12.1). Since each parameter could only be measured from different lithologies and stratigraphic levels the data obtained are complementary. Furthermore, the type of data derived by the two methods reflect different, although not well understood, reaction kinetics. Thus, the dual approach provided improved constraints in developing thermal models.

Geological Setting

The Ramon 1, Makhtesh Qatan 2, and Kurnub 1 boreholes were drilled in three breached anticlines in the Negev area, southern Israel (Fig. 12.1). The three anticlines named Ramon, Hazera, and Hatira, respectively, are related to a system of northeast to east-northeast trending, elongate, asymmetric folds, that comprise the southeastern Negev-Sinai branch of the S-shaped Syrian Arc (Bentor and Vroman, 1954; de Sitter, 1962).

The stratigraphy of the studied area (Fig. 12.2) reveals a discontinuous Early Permian through Tertiary succession that unconformably overlies arkosic sandstones and igneous rocks of the Pre-

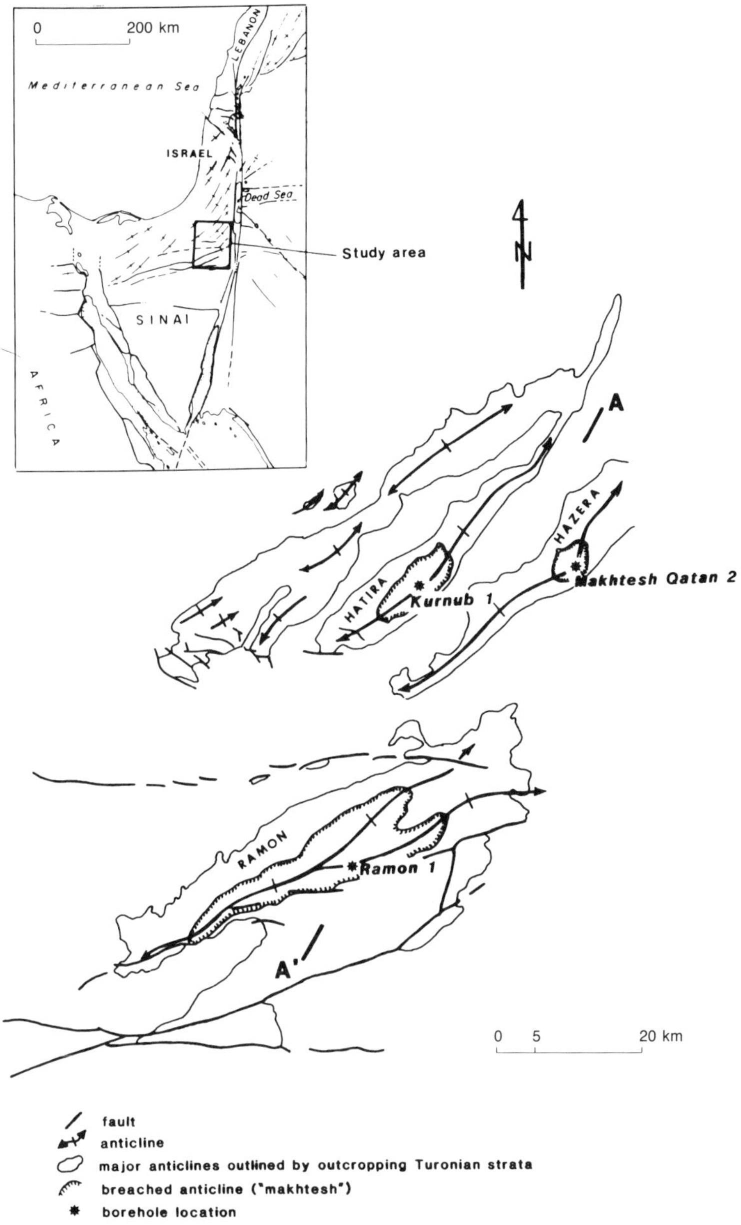

FIGURE 12.1. Locality map and regional structures. A-A' shows line of cross section in Figure 12.7.

cambrian Zenifim Formation (Weissbrod, 1981). The Permian to Late Cretaceous sedimentary succession thickens gradually to the west and mainly comprises carbonates and sandstones deposited over a wide shelf. A series of transgressive-regressive phases during this time has resulted in some cycles of lagoonal deposits and more pronounced continental facies of blanket sandstones and shales (Goldberg, 1970; Bartov et al., 1972; Druckman, 1974; Bartov and Steinitz, 1977; Weissbrod, 1981). Field evidence indicates that folding of the broad sedimentary wedge into a series of elongate anticlines and synclines commenced in late Turonian-early Senonian time (Bentor and Vroman, 1951, 1954; Freund, 1965).

Thickness of stratigraphic units for each of the three investigated sections, including estimates of amount of eroded section, is given in Feinstein (1985).

Igneous rocks crop out only in the Ramon anticline (Bentor, 1952, 1963; Mazor, 1955). They are exposed over a wide area and comprise both extru-

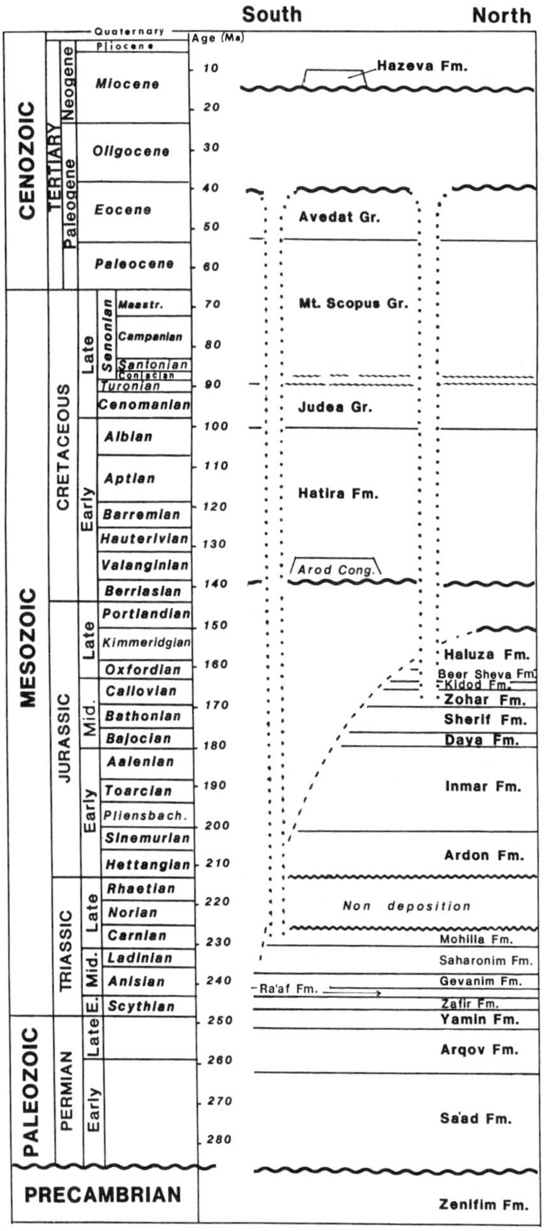

FIGURE 12.2. Stratigraphic table for the Har HaNegev area compiled from the studies of Bartov et al. (1981), Derin and Gerry (1981), Flexer et al. (1981), Eshet (1983), Lewy (1983), and Lang and Mimran (1985). Dashed line represents regional Early Cretaceous erosion surface; dotted lines represent local Neogene erosion surface in breached anticlines. Note deeper Early Cretaceous and Neogene erosion in south (Ramon anticline) compared to north (Hazera and Hatira anticlines). Chronostratigraphy follows Harland et al. (1982).

sive (mainly basalts and associated differentiates) and intrusive (mainly essexites and nordmarkites) phases (Bentor, 1952, 1963; Weissbrod, 1962; Garfunkel and Katz, 1967; Bonen, 1980). In general, the intrusive phases predate the extrusives. The majority of igneous rocks exposed in the Ramon anticline were mapped as Early Cretaceous whereas others were assigned a Jurassic age (Bentor and Vroman, 1951; Garfunkel and Katz, 1967). Starinsky et al. (1980) and Lang and Steinitz (1985) reported whole-rock K-Ar ages for the Early Cretaceous rocks ranging from 145 Ma to 100 Ma. Subsurface igneous rocks, in pre-Late Jurassic strata, have been documented from the three drilled anticlines (Weissbrod, 1969; Goldberg, 1970; Druckman, 1974). In Ramon 1, an Early Cretaceous K-Ar age was reported from an igneous rock of intermediate composition intruding the Lower Triassic section (Recanati, 1985).

Five regional unconformities truncate the Phanerozoic sedimentary succession studied (Fig. 12.2). Three of the unconformities (basal Permian, basal Cretaceous, and Neogene) were associated with deep erosion. The basal Permian unconformity separates the Phanerozoic sedimentary succession from the Precambrian rocks. The basal Cretaceous unconformity (Fig. 12.2) coincides with a period of relatively intense magmatism, regional uplift, and increasing erosion to the south and east (Goldberg, 1970; Freund et al., 1975). By contrast, Neogene erosion was most pronounced in the anticlines, particularly in the three investigated (Fig. 12.2). The remaining two unconformities, at the Triassic-Jurassic boundary and at the basal Senonian, record relatively little erosion (Goldberg, 1970; Lewy, 1973; Druckman, 1974). For the reconstruction of burial history these unconformities represent time breaks of nondeposition.

Burial History

Burial history can be reconstructed on the basis of thickness and chronology of the stratigraphic units (e.g., van Hinte, 1978; Sclater and Christie, 1980; Feinstein, 1981). Burial curves for the investigated sections are illustrated in Figures 12.3 to 12.5.

The study wells were spud in the exposed cores of the anticlines, and thus drilling commenced in Middle Triassic strata in the Ramon and in Middle-

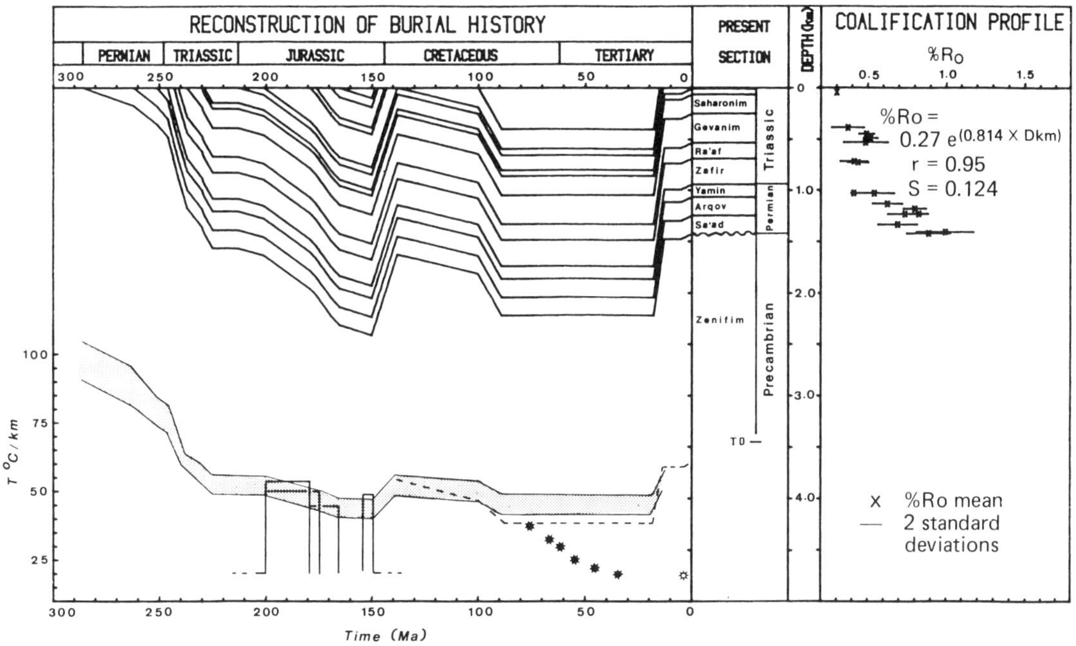

Alternative possible paleothermal gradients, which, in combination with the given burial history, could account for the evolution of the measured coalification profile.

– – – Upper limit of thermal gradient at postcoalification time based on vitrinite reflectance measurements.

* Maximum possible paleothermal gradient derived from apatite fission-track age assuming total annealing at 110°C.

Limit of possible paleothermal gradient derived from zircon fission-track age (deepest sample) assuming annealing temperatures between 200°-230°C.

FIGURE 12.3. Ramon 1 drill hole. Coalification profile (right side), burial curve (upper left), and thermal gradient history $(\Delta T/\text{km})_t$ as defined by vitrinite reflectance measurements and fission-track dating (lower left). TD = total depth drilled. Open star indicates present-day thermal gradient. *Note*: 1) high thermal gradient during Jurassic required to account for coalification gradient in Permo-Triassic section; 2) maximum possible paleothermal gradients derived from zircon fission-track data slightly overlap with those derived by time-temperature coalification calculations; 3) maximum paleothermal gradients derived from apatite fission-track ages considerably constrain those calculated for postcoalification time from vitrinite reflectance data; and 4) maximum paleothermal gradients defined by apatite fission-track ages indicate that present-day thermal gradient has prevailed since at least Early Oligocene time.

Late Jurassic strata in the Hazera and Hatira anticlines. Therefore, only the older portion of the sedimentary succession, as well as a considerable part of the underlying Precambrian Zenifim complex, was penetrated. Burial history for the rest of the sedimentary succession was reconstructed from outcrops preserved on the peripheral escarpments of the breached anticlines and surrounding area. Tertiary burial history, which involved relatively minor deposition on the anticlines, was inferred from regional stratigraphic considerations.

Sediment accumulation and burial rate from Early Permian through Jurassic in the Ramon anticline is virtually identical to that of the Hazera anticline (Makhtesh Qatan 2) and only slightly less than that of the Hatira anticline (Kurnub 1).

On the northern flank of the Ramon anticline the basal Cretaceous Arod Conglomerate unconformably overlies Middle Jurassic formations. However, south of the Ramon 1 borehole (Fig. 12.1), due to deeper erosion, the conglomerate directly overlies the Middle Triassic Saharonim Formation (Garfunkel, 1964), indicating complete truncation of the Jurassic-Upper Triassic section by Early Cretaceous time. This truncation was somewhat less at the well site (approximately 750

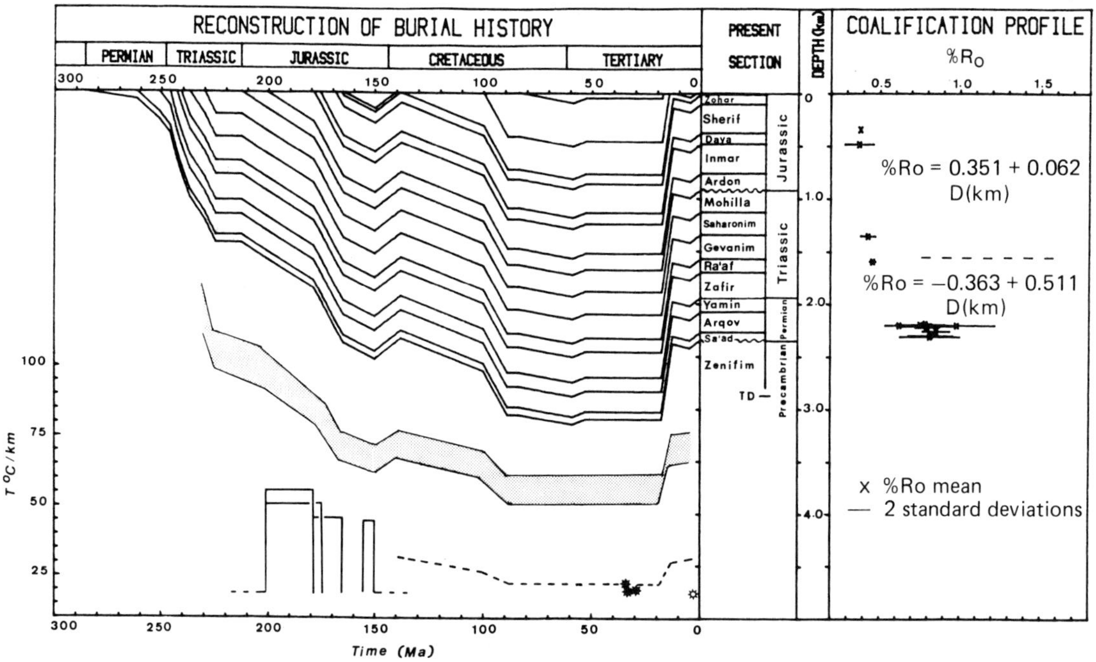

Alternative possible paleothermal gradients, which, in combination with the given burial history, could account for the evolution of the measured coalification profile.

– – – Upper limit of thermal gradient at postcoalification time based on vitrinite reflectance measurements.

* Maximum possible paleothermal gradient derived from apatite fission-track age assuming total annealing at 110°C.

Limit of possible paleothermal gradient derived from zircon fission-track age (deepest sample) assuming annealing temperatures between 200°-230°C.

FIGURE 12.4. Makhtesh Qatan 2 drill hole. Coalification profile (right side), burial curve (upper left), and thermal gradient history $(\Delta T/\text{km})_t$ as defined by vitrinite reflectance measurements and fission-track dating (lower left). TD = total depth drilled. Open star indicates present-day thermal gradient. *Note*: 1) high thermal gradient during Jurassic required to account for coalification gradient in Permo-Triassic section; 2) maximum possible paleothermal gradients derived from zircon fission-track data do not provide improved constraints on time-temperature coalification calculations; 3) maximum paleothermal gradients for Early Oligocene time, derived from apatite fission-track ages, are virtually identical to those calculated from vitrinite reflectance data; and 4) maximum paleothermal gradients calculated from vitrinite reflectance data for postcoalification time indicate that the present-day thermal gradient has prevailed since Late Cretaceous time.

to 800 m) where part of the Late Triassic Mohilla Formation has been preserved. In the Hazera and the Hatira anticlines the erosion was considerably less (approximately 150 m) with only the uppermost portion of the Jurassic section removed. Therefore, despite the similarity of Permian to Jurassic burial, the post-Jurassic burial history in the Ramon anticline was markedly different from that of the Hazera and Hatira anticlines and indicates considerably shallower burial depths (cf. Figs. 12.3 to 12.5).

The relatively thick Early Cretaceous Hatira Formation (predominantly clastics) and Cenomanian-Turonian Judea Group (predominantly carbonates) represent regional phases of rapid sedimentation and burial which followed the basal Cretaceous truncation. This phase of sedimentation was more pronounced in the area of the Hazera and Hatira anticlines (cf. Figs. 12.3 to 12.5). Following initiation of Syrian Arc folding in late Turonian-Senonian time, sedimentation was characterized by minor cycles of accumulation and erosion with no significant burial increase on structural highs (Bartov and Steinitz, 1977).

Deep erosion in Neogene time resulted in a decrease of burial depth (Figs. 12.2 to 12.5) and

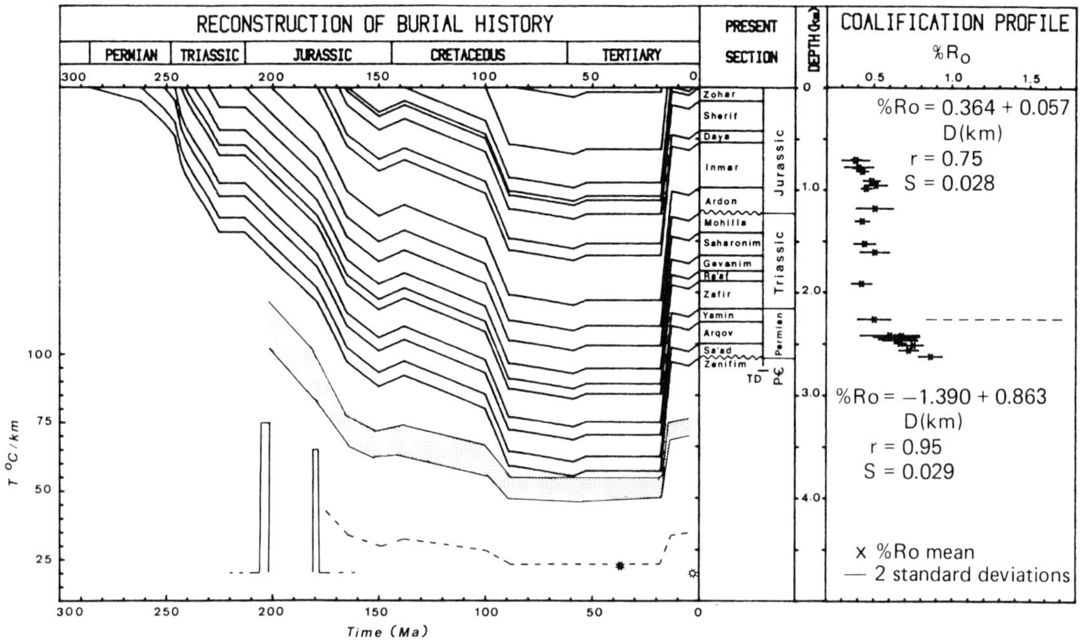

Alternative possible paleothermal gradients, which, in combination with the given burial history, could account for the evolution of the measured coalification profile.

– – – Upper limit of thermal gradient at postcoalification time based on vitrinite reflectance measurements.

* Maximum possible paleothermal gradient derived from apatite fission-track age assuming total annealing at 110°C.

Limit of possible paleothermal gradient derived from zircon fission-track age (deepest sample) assuming annealing temperatures between 200°-230°C.

FIGURE 12.5. Kurnub 1 drill hole. Coalification profile (right side), burial curve (upper left), and thermal gradient history $(\Delta T/\mathrm{km})_t$ as defined by vitrinite reflectance measurements and fission-track dating (lower left). TD = total depth drilled. Open star indicates present-day thermal gradient. *Note*: 1) markedly high thermal gradient during Jurassic required to account for coalification gradient in Permo-Triassic section; 2) maximum possible paleothermal gradients derived from zircon fission-track data do not provide improved constraints on time-temperature coalification calculations; 3) maximum paleothermal gradient derived from apatite fission-track age is virtually identical to that calculated from the vitrinite reflectance data; and 4) maximum paleothermal gradients calculated from vitrinite reflectance data for postcoalification time indicate that the present-day thermal gradient has prevailed since Late Cretaceous time.

exposed Middle Triassic rocks in the core of the Ramon anticline and Middle-Late Jurassic rocks in the core of the Hazera and Hatira anticlines.

Analytical Methods and Results

Vitrinite Reflectance

Vitrinite reflectance measurements were carried out on finely polished (0.3-µm alumina) rock chip mounts of borehole cuttings and core samples. Analytical methods and data processing followed the procedures described by Stach et al. (1975, pp. 240–243 and 263–273) and Zeiss publication A 41-825.8-e. Due to the relatively low ranges of reflectance encountered ($<1\%$ R_o) and the frequent small size of particles, reflection was measured on randomly oriented organic particles under nonpolarized light.

Suitable material for measurement was found in 15 samples out of 30 examined from the Permo-Triassic section of Ramon 1, 14 out of 45 examined from the Permian-Jurassic section of Makhtesh Qatan 2, and 23 out of 46 examined from the equivalent section in Kurnub 1. Data obtained on all samples measured are given in Tables 12.1 to

TABLE 12.1. Vitrinite reflectance of whole-rock samples from Ramon 1 drill hole.

Sample no.	Depth (m)	Formation	Rock type	No. of meas.	% R_o Min.	Max.	Mean	SD
1	46–49	Saharonim	Shale	2	–	–	0.30	–
13	388–390	Gevanim	Shale	4	0.31	0.42	0.37	0.052
15	450–452	Gevanim	Coal	13	0.46	0.53	0.49	0.024
16	490	Gevanim	Shale	4	0.50	0.55	0.52	0.022
43a	532–536	Gevanim	Coal and shale	61	0.41	0.61	0.49	0.071
43b	532–536	Gevanim	Coal and shale	29	0.65	0.93	0.86*	0.075
18	718	Zafir	Shale	5	0.36	0.47	0.41	0.044
39	728–732	Zafir	Shale	12	0.41	0.52	0.44	0.033
7	1,030	Yamin	Shale	4	0.40	0.42	0.41	0.008
32	1,030	Yamin	Shale	12	0.44	0.64	0.54	0.063
31	1,132	Arqov	Shale	17	0.55	0.69	0.63	0.047
30a	1,182–1,184	Arqov	Coal	10	0.75	0.85	0.80	0.035
30b	1,182–1,184	Arqov	Coal	53	0.89	1.15	1.05*	0.082
30c	1,182–1,184	Arqov	Coal	3	1.23	1.34	1.27*	–
29a	1,233	Arqov	Shale	10	0.44	0.52	0.48†	0.033
29b	1,233	Arqov	Shale	8	0.67	0.86	0.74	0.055
29c	1,233	Arqov	Coal	56	0.75	0.88	0.83	0.029
27	1,336–1,338	Saad	Shale	25	0.60	0.80	0.69	0.062
24	1,410–1,414	Saad	Coal	20	0.88	1.15	0.99	0.091
23	1,423–1,428	Saad	Shale	13	0.78	1.04	0.89	0.069

All samples were taken from cuttings. Identical sample numbers followed by a, b, or c indicate that the reflectance measurements are distributed into three subpopulations.
*Indication of oxidation (includes possibility of "recycling").
†Measurement obtained from unsuitable material (e.g., ulminite A or "pseudo-vitrinite").

TABLE 12.2. Vitrinite reflectance of whole-rock samples from Makhtesh Qatan 2 drill hole.

Sample no.	Depth (m)	Formation	Rock type	No. of meas.	% R_o Min.	Max.	Mean	SD
1	342	Sherif	Shale	7	0.31	0.44	0.38	–
C1	483–492	Daya	Shale	64	0.25	0.47	0.37	0.044
50a	1,353	Gevanim	Shale	33	0.38	0.47	0.43	0.022
50b	1,353	Gevanim	Shale	5	0.61	0.72	0.65*	–
68	1,594–1,597	Raaf	Coal	30	0.44	0.47	0.46	0.009
29a	2,186–2,188	Arqov	Shale	18	0.47	0.72	0.59†	0.082
29b	2,186–2,188	Arqov	Shale	10	0.75	0.82	0.79	0.019
28	2,196–2,198	Arqov	Shale	15	0.64	0.89	0.76	0.076
27a	2,202–2,203	Arqov	Shale	21	0.58	0.70	0.62	0.043
27b	2,202–2,203	Arqov	Shale	10	0.82	1.17	0.98	0.118
C13	2,210–2,216	Arqov	Shale and coal	48	0.65	0.95	0.79	0.092
26a	2,222–2,224	Arqov	Shale and coal	52	0.71	0.89	0.80	0.037
26b	2,222–2,224	Arqov	Shale and coal	18	0.97	1.22	1.05*	0.067
25	2,260–2,262	Arqov	Shale and coal	36	0.76	0.92	0.85	0.042
24	2,302–2,306	Saad	Shale	11	0.73	0.96	0.82	0.069
23a	2,310–2,312	Saad	Shale	46	0.61	0.99	0.81	0.093
23b	2,310–2,312	Saad	Shale	3	1.20	1.23	1.21*	–
21a	2,342–2,344	Zenifim	Shale	12	0.55	0.75	0.63‡	0.071
21b	2,342–2,344	Zenifim	Shale	8	0.77	0.95	0.83‡	–
20a	2,346–2,348	Zenifim	Shale	43	0.50	0.74	0.62‡	0.070
20b	2,346–2,348	Zenifim	Shale	12	0.76	0.86	0.81‡	0.038

All samples were taken from cuttings except those numbers prefixed by C (core material).
Identical sample numbers followed by a or b indicate the reflectance measurements are distributed into two subpopulations.
*Indication of oxidation (includes possibility of "recycling").
†Measurement obtained from unsuitable material (e.g., ulminite A or "pseudo-vitrinite").
‡Cavings.

TABLE 12.3. Vitrinite reflectance of whole-rock samples from Kurnub 1 drill hole.

Sample no.	Depth (m)	Formation	Rock type	No. of meas.	% R_o Min.	Max.	Mean	SD
39a	707–709	Inmar	Coal and shale	54	0.35	0.41	0.39	0.016
39b	707–709	Inmar	Coal and shale	11	0.52	0.54	0.53*	0.009
40a	713–715	Inmar	Coal	56	0.32	0.47	0.39	0.043
40b	713–715	Inmar	Coal	7	0.49	0.52	0.51*	0.014
41	779–782	Inmar	Shale	92	0.34	0.50	0.41	0.045
42a	788–790	Inmar	Coal	49	0.37	0.43	0.41	0.018
42b	788–790	Inmar	Coal	18	0.50	0.55	0.53*	0.015
44a	820–825	Inmar	Coal	22	0.40	0.45	0.43	0.017
44b	820–825	Inmar	Coal	64	0.49	0.58	0.53*	0.023
45	910–916	Inmar	Coal and shale	67	0.44	0.53	0.49	0.025
35	954–957	Inmar	Coal	99	0.43	0.60	0.51	0.035
36a	986–989	Ardon	Shale	30	0.41	0.48	0.45	0.015
36b	986–989	Ardon	Shale	34	0.50	0.57	0.53*	0.010
3	1,178–1,182	Ardon	Shale	6	0.43	0.58	0.51	0.057
31	1,303–1,306	Mohilla	Shale	10	0.40	0.47	0.43	0.021
22	1,524–1,527	Saharonim	Shale	21	0.40	0.52	0.44	0.033
28	1,606–1,608	Saharonim	Shale and coal	85	0.43	0.61	0.51	0.045
18	1,913–1,914	Zafir	Shale	41	0.37	0.49	0.43	0.031
14	2,259–2,260	Yamin	Shale and coal	90	0.44	0.61	0.50	0.051
10	2,415–2,419	Arqov	Shale	17	0.50	0.72	0.60	0.092
26a	2,419–2,422	Arqov	Shale	9	0.61	0.76	0.68	0.055
26b	2,419–2,422	Arqov	Shale	35	0.78	0.97	0.90*	0.058
11	2,431–2,434	Arqov	Shale	18	0.55	0.75	0.65	0.072
9	2,454–2,455	Arqov	Shale	17	0.54	0.72	0.65	0.059
8	2,465–2,466	Arqov	Shale	6	0.55	0.71	0.67	0.054
7a	2,492–2,500	Arqov	Shale	29	0.63	0.70	0.67	0.018
7b	2,492–2,500	Arqov	Shale	65	0.73	0.83	0.78*	0.036
6	2,512–2,513	Saad	Shale	20	0.68	0.80	0.75	0.032
5	2,561–2,564	Saad	Shale	41	0.65	0.79	0.72	0.030
4	2,622–2,628	Saad	Coal and shale	23	0.80	0.93	0.86	0.037

All samples were taken from cuttings. Identical sample numbers followed by a or b indicate that the reflectance measurements are distributed into two subpopulations.
*Indication of oxidation (includes possibility of "recycling").

12.3. Coalification profiles for the three boreholes are plotted in Figures 12.3 to 12.5.

In some of the samples percent R_o values are distributed into two or three frequency populations (Tables 12.1 to 12.3). Factors that may affect the recorded results, such as oxidation, "vitrinite B", and cavings are indicated by footnotes. Measurements on such material do not reflect the true coalification and were therefore omitted in the construction of the coalification curves shown in Figures 12.3 to 12.5.

Vitrinite reflectance data in the Jurassic section of Makhtesh Qatan 2 (eight measurements) and Kurnub 1 (three measurements) were reported by Kisch (1978). The reflectance measurements presented here are in accordance with or slightly lower than those measured by Kisch. Feinstein (1985) reexamined those samples measured by Kisch and suggested that some may have been affected by oxidation. For most of the Triassic succession, vitrinite reflectance measurements reported here are in broad agreement with palynomorph color alteration data presented by Eshet (1983). However, for the lowermost Triassic Eshet's palynomorph data reveal a considerably higher level of organic maturation.

Fission-Track Dating

For fission-track dating, apatite, zircon, and sphene splits in the size range of 63 to 250 μm were recovered from crushed and ground cores using

TABLE 12.4. Fission-track age of apatite, zircon, and sphene from cores in investigated drill holes.

Core no.	Drill-hole depth (m)	Fossil track density ($\times 10^6 t/cm^2$)	Induced track density ($\times 10^6 t/cm^2$)	Neutron dose ($\times 10^{15} n/cm^2$)	No. of grains	r	Age* ± 1σ (Ma)
			Apatite				
Ramon-8†	1,475–1,487	0.287 (498)	0.796 (690)	3.51 (2,693)	8	0.993	75.3±4.4
Ramon-9†	1,764–1,780	1.170 (1,042)	4.088 (1,820)	4.04 (1,009)	10	0.993	68.5±2.7
Ramon-11†	1,984–1,988	1.291 (1,185)	4.379 (2,010)	3.54 (2,693)	9	0.974	62.2±2.3
Ramon-12†	2,579–2,587	0.243 (714)	0.942 (1,384)	3.58 (2,693)	12	0.994	54.0±2.5
Ramon-14†	3,045–3,057	0.180 (605)	0.915 (1,540)	3.66 (1,090)	11	0.975	42.7±2.1
Ramon-19†	3,434–3,439	0.082 (451)	0.541 (1,489)	3.72 (1,090)	10	0.982	33.7±1.9
M.Qatan-32†	2,683–2,700	0.070 (107)	0.444 (340)	3.44 (2,693)	2	0.917	32.4±3.6
M.Qatan-34†	2,739–2,754	0.186 (103)	1.261 (350)	4.10 (1,009)	7	0.954	36.1±4.1
M.Qatan-35‡	2,811–2,813	0.160 (409)	1.150 (1,467)	3.53 (1,090)	5	0.995	29.3±1.6
Kurnub-8‡	2,658–2,659	0.239 (122)	1,384 (353)	3.65 (2,693)	2	–	37.6±4.0
			Zircon				
Ramon-7§	1,342–1,352	8.319 (2,461)	2.711 (401)	1.83 (3,029)	8	0.929	328±18
Ramon-12†	2,579–2,587	12.98 (1,721)	3.876 (514)	1.76 (3,029)	5	0.939	343±23
Ramon-19†	3,434–3,439	11.97 (1,880)	3.718 (292)	1.80 (3,029)	5	0.912	338±21
M.Qatan-16§	2,274–2,290	9.679 (1,481)	3.503 (268)	1.90 (3,029)	5	0.996	307±20
M.Qatan-26¶	2,425–2,428	9.442 (2,273)	2.833 (341)	1.75 (1,132)	8	0.972	340±20
M.Qatan-32†	2,682–2,700	9.891 (1,231)	3.182 (198)	1.81 (3,029)	4	0.998	328±25
M.Qatan-34†	2,739–2,754	3.605 (1,368)	0.559 (106)	0.91 (1,489)	6	0.997	342±35
Kurnub-11‡	2,741–2,743	1.996 (448)	0.321 (36)	0.98 (1,079)	2	–	355±85
			Sphene				
Ramon-11†	1,984–1,988	9.257 (2,219)	4.664 (559)	3.87 (4,322),	6	0.997	444±21
Ramon-19†	3,434–3,439	21.00 1,585)	10.84 (409)	3.99 (4,322)	5	0.993	447±25

Number in parentheses = number of tracks counted. All track densities are quoted for internal surfaces (g = 0.5). r = Correlation coefficient.

*$\lambda_f = 7.03 \times 10^{-17}$/yr, $\lambda_D = 1.551 \times 10^{-10}$/yr, $^{235}\sigma = 5.802 \times 10^{-22}$ cm², I = 7.252×10^{-3}.

Details of cores:

† Zenifim Formation (arkose) – Precambrian.

‡ Zenifim Formation (andesite) – Precambrian.

§ Saad Formation (sandstone) – Early Permian.

¶ Zenifim Formation (diabase) – Precambrian.

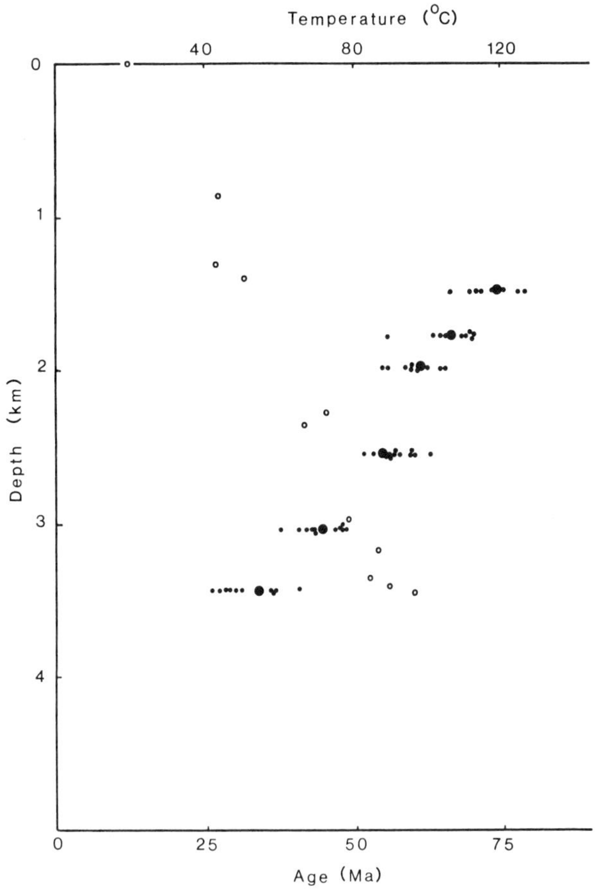

FIGURE 12.6. Apatite fission-track ages and measured temperature (open circles) as a function of depth in Ramon 1 drill hole. The weighted average age for each sample (large solid circle) was calculated from ages obtained on several detrital apatite grains (small solid circles).

heavy liquids and a Frantz isodynamic magnetic separator. Apatite was extracted only from six cores out of nine cores and five samples of cuttings examined in Ramon 1, from three cores out of 11 examined from Makhtesh Qatan 2, and in one core out of two examined from Kurnub 1. Relatively abundant apatite was found in nearly all samples of arkose and igneous rocks of the Precambrian Zenifim Formation that were examined, but it occurred only in trace amounts in upper Paleozoic sediments. Zircon was ubiquitous in all samples of Precambrian to Mesozoic arkose and sandstone examined. The presence of these minerals is in accordance with the findings of Weissbrod and Nachmias (1985), who studied the heavy mineral composition of the Lower Cambrian-Mesozoic clastic sequence ("Nubian Sandstone") in southern Israel, Sinai, and southern Jordan. Sphene was only found in two core samples of arkose from the Zenifim Formation in Ramon 1.

Fission-track ages were obtained for individual grains using the external detector method (Gleadow, 1981) with Brazil Ruby muscovite detectors. All samples were irradiated in the RT-4 facility of the National Bureau of Standards reactor in Gaithersburg, Maryland. Methodology and standardization employed was identical to that described by Kohn, Shagam, et al. (1984). Fission-track age data and sample information are given in Table 12.4.

Average fission-track ages of apatite grains from the Precambrian Zenifim Formation in Ramon 1 show a steady decrease from approximately 75 Ma to 34 Ma with increasing depth (Fig. 12.6). For Makhtesh Qatan 2 average apatite fission-track ages ranging between approximately 35 Ma and 32 Ma were determined from three samples within 28 to 150 m of the maximum depth (2,840 m) drilled. In Kurnub 1, an apatite fission-track age of 37 Ma was determined from a sample within 85 m of the maximum depth (2,743 m) drilled.

Fission-track ages on 43 detrital zircon grains from six samples of Precambrian Zenifim Formation and two samples of Early Permian Saad Formation at different depths within the three boreholes yield weighted average ages in the range 307 to 355 Ma (Table 12.4).

Sphene ages derived from 11 detrital grains in two cores of the Zenifim Formation of Ramon 1 yield older average fission-track ages (428 to 431 Ma) than those derived from zircons in the same strata (Table 12.4).

Discussion

Time of Coalification

A schematic cross section from Ramon 1 through Kurnub 1 to Makhtesh Qatan 2 using the top of the Judea Group (latest Turonian) as a datum (Fig. 12.7) shows the angular unconformity between Cretaceous and underlying strata. Isoreflectance values in the Permo-Triassic section were encountered at the same stratigraphic position in Ramon 1 and Makhtesh Qatan 2 and at slightly deeper stratigraphic position in Kurnub 1 (Figs. 12.3 to 12.5). Isoreflectance contours (Fig. 12.7) are therefore approximately concordant with the deformed Permian to Jurassic formations and markedly discordant with the overlying Cretaceous beds. This clearly indicates that the recorded coalification predates the basal Cretaceous unconformity.

Thermal History as Revealed by Coal Ranks

The burial curves (Figs. 12.3 to 12.5) and empirical time-temperature models (e.g., Hood et al., 1975; Waples, 1980) permit use of the measured coalification profiles (Figs. 12.3 to 12.5) for quantitative paleothermometric analysis. The time-temperature coalification model used herein is a slightly modified version of the Lopatin (1971) Time-Temperature Index (TTI), as presented by Waples (1980). Modifications are as follows: 1) the minimum temperature for initiation of thermally driven coalification was set at 50°C (Teichmueller and Teichmueller, 1966; Vassoyevich et al., 1969), and 2) an option was inserted for placing a limit on the amount of cooling that occurs before the coalification rate decays to zero. In addition to the first-order kinetic considerations, restriction 2 above is essential for cases where coalification ceased at a relatively early stage in the geological history and did not progress, despite the long time interval elapsed (e.g., Patteisky et al., 1962; Teichmueller and Teichmueller, 1966; Hacquebard, 1975; Cook and Kantsler, 1980).

Feinstein (1985) proposed that the amount of cooling required to freeze coalification is not constant but may vary according to coal rank and the temperature under which it evolved. At present there is no suitable scale to account for Feinstein's proposal. Hence, in accordance with the effective thermal range of the LOM (Level of Organic Metamorphism) model of Hood et al. (1975), cooling of 15°C below the maximum temperature attained is here assumed as the approximate value required for cessation of coalification.

In the absence of other constraints, time-temperature calculations based on single coal rank measurements in a stratigraphic section may yield an infinite number of solutions. The range of possible models for thermal gradient with time can be considerably restricted if the coalification profile rather than a discrete point on it is considered. A relationship between slope of the coalification profile and the thermal gradient under which it evolved has been suggested by Patteisky et al. (1962), Teichmueller and Teichmueller (1966), Buntebarth (1978), and Cook and Kantsler (1980).

The following steps were used here for derivation of paleothermal gradients: 1) assumption of hypothetical models for thermal gradient with time, $(\Delta T/\text{km})_t$; 2) integration of these models with the burial curve to obtain variation of temperature with time; 3) Lopatin's calculation of the theoretical TTI (modified after Waples, 1980) produced by each model; and 4) selection of appropriate model on the basis of closest match between calculated TTI values with coalification profile. Possible solutions derived are presented in Tables 12.5 to 12.7 and Figures 12.3 to 12.5. For all calculations surface temperatures were assumed to be 20°C and constant with time. In light of the numerous uncertainties involved, the various paleothermal models tested should be viewed as a schematic framework outlined by age, duration, and average thermal gradient. In accordance with constraints on time of coalification

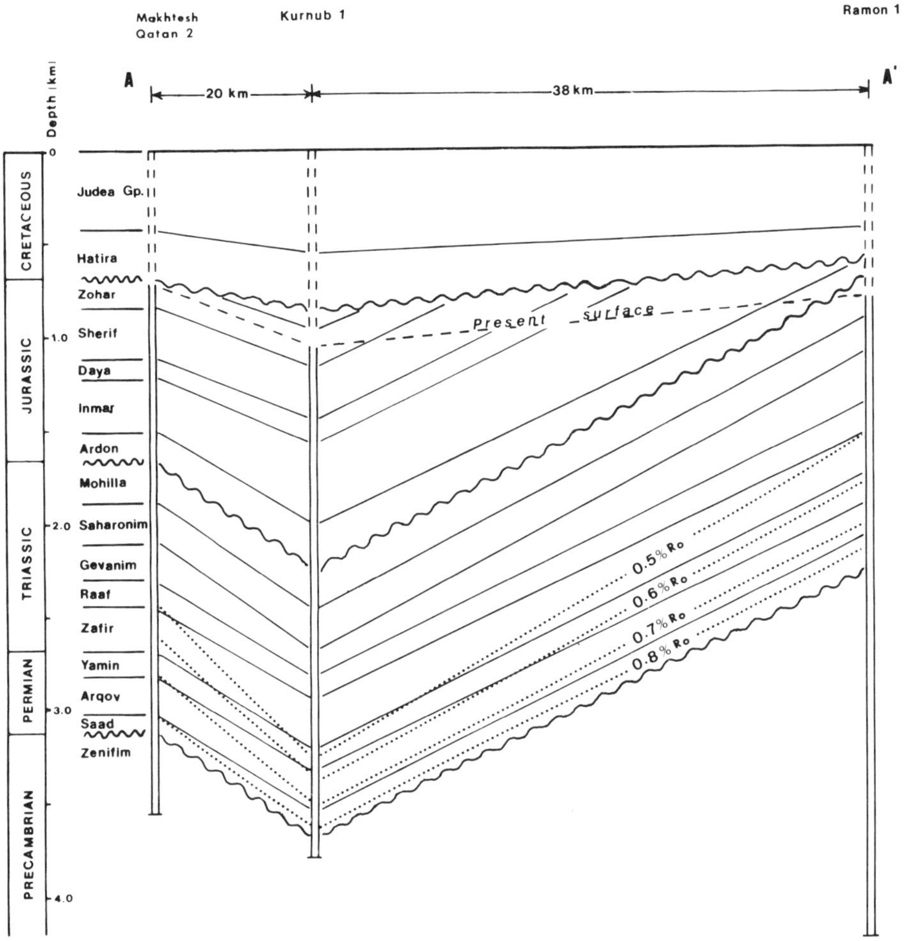

FIGURE 12.7. Schematic NNE-SSW cross section (along A-A′, Fig. 12.1) through investigated boreholes, with top of Judea Group strata (late Turonian) as a datum. Dotted lines illustrate measured coal-rank isograds (0.5% R_o and higher). Isorank contours are generally parallel to subparallel with Permo-Triassic strata and show strong angular relationships with the basal Cretaceous unconformity. These relationships imply that most of the coalification predated the Early Cretaceous.

(Fig. 12.7), age of possible thermal events under which coalification could have evolved is restricted to Jurassic time. This restriction is further supported by TTI calculations, but still leaves a relatively large time range for possible solutions within the Jurassic. Ages for "thermal events" tested have arbitrarily been assigned to coincide with Jurassic formation boundaries.

Similarity of the calculated paleothermal gradients presented in Figures 12.3 and 12.4 and Tables 12.5 and 12.6 implies that the coalification profiles in Ramon 1 and Makhtesh Qatan 2 evolved under the same thermal conditions, namely a relatively short (approximately 10 to 25 my) Jurassic thermal event with an average thermal gradient of 45° to 55°C/km. In Kurnub 1, calculated solutions (Table 12.7, Fig. 12.5) indicate an even shorter thermal event (<5 my) with a higher thermal gradient (65° to 75°C/km). The magnitude of the thermal gradients revealed by these calculations, particularly for Kurnub 1, may appear unrealistically high. However, similar paleothermal gradients based on coalification profiles of the same slope developed over similar time spans have been reported by Teichmueller and Teichmueller (1966) and Buntebarth (1978). In light of the above, it is

TABLE 12.5. Ramon 1 drill hole: Theoretical time-thermal gradient conditions that best account for evolution of the measured coalification profile.

Data of coalification profile				Possible thermal events and corresponding calculated coal ranks			
				Inmar Fm. 201–179 Ma (22 my) 54°C/km	Inmar-Daya Fms. 201–176 Ma (25 my) 50°C/km	Sherif-Zohar Fms. 176–165 Ma (11 my) 47°C/km	Halutza Fm. 155–150 Ma (5 my) 48°C/km
Formation	Present depth (base)(m)	Mean $\%R_o$	Equiv. TTI	TTI	TTI	TTI	TTI
Gevanim	535	0.42	<3	1.7	2.1	2.8	2.2
Raaf	690	0.47	<3	3.1	3.6	4.6	4.1
Zafir	944	0.58	8	7.9	8.7	10.2	9.6
Yamin	1,067	0.64	13	12.6	13.3	15.0	14.2
Arqov	1,249	0.75	26	24.8	24.9	26.8	25.4
Saad	1,425	0.86	44	48.6	46.5	46.7	45.2

TTI values derived by applying Waples' (1980) TTI/R_o plot to vitrinite reflectance measurements (left side) are compared with theoretical TTI values (right side) calculated by Lopatin's method (in Waples, 1980) together with the constructed burial curve (Fig. 12.3). Theoretical time-thermal gradient conditions presented, out of numerous tested, represent the closest match between empirical and calculated TTI data sets.

concluded that the coalification profile in the deeper part of the sedimentary section in Har HaNegev region reflects a relatively short, intense Jurassic thermal event.

An intensive phase of volcanism in northern Israel, the Asher Volcanics, is assigned an Early Jurassic age (Bonen, 1980; Gvirtzman and Steinitz, 1983; Steinitz et al., 1983). It is tempting to postulate that the Jurassic magmatism reported in the Har HaNegev by Bentor and Vroman (1951), Garfunkel and Katz (1967), Bonen (1980), and Lang and Steinitz (1985), together with the Asher Volcanics, are an expression of a regional Jurassic thermal event. However, there is not enough evidence available at present to support such a correlation. The particularly high paleothermal gradients inferred for Kurnub 1 may be indicative of proximity to localized magmatic activity.

Predeformation cessation of the coalification process as reported here from the Early Cretaceous

TABLE 12.6. Makhtesh Qatan 2 drill hole: Theoretical time-thermal gradient conditions that best account for evolution of the measured coalification profile.

Data of coalification profile				Possible thermal events and corresponding calculated coal ranks			
				Inmar Fm. 201–179 Ma (22 my) 55°C/km	Inmar-Daya Fms. 201–176 Ma (25 my) 51°C/km	Sherif-Zohar Fms. 176–165 Ma (11 my) 46°C/km	Halutza Fm. 155–150 Ma (5 my) 45°C/km
Formation	Present depth (base)(m)	Mean $\%R_o$	Equiv. TTI	TTI	TTI	TTI	TTI
Gevanim	1,570	0.45	<3	2.1	2.7	3.7	3.8
Raaf	1,698	0.5	3	3.5	4.2	5.5	5.8
Zafir	1,940	0.63	12	8.7	10.1	11.7	12.2
Yamin	2,074	0.70	20	14.5	16.2	17.8	18.4
Arqov	2,268	0.80	33	30.5	32.5	32.2	33.4
Saad	2,341	0.83	39	40.3	42.2	40.2	41.6

TTI values derived by applying Waples' (1980) TTI/R_o plot to vitrinite reflectance measurements (left side) are compared with theoretical TTI values (right side) calculated by Lopatin's method (in Waples, 1980) together with the constructed burial curve (Fig. 12.4). Theoretical time-thermal gradient conditions presented, out of numerous tested, represent the closest match between empirical and calculated TTI data sets.

TABLE 12.7. Kurnub 1 drill hole: Theoretical time-thermal gradient conditions that best account for evolution of the measured coalification profile.

Data of coalification profile				Possible thermal events and corresponding calculated coal ranks	
				Ardon (part) 204–201 Ma (3 my) 75°C/km	Inmar (part) 180–179 Ma (1 my) 65°C/km
Formation	Present depth (base) (m)	Mean $\%R_o$	Equiv. TTI	TTI	TTI
Raaf	1,887	0.47	<3	<3.0	<3.0
Zafir	2,158	0.49	<3	3.0	4.0
Yamin	2,286	0.52	4	6.0	7.5
Arqov	2,495	0.70	20	18.5	19.0
Saad	2,634	0.81	36	38.5	35.5

TTI values derived by applying Waples' (1980) TTI/R_o plot to vitrinite reflectance measurements (left side) are compared with theoretical TTI values (right side) calculated by Lopatin's method (in Waples, 1980) together with the constructed burial curve (Fig. 12.5). Theoretical time-thermal gradient conditions presented, out of numerous tested, represent the closest match between empirical and calculated TTI data sets.

in the Har HaNegev and from even earlier times in other cases (e.g., Carboniferous in Patteisky et al., 1962; Teichmueller and Teichmueller, 1966) implies that postdeformation subsurface temperatures have never exceeded those under which the previous coalification evolved.

In Ramon 1, "freezing" of coalification could be attributed to the deep basal Cretaceous erosion that removed almost the entire Jurassic section (approximately 800 m) followed by only 600 m of Cretaceous-Paleogene sedimentation. It would be more difficult, however, to invoke this explanation for Makhtesh Qatan 2 and Kurnub 1 boreholes where this erosion was limited to the upper Jurassic (< approximately 200 m) but Cretaceous-Paleogene sedimentation attained thicknesses of 800 to 1000 m prior to Neogene erosion. Despite the increased burial, coalification did not progress. Hence, post-Jurassic cooling was mainly the result of a regional decay of the thermal gradient rather than erosion.

Postcoalification Paleothermal Gradients

In order to constrain postcoalification thermal gradients it is stressed that post-Jurassic temperatures could not have attained the thermal range under which preexisting coalification could have been destabilized. Thus, the time factor could not have effectively influenced the coalification process. Maximum temperatures attained during coalification are obtained from time-temperature calculations (represented graphically as paleothermal gradients in Figs. 12.3 to 12.5). Following the discussion above and the LOM model of Hood et al. (1975) it is inferred to a first approximation that post-Jurassic temperatures were at least 15°C lower than the maximum temperatures that prevailed during coalification. Using this inference together with the known burial history, maximum possible postcoalification thermal gradients have been calculated. These data for the three boreholes are plotted in Figures 12.3 to 12.5.

Fission-Track Ages and Their Bearing on Thermal History

Apatite fission-track ages from Ramon 1 (Table 12.4) are considerably younger than the Precambrian Zenifim arkose and the crystallization ages of its source rocks, and therefore record cooling following thermal resetting. The shape of the apatite age profile (Fig. 12.6) is similar to parts of those reported from borehole studies elsewhere (e.g., Naeser, 1979, 1981; Briggs et al., 1981; Gleadow and Duddy, 1981; Gleadow et al., 1983).

The burial history of Ramon 1 (Fig. 12.3) indicates that the last phase of relatively deep burial extended from late Turonian to Neogene time. In order to erase all tracks during such an interval (i.e., 6×10^7 to 7×10^7 yr), the rock must be held at a temperature between about 105°C and 125°C (Gleadow and Duddy, 1981; Naeser, 1981;

Gleadow et al., 1983). During the last period of deep burial, rocks from which apatite grains were dated attained depths of about 2,075 to 4,035 m. Despite the present-day relatively low thermal gradients (averaging approximately 20°C/km, Levitte and Olshina, 1985), it would be reasonable to assume that most samples underwent at least partial track annealing and that the deeper samples underwent total track annealing. The temperature for total annealing of apatites (here arbitrarily assumed at 110°C for the Ramon 1 burial history) and the reconstructed depth of the sampled rocks at the ages recorded can be used to indicate the last time that the host rocks could have been exposed to gradients greater than those required for total track annealing. Such maximum possible paleothermal gradients obtained for Ramon 1 have been plotted in Figure 12.3. The maximum possible gradient decreases from Late Cretaceous, and since Early Oligocene overlaps with the present-day gradient. The calculated maximum paleothermal gradients thus considerably constrain the limits placed on maximum possible postcoalification thermal gradients calculated from vitrinite reflectance alone.

The cooling recorded by the apatite age pattern could result from either erosion or decay of the regional thermal gradient. As previously indicated, however, the magnitude of early Senonian to Paleogene erosion was insufficient by itself to account for the cooling. Thus, cooling is mainly attributed to a decrease in the regional thermal gradient. Whether the fission-track ages record cooling following a post-Jurassic thermal event (e.g., Cretaceous magmatism in the Ramon area) that did not affect coalification, or decay of a relatively high Jurassic thermal gradient (as previously inferred from the coalification profile) to its present-day relatively low level requires further investigation.

As for Ramon 1, the Makhtesh Qatan 2 and Kurnub 1 apatite ages (Table 12.4) are interpreted as having been reset, and reconstruction of burial depth (i.e., approximately 3,490 to 3,610 m and 3,850 m, respectively) at times given by apatite ages allows a maximum paleothermal gradient to be calculated. The data indicate that since Early Oligocene time thermal gradients never exceeded approximately 22° to 25°C/km.

The deepest apatite sample in Ramon 1 and those dated in Makhtesh Qatan 2 and Kurnub 1 all yield Early Oligocene fission-track ages. Since the samples are all at approximately the same reconstructed burial depths, they indicate similar paleothermal gradients (approximately 22° to 25°C/km). It is therefore concluded that: 1) the relatively low present-day regional gradients in the study area have prevailed since at least Early Oligocene time, and 2) under the Early Oligocene depth-thermal gradient regime the apatite ages probably indicate cooling from the zone of total track annealing. Thus, the Early Oligocene ages not only constrain the maximum paleothermal gradients but they also provide fixed points in time on the regional cooling curve (Figs. 12.3 to 12.5). These cooling ages could either be interpreted as transient points on the decay curve of regional thermal gradient or as being due to relatively deep Neogene erosion. At this stage we cannot distinguish between the two possibilities.

Zircon fission-track ages from the three boreholes (Table 12.4) show no consistent age trend with depth, and the overlap of errors quoted indicates that the average ages are virtually indistinguishable. The 307 to 355 Ma zircon ages have been previously interpreted (Kohn, Eyal, et al., 1984) as probably reflecting cooling related to major regional Early Carboniferous uplift and deep truncation probably associated with the basal Permian-Precambrian unconformity (Fig. 12.2).

The preservation of Early Carboniferous fission-track ages indicates that zircons have not been subjected to temperatures required to cause total or severe partial annealing during Permian to Tertiary time. Since zircons are more resistant to track annealing than apatites, they enable higher limits to be placed on maximum paleothermal gradients encountered in the strata studied and on the variation of the gradients with time. Maximum paleogradients calculated for zircons from the deepest samples studied in each borehole, based on the burial evolution curves for the boreholes and track retention temperatures for zircon, are plotted in Figures 12.3 to 12.5. For Makhtesh Qatan 2 and Kurnub 1 boreholes, maximum possible paleothermal gradients are considerably greater than those derived from vitrinite reflectance measurements. In Ramon 1, the sample analyzed represents a much deeper level below the basal Permian unconformity. Hence, calculated paleogradients provide a more rigorous constraint on thermal history. The limits defined during Jurassic time slightly overlap with the possible thermal gradients revealed by time-temperature coalification calculations. This may indicate that either the zircon closure temperature used here should be slightly higher or that Jurassic thermal gradients calculated from vitrinite measurements should be slightly less intense (and accordingly of longer duration). Limits placed on thermal gradients by zircons during postcoalification time closely follow maximum thermal gradients derived from coal rank measurements. However, as demonstrated above, apatite ages better constrain the paleothermal gradients for most of this period.

The older sphene ages (Table 12.4) are in accord with the higher closure temperatures reported for sphene. As for zircon, sphene dates probably record pre-Permian events, and hence the potential of sphene as a constraining parameter for thermal history in the Permian-Tertiary section of the Har HaNegev is rather limited.

Conclusions

Vitrinite reflectance and fission-track measurements together with reconstruction of burial history in the three investigated boreholes of the Har HaNegev region indicate the following:

1. Coalification predated the Early Cretaceous regional unconformity.
2. Paleothermal gradient models yielding calculated TTI that match the measured coalification values for the deeper part of the sedimentary section reflect a relatively short, intense, Jurassic thermal event: for Ramon 1 and Makhtesh Qatan 2, 45° to 55°C/km over a period of approximately 10 to 25 my, and for Kurnub 1, 65° to 75°C/km over a period of <5 my.
3. Coal ranks have remained "frozen" since Early Cretaceous time implying that temperatures over the last approximately 140 my were lower than those that prevailed during coalification. The recorded cooling, despite an increase in burial depth, strongly suggests decay of the regional thermal gradient.
4. Apatite fission-track dates:
   a. are considerably younger than those of host strata and are interpreted as cooling ages following partial to total thermal resetting;
   b. reveal maximum paleothermal gradients that indicate cooling through Tertiary time, confirming statement 3 above;
   c. in Ramon 1, yield maximum paleothermal gradients that are lower than, and considerably constrain, the limits placed on maximum possible postcoalification thermal gradients derived from vitrinite reflectance alone;
   d. in Makhtesh Qatan 2 and Kurnub 1, yield maximum paleothermal gradients that are in good agreement with maximum possible postcoalification thermal gradients derived from vitrinite reflectance;
   e. in the three boreholes suggest that the present-day relatively low thermal gradient (approximately 20°C/km) has prevailed in the study area since at least Early Oligocene time; and
   f. from samples within 150 m of total depth of the drill holes indicate cooling from the zone of total track annealing under the Early Oligocene depth-thermal gradient regime and therefore provide fixed points in time on the regional cooling curve.
5. Zircon fission-track dates probably record cooling resulting from major regional Early Carboniferous uplift prior to subsidence and deposition of the Permian section. The preservation of these ages thus limits the maximum possible Permian to present-day paleotemperatures in the investigated section. Similarly, sphene ages most likely record pre-Permian thermal history, but at higher temperatures.

In the present study the integration of vitrinite reflection and fission-track measurements has provided a powerful means of thermal history analysis. The measurement of two parameters with different reaction kinetics from varied lithologies in different parts of the section provided complementary data and considerably improved constraints in developing paleothermal models.

Acknowledgments. We are deeply indebted to B.S. Carpenter at National Bureau of Standards, Gaithersburg, Maryland, for irradiating our samples. D. Koshashvili provided technical assistance and D. Elazar assisted with various aspects of computer programming. Constructive criticism of an earlier version of this work was provided by N.H. Bostick, N.D. Naeser, R. Shagam, and an anonymous reviewer. This study was partly supported by grants from Oil Exploration (Investments) Ltd. and the Israel Ministry of Energy and Infrastructure.

References

Bartov, Y., Arkin, Y., Lewy, Z., and Mimran, Y. 1981. Regional stratigraphy of Israel: A guide to geological mapping. Jerusalem, Israel Geological Survey, 1 p.

Bartov, Y., Eyal, Y., Garfunkel, Z., and Steinitz, G. 1972. Late Cretaceous and Tertiary stratigraphy and paleogeography of sourthern Israel. Israel Journal of Earth Sciences 21:69–97.

Bartov, Y., and Steinitz, G. 1977. The Judea and Mount Scopus Groups in the Negev and Sinai with trend surface analysis of the thickness data. Israel Journal of Earth Sciences 26:119–148.

Bentor, Y.K. 1952. Magmatic intrusions and lava sheets in the Ramon area of the Negev. Geological Magazine 89:129–140.

Bentor, Y.K. 1963. The magmatic petrology of Makhtesh Ramon. Israel Journal of Earth Sciences 12:85.

Bentor, Y.K., and Vroman, A. 1951. The Geological Map of Israel, Series 1, Sheet 18, Ovdat. Jerusalem, Geological Survey of Israel, scale 1:100,000.

Bentor, Y.K., and Vroman, A. 1954. A structural map of Israel (1:250,000) with remarks on its dynamic interpretation. Bulletin Research Council of Israel 4:125–135.

Bonen, D. 1980. The Mesozoic basalts of Israel (in Hebrew with English summary). Ph.D. thesis, The Hebrew University, Jerusalem, 158 pp.

Bostick, N.H. 1979. Microscopic measurement of the level of catagenesis of solid organic matter in sedimentary rocks to aid exploration for petroleum and to determine former burial temperature: A review. In: Scholle, P.A., and Schluger, P.R. (eds.): Aspects of Diagenesis. Society of Economic Paleontologists and Mineralogists Special Publication 26, pp. 17–43.

Briggs, N.D., Naeser, C.W., and McCulloh, T.H. 1981. Thermal history of sedimentary basins by fission-track dating. Nuclear Tracks 5:235–237.

Buntebarth, G.V. 1978. The degree of metamorphism of organic matter in sedimentary rocks as a paleo-geothermometer, applied to the upper Rhine Graben. Pure Applied Geophysics 117:83–91.

Cook, A.C., and Kantsler, A.J. 1980. The maturation history of the epicontinental basins of western Australia. United Nations Economic and Social Commission for Asia and the Pacific, CCOP/SOPAC Technical Bulletin 3, pp. 171–195.

Derin, B., and Gerry, E. 1981. Late Permian-Late Triassic stratigraphy in Israel and its significance to oil exploration. Oil Exploration in Israel Symposium, Jerusalem, pp. 9–11.

de Sitter, L.U. 1962. Structural development of the Arabian Shield in Palestine. Geologie en Mijnbouw 41:116–124.

Dow, W.G. 1977. Kerogen studies and geological interpretations. Journal of Geochemical Exploration 7:79–99.

Druckman, Y. 1974. The stratigraphy of the Triassic sequence in southern Israel. Israel Geological Survey Bulletin 64, 92 pp.

Eshet, Y. 1983. Palynostratigraphy, thermal alteration index and kerogen analysis of the Permo-Triassic succession in Makhtesh Katan 2 well, Negev, Israel. Israel Geological Survey Report S/12/83, 164 pp.

Feinstein, S. 1981. Subsidence and thermal history of southern Oklahoma aulacogen: Implications for petroleum exploration. American Association of Petroleum Geologists Bulletin 65:2521–2533.

Feinstein, S. 1985. Coal rank and the thermal history of the sedimentary succession in southern Israel (in Hebrew with English abstract). Ph.D. thesis, Ben Gurion University of the Negev, Beer Sheva, Israel, 138 pp.

Flexer, A., Livnat, A., Shafran, N., and Gill, D. 1981. Stratigraphic table of Israel: Outcrops and subsurface. In: Atlas Project. Tel Aviv, Oil Exploration (Investments) Ltd. (unpublished).

Freund, R. 1965. A model of the structural development of Israel and adjacent areas since Upper Cretaceous times. Geological Magazine 102:189–205.

Freund, R., Goldberg, M., Weissbrod, T., Druckman, Y., and Derin, B. 1975. The Triassic-Jurassic structure of Israel and its relation to the origin of the eastern Mediterranean. Israel Geological Survey Bulletin 65, 26 pp.

Garfunkel, Z. 1964. Tectonic problems on the Ramon lineament (in Hebrew). M.Sc. Thesis, The Hebrew University, Jerusalem, 68 pp.

Garfunkel, Z., and Katz, A. 1967. New magmatic features in Makhtesh Ramon, southern Israel. Geological Magazine 104:608–629.

Gleadow, A.J.W. 1981. Fission-track dating methods: What are the real alternatives? Nuclear Tracks 5:3–14.

Gleadow, A.J.W., and Brooks, C.K. 1979. Fission track dating, thermal histories and tectonics of igneous intrusions in east Greenland. Contributions to Mineralogy and Petrology 71:45–60.

Gleadow, A.J.W., and Duddy, I.R. 1981. A natural long-term track annealing experiment for apatite. Nuclear Tracks 5:169–174.

Gleadow, A.J.W., Duddy, I.R., and Lovering, J.F. 1983. Fission track analysis: A new tool for the evaluation of thermal histories and hydrocarbon potential. Australian Petroleum Exploration Association Journal 23:93–102.

Gleadow, A.J.W., Duddy, I.R., and Lovering, J.F. 1984. Fission track thermochronology of sedimentary basins. Abstracts, Fourth International Fission Track Dating Workshop, Troy, NY, p. 14.

Goldberg, M. 1970. The lithostratigraphy of Arad Group (Jurassic) in the northern Negev (in Hebrew with English abstract). Ph.D. thesis, The Hebrew University, Jerusalem, 137 pp.

Gvirtzman, G., and Steinitz, G. 1983. The Asher volcanics: An Early Jurassic event in northern Israel. Israel Geological Survey Current Research 1982, pp. 28–33.

Hacquebard, P.A. 1975. Pre- and postdeformational coalification and its significance for oil and gas exploration. In: Alpern, B. (ed.): Petrographie de la Matiere Organique des Sediments, Relations avec la Paleotemperature et le Potential Petrolier. Paris, Centre National de la Recherche Scientifique, pp. 225–241.

Harland, W.B., Cox, A.V., Llewellyn, P.G., Pickton, C.A.G., Smith, A.G., and Walters, R. 1982. A Geologic Time Scale. Cambridge, Cambridge University Press, 131 pp.

Harrison, T.M., Armstrong, R.L., Naeser, C.W., and Harakal, J.E. 1979. Geochronology and thermal history of the Coast Plutonic Complex, near Prince Rupert, British Columbia. Canadian Journal of Earth Sciences 16:400–410.

Hood, A., Gutjahr, C.C., and Heacock, R.L. 1975. Organic metamorphism and the generation of petroleum. American Association of Petroleum Geologists Bulletin 59:986–996.

Hurford, A.J. 1986. Cooling and uplift patterns in the Lepontine Alps south central Switzerland and an age of vertical movement on the Insubric line. Contributions to Mineralogy and Petrology 92:413–427.

Kisch, H.J. 1978. Coal ranks in Jurassic and Lower Cretaceous clastic rocks of the northern Negev and the southern coastal plain of Israel: Interpretation and implication for the maturity of potential petroleum source rocks. Israel Journal of Earth Sciences 27: 23–35.

Kohn, B.P., Eyal, M., and Feinstein, S. 1984. Early Carboniferous fission-track ages from southern Israel: Evidence for major uplift and erosion of a thick lower Paleozoic section. Abstracts, Annual Meeting Israel Geological Society, Arad, p. 56.

Kohn, B.P., Shagam, R., Banks, P.O., and Burkley, L.A. 1984. Mesozoic-Pleistocene fission-track ages on rocks of the Venezuelan Andes and their tectonic implications. Geological Society of America Memoir 162, pp. 365–384.

Lang, B., and Mimran, Y. 1985. An early Cretaceous volcanic sequence in central Israel and its significance to the absolute date of the base of the Cretaceous. Journal of Geology 93:179–184.

Lang, B., and Steinitz, G. 1985. New K-Ar ages of Mesozoic magmatic rocks in Makhtesh Ramon. Abstracts, Annual Meeting Israel Geological Society, Yotvata, p. 56.

Levitte, D., and Olshina, A. 1985. Isotherm and geothermal gradient maps of Israel. Israel Geological Survey Report 60/84, 94 pp.

Lewy, Z. 1973. The geological history of Sinai and southern Israel during the Coniacian (in Hebrew with English abstract). Ph.D. thesis, The Hebrew University, Jerusalem, 189 pp.

Lewy, Z. 1983. Upper Callovian ammonites and Middle Jurassic geological history of the Middle East. Israel Geological Survey Bulletin 76, 56 pp.

Lopatin, N.V. 1971. Temperature and geological time as factors of carbonification. Akademiya Nauk SSSR Series Geologicheskaya Izvestiya, no. 3, pp. 95–106.

Mazor (Pozner), E. 1955. The magmatic occurrences in Makhtesh Ramon (in Hebrew). M.Sc. thesis, The Hebrew University, Jerusalem, 29 pp.

Naeser, C.W. 1979. Fission track dating and geologic annealing of fission tracks. In: Jäger, E., and Hunziker, J.C. (eds.): Lectures in Isotope Geology. Berlin, Springer-Verlag, pp. 154–169.

Naeser, C.W. 1981. The fading of fission tracks in the geological environment: Data from deep drill holes. Nuclear Tracks 5:248–250.

Naeser, N.D. 1984. Thermal history determined by fission track dating for three sedimentary basins in California and Wyoming. Abstracts, Fourth International Fission Track Dating Workshop, Troy, NY, p. 37.

Patteisky, V.K., Teichmueller, M., and Teichmueller, R. 1962. Das Inkohlungsbild des Steinkohlengebirges an Rhein und Ruhr, dargestellt im Niveau von Floz Sonnenschein. Fortschrift Geologisches Rheinland und Westfalen 3:687–700.

Recanati, P. 1985. Lower Cretaceous intrusion in the Triassic subsurface sequence of the Ramon-1 drillhole. Abstracts, Annual Meeting Israel Geological Society, Yotvata, p. 83.

Robert, P. 1980. The optical evolution of kerogen and geothermal histories applied to oil and gas exploration. In: Durand, B. (ed.): Kerogen: Insoluble Organic Matter from Sedimentary Rocks. Paris, Editions Technip, pp. 385–414.

Sclater, J.G., and Christie, P.A.F. 1980. Continental stretching: An explanation of the post-mid-Cretaceous subsidence of the central North Sea Basin. Journal of Geophysical Research 85:3711–3739.

Stach, E., Mackowsky, M.T., Teichmueller, M., Taylor, G.H., Chandra, D., and Teichmueller, R. 1975. Stach's Textbook of Coal Petrology. Berlin, Gebruder Borntraeger, 428 pp.

Starinsky, A., Bielski, M., Bonen, D., and Steinitz, G. 1980. Rb-Sr whole rock age of the syenitic intrusions

(Shen Ramon and Gevanim) in the Ramon area, southern Israel. Israel Journal of Earth Sciences 29:177–181.

Steinitz, G., Gvirtzman, G., and Lang, B. 1983. Evaluation of K-Ar ages of the Asher volcanics. Israel Geological Survey Current Research 1982, pp. 34–38.

Teichmueller, M., and Teichmueller, R. 1966. Die Inkohlung im saar-lothringischen Karbon, verglichen mit der im Ruhrkarbon. Zeitschrift der Deutschen Geologischen Gesellschaft 117:243–279.

Tissot, B., and Welte, D.H. 1978. Petroleum Formation and Occurrence. Berlin, Springer-Verlag, 538 pp.

van Hinte, J.E. 1978. Geohistory analysis: Application of micropaleontology in exploration geology. American Association of Petroleum Geologists Bulletin 62: 201–222.

Vassoyevich, N.B., Korchogina, Y.I., Lopatin, N.V., and Chernyshev, V.V. 1969. Principal phase of oil formation. Vestnik Moskov University 6:3–27. (English translation: International Geological Review, 1970, 12:1276–1296.)

Waples, D.W. 1980. Time and temperature in petroleum formation: Application of Lopatin's method to petroleum exploration. American Association of Petroleum Geologists Bulletin 64:916–926.

Weissbrod, T. 1962. Feldspar rocks in Shen Ramon and Har Gvanim, Makhtesh Ramon. Geological Survey of Israel Report M.P. 115/61, 16 pp.

Weissbrod, T. 1969. The Paleozoic of Israel and adjacent countries: Part I. The subsurface Paleozoic stratigraphy of southern Israel. Geological Survey of Israel Bulletin 47, 34 pp.

Weissbrod, T. 1981. The Paleozoic of Israel and adjacent countries (lithostratigraphic study) (in Hebrew with English summary). Geological Survey of Israel Report M.P. 600/81, 276 pp.

Weissbrod, T., and Nachmias, J. 1985. Stratigraphic significance of heavy minerals in the late Precambrian-Mesozoic clastic sequence ("Nubian Sandstone") in the near East. Abstracts, Annual Meeting Israel Geological Society, Yotvata, pp. 105–106.

Zaun, P.E., and Wagner, G.A. 1985. Fission-track stability in zircons under geological conditions. Nuclear Tracks and Radiation Measurements 10:303–307.

Zeiss Publication (undated). Reflection microscope photometry, application: Measurements on coal samples. Carl Zeiss Publication A 41-825.8-e, Uberkochen, West Germany, Carl Zeiss, 36 pp.

Zeitler, P.K. 1984. Closure temperature implications of concordant $^{40}Ar/^{39}Ar$ potassium feldspar and zircon fission track ages from high-grade terranes. Abstracts, Fourth International Fission Track Dating Workshop, Troy, NY, p. 60.

13
Estimating the Thickness of Sediment Removed at an Unconformity Using Vitrinite Reflectance Data

Charlene Armagnac, James Bucci, Christopher G. St. C. Kendall, and Ian Lerche

Abstract

When variations in vitrinite reflectance with depth for a well are combined with burial-history data, it is possible to estimate how the heat flux at that location varied with time and thus understand how vitrinite reflectance behaves as a function of depth through time. If an unconformity representing a large amount of erosional removal occurs in the well, the trend of measured vitrinite reflectance values with depth shows a discontinuity at that depth. Both the thickness of sediment removed at an unconformity and the heat flux with time can be determined simultaneously by varying the sediment thickness until a minimum is reached in the mean-square-residual fit of the match between the measured vitrinite reflectance values and predicted vitrinite reflectance values. This scheme has been tested successfully on unconformities in wells drilled in the National Petroleum Reserve of Alaska (NPRA) and in one well from the Rharb basin of Morocco. The predicted magnitudes of the erosional events determined from vitrinite reflectance match the regional geology, suggesting that the system may be useful in areas where well and vitrinite reflectance data exist but regional geologic control is poor.

Introduction

The stratigrapher tracking the history of basins is constantly confronted with the problem of determining the amount of sediment deposition and erosion at unconformities. Similarly, in some sedimentary basins, the geologist is asked to determine the timing of listric faulting, the emplacement of igneous intrusions and salt tongues, and the onset of overpressuring and overthrusting. In this chapter we focus on unconformities, but in later papers we will consider how to determine the parameters that describe the onset and magnitude of these other phenomena that also affect the thermal response of sedimentary sequences.

The classical approach to determining when erosional events were initiated and in estimating how much sediment was eroded is to extract this information from the regional stratigraphy. Some of these techniques are outlined by Van Hinte (1978) in an article in which he describes the use of burial-history or geohistory algorithms. These techniques include establishment of the age of the youngest rocks beneath the unconformity at specific locations. The stratigrapher then locates the oldest rocks resting on the unconformity and notes their age. This approach does not handle the problem of how much erosion took place prior to the development of the present unconformity surface or when erosion began. One possible solution is to assume that the rate of sedimentation below the unconformity continued into it, and that the rate of deposition following the unconformity matched the rate of erosion (Van Hinte, 1978; Guidish et al., 1985). One can assume that erosion began at the time where the slopes of these lines intersect on a burial-history diagram (Fig. 13.1). Another method is to: 1) assume that the erosion associated with the unconformity began at the most obvious major sea-level change on the coastal onlap chart of Vail et al. (1977), and then 2) extrapolate the rate of sediment accumulation prior to the unconformity to this time, and 3) erode to the age of the overlying sediment (Fig. 13.2). Undoubtedly there are other ways of deriving the timing and erosion involved

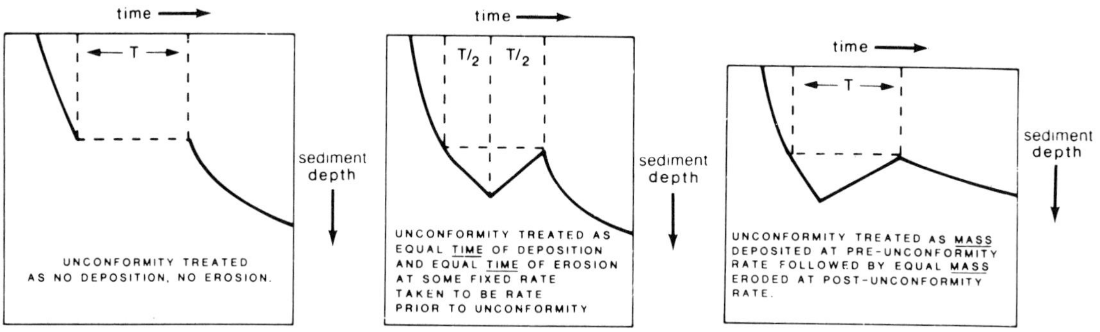

FIGURE 13.1. Burial-history diagram of hypothetical well with unconformities handled in different ways.

during an unconformity (Moshier and Waples, 1985), but all carry similar assumption to the approaches listed above.

Once we have estimated the time of the onset of erosion we need an independent means to establish how much sediment was eroded. Magara (1976) used sonic logs to plot shale transit time with depth and estimated the amount of erosion at unconformities by extrapolating the rate of change in the transit time into the unconformity. The sediment removed from the unconformity is added back until the transit times on either side of the unconformity match. Dow (1977) outlined a method of estimating the amount of material removed at an unconformity by examining vitrinite maturation on a logarithmic scale. The method entails the development of a maturation profile, the slope of which is influenced by the geothermal gradient, the geological age, and sedimentation rate of the section under investigation. With the presence of a major unconformity, the slope of the maturation profile is offset and the thickness of the sediment removed during this event can be estimated by projecting the trend of the earlier vitrinite reflectance values into the slope of later values (Fig. 13.3). We use a similar scheme to those of Magara (1976) and Dow (1977) but, instead of invoking a discontinuity in a trend in transit time or vitrinite reflectance with depth, we tie thermal history to the vitrinite reflectance (Lerche et al., 1984).

The method, like that of Dow (1977), uses the fact that if vitrinite reflectance is traced across an unconformity in a well or a measured section then the trend of the vitrinite reflectance data (Fig. 13.4A) will show a "jump" at the unconformity (Fig. 13.5). Thus between unconformities there is a progressive upward decrease in vitrinite reflectance, reflecting a shorter temperature history. At the unconformity, this trend is interrupted, and the vitrinite reflectance values above the unconformity are markedly less than those below the "jump." The trend with depth of the younger vitrinite reflectances may or may not parallel the trend of the vitrinite reflectance values below the unconformity. We determine the heat flow during the unconformity period by using the vitrinite reflectance data above and below it. Then we vary the size of the unconformity, in terms of its timing and the amount of sediment removed, until the predicted vitrinite reflectance values follow the trend of the vitrinite reflectance measurements on either side of the unconformity. In other words, as we have noted elsewhere (Lerche et al., 1984), the cumulative thermal history is retained by the vitrinite reflectance throughout time, recording the thermal maturation of the sediments, including evidence of what happened during any unconformities. This information can be extracted to determine the total sediment thickness stripped off the sedimentary column during an unconformity.

In Lerche et al. (1984) we showed how paleoheat flux was determined using vitrinite reflectance based on the assumptions: 1) that the burial history of the sediments is precisely known, 2) that the dominant source of heat to the sediments was from the basement, and 3) that the dominant mode of heat transport was by thermal conduction.

Our purpose in this chapter is to provide quantitative formulae that detail how the thermal history of sediments is influenced when one or more of the above assumptions is not met. These formulae are then used to determine not only the

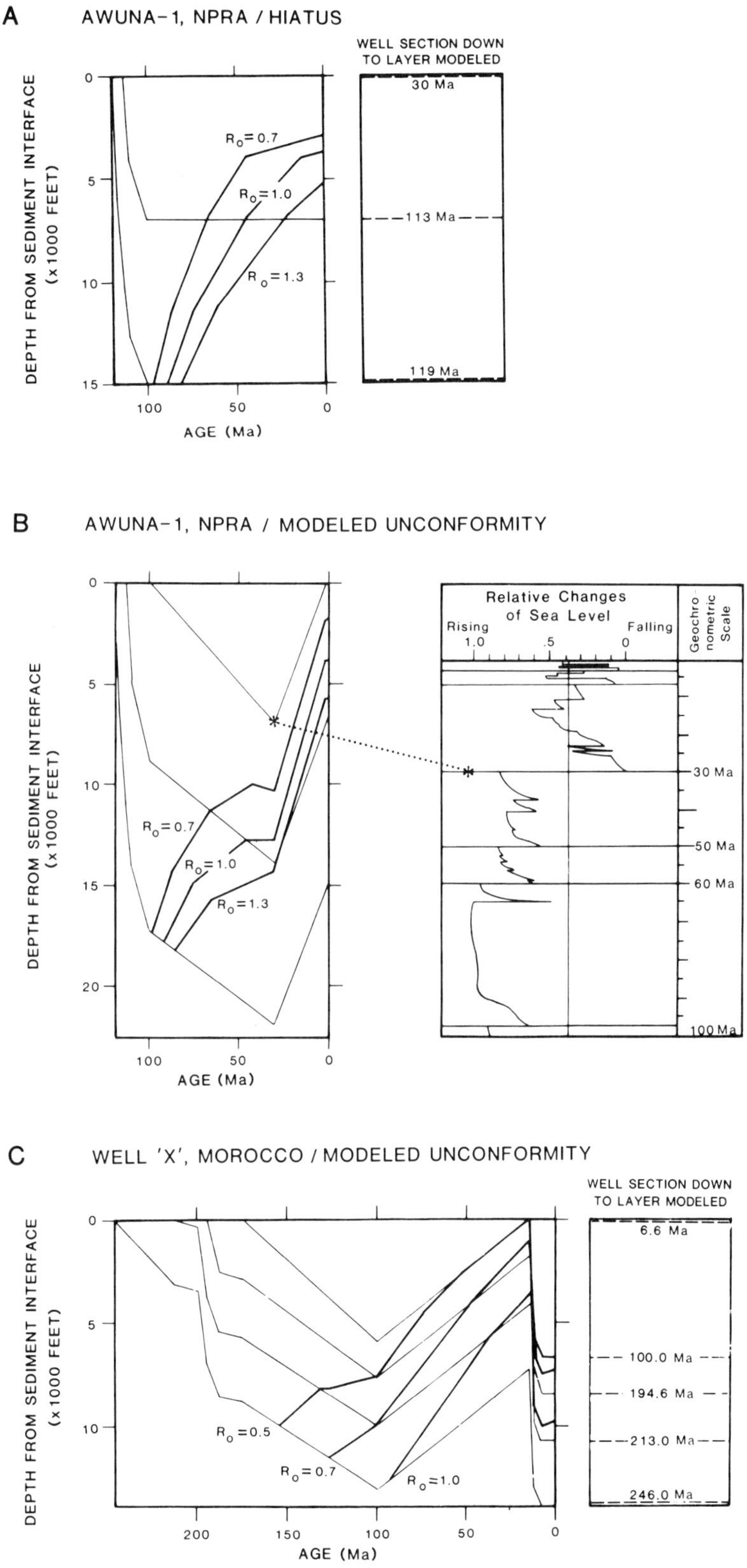

FIGURE 13.2. Combined geohistory plots for NPRA well Awuna-1 with the Tertiary unconformity shown, as a hiatus (A) and with the onset of erosion coinciding with the Oligocene sea-level low from the Vail et al. (1977) coastal onlap chart (B). C, Plot for Moroccan Well "X" showing the onset of erosion coinciding with the mid-Cenomanian sea-level low set to 100 Ma; note that from Vail et al. (1977) it might be argued that the low is at 95 Ma.

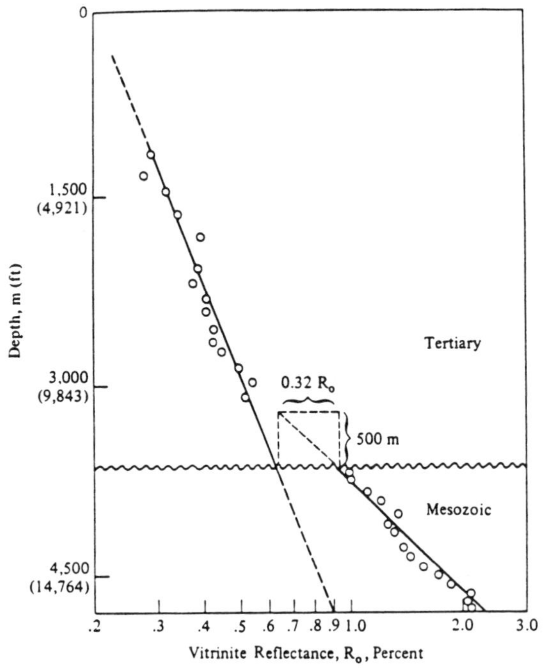

FIGURE 13.3. Dow's (1977) method for estimating unconformity size. (After Dow, 1977.)

temporal variation of paleoheat flux but also the size of an erosional unconformity. First, however, we need to formulate quantitatively how various factors that violate these assumptions can influence the thermal history of sedimentary material.

Two factors are apparent. First, direct changes in the burial path of sediments mean that, even if the paleoheat flux were constant at its present-day value, there would be a direct change in the temperature history of the sediments resulting from the altered burial path. Thus vitrinite reflectance, or any other thermal indicator, would be directly influenced. Second, a direct change in the temperature can also be caused by igneous intrusives, overthrusting, listric faulting, variable paleo-overpressuring and fluid flow with time, and proximity to dynamically evolving salt ridges, diapirs, and "tongues." Thus though the burial path may be very precisely known, both the thermal conductivity and the sediment temperature along the path are influenced directly by the above events.

Method for Determining Size of Unconformity

As detailed in Guidish et al. (1985), if isostatically compensated burial-history models are used, it is a simple matter to ascribe any number of variations to an unconformity, including a depositional hiatus as the limiting form. The cumulative record of thermal history carried by a thermal indicator, like vitrinite reflectance, provides the precise information needed to determine the unconformity thickness, if the timing of the unconformity can be defined.

Vitrinite in sedimentary layers deposited later than the unconformity has no means of recording the presence of that unconformity. However, vitrinite occurring in sedimentary layers deposited earlier than the unconformity records not only the thermal history before the unconformity period but also during and after. This is because the combination of vitrinite reflectance theory (Lerche et al., 1984) together with isostatic burial-history "backstripping" (Guidish et al., 1985) allows a determination of the paleoheat flux with time. Usually one assumes a linear dependence $Q(t) = Q_0(1 + \beta t)$ where Q_0 is present-day heat flux, t is time, and β is the linear constant coefficient to be determined (Fig. 13.4B). This simple formulation is the best that can be ruggedly extracted at this time from most present-day vitrinite reflectance measurements with depth, because of both the scatter in present-day vitrinite reflectance measurements and the degree of uncertainty in the biostratigraphic markers. However, we are currently working on a nonlinear method based on tomography, which appears to be a promising way of circumventing such simplistic formulations of heat-flux variations with time, despite data scatter.

In Lerche et al. (1984) the degree of accuracy is provided by the logarithm of the mean-square-residual (MSR), which is a measure of the degree of mismatch between the theoretically predicted vitrinite reflectance with depth and the observed values (Fig. 13.4C). As β is varied, so the MSR varies, and a minimum in MSR indicates the best fit. By such a procedure, β is relatively tightly constrained and the fit of theoretical to observed vitrinite reflectance with depth for the "best" β

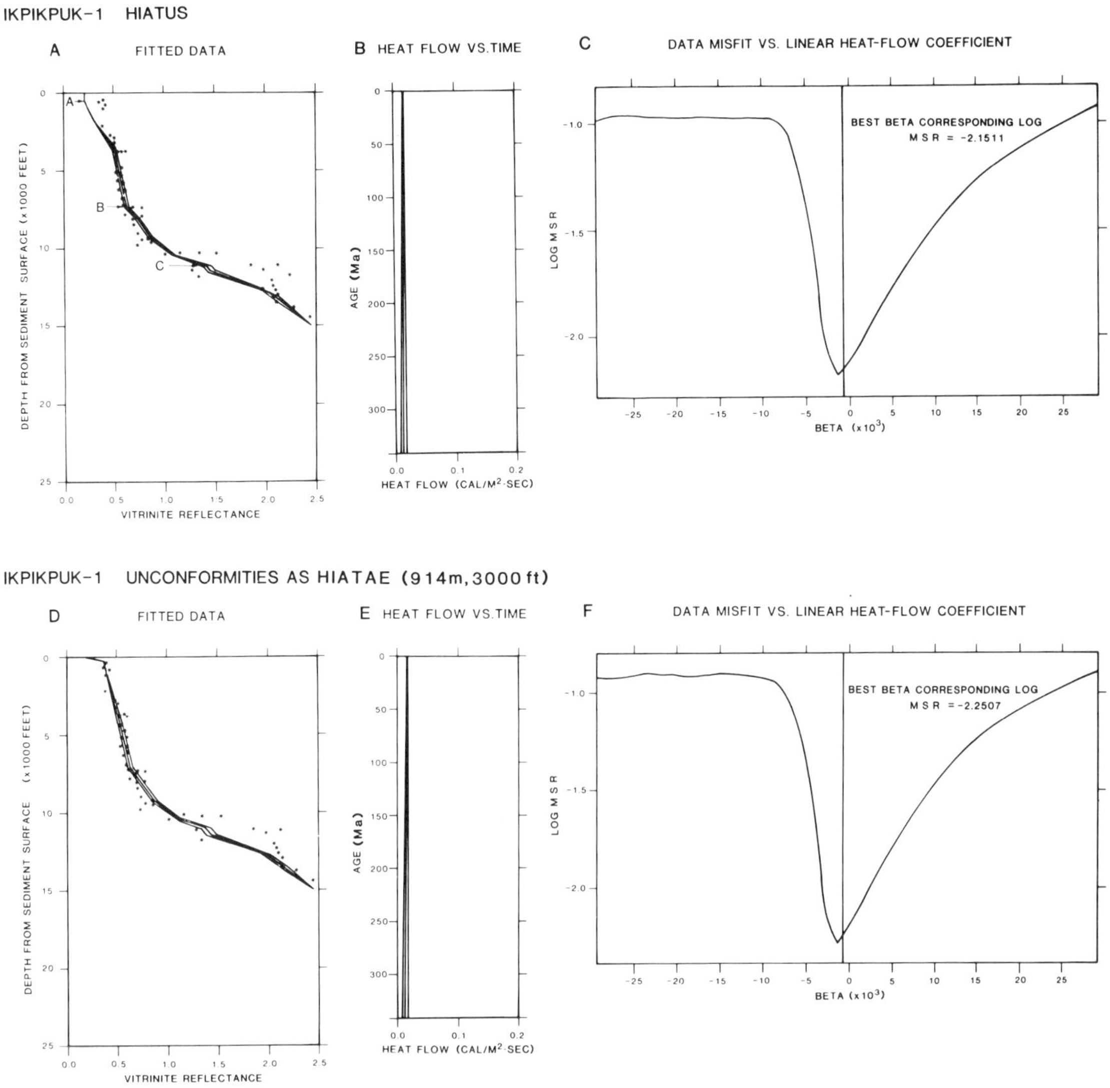

FIGURE 13.4. A, Predicted and observed vitrinite reflectance data (Lerche et al., 1984) as a function of depth for Ikpikpuk-1. Note the "jumps" in the near surface (arrow A), at 2,270 m (7,460 ft) (arrow B), and at 3,500 m (11,450 ft) (arrow C). These "jumps" coincide with unconformities (see Fig. 13.5 for burial-history plot that shows these unconformities). The central line enclosed by the outer error envelope is predicted vitrinite reflectance for the best β for a linear heat flow. B, Plot of "best" heat flow for best "β" used to predict vitrinite reflectance line of A for Ikpikpuk-1 with unconformities treated as hiatae. C, Plot of log MSR (of the degree of mismatch between the predicted vitrinite with depth to the observed values) and the corresponding β for Ikpikpuk-1 with unconformities treated as hiatae. D, Predicted and observed vitrinite reflectance as functions of depth for Ikpikpuk-1 when the unconformity is modeled with 914 m (3,000 ft) of deposition before the onset of erosion at 30 Ma. Central line and error envelope as in A. E, Plot of "best" heat flow for best "β" for Ikpikpuk-1 with 914 m (3,000 ft) of deposition modeled for unconformity. F, Plot of MSR against β for 914 m (3,000 ft) of deposition at 30 Ma unconformity for Ikpikpuk-1.

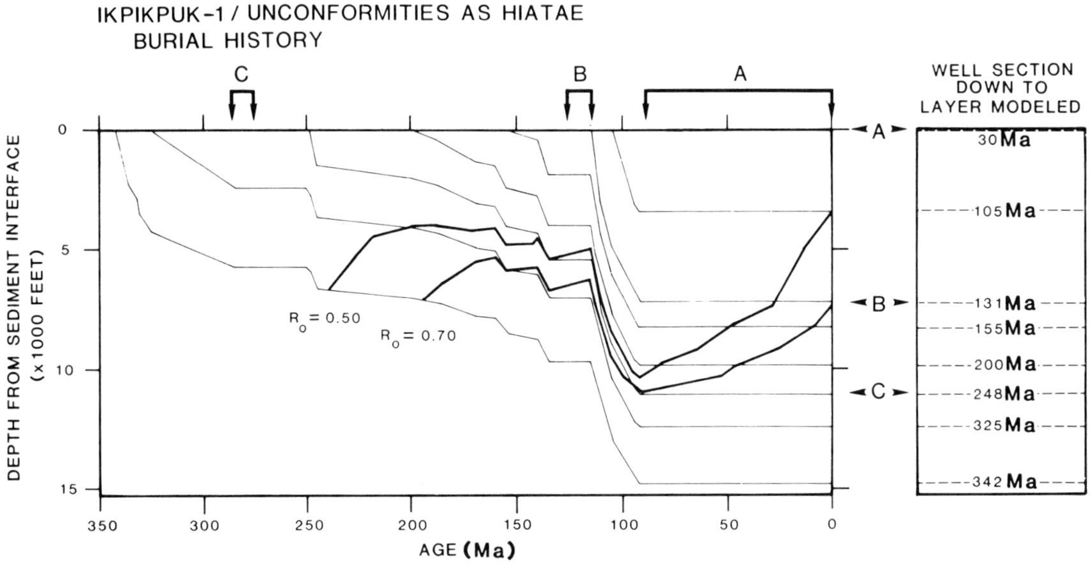

FIGURE 13.5. Burial-history plot for Ikpikpuk-1 showing unconformities at 92 Ma at a depth of 30 m (100 ft) (arrow A), at 135 Ma at a depth of 2,270 m (7,460 ft) (arrow B), and at 285 Ma at 3,500 m (11,450 ft) (arrow C). The unconformity we concentrate on in this chapter is at the near surface (arrow A). The unconformities are shown as hiatae in this diagram and can be recognized as jumps on the vitrinite versus depth plot (Fig. 4A).

value is very good. Further details of this scheme are given in Appendix 13.A.

Unconformities add another dimension to the problem. The total amount of material deposited

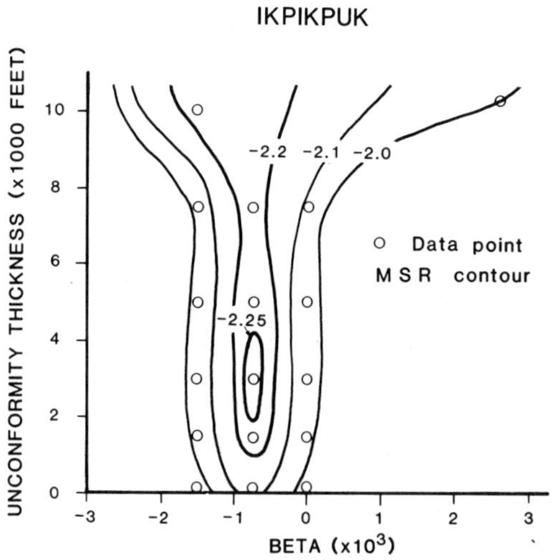

FIGURE 13.6. Plot of unconformity thickness versus best "β" for Ikpikpuk-1. The lowest MSR is at −2.25, which coincides with the Oligocene unconformity eroding 914 m (3,000 ft) and a best "β" of −0.0007/my.

and then eroded is unknown. However, for a given total amount of assumed erosion we find the best β, and record the MSR (Fig. 13.6). We then increase the amount of erosion in stepped increments and at each step we record the best β and the MSR. With too little or too much erosion the degree of mismatch between predicted and observed vitrinite reflectances with depth will be large. With too little erosion, the discontinuity in the reflectance profiles at the unconformity surface will not be large enough to match the observed jump while, with too great an amount of erosion, the predicted jump will be too large relative to observations. There is, then, a "best" amount of erosional removal during the unconformity that equates with both the smallest MSR and the best β value.

One caveat needs to be mentioned. When the unconformity occurs near the present-day sedimentary surface, as occurs in several wells in the NPRA, we do not have vitrinite reflectance on the upper side of the unconformity to guide us. In this case we take the vitrinite reflectance to have a specific value at the depositional surface. Our investigation of nearly 200 wells has narrowly constrained this reflectance value to be $R_o = 0.18 \pm 0.05$, as do other workers like Hunt (1979). When

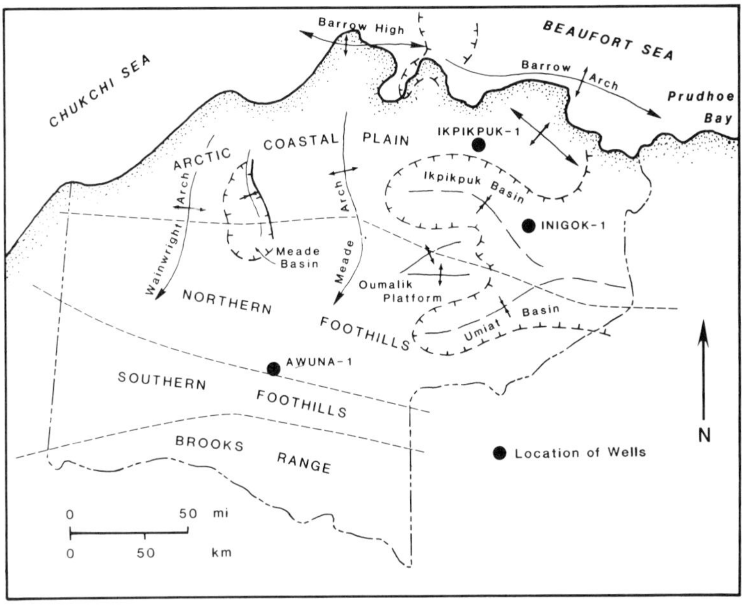

FIGURE 13.7. Location of Awuna-1, Inigok-1, and Ikpikpuk-1 in the National Petroleum Reserve of Alaska (NPRA) and of the major physiographic terrains described in the text. (Modified from Kirschner et al., 1983.)

this depositional value is so constrained, we can go through the previous sequence of operations with wells with present-day surface unconformities until the best thickness, h, and best β value are determined. We have followed this procedure for a number of wells in the NPRA and one well from Morocco as described in the following sections.

Mid-Tertiary Unconformity in Alaskan Wells from the National Petroleum Reserve

Regional Geology

Using the above scheme we have estimated the thicknesses of unconformities for a number of wells in the NPRA to see if we could improve our understanding of the regional geology. We used the NPRA because it provides an excellent data base that has already been studied by many geologists using seismic cross sections, outcrop, and well information to interpret the structural and stratigraphic history (Grantz et al., 1979; Churkin et al., 1979, 1980; Mull, 1980; Bushnell, 1981; Molenaar, 1981a, 1981b, among others). The wells we used are in the southern, eastern, and northern NPRA and include Awuna-1, Inigok-1, and Ikpikpuk-1, respectively (Fig. 13.7). In the northern area along the Barrow Arch, a mainly Cretaceous section is penetrated, but at Inigok-1 to the southeast, sediments of Mississippian age have also been drilled. The wells cut across several of the unconformities seen on Figures 13.8 and 13.9.

A major unconformity erodes down to Brookian rocks of middle Cretaceous age in the western NPRA and down to Tertiary-aged rocks in the eastern NPRA. The beds beneath the unconformity have been dated from Eocene to Maastrichtian (Nelson, 1981, citing both Carter et al., 1977, and Wolfe, personal communication, 1979), suggesting that a period of major erosion began after the Eocene.

The global coastal onlap curves of Vail et al. (1977) show a major drop in eustatic sea level at approximately 30 Ma, in the Oligocene (Fig. 13.2B). This sea-level change apparently halted deposition in the NPRA and coincided with the period of marked erosion. We believe it is responsible for having initiated erosion in the vicinity of all the wells we studied. Thus we use the time of this sea-level drop to begin a period of erosion in the computer runs we made of our burial-history model for these wells. At the same time we varied the thicknesses of sediment deposited before the onset of erosion to test how much sediment could have been deposited in the late Cretaceous to early Tertiary. We plotted beta (β), the fractional rate of

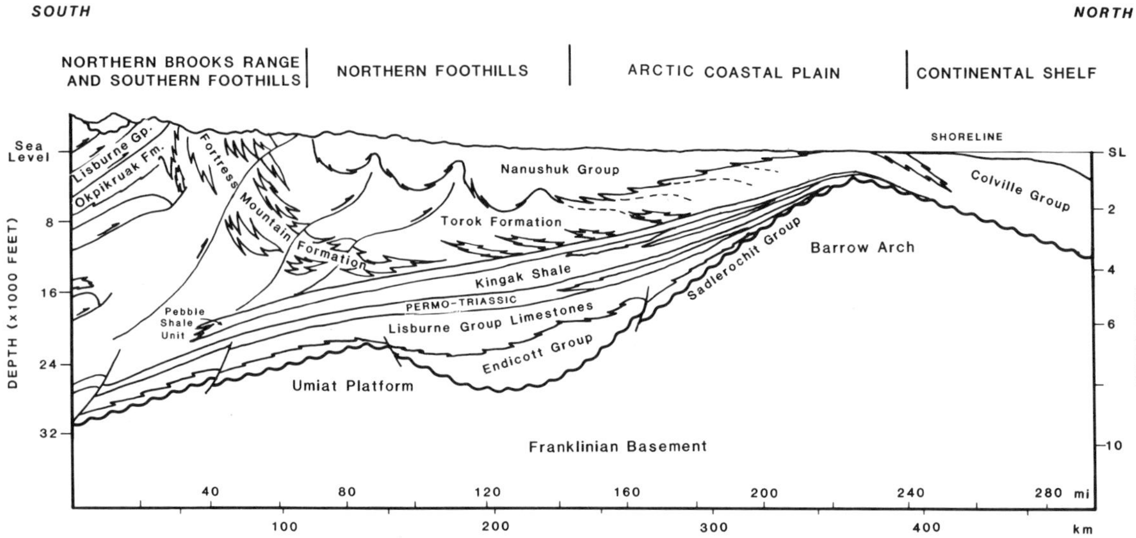

FIGURE 13.8. Schematic north-south cross section through NPRA showing general stratigraphy and structure. Note the decreasing complexity northward. (Modified from Bird, 1981.)

rate of change in paleoheat flux, versus unconformity thickness (h) (Figs. 13.6, 13.10, 13.11). These plots were contoured for values of equal log MSR and show ranges of unconformity thickness centered around a minimum value of the log MSR for that best β as well as ranges in β for a given unconformity thickness. The "bullseye" area of the plot shows the best range of fit of unconformity thickness and of the linear variation in heat flux for that well.

Input Data

The input data used in our burial-history computer runs consist of elevation, Kelly-Bushing height, bottom-hole temperature, and well location plus the lithology, water depth at deposition, age, and vitrinite reflectance values for selected layers from each well.

For the NPRA, the elevations, Kelly-Bushing heights, and the well locations were obtained from well-log headers. Where possible, temperature information was obtained from Blanchard and Tailleur (1982) and from log-header information where necessary. The data base used for all the burial and geohistory calculations in this chapter have depth expressed in feet and temperature in Fahrenheit, the convention of the North American oil industry. In the text we refer to depths in both meters and feet; however in the illustrations and tables we have used feet and Fahrenheit since this way we reduce the chance of error and prevent any confusion for scientists referring back to the original well data and logs.

Biostratigraphy provided the unit/time markers (layers) and the water depths. The relationship of paleobathymetry to depositional setting was obtained from Van Hinte (1978). Witmer (1981) and McClelland Engineers, Inc. (1982) were the sources for biostratigraphic information. The time markers were correlated to the Geologic Time Scale of Palmer (1983), which we slightly modified in our age determinations. Well logs were used to obtain the lithology information for each layer. The logs came from the National Geophysical and Solar-Terrestrial Data Center of the National Oceanic and Atmospheric Administration in Boulder, Colorado. The vitrinite reflectance data are from a geochemical data base for NPRA released by Petroleum Information (1981).

We have taken these data "as is" and have not attempted to cull them or improve their quality. We appreciate that the temperature data, the vitrinite reflectance values, and the time markers are all extremely noisy. However, despite the data quality, the quantitative results obtained match well with the qualitative behavior inferred from the regional geology.

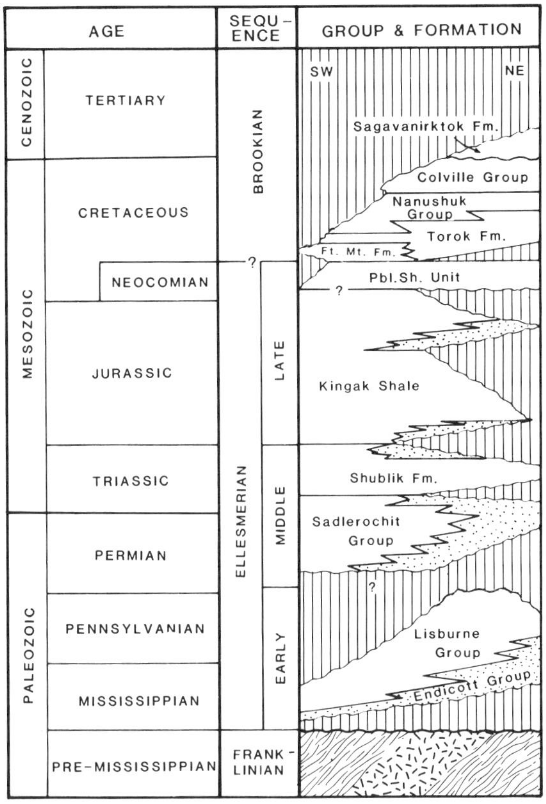

FIGURE 13.9. General stratigraphic terminology and relationships for the NPRA. Ft. Mt. Fm = Fortress Mountain Formation. (After Kirschner et al., 1983.)

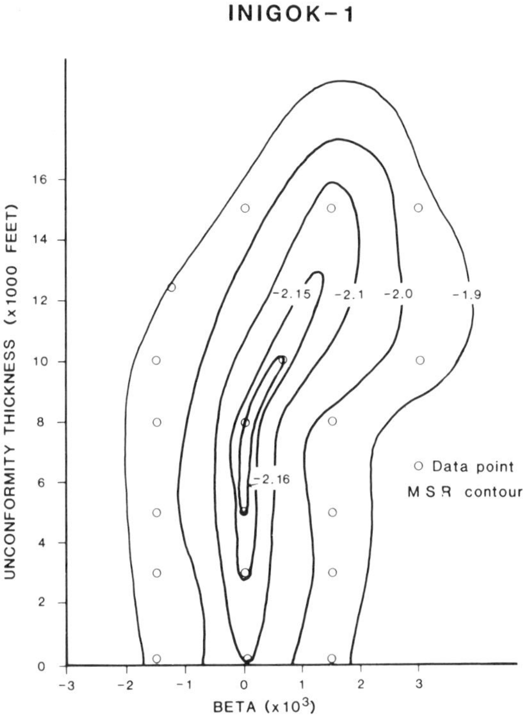

FIGURE 13.10. Plot of unconformity thickness versus best "β" for Inigok-1. The lowest MSR is at −2.165, which coincides with the Oligocene unconformity eroding 2,440 m (8,000 ft) and a best "β" of 0/my.

Runs and Results

We input the general data for our wells as listed in Appendix 13.B. We varied the thickness of the Oligocene unconformity in the manner described above. Figure 13.4 includes the results of such manipulations for Ikpikpuk-1 for both a hiatus (Fig. 13.4, A to C) and a 914-m (3,000-ft) depositional and erosional event (Fig. 13.4, D to F). The outputs include a series of plots showing the resulting predicted values of vitrinite reflectance against measured reflectance values, predicted heat flux, and the MSR versus β for the two unconformity models. Figure 13.6 shows how we obtained the best β and the thickness of sediment deposited and eroded in the unconformity.

For Ikpikpuk-1 (Figs. 13.6 and 13.12), a minimum log MSR value corresponds to a thickness of approximately 914 m (3,000 ft) and a β of −0.0007/my. Hereafter all values of β are quoted

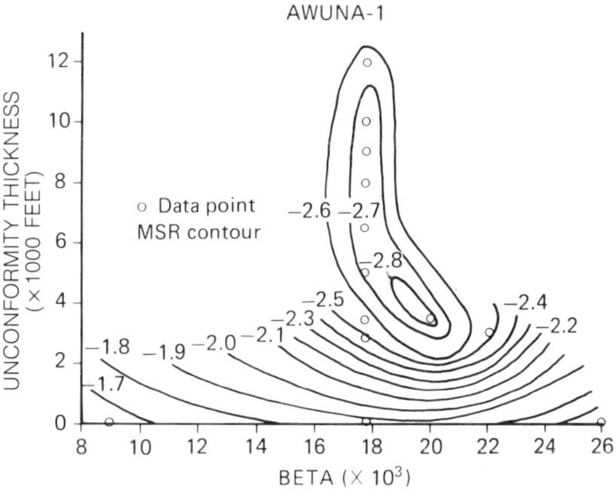

FIGURE 13.11. Plot of unconformity thickness versus "β" for Awuna-1. The lowest MSR is at −2.8, which coincides with the Oligocene unconformity eroding 1,070 m (3,500 ft) and a best "β" of 0.02/my.

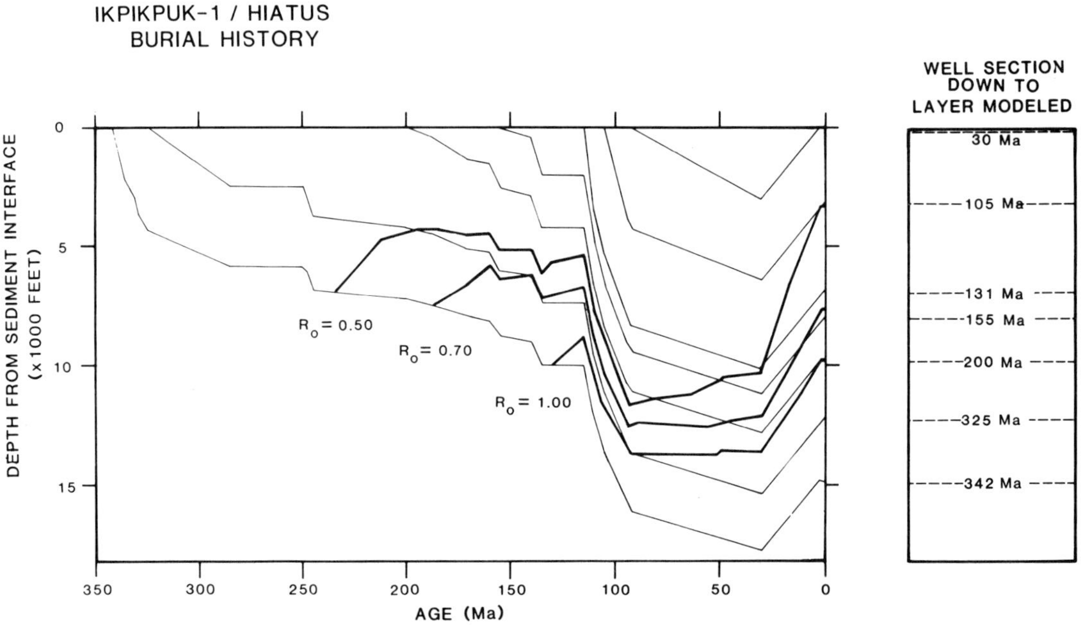

FIGURE 13.12. Burial-history plot of best unconformity thickness of 914 m (3,000 ft) for Ikpikpuk-1.

in my^{-1}. Calculated minimum and maximum values for β at this thickness are −0.0015 and 0, respectively.

In well Inigok-1 (Fig. 13.10), a minimum log MRS value of −2.16 is found using thicknesses of 1,525 to 3,050 m (5,000 to 10,000 ft), with the best value of −2.1643 occurring at 2,440 m (8,000 ft) (Figs. 13.13 and 13.14). The respective minimum, best, and maximum β at this fit are −0.0015, 0, and 0.0015.

Well Awuna-1 (Fig. 13.11) has the best fit at approximately 1,070 m (3,500 ft) with a minimum log MSR value of −2.8077 and a best β of 0.0177.

It is interesting to note that Inigok-1, located on a horst block adjacent to the Ikpikpuk Basin, shows the greatest amount of sediment removed at the unconformity. This may imply that rejuvenation of the horst structure occurred in the late Cretaceous-middle Tertiary. A study integrating the results from the thermal-history program with the regional geological history is presently underway.

In addition to unconformity thickness estimates using the burial-history program of Lerche et al. (1984) we also used Dow's (1977) method to estimate the amount of material deposited and eroded at the sub-Gubik unconformity for eight of the NPRA wells. As we explained earlier, in this method a line is fitted to the trend of the vitrinite reflectance values and is extended through an unconformable layer until it intersects the trend of the vitrinite reflectance above that unconformity (Fig. 13.3). Because no reflectance values were available above the shallow sub-Gubik unconformity, the trend of the line was extended until it reached a surface reflectance value of 0.2%. The thickness at which the surface reflectance value is met is assumed to match the amount of material deposited and eroded at the unconformity. A certain amount of variability is readily apparent in the best fit value, and therefore the range of possible thicknesses was measured. The wells that do not contain estimates of thickness by this method had too much scatter in the data to fit any reasonable line. Dow's method gave a smaller range of values than ours, but there was no consistent behavior in the differences between the two methods. However with Dow's method 1) decompaction is not modeled; 2) thermal conductivity variations with porosity, depth, lithology, and time are not accounted for; and 3) paleoheat flow is assumed to be constant so that it is not clear to what extent systematic and statistical errors are modifying results

FIGURE 13.13. A, Predicted and observed vitrinite reflectance as functions of depth for Inigok-1. The unconformity is modeled with 2,440 m (8,000 ft) of deposition before the onset of erosion at 30 Ma. Line and error envelope as in Figure 4A. B, Plot of "best" heat flow for best "β" for Inigok-1 with 2,440 m (8,000 ft) of deposition modeled for unconformity. C, Plot of MSR against β for 2,440 m (8,000 ft) of deposition at 30 Ma unconformity for Inigok-1.

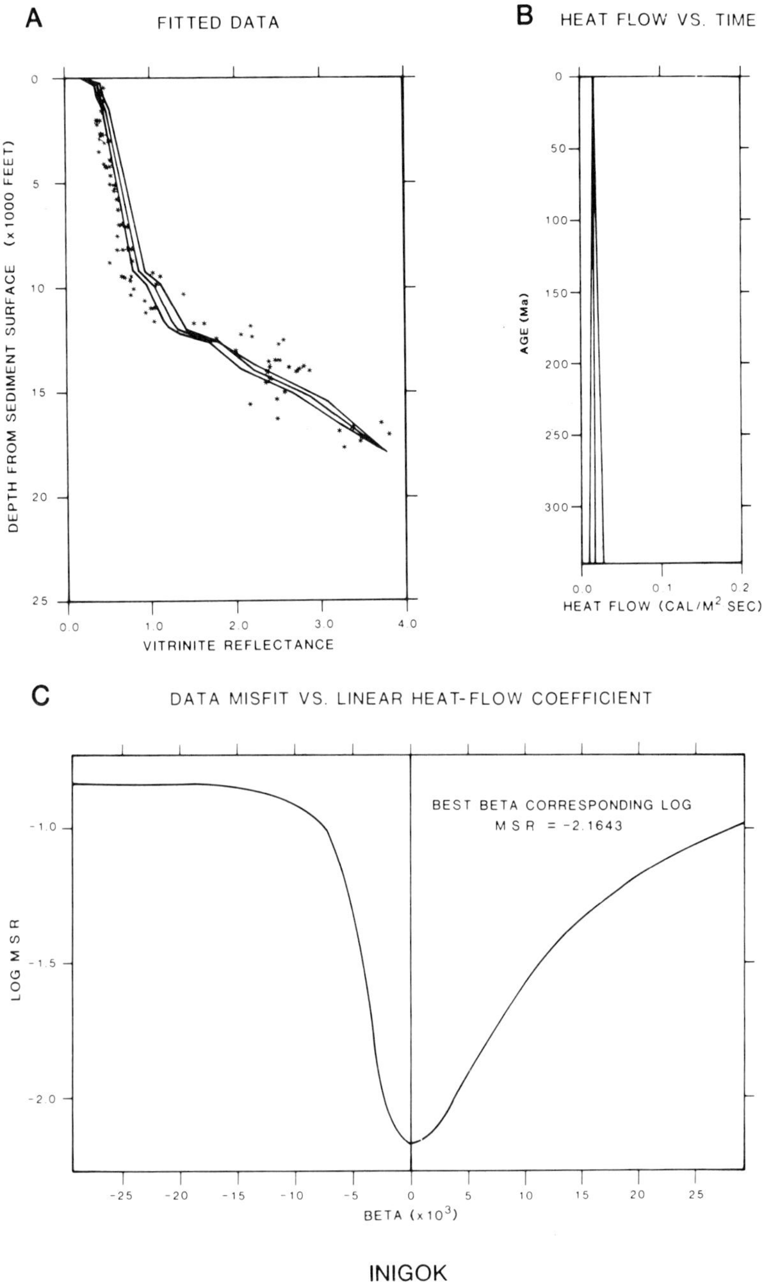

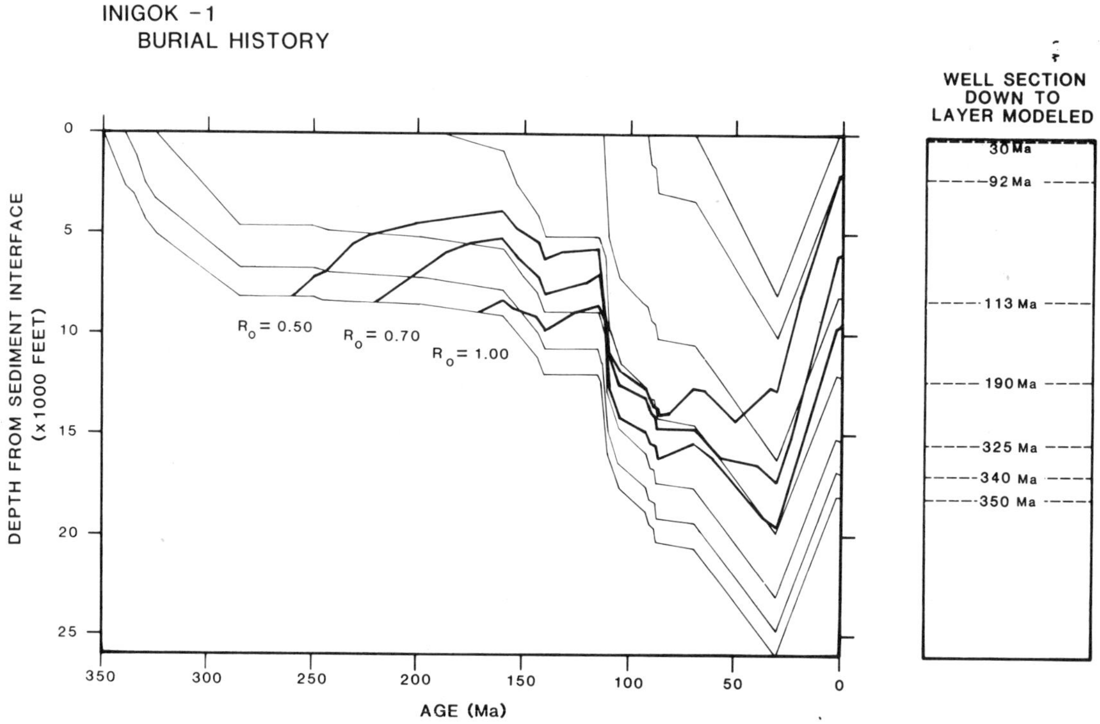

FIGURE 13.14. Burial-history plot of best unconformity thickness of 2,440 m (8,000 ft) for Inigok-1.

based on this method. A comparison of the unconformity size estimated by the two methods is shown in Table 13.1.

Late Cretaceous/Early Tertiary Unconformity in Moroccan Well "X" from the Rharb Basin

Regional Geology

Using the methodology set out above for the NPRA, we estimated the size of the late Cretaceous/early Tertiary unconformity that can be seen in Well "X" from the Rharb basin. This basin, as Bucci (1985) describes it, is located in Morocco just south of the Rif overthrust zone, at the north end of the Moroccan Mesata (Fig. 13.15). The Mesozoic and Tertiary rocks of the basin lie unconformably on a structurally deformed, slightly metamorphosed Hercynian basement with some intrusive granites (Fig. 13.16). The extension associated with the initiation of the Atlantic in the Triassic developed a series of normal faults that strike N. 45° E. in the Rharb basin area, creating a series of horsts and grabens. The resulting basin is filled by a Mesozoic succession related to rifting and postrift sedimentation, and is followed by a Tertiary section related to Alpine deformation and foredeep development.

The earliest rift fill is a presumed Triassic breccia composed of fragments of Paleozoic rock (too thin to mark on Fig. 13.16), which lies unconformably on the Hercynian basement. Overlying these breccias is a sequence of continental red beds containing anhydrite, gypsum, and salt beds. Basalts, 100 m (330 ft) thick, are interbedded with the red beds. The Triassic red beds terminate with a thick succession of dolomites and anhydrites with minor shales. The postrift fill began with Liassic shallow-marine carbonates, followed by deep-water marls. The Jurassic deposition was terminated in the Dogger by arkosic sands that were probably sourced from nearby horst blocks. No upper Jurassic is present in the Rharb basin but thin onlapping Cenomanian and very thin Turonian

TABLE 13.1. Comparison of Estimates of Missing Section in Selected Alaskan Wells

	Range in thickness (× 10^3 ft)		
Well name	Dow's (1977) method	Vitrinite inversion method*	β (my^{-1})†
West Dease-1	–	9 ± 8	10^{-2} (1.33 ± 0.58)
East Simpson-2	5 ± 2	9 ± 8	10^{-3} (6.5 ± 6.5)
Drew Point-1	4.5 ± 2.5	9 ± 9	10^{-2} (1.52 ± 0.65)
North Kalikpik-1	5 ± 3	9 ± 8	10^{-3} (4.3 ± 4.3)
Peard-1	8.5 ± 1.5	10 ± 10	10^{-3} (4.4 ± 4.4)
Kugrua-1	8 ± 4	10 ± 10	10^{-3} (3.8 ± 7.7)
Ikpikpuk-1	7 ± 2	9 ± 8	10^{-3} (2.20 ± 6.5)
Atigaru Point-1	–	10 ± 10	10^{-2} (−2.3 ± 0.47)
South Meade-1	–	10 ± 10	10^{-2} (−1.66 ± 0.63)
S. Harrison Bay-1	–	10 ± 10	10^{-3} (1.15 ± 1.15)
West Fish Creek-1	5 ± 2.5	9 ± 8	10^{-3} (0 ± 3.8)
Inigok-1	7 ± 3	10 ± 5	10^{-2} (1.10 ± 0.22)
Tunalik-1	–	10 ± 10	10^{-2} (2.0 ± 0.4)

*Error within 10% of minimum log MSR.
†Beta values based on data above Cretaceous unconformity.

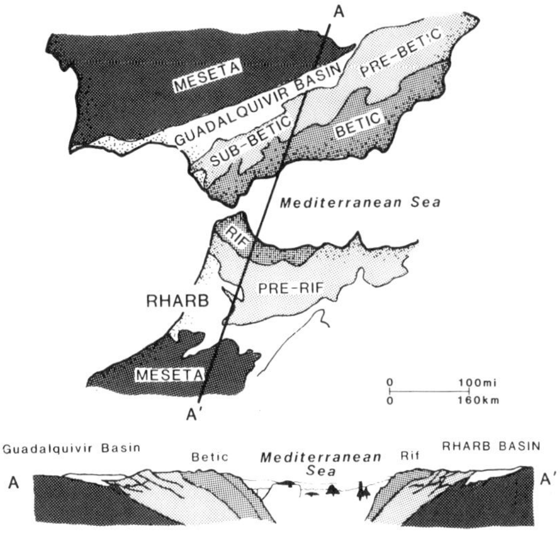

FIGURE 13.15. Location map for the Rharb basin with a schematic crustal cross section across northern Morocco. (After Biju-Duval and Montadert, 1977.)

shallow-water carbonates and marls are present on the western flank. A thick Cretaceous sequence may have been deposited in the basin itself, but if so, it was eroded in the late Cretaceous to early Eocene (Arana and Vegas, 1974). Bucci (1985) discusses the possibility that the western Rharb basin may have been a positive feature until the Tertiary, although no conclusive evidence for this is available. As we shall show, within the basin itself, the vitrinite reflectance data from Well "X" in the Sidi Fili Field can be used to argue that an extra 1,830 m (6,000 ft) or more of Mesozoic sediment was present before erosion was initiated at a Cenomanian sea-level low some 96 to 100 Ma ago (Fig. 13.2) (Vail et al., 1977).

In the early Miocene, overthrusting from the north in the Rif extended this nappe complex to the northern edge of the Rharb basin. Deposition began with Aquitanian and Burdigalian sands, which were succeeded by shaley and marly sediments of Serravalian to Tortonian age. These deep-water sediments (Bucci, 1985) are capped by Pliocene shallow-water sands and marly sands.

Bucci (1985) reports an abrupt deflection in the basement and an increase in sediment thickness adjacent to the Rif nappe system, suggesting that basin subsidence in the Rharb basin at the time of nappe implacement was at least partially related to crustal loading.

Input Data for Well "X"

The input data used for modeling the unconformity were the same kinds as we used for the NPRA wells (Appendix 13.B). The source of our information came from Bucci (1985), who gathered elevation, Kelly-Bushing height, and temperatures from well-log headings. Biostratigraphy and lithologies came from final logs and well reports from the Bureau de Recherche et de Participations Minieres, Rabat, Morocco. The vitrinite samples were assembled by James Bucci and their reflectances were measured by Iris Hubbard of the Earth Sciences Resources Institute at the University of South Carolina.

Runs and Results

On the basis of the work of Arana and Vegas (1974) and the coastal onlap charts of Vail et al. (1977), which show a Cenomanian sea-level low, we estimate that the unconformity that extends from 175 Ma to 13.5 Ma in Well "X" began at around 96 to 100 Ma (Fig. 13.2). This sea-level event coincides with an early-Alpine orogenic event during which regional uplift is believed to have occurred (Arana and Vegas, 1974). Several examples with varying amounts of erosion beginning at 100 Ma were used. At the same time the linear heat-flux coefficient β

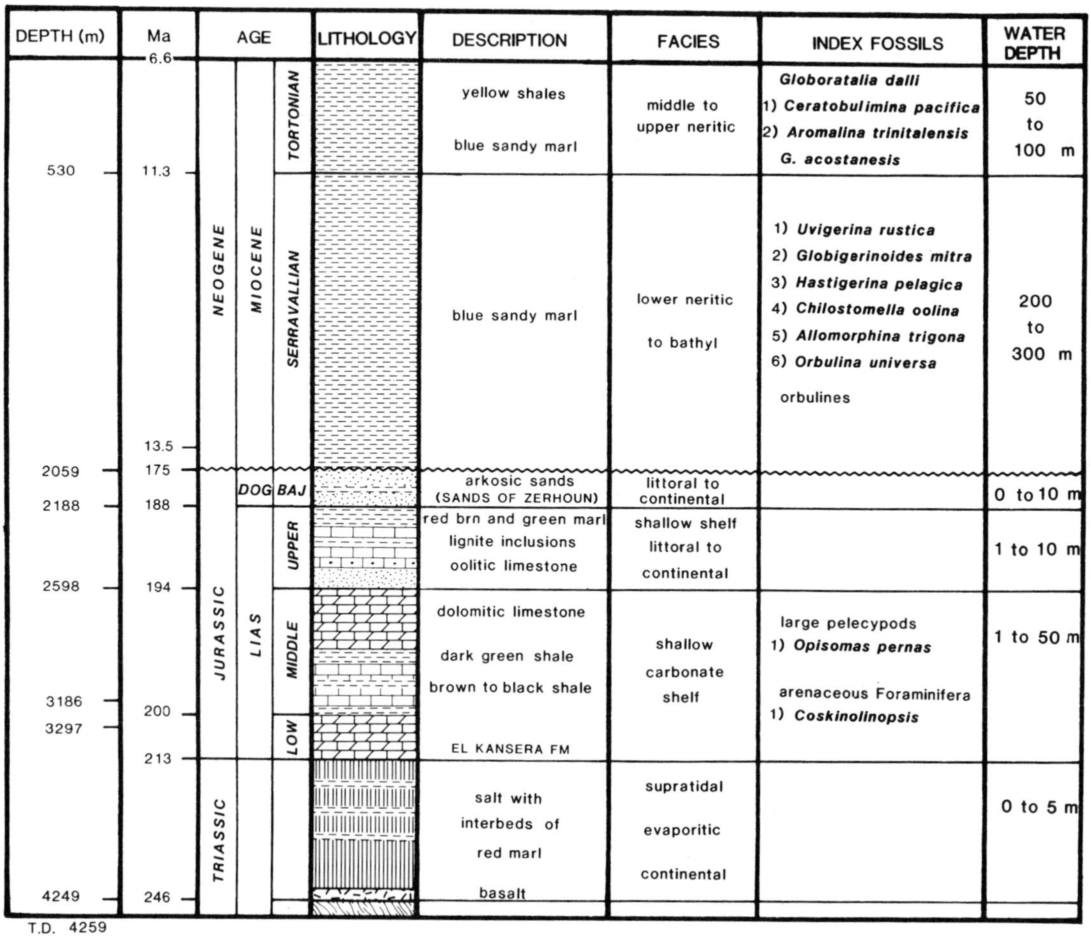

FIGURE 13.16. Stratigraphy of Well "X" from the Sidi Fili Field of the Rharb basin, Morocco. (After Bucci, 1985.)

was plotted as a function of the log MSR (Fig. 13.17 C). The "best" β values were then plotted as a function of 20 different thicknesses (h), and the log MSR values were contoured as they were for the Alaskan wells (Fig. 13.18). The minimum MSR contour occurs when approximately 1,830 m (6,000 ft) of section are eroded at the unconformity. Additionally, the best range of h (smallest contour interval) occurs between 1,470 m and 2,170 m (4,800 ft and 7,100 ft) with the β range of 0.0107 and 0.0078 (Fig. 13.18).

The implications of this study on the regional stratigraphy and on hydrocarbon generation modeling are important. In the case of the regional stratigraphy this study suggests that the late Cretaceous/early Tertiary unconformity seen in the region was more than just a long-lived hiatus but actually removed some 1,830 m (6,000 ft) or so of upper Cretaceous sediment. This result opposes the current stratigraphic model, which would have a long-lived hiatus instead. From the hydrocarbon generation viewpoint, the extra 1,830 m (6,000 ft) of rock means that potential source rocks from the Liassic could have entered the oil window earlier, remained there longer, and hence generated more oil.

Discussion and Conclusion

While the isostatic burial history is an equilibrium calculation and so contains no record of the evolutionary paths of the sediments, the cumulative

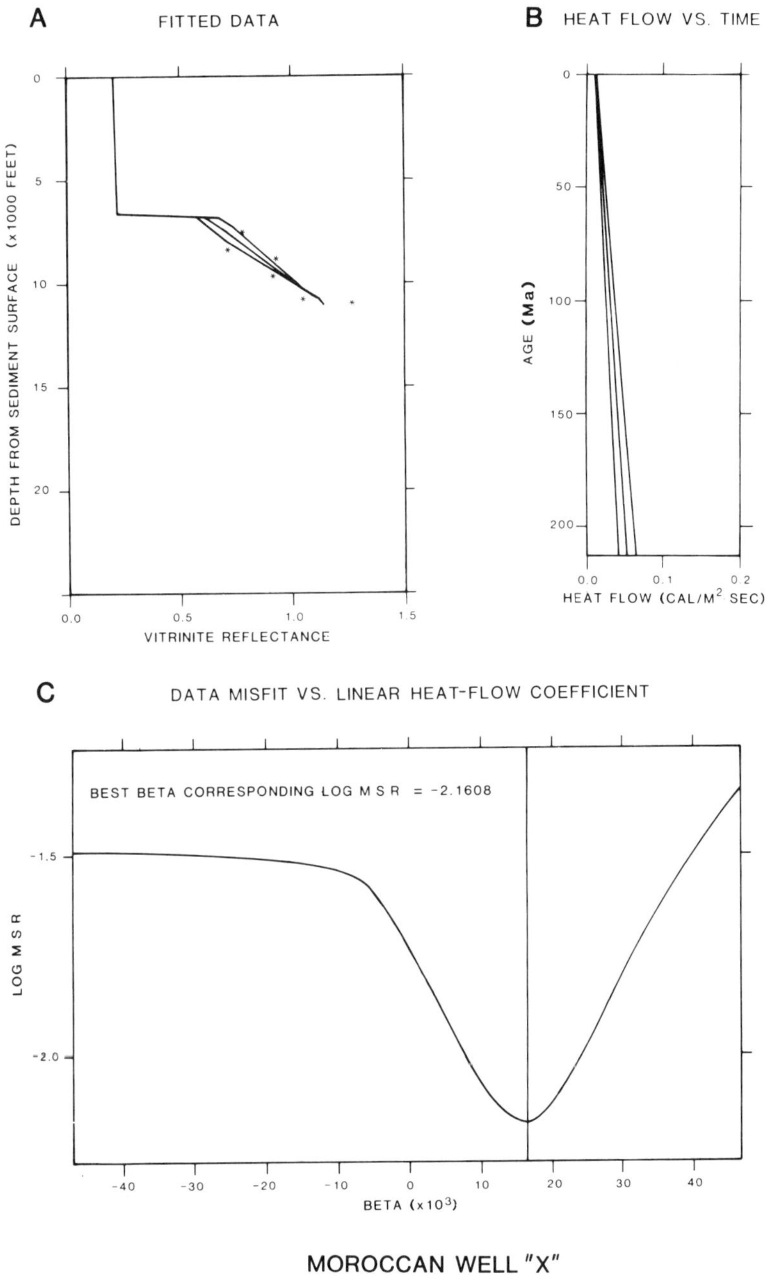

FIGURE 13.17. A, Predicted and observed vitrinite reflectance as functions of depth for Moroccan Well "X" when the unconformity is modeled with 1,830 m (6,000 ft) of deposition before the onset of erosion at 30 Ma. Line and error envelope as in Figure 4A. B, Plot of "best" heat flow for best "β" for Moroccan Well "X" with 1,830 m (6,000 ft) of deposition modeled for unconformity. C, Plot of MSR against β for 1,830 m (6,000 ft) of deposition at 30 Ma unconformity for Moroccan Well "X".

memory of the vitrinite reflectance does contain knowledge of the burial paths of the sediments because it tracks the thermal history along those paths. We have seen by explicit example that the temporal information recorded in the vitrinite reflectance is sufficient to permit us to estimate the thickness of unconformities, be they currently at depth or at the sediment surface.

In recording vitrinite reflectance with depth, we suggest that considerably more care, and more vitrinite samples from each depth, be taken than has heretofore been deemed necessary. We believe

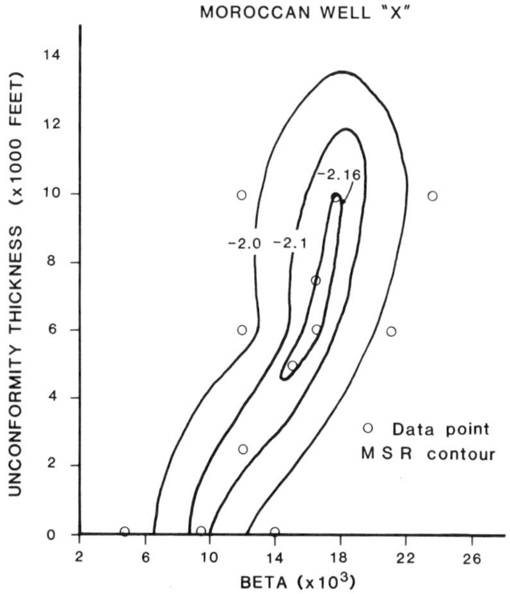

FIGURE 13.18. Plot of unconformity thickness versus "β" for Moroccan Well "X" from the Rharb basin. The lowest MSR is at −2.16, which coincides with the Cenomanian unconformity eroding 1,830 m (6,000 ft) and a best β of + 0.0164/my.

that the minimization of statistical and systematic errors in the measurements of vitrinite reflectance at each depth, and the consequent improvement of the quality of the vitrinite data, will go a long way to enabling us to improve not only our resolution of the temporal behavior of the heat flux with time, but also our understanding and determination of the evolution of unconformities. A similar plea is addressed to improving temperature data gathered at wells and the quality of the age markers. If all these data are improved, our understanding of the evolutionary behavior of heat flux will improve so that we will be able to determine more accurately and correctly the hydrocarbon maturation in a given area and the stratigraphic and structural evolution of potential hydrocarbon traps within that area.

We are currently programming and testing other thermal indicators to improve on the resolution of the unconformities that we achieve with vitrinite reflectance. These indicators include sterane and hopane isomerization and aromatization, apatite fission-track annealing, $^{40}Ar/^{39}Ar$, pyrotically derived carbon-isotope compositions as kerogen maturity indicators, and carya pollen translucency. As a result we anticipate being able to determine more tightly the magnitude and timing of unconformities. Preliminary results are very encouraging and will be reported in future papers. We are also working on a tomographic system for deriving heat flux through serial unconformities that will allow us to see variations in the rate of change as well as magnitude of heat flux.

Acknowledgments. We should like to thank Chevron, Cities Service, Core Labs, Western Geophysical, Marathon, Norsk Hydro, Saga, Statoil, Sun, and Union who have all generously provided financial support for this research, and the U.S. Geological Survey for releasing to the public domain most of the data we used.

The results for Alaska are from Charleen Armagnac's M.S. thesis and those for Morocco are from James Bucci's M.S. thesis. We thank James Edmund Iliffe for reviewing this paper and Iris Hubbard of the Earth Science Institute of the University of South Carolina for measuring the vitrinite reflectance values for Well "X" of the Rharb basin.

References

Arana, V., and Vegas, R. 1974. Plate tectonics and volcanism in the Gibralter arc. Tectonophysics 54:197–212.

Biju-Duval, B., and Montadert, L. 1977. Introduction to the structural history of the Mediterranean basins. In: Biju-Duval, B., and Montadert, L. (eds.): International Symposium on the Structural History of the Mediterranean Basins, Split, Yugoslavia, 1976. Paris, Editions Technip, pp. 1–12.

Bird, K.J. 1981. Petroleum exploration of the North slope, Alaska. In: Mason, J.F. (ed.): Petroleum Geology in China: Principal Lectures Presented to the United Nations International Meeting on Petroleum Geology, Beijing, China, 18–25 March 1980. Tulsa, OK, Penn Well Books, pp. 233–248.

Blanchard, D.C., and Tailleur, I.L. 1982. Temperatures and interval geothermal-gradient determinations from wells in National Petroleum Reserve in Alaska. U.S. Geological Survey Open-File Report 82–391, 79 pp.

Bucci, J. 1985. Basin analysis, thermal history and hydrocarbon generation of the Rharb Basin, Morocco. M.S. thesis, University of South Carolina, Columbia, 90 pp.

Bushnell, H. 1981. Unconformities: Key to North Slope oil. Oil and Gas Journal 79(2):114–118.

Carter, R.D., Mull, C.G., Bird, K.J., and Powers, R.B. 1977. The petroleum geology and hydrocarbon potential of Naval Petroleum Reserve No. 4, North Slope, Alaska. U.S. Geological Survey Open-File Report 77-475, 61 pp.

Churkin, M., Jr., Carter, C., and Trexler, J.H., Jr. 1980. Collision-deformed Paleozoic continental margin of Alaska: Foundation for microplate accretion. Geological Society of America Bulletin 91:648–654.

Churkin, M., Jr., Nokleberg, W.J., and Huie, C. 1979. Collision-deformed Paleozoic continental margin, western Brooks Range, Alaska. Geology 7:379–383.

Dow, W.G. 1977. Kerogen studies and geological interpretations. Journal of Geochemical Exploration 7: 79–99.

Grantz, A., Eittreim, S., and Dintar, D.A. 1979. Geology and tectonic development of the continental margin north of Alaska. Tectonophysics 59:263–291.

Guidish, T.M., Kendall, C.G.St.C., Lerche, I., Toth, D.J., and Yarzab, R.F. 1985. Basin evaluation using burial history calculations: An overview. American Association of Petroleum Geologists Bulletin 69: 92–105.

Hunt, J.M. 1979. Petroleum Geochemistry and Geology. San Francisco, W.H. Freeman, 617 pp.

Kirschner, C.E., Gryc, G., and Molenaar, C. 1983. Regional seismic lines in the National Petroleum Reserve in Alaska. In: Bally, A.W. (ed.): Seismic Expression of Structural Styles: Vol. 1. American Association of Petroleum Geologists Studies in Geology, Vol. 15, pp. 1.2.5-1 – 1.2.5-14.

Lerche, I., Yarzab, R.F., and Kendall, C.G.St.C. 1984. Determination of paleoheatflux from vitrinite reflectance data. American Association of Petroleum Geologists Bulletin 68:1704–1717.

Magara, K. 1976. Thickness of removed sedimentary rocks, paleopore-pressure and paleotemperature, southwestern part of Western Canada Basin. American Association of Petroleum Geologists Bulletin 60: 554–565.

McClelland Engineers, Inc. 1982. Alaskan North Slope Generalized Subsurface Biostratigraphic Chart. San Diego, CA, McClelland Engineers, Inc., 1 p.

Molenaar, C.M. 1981a. Depositional history of the Narushuk Group and related strata. In: Albert, N.R.D., and Hudson, T. (eds.): The United States Geological Survey in Alaska: Accomplishments during 1979. U.S. Geological Survey Circular 823-B, pp. B4-B5.

Molenaar, C.M. 1981b. Depositional history and seismic stratigraphy of Lower Cretaceous rocks, National Petroleum Reserve in Alaska and adjacent areas. U.S. Geological Survey Open-File Report 81–1084, 42 pp.

Moshier, S.O., and Waples, D.W. 1985. Quantitative elevation of Lower Cretaceous Mannville Group as source rock for Alberta's oil sands. American Association of Petroleum Geologists Bulletin 69:161–172.

Mull, C.C. 1980. Evolution of Brooks Range thrust belt and Arctic Slope, Alaska (abst.). American Association of Petroleum Geologists Bulletin 64:754.

Nelson, R.E. 1981. Paleoenvironments during deposition at a section of the Gubik Formation exposed along the lower Colville River, North Slope. In: Albert, N.R.D., and Hudson, T. (eds.): The United States Geological Survey in Alaska: Accomplishments during 1979. U.S. Geological Survey Circular 823-B, pp. B9-B11.

Palmer, A.R. 1983. The Decade of North American Geology 1983 Geologic Time Scale. Geology 11: 503–504.

Petroleum Information. 1981. Geochemical Data File Developed by Petroleum Information for N.P.R.A., North Slope (2 vols): Denver, Petroleum Information.

Toth, D.J., Lerche, I., Petroy, D.E., Meyer, R.J., and Kendall, C.G.St.C. 1981. Vitrinite reflectance and the derivation of heat flow changes with time. In: Bjoroy, M., et al. (eds.): Advances in Organic Geochemistry 1981. New York, Wiley, pp. 588–596.

Vail, P.R., Mitchum, R.M., Jr., and Thompson, S., III. 1977. Seismic stratigraphy and global changes of sea level: Part 4. Global cycles of relative changes of sea level. In: Payton, C.E. (ed.): Seismic Stratigraphy: Applications to Hydrocarbon Exploration. American Association of Petroleum Geologists Memoir 26, pp. 83–97.

Van Hinte, J.E. 1978. Geohistory analysis: Applications of micropaleontology in exploration geology. American Association of Petroleum Geologists Bulletin 62: 201–222.

Witmer, R.J. 1981. Biostratigraphic correlations of selected test wells of National Petroleum Reserve in Alaska. U.S. Geological Survey Open-File Report 81-1165, 89 pp.

Appendix 13.A. Vitrinite Reflectance as a Thermal Indicator

The physical laws of reflection and refraction show that the fractional intensity, R, of monochromatic light which is normally reflected from a plane interface of vitrinite of refractive index, n, surrounded by immersion oil of refractive index, n_o, is given by

$$R = \frac{(n - n_o)^2}{(n + n_o)^2} \quad \text{(A1)}$$

The classical Lorentz-Lorenz formula for the refractive index of a solid gives

$$n^2 = [1 + 2N\,g(\omega)]\,[1 - N\,g(\omega)]^{-1} \quad \text{(A2)}$$

where N is the number density of molecules, and $g(\omega)$ is the molecular structure factor that may be either positive or negative, depending on the frequency, ω.

Thus, from Equations A1 and A2 we see that the change in vitrinite reflectance, R, with depth is due to a change in N, the number density of molecules, with depth. The change in number density is due to the thermochemical decrease in number density described by a first-order reaction process as given by

$$\left(\frac{dN}{dt}\right)_{\text{chem}} = -k_0\,N \exp[-E(T)/\mathbf{R}T] \quad \text{(A3)}$$

where k_0 is the rate constant in inverse seconds, $\mathbf{R}$ is the gas constant, $T = T(t,z)$ is the subsurface temperature at time t and depth z, and $E(T)$ is the activation energy.

It is customary to write $E(T)$ in the form

$$E(T) = E_M - \mathbf{R}T(T/T_0) \quad \text{(A4)}$$

E_M, a constant, describes the basic molecular binding energy and the term $T(T/T_0)$ describes the exponentiation of the reaction rate for every T_0 degrees increase in temperature (due physically to excitation of phonons and umklapp processes weakening bond strengths in the solid vitrinite as the temperature rises).

If we assume all samples of vitrinite start with the same number density, N_0, then we can determine the parameter, $N_0\,g(\omega)$, in terms of vitrinite measurements made on recently deposited samples (i.e., close to $z = 0$). We do this by first determining the refractive index, n_*, of a vitrinite sample on the surface from its value, R_*, of vitrinite reflectance from

$$n_* = n_o \frac{(1 \mp R_*^{1/2})}{(1 \pm R_*^{1/2})} \quad \text{(A5)}$$

The quantity $N_0\,g(\omega)$ is given by

$$N_0\,g(\omega) = \frac{(n_*^2 - 1)}{(n_*^2 + 2)} \quad \text{(A6)}$$

where n_* is given by Equation A5. The quantity, N, in Equation A2 is then related to N_0 by

$$N(t) = N_0\,f(t,t_i) \quad \text{(A7)}$$

where f is the fractional number density. f satisfies the differential equation

$$\frac{df}{dt} = -k_0\,f \exp[T(t,z)/T_0 - E_M/\mathbf{R}T(t,z)] \quad \text{(A8)}$$

The solution for the fractional number density as a function of time is given by

$$f(t,t_i)\ ½ \exp\Big[-k_0 \int_{-t_i}^{t} dt^1 \exp[T(t^1,z(t^1))/T_0 - E_M/\mathbf{R}T(t^1,z(t^1))]\Big] \quad \text{(A9)}$$

With the time-temperature index, TTI, defined as

$$\text{TTI}(t,t_i) = \int_{-t_i}^{t} dt^1 \exp[T(t^1,z(t^1))/T_0 - E_M/\mathbf{R}T(t^1,z(t^1))] \quad \text{(A10)}$$

then the fractional number density as a function of time is given by

$$f(t,t_i) = \exp[-k_0\,\text{TTI}(t,t_i)] \quad \text{(A11)}$$

The fractional number density given in Equation A11 is connected to the measured vitrinite reflectance as a function of depth through Equations A7, A2, and A1. The small vitrinite reflectance values observed ($R < 5\%$) allows us to approximate

$$k_0\,\text{TTI}(0,t_i) = 12n_o^2[(2 + n_o^2)(n_o^2 - 1)]^{-1} [R(Z_i)^{1/2} - R_*^{1/2}] \quad \text{(A12)}$$

In practice, we cannot measure R_* in Equation A12 since the vitrinite reflectance cannot be measured on the surface, as assumed in the derivation of Equation A12. However, we can measure a value of vitrinite reflectance R_1 at a relatively shallow depth Z_1. This reflectance R_1 is related to R_* by

$$k_o\,\text{TTI}(0,t_1) = 12n_o^2[(2 + n_o^2)(n_o^2 - 1)]^{-1} [R_1(Z_1)^{1/2} - R_*^{1/2}] \quad \text{(A13)}$$

R_* is then easily expressed in terms of the smallest measured vitrinite reflectance R_1. Thus,

$$k_o\,[\mathrm{TTI}(0,t_i) - \mathrm{TTI}(0,t_1)] =$$

$$12n_o^2\,(2 + n_o^2)^{-1}\,(n_o^2 - 1)^{-1}\,[R(Z_i)^{1/2} - R_1^{1/2}] \tag{A14}$$

If interest is taken to center primarily on the shape of the TTI curve with respect to time and not its magnitude, we can obtain that structure without having to know the thermochemical reaction constant, k_0, as follows. Take the largest measured $R(Z_i)$ value, say R_{max}, occurring at depth $Z = Z_{max}$. Define a normalized TTI index, VITTI, by

1. dividing the left-hand side of Equation A14 by the corresponding value with TTI_{max} substituted for TTI $(0,t_i)$, and
2. dividing the right-hand side by the corresponding terms in R_{max}.

In this case $0 < \mathrm{VITTI} < 1$ in the range $-t_{max} < t_i < 0$. The use of this renormalized TTI corresponds to a scale change based on measured vitrinite values and not based on the thermochemical rate constant, k_0, Thus, one of the parameters defining the maturation index is normalized out of the problem:

$$\mathrm{VITTI}(t_i) = [\mathrm{TTI}(0,t_i) - \mathrm{TTI}(0,t_1)]$$

$$[\mathrm{TTI}(0,t_{max}) - \mathrm{TTI}(0,t_1)]^{-1} =$$

$$[R(Z_i)^{1/2} - R_1^{\,1/2}]\;[R_{max}(Z_{max})^{1/2} - R_1^{\,1/2}]^{-1} \tag{A15}$$

As detailed in Lerche et al. (1984) in order to resolve certain technical difficulties a critical temperature, T_c, had to be introduced, and the final form of the time-temperature integral became

$$\mathrm{TTI}(0,t_i) = I^{[0]}\,(i) =$$

$$\int_{-t_i}^{0} dt\,\exp(T - T_c)/T_0),\; T > T_c \tag{A16}$$

with the integrand zero in $T < T_c$.

The role of the critical temperature is simple: below $T = T_c$, vitrinite does not mature. Also as detailed in Lerche et al. (1984) the value of $T_A \equiv E_M/\mathbf{R}$ is so small as to be ignored. D.W. Waples (personal communication, 1983) has noted that the low value of T_A, suggestive of van der Waals-type interactions, itself suggests a critical temperature effect since, viewed in the light of transition-state theory, T_c would represent a minimum in the change in entropy between reactant and activated complex. He also notes that, "The 'threshold' for vitrinite maturation might even involve the formation of true vitrinite (starting at $R_o = 0.35$ or 0.4%), and might thus be dependent on maturity rather than temperature as T_c indicates."

The introduction of a fixed critical temperature made it possible to fit *all* sets of vitrinite data in the geohistory data bases we have built for wells from Australia, Indonesia, the North Sea, and Alaska.

The final algorithm implemented by Lerche et al. (1984) is essentially an implementation of two equations. Paleoheat flux varies linearly with time

$$Q\,(t) = Q_0(1 + \beta\,t) \tag{A17}$$

and the theoretical maturation integral possesses a critical temperature:

$$I^{[0]}\,(i) = \int_{-t_i}^{0} \exp((T - T_c)/T_0)\,dt \text{ in } T > T_c \tag{A18}$$

with the integrand set to zero in $T < T_c$.

Recalibration of lab data (see Toth et al., 1981) with the final integral form led us to *fix* T_0 at 200°K; the best value of $T_c = 295° \pm 20°$K was determined by exhaustive trial-and-error searching on over 300 wells; grid searching in a $T_0 - \beta$ space with T_c *fixed* at 295°K enforced the reasonableness of this value for T_0. With T_0 and T_c *fixed*, it proved possible to do an exhaustive trial-and-error search for the best β for each well. It seems worth commenting here that with the burial-history parameters determined, and with T_0 and T_c fixed for all wells, we have only the parameter β to be determined for each suite of vitrinite data for each well.

With the provision of high-quality vitrinite reflectance data, and with the novel inversion schemes now available based on tomographic methods, we can relax the constraint that the heat flux varies linearly with time. Instead we can now take the heat flux to vary piecewise linearly with time over selected intervals of time so that β varies with the time interval.

Thus if the time intervals are labeled $t_1, t_2, \ldots$ then we can write

$$Q(t) = Q_0\,[1 + \beta_n(t - t_n)]\prod_{i=1}^{n=1}[1 + \beta_i(t_i - t_{i-1})] \tag{A19}$$

in $t_{n+1} > t > t_n$ where the $\beta_1, \beta_2, \ldots \beta_n \ldots$ and $t_1, t_2 \ldots t_n$ are to be determined using the inversion technique outlined in the paper we have in preparation based on thermal indicator tomography.

The form of $Q(t)$ chosen above is linear in each time interval and the functional form allows $Q(t)$ to be continuous across each time line at $t = t_{n-1}, t_{n-2}, \ldots t_1$.

Appendix 13.B. Input Data

Ikpikpuk-1

TABLE 13.B1. Depth, age, paleobathymetry, and lithology of selected samples from Ikpikpuk-1, National Petroleum Reserve of Alaska

Depth* (ft)	Age (Ma)	Paleobathymetry (ft)	Lithology†
100	1.8	65	1
100	92	65	2
550	94	10	2
3,485	105	65	1
5,180	110	200	1
7,237	115	300	1
7,460	131	450	1
7,460	135	200	1
8,100	140	550	1
8,290	155	550	1
9,100	160	550	2
9,250	170	550	2
9,730	188	65	1
9,898	200	65	1
10,370	245	200	2
11,098	248	10	1
11,290	250	10	1
11,450	278	10	1
11,450	285	120	3
12,480	325	120	3
12,930	330	120	3
13,450	332	120	3
13,760	336	100	3
14,850	342	100	1
15,310	350	100	1

Well data: Kelly-Bushing height = 20 ft; ground elevation = 32 ft; sediment surface temperature = 30°F; BHT = 288°F (at 14,160 ft).

*Unconformity noted by repeating the same depth twice.

† 1 = shale; 2 = sandstone; 3 = carbonate.

TABLE 13.B2. Vitrinite reflectance data for Ikpikpuk-1

Depth (ft)	R_o	Depth (ft)	R_o
475	0.39	8,465	0.71
655	0.37	9,065	0.74
895	0.41	9,365	0.86
1,075	0.40	9,485	0.78
2,155	0.38	9,545	0.87
2,755	0.48	9,825	0.73
2,930	0.52	10,283	1.16
2,950	0.51	10,288	1.52
3,325	0.50	10,297	1.34
3,784	0.54	10,445	1.01
3,800	0.57	11,075	1.29
3,805	0.61	11,112	2.12
3,935	0.52	11,118	1.85
4,535	0.55	11,405	1.98
4,895	0.57	11,723	2.25
5,120	0.54	11,775	1.34
5,315	0.59	12,075	2.06
5,695	0.60	12,375	2.08
5,735	0.54	12,645	2.09
6,155	0.61	12,745	1.98
6,335	0.56	12,975	2.13
6,935	0.60	13,275	2.12
7,235	0.61	13,545	2.14
7,369	0.70	13,845	2.28
7,373	0.78	14,495	2.45
7,565	0.67	14,795	2.56
7,865	0.62	15,095	2.69
7,985	0.77	15,395	2.55
8,165	0.69		

Inigok-1

TABLE 13.B3. Depth, age, paleobathymetry, and lithology of selected samples from Inigok-1, National Petroleum Reserve of Alaska

Depth* (ft)	Age (Ma)	Paleobathymetry (ft)	Lithology†
110	1.8	65	2
110	69	65	2
560	87	65	2
1,390	88	15	1
1,490	90	15	1
2,240	92	10	1
2,390	94	65	2
3,520	105	65	1
5,180	110	200	1
8,310	113	200	1
9,040	115	200	2
9,040	131	450	1
9,270	140	200	1
9,810	143	200	1
11,006	155	550	1
11,670	160	450	1

Table 13.B3. *Continued.*

Depth* (ft)	Age (Ma)	Paleobathymetry (ft)	Lithology†
11,795	170	200	1
12,045	190	200	1
12,170	200	200	1
12,400	245	200	1
12,625	250	10	1
13,890	278	120	3
13,890	285	120	3
15,215	325	120	3
15,740	330	100	3
16,220	333	100	3
16,490	335	100	3
16,890	340	100	3
18,110	350	10	3

Well data: Kelly-Bushing height = 43 ft; ground elevation = 120 ft; sediment surface temperature = 30°F; BHT = 343°F (at 17,300 ft).

*Unconformity noted by repeating the same depth twice.

† 1 = shale; 2 = sandstone; 3 = carbonate.

Table 13.B4. Vitrinite reflectance data for Inigok-1

Depth (ft)	R_o	Depth (ft)	R_o
410	0.42	9,814	1.06
725	0.42	10,045	0.80
905	0.42	10,295	0.76
1,025	0.41	10,306	1.37
1,625	0.43	10,645	0.94
1,985	0.44	10,998	1.03
2,045	0.40	11,008	1.02
2,165	0.41	11,245	0.93
2,225	0.39	11,704	1.05
2,632	0.45	11,714	1.49
2,642	0.47	11,776	1.64
2,652	0.46	11,893	2.18
2,661	0.45	12,235	2.05
2,835	0.40	12,355	2.17
3,072	0.49	12,415	1.75
3,080	0.46	12,515	2.56
3,410	0.42	12,720	2.51
4,015	0.51	13,015	1.99
4,206	0.47	13,483	2.52
4,211	0.51	13,493	2.53
4,216	0.50	13,585	2.39
4,615	0.55	13,848	2.64
5,000	0.52	13,873	2.77
5,010	0.58	13,880	2.72
5,215	0.58	13,885	2.40
5,815	0.60	14,033	2.84
6,415	0.60	14,058	2.73
7,015	0.65	14,125	2.73
7,054	0.63	14,185	2.40

Table 13.B4. *Continued.*

Depth (ft)	R_o	Depth (ft)	R_o
7,064	0.70	14,485	2.37
7,615	0.61	14,785	2.38
8,210	0.68	15,085	2.58
8,215	0.62	15,385	2.48
8,220	0.72	15,685	2.16
8,230	0.69	16,285	2.51
8,240	0.77	16,555	3.71
8,815	0.78	16,735	3.37
8,842	0.53	16,855	3.21
8,852	0.52	17,065	3.82
9,343	1.04	17,155	4.01
9,445	0.70	17,455	3.46
9,448	0.66	17,545	4.30
9,453	1.11	17,725	3.28
9,458	0.73	18,025	4.36
9,564	0.77	18,295	4.14

Awuna-1

Table 13.B5. Depth, age, paleobathymetry, and lithology of selected samples from Awuna-1, National Petroleum Reserve of Alaska

Depth* (ft)	Age (Ma)	Paleobathymetry (ft)	Lithology†
50	1.6	0	2
50	100	20	1
4,000	110	300	1
7,000	113	1,200	1
11,200	117	2,900	1
15,000	119	3,500	1

Well data: Kelly-Bushing height = 26 ft; ground elevation = 1,103 ft; sediment surface temperature = 40°F; BHT = 222°F (at 11,140 ft).

*Unconformity noted by repeating the same depth twice.

† 1 = shale; 2 = sandstone.

Table 13.B6. Vitrinite reflectance data for Awuna-1

Depth (ft)	R_o	Depth (ft)	R_o
640	0.66	5,070	1.13
1,240	0.70	5,340	1.34
1,920	0.64	5,940	1.39
2,457	0.82	6,220	1.47
2,700	0.87	6,850	1.56
3,300	0.90	7,490	1.62
3,870	0.93	8,110	1.86
4,470	1.15		

Morocco Well "X"

TABLE 13.B7. Depth, age, paleobathymetry, and lithology of selected samples from Well "X", Morocco

Depth* (ft)	Age (Ma)	Paleobathymetry (ft)	Lithology†
0	6.6	200	1
1,739	11.3	200	1
6,755	13.5	500	1
6,760	100	200	1
6,760	175	500	2
7,179	18.8	25	2
8,524	194.6	150	3
10,453	200	150	3
10,817	213	15	4
13,940	246	15	4

Well data: Kelly-Bushing height = 20 ft; ground elevation = 421 ft; sediment surface temperature = 86°F; BHT = 282°F (at 13,940 ft).

*Unconformity noted by repeating the same depth twice.

†1 = shale; 2 = sandstone; 3 = carbonate; 4 = salt.

TABLE 13.B8. Vitrinite reflectance data for Well "X", Morocco

Depth (ft)	R_o (mean)	R_o (predicted)	Standard deviation
7,536	0.88	0.68	0.16
8,360	0.72	0.78	0.07
8,878	0.94	0.90	0.15
9,698	0.92	1.03	0.10
10,876	1.05	1.20	0.16
11,021	1.27	1.24	0.19

14

A Finite Element Model of the Subsidence and Thermal Evolution of Extensional Basins: Application to the Labrador Continental Margin

Dale R. Issler and Christopher Beaumont

Abstract

A one-dimensional finite element model has been developed to study postrift thermal and subsidence histories of basins formed by lithospheric stretching. The model includes depth-dependent extension, radiogenic heat production, and variations in the sediment thermal properties. Inputs to the model are a rifting age, a sediment budget, and the stretching parameters β and δ, for the crust and subcrustal lithosphere, derived from backstripping analysis.

Thermal histories of sediments are computed assuming conductive heat transport and model-derived temperatures are used to make predictions of organic maturity using chemical reaction kinetic theory. In addition to vitrinite reflectance, a promising technique of measuring organic maturity involves aromatization-isomerization (A-I) reactions associated with biological marker compounds common to most organic-rich sediments. These are unimolecular, first-order reactions that precede the main phase of oil generation and therefore can be used to locate the top of the petroleum generation zone. Model results can be compared with organic maturity measurements (A-I products, vitrinite reflectance), crustal thickness estimates from seismic refraction experiments, corrected bottom-hole temperatures and heat-flow measurements, and paleobathymetry or stratigraphy at the modeled location.

Preliminary results from a study of seven wells from the Cretaceous age Labrador continental margin, northeastern Canada, illustrate use of the model. The results show good agreement between theory and observations when measured thermal properties of the crust and sediments are incorporated into the model. The combined effects of increased sediment accumulation and decreased heat flow with time means that organic maturation is most sensitive to present thermal conditions where sediments are at their maximum temperature. Synrift and early postrift sediments are too thin to record the effects of the early temperature history when predicted heat flow was much higher. Therefore, although the margin has experienced a complex rifting history, a simple model in which the extension was effectively instantaneous can account for subsidence, present-day temperatures, and organic maturity at the modeled well locations. Although the resolution of the observations is poor and interpretations conflicting, the model generally predicts paleobathymetries greater than those inferred from biostratigraphic data, suggesting that refinements to the model are necessary to improve on subsidence predictions.

Introduction

In recent years, significant progress has been achieved in quantitative modeling of sedimentary basin evolution. Advances in understanding the mechanisms of basin formation have permitted the construction of models based on known physical principles. Two processes believed to produce sedimentary basins are flexure of the lithosphere in response to surface loading by overthrusts (Beaumont, 1981; T.E. Jordan, 1981) and thinning of the lithosphere by stretching (McKenzie, 1978). This chapter is concerned with the second process and concentrates on the subsidence and thermal history of the sediments.

Continental margins are a type of extensional basin receiving great attention because of their potential for hydrocarbon resources. In McKenzie's (1978) model, the crust and subcrustal lithosphere are thinned by stretching during the rift phase of continental breakup (Fig. 14.1A). Stretching produces isostatic uplift or subsidence of the surface. Following rifting, the attenuated

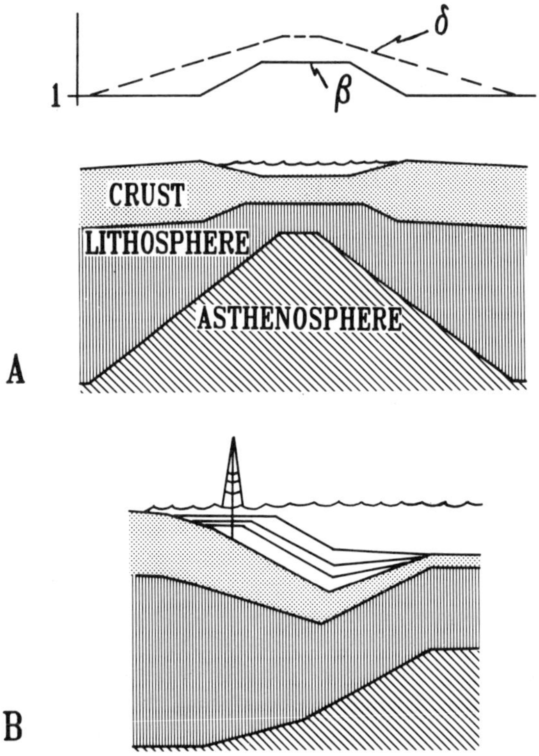

FIGURE 14.1. The development of a passive continental margin according to a modified McKenzie (1978) stretching model. A, During continental rifting, the crust and subcrustal lithosphere undergo depth-dependent extensional thinning and the asthenosphere passively upwells to replace stretched lithosphere. Isostatic adjustments produce elevation changes within the rift zone. The stretching factors, β and δ, represent the ratio of the initial thickness to final stretched thickness for the crust and subcrustal lithosphere, respectively. The variation of β and δ across this region shows that the subcrustal lithosphere has been thinned more than the crust. B, Following rifting, the attenuated region subsides as a result of conductive cooling and thermal contraction of the lithosphere. This thermal subsidence is modified by surface loading of water and sediments that fill the space created by lithospheric contraction.

region subsides over a period of time due to thermal contraction of the lithosphere as it cools. The overall tectonic subsidence is amplified by sediment and water loading (Fig. 14.1B). McKenzie's model assumes that rifting is an instantaneous event and that extension by an amount β is uniform with depth in the lithosphere, but that β varies laterally. Figure 14.1A illustrates the concept of depth-dependent extension where the hotter, ductile, lower lithosphere extends by an amount greater than the cooler, brittle part of the crust. The stretching parameters, δ and β, represent the ratio of initial thickness to final stretched thickness of the lower and upper lithosphere, respectively. In this example, decoupling of β-δ stretching occurs at the crust-mantle boundary. (Note that we follow the notation of Beaumont et al. [1982] with the δ-β convention reversed with respect to that of Royden and Keen [1980] and Sclater et al. [1980]).

Extension models have been used to study subsidence at point locations within a basin in which local isostasy is assumed (Royden and Keen, 1980; Royden et al., 1980; Sclater and Christie, 1980; Sclater et al., 1980; Foucher et al., 1982; Wood, 1982) and to represent cross sections through a basin that allows lithospheric flexure in response to sediment and water loading to be included (Steckler, 1981; Beaumont et al., 1982; Sawyer et al., 1983). These models successfully predict the first-order subsidence characteristics of extensional basins. The associated temperature predictions can be used to compute organic maturation and, therefore, the timing and location of petroleum generation in sedimentary basins (e.g., Turcotte and McAdoo, 1979; Royden et al., 1980). With few exceptions (Mackenzie and McKenzie, 1983; Mackenzie et al., 1985), most model predictions have not been adequately tested. In this chapter, we show that a model based on simple assumptions can account for subsidence, organic maturity, and present-day temperatures within a basin when the model is properly calibrated with measured thermal properties of the sediment and crust.

This chapter is divided into three parts. The first outlines modeling procedures, discusses limitations of the model, and shows how model predictions can be tested. The second illustrates how well data from the Labrador continental margin can be modeled. The third contains a discussion of the Labrador results and the uncertainties associated with the modeling.

Modeling Procedure

A summary of the essential elements of the modeling process is given in Figure 14.2. There are three major steps: 1) acquisition of the raw data necessary for choosing input parameter values, 2) model construction and solution of the equations using the finite element method, and 3) comparison of model-derived temperature histories with observations, and the prediction of temperature-related geological and geochemical properties for comparison with observations.

Raw Input Data

Basic raw data include biostratigraphic, lithological and porosity information, seismic reflection and refraction data, and thermal property measurements on samples of sediment and crust. The first three items can be obtained from well-history reports, micropaleontology and palynology reports, sonic logs, and composite lithology logs.

Geological Data

Biostratigraphic information is needed for both age dating of sediment units and paleobathymetry interpretations. A qualitative comparison can be made between computed water depths of deposition and these paleobathymetry estimates. Sediment ages and thicknesses are used to determine how the sediment budget changes with time during subsidence at a particular location. We use the time scale of Harland et al. (1982) to obtain numerical values for paleontological ages.

Lithology and porosity information is necessary in order that sediment compaction can be calculated during forward modeling and that decompaction can be estimated during backstripping (Sclater and Christie, 1980; Falvey and Middleton, 1981). These data are also needed for calculating sediment thermal transport properties. Porosity reduction in clays and shales is primarily the result of mechanical grain compaction with burial depth, geological age and clay (shale) content being the dominant controlling factors (Hedberg, 1936; Dzevanshir et al., 1986). With quartzose sandstones, however, decreases in porosity below a depth of about 1.5 km are largely governed by chemical processes (pressure solution, cementation) which strongly depend on temperature as well as other factors (Maxwell, 1964; Angevine and Turcotte, 1983). Consequently, we might expect to observe more variability in sandstone porosity with depth because changes in pore water chemistry have important effects on porosity development.

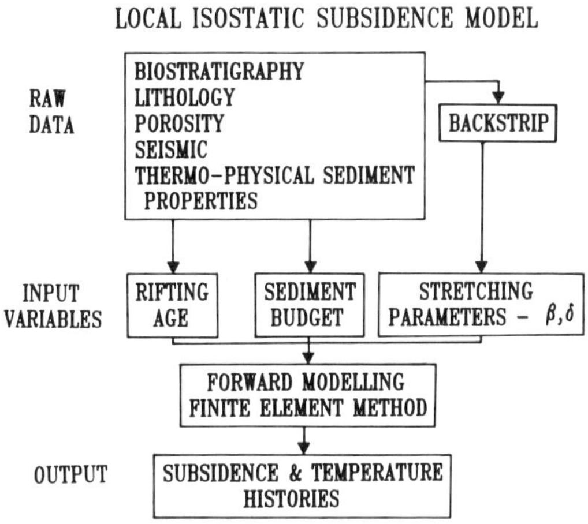

FIGURE 14.2. A summary of the steps involved in the modeling procedure (see text for more details).

Spot measurements of porosity on conventional and sidewall cores of sandstone from Labrador Shelf wells are highly variable and show a poor correlation with depth and sonic transit time data from logs. In contrast, there is an excellent correlation ($R = -0.94$) between sonic transit time and depth for these sandstones. We lack sufficient shale porosity measurements to define a porosity-depth relation but there is also a good correlation between shale sonic transit time and depth ($R = -0.87$). The equation relating sonic transit time with depth for sandstone is

$$\Delta t = 538\exp(-2.649\times10^{-4}\,z)$$

and for shale is

$$\Delta t = 515\exp(-1.832\times10^{-4}\,z)$$

where z is the depth from sea floor in meters and Δt is sonic transit time in μs/m. Sonic log measurements are sensitive to changes in porosity but because of the scale over which measurements are made, small-scale fluctuations in porosity should be averaged out.

Our approach was to use sonic log transit time measurements on sandstones and shales and convert these to porosity values (Raymer et al., 1980) in order to obtain porosity-depth functions for these lithologies. The Labrador Shelf sediments were best described using an exponential relation of the form,

$$\Phi = \Phi_0 \exp(-cz)$$

for shale porosity and a linear equation of the form,

$$\Phi = \Phi_0 - cz$$

for sandstone porosity where Φ_0 is the surface porosity, z is the depth measured from sea floor, and c is the compaction constant (see Appendix 14.A for parameter values). These results are consistent with observations in other areas (Maxwell, 1964; Selley, 1978; Magara, 1980). The equations are used to predict changes in thickness and porosity as a sediment unit undergoes compaction during burial. The main lithologies on the Labrador Shelf are sandstone and shale, with shale predominating. Lithological variations can adequately be expressed as a variable mixture between these two end members. Post-Paleozoic carbonate sediments are negligible at this margin.

There are difficulties with converting sonic transit time values into porosity for shales because little has been published on the subject. Previous studies (Perrier and Quiblier, 1974; Magara, 1976) have correlated sonic transit time with shale bulk density or directly with porosity measured on mudstone cores (Magara, 1968) and porosities predicted from these relations agree to within about 5% porosity. Porosity values computed from bulk density measurements depend on the assumed values for the matrix and pore fluid and uncertainties in these parameters could alter estimates by 5 to 10% porosity. Sandstone porosities computed from sonic transit time data fall within the range of measured porosities and are in excellent agreement with porosity determinations on Paleocene sandstones of the North Sea (Selley, 1978). Shale porosity is not well determined and our computed values lie outside the range of most literature values (Rieke and Chilingarian, 1974). It is possible our assumed value of matrix velocity is too high and therefore gives an upper bound to shale porosity. There is some evidence for undercompaction in some of the deeper shale-rich sections. For example, at North Leif I-05 (Fig. 14.7), hole cleaning problems caused by splintery shales, formation pressure tests, and sonic log data indicate moderate overpressures below 2,000 m.

Seismic reflection profiles provide vital information for interpreting subsidence histories in the neighborhood of well locations. Most of the Labrador Shelf wells were drilled near the crests of rotated basement fault blocks or horsts and thus sample a condensed section, often lacking sediment deposited during the earliest phase of subsidence. Seismic reflection profiles passing near well locations allow us to trace the stratigraphy off structure into less disturbed, more representative regions and repeat the modeling. They also enable us to model along strike sections and cross sections to understand how stretching varies within the basin.

Seismic refraction data enable us to estimate prestretched crustal thicknesses in adjacent unextended regions and to test model predictions of crustal stretching using the assumed initial thickness.

Thermal Property Data

Computed temperature histories are highly sensitive to thermal properties assigned to the sediment and crust. Therefore, measurements of thermal conductivity and radiogenic heat production were made on samples from sediment and Precambrian basement rocks of the Labrador Shelf. Heat production was measured for 28 pulverized drill cuttings samples using a gamma ray spectrometer with a solid state Ge (Li) detector (Lewis, 1974; Keen and Lewis, 1982). A standard divided-bar apparatus was used to measure the thermal conductivity of 46 water-saturated chip samples (Sass et al., 1971; Sass et al., 1984). The samples were described in detail with the aid of a binocular microscope, diagrams for visual estimation of the percentage of various constituents, and well-history reports. Most of the samples contained greater than 80% quartz or more than 90% shale. Error estimates of the composition are difficult to quantify but probably increase for samples in the middle range. We believe they are, however, accurate to within approximately 10%.

Within the samples, shales, mudstones, and greywackes are underrepresented (due to decomposition on washing) and sand grains are overrepresented (due to decomposition of sandy shales and greywackes; Petro-Canada, personal communication, 1984). Also, caving of borehole material, disaggregation of friable sandstones into loose quartz grains, and facies change caused variations in measured thermal properties. These effects are best characterized by correlating the measured

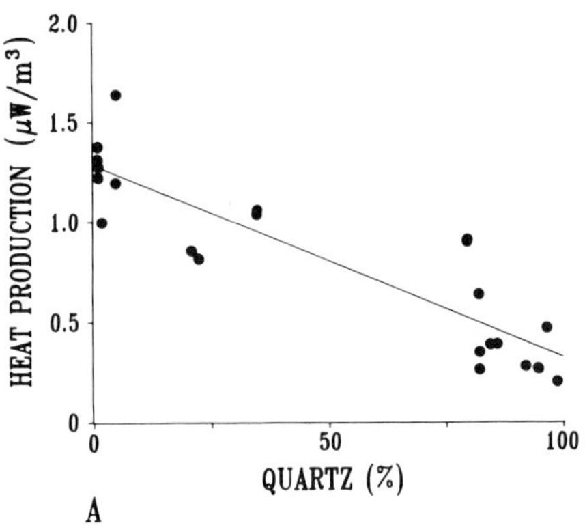

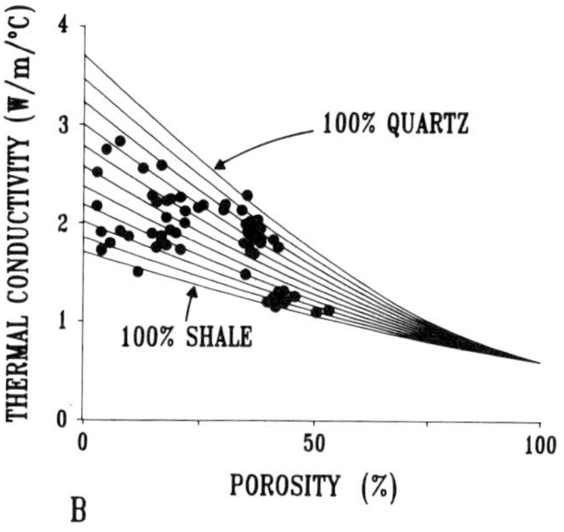

FIGURE 14.3. Thermal property measurements for borehole samples from the Labrador Shelf. A, Correlation between radiogenic heat production and quartz content for sediments consisting of quartz and shale. A linear regression gives the relation,

$$HP = -0.96(FQ) + 1.29$$

where HP is heat production and FQ is the fractional percent quartz; correlation coefficient, −0.89. B, Thermal conductivity versus porosity for various combinations of quartz, shale, and water. Each curve represents conductivity as a function of water content for 10% changes in quartz (shale) content over the range 0% quartz (100% shale) to 100% quartz (0% shale). The dots represent measured values of thermal conductivity for Labrador Shelf core samples (<30% porosity; J. Wright, personal communication, 1984) and chip samples (>30% porosity).

properties with the percentage of sand and shale in each of the samples.

There is a significant correlation between heat production and quartz content in Figure 14.3A. Shales, therefore, have a higher rate of heat production than sandstones. The relationship determines matrix heat production values for shale and sandstone as well as for mixtures of the two. A quartz content of 75% was used when assigning thermal properties to Labrador Shelf sandstones (see Appendix 14.A).

Figure 14.3B is a plot of thermal conductivity against porosity for different combinations of quartz, shale, and water. The curved lines range from 0% quartz (100% shale) to 100% quartz in increments of 10% and were computed using the self-consistent approximation of Budiansky (1970) which was adapted for a three-phase composite material (Mackenzie and McKenzie, 1983). The dots represent measured values obtained from 38 cuttings samples and 30 disk samples taken from cores (Petro-Canada, personal communication, 1984). Unfortunately, detailed lithological descriptions were not available for the cores but these data are included to show that they fall within the range of predictions. Matrix conductivity values were computed for shale and sandstone based on the zero porosity intercepts of curves that bound the observations and the inference that Labrador sandstones are 75% quartz (Appendix 14.A). Using the matrix values, we can approximate in situ thermal properties for various combinations of sediment and water.

Changes in the matrix values of specific heats and thermal conductivities in response to increasing temperature and pressure during burial were estimated by fitting polynomials to tabulated data. The specific heat and thermal conductivity of water are computed as a function of temperature and pressure and adjusted to a 3.5% saline solution (Keenan et al., 1978; Von Driska, 1982; Weast et al., 1983). These properties are relatively insensitive to realistic changes in salinity. Similarly, the conductivity and heat capacity of sandstone, shale, and basalt are corrected for temperature (Birch and Clark, 1940; Kappelmeyer and Haenel, 1974; Roy et al., 1981; Weast et al., 1983). Variations in pressure are not important for these rocks. Its effect is to increase thermal conductivity due to closing of microcracks but this is minimized in measuring water-saturated samples (Walsh and Decker, 1966; Hurtig and Brugger, 1970).

Radiogenic heat production from crustal rocks was computed using an average of four measure-

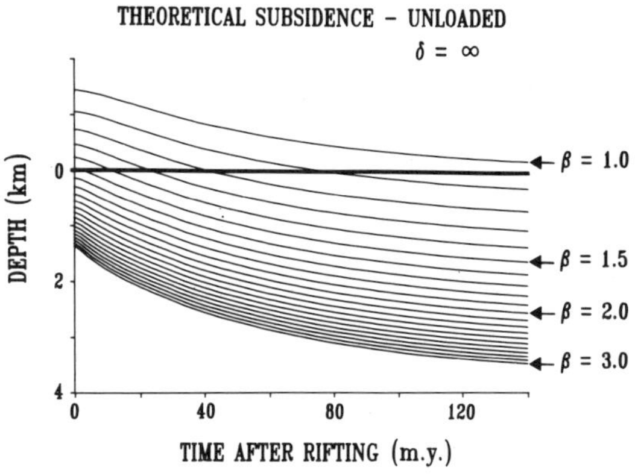

FIGURE 14.4. Theoretical basement subsidence curves for various amounts of crustal stretching (β = 1.0–3.0) and constant subcrustal lithosphere attenuation ($\delta = \infty$) generated using the depth-dependent stretching model of Royden and Keen (1980). Extension, which is assumed to be instantaneous, produces initial uplift or subsidence depending on the ratio of β to δ and is followed by exponentially declining thermal subsidence. These curves are for basement subsidence unmodified by sediment and water loads (see text for more details).

ments from basement samples and assuming this was distributed over a crustal layer of uniform heat production. Due to the absence of heat-flow and heat-production measurements at this margin (Roy et al., 1968; Hyndman et al., 1979; Wright et al., 1980), a heat-producing layer of 7.5 km (unstretched thickness) was chosen and is the same value used for the Scotian Shelf to the south (Beaumont et al., 1982). The remaining properties of the crust, subcrustal lithosphere, and asthenosphere are similar to those commonly quoted in the literature (Parsons and Sclater, 1977; Beaumont et al., 1982; see Appendix 14.A).

Model Input Variables

The remaining input variables, once the material property values have been determined, are the rifting age, the sediment budget, and the stretching parameters, β and δ, for the upper and lower lithospheric layers, respectively. The first two inputs can be deduced from available geological data. β and δ can then be estimated by the "backstripping" technique (Steckler and Watts, 1978; Sclater and Christie, 1980; Wood, 1982). This involves stepping backwards in time, sequentially removing successive sediment layers in each time interval and restoring basement to its unloaded elevation assuming Airy or local isostasy. The result is a tectonic subsidence curve, from which the effects of sedimentation and water loading have been removed.

Figure 14.4 shows the forward theoretical prediction of basement subsidence for different values of β and $\delta = \infty$ using the depth-dependent extension model of Royden and Keen (1980). Rifting is assumed to be instantaneous at 0 my and the crust and total lithosphere are, respectively, 35 and 125 km thick. Decoupling of the β and δ stretching is at the crust-mantle boundary. Stretching results in initial uplift ($\beta < \approx 1.5$) or subsidence ($\beta > \approx 1.5$) and depends on a balance between crustal thinning and replacement of cold lithosphere by its hotter equivalent or asthenosphere. The postextension exponentially decaying subsidence results from thermal contraction of the cooling lithosphere.

Basement tectonic subsidence curves can be compared with theoretical curves to obtain estimates of β and δ. These are only approximate values because this analysis does not include flexure of the lithosphere, due to partial lateral support of surface loads, which decreases subsidence in the deepest parts of basins and increases subsidence near the hinge zones (Beaumont et al., 1982; Karner and Watts, 1982). Other factors not included are the contribution of radiogenic heat sources, thermal blanketing by sediments (Lucazeau and Le Douaran, 1985), and, to a much lesser degree, enhanced cooling and subsidence that accompanies self-contraction of the lithosphere. With the exception of flexure, these factors are considered in the forward modeling. Their overall effect is to reduce the total amount of subsidence and, therefore, it is necessary to use values of β that are slightly greater than those estimated from "backstripping."

Forward Modeling

Our model is based on a variation of the McKenzie (1978) stretching model and allows for depth-dependent extension, radiogenic heat production in the crust and sediments, and changes in the thermal characteristics of the sediments with temperature and compaction. The approach is similar to that of Beaumont et al. (1982) except that the finite element method (Strang and Fix, 1973) is used to solve the one-dimensional, heat diffusion equation, permitting a more detailed treatment of the sediment thermal properties. Beaumont et al. (1982) used the finite difference method for thermal calculations and averaged thermal properties over the entire sediment column in each model timestep. Also, their model is a two-dimensional, thermomechanical model used to study how profiles across a continental margin evolve with time. The model presented here is one-dimensional and used to study subsidence at well sites or point locations within a basin. The model and its limitations are discussed below.

The stretching parameters β and δ allow for different amounts of extension in the upper and lower lithosphere, thereby approximating the change in extension controlled by the rheology with depth. The dependence of the rheology of rocks on temperature means that, under an extensional stress regime, upper cooler regions of the lithosphere are rigid and deform by brittle failure, whereas, hotter, deeper regions will deform by ductile flow. Supporting evidence comes from laboratory experiments on materials thought to be representative of the lithosphere (Brace and Kohlstedt, 1980), observations of an aseismic lower crust in intraplate regions (Chen and Molnar, 1983), listric faulting in the upper crust (Montadert et al., 1979), and rheological models examining the effect of ductile deformation on normal faulting in an extending crust (Smith and Bruhn, 1984). Decoupling between β-δ stretching may occur at midcrustal levels along a shear zone where overlying faults of the brittle region are thought to sole out (e.g., Rehrig and Reynolds, 1980; Gans et al., 1985) or within the inferred brittle region of the upper mantle (Sawyer, 1985). For simplicity and lack of data, we choose the crust-mantle boundary as the decoupling surface. This should be a good approximation because thinning of the crust and heat input to the base of the subcrustal layer are the dominant controls on subsidence.

Space problems occur when the total amounts of extension of the upper and lower layers across the rift zone are not equal. Greater extension of the lower layer, $\delta > \beta$, can be accommodated by magmatic intrusion into the brittle region (Royden and Keen, 1980) or by proposing that the stretched lower lithosphere originally occupied a narrower region than the upper lithosphere (Beaumont et al., 1982). Depth-dependent extension for a pure shear model has been criticized on the basis of strain incompatibility problems (Kligfield et al., 1984). This difficulty is avoided by increasing the initial width of the stretched region with depth (Rowley and Sahagian, 1986) or by varying the geometry to accommodate both pure shear and simple shear extension (Dokka and Pilger, 1983).

More importantly, the model is limited by problems in estimating δ because the lithosphere cools and thickens with time since rifting. However, observations of rift flank uplifts (Illies and Greiner, 1978; Zorin, 1981; Morgan, 1983; Steckler, 1985) and evidence for broad regions of attenuated lithosphere corresponding to $\delta > \beta$ (Fairhead and Reeves, 1977; Brown and Girdler, 1980; Panza et al., 1980; Gough, 1986) are consistent with the model.

The Model

Consider a column of crust and subcrustal lithosphere of thickness a, initially in thermal equilibrium, with its base defined by the solidus temperature, T_m (Fig. 14.5 and Appendix 14.A for parameter values). There is an upper crustal region of thickness b, uniformly enriched in radiogenic heat-producing elements, which causes nonlinearity in the equilibrium temperature configuration (solid curve, Fig. 14.5). To simulate rifting, the crust and subcrustal lithosphere are thinned kinematically by stretching. Isothermal mantle material at temperature T_m is assumed to passively upwell and fill the void created by stretching.

Changes in basement elevation that accompany depth-dependent extension depend on how various combinations of β and δ affect the density structure of the stretched column. Local isostatic adjustment of a column of depth a predicts either uplift or subsidence (see Beaumont et al., 1982, for pertinent equations). It is assumed that stretching is sufficiently rapid to conserve heat.

At some time t after stretching, the column has conductively cooled and contracted. The base of the lithosphere has migrated to a new depth and the basement has subsided, creating space for sediments. Additional isostatic subsidence occurs in response to sediment loading and eventually, with

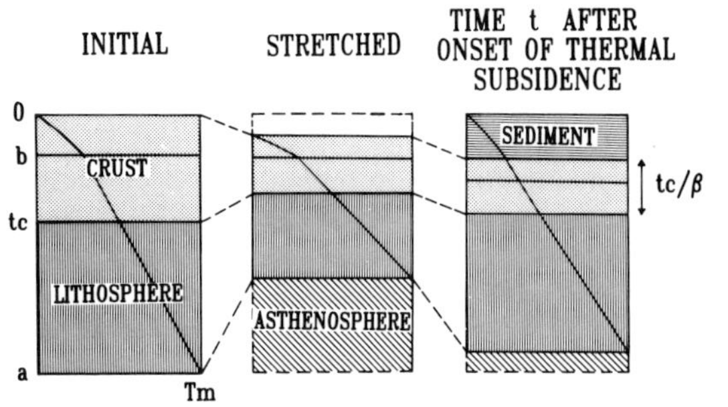

FIGURE 14.5. Diagrams illustrating the basic concepts underlying the model. Initially, we have a column of lithosphere with crustal thickness, t_c, and of total thickness, a, which is in thermal equilibrium with its base defined by the solidus temperature, T_m. An upper crustal layer of thickness, b, enriched in radiogenic heat-producing elements, produces curvature in the equilibrium temperature gradient (solid curve). The crust and subcrustal lithosphere are stretched and thinned by β and δ, respectively, and the asthenosphere fills the space created by stretching. Extension produces a change in surface elevation which, in this example, is initial subsidence. The new stretched crustal thickness is given by the ratio, t_c/β. At some time after stretching, the column cools and thermally contracts and the base of the lithosphere migrates to a new depth. Sediments with variable thermal properties are added to the column and conductively heat up as the column cools and contracts (see text for more details).

sufficient cooling, subsidence approaches some final equilibrium value. It is assumed that the base of the lithosphere is maintained at depth a and that lithosphere depressed below this depth is converted to asthenosphere.

Computational Steps

Progressive subsidence is predicted as a function of β and δ, the rifting age, and sediment budget by timestepping the finite element cooling and contraction model. Typical Labrador Shelf models use 10 model timesteps with each time interval corresponding to the deposition of an individual formation. Our approach is to interface two separate Fortran programs. The IMSL finite element program, TWODEPEP (Sewell, 1981), solves the one-dimensional heat diffusion equation with heat production using initial and boundary conditions supplied for each timestep. The difference between the initial and final temperature distributions for each timestep, $\Delta T(z)$, is used to calculate the thermal contraction assuming

$$\Delta\rho(z) = \rho_0 \alpha \Delta T(z)$$

where $\Delta\rho$ is the associated density change, ρ_0 is the reference density at 0°C, and α is the volume coefficient of thermal expansion. Sediment is then added, the isostatic adjustment calculated, and the problem is reformulated for the next finite element timestep.

The finite element grid is selected using a temperature-dependent triangulation scheme. The new layer of sediment is assumed to be at 0°C at the start of each timestep. Thus, we maximize the effects of sediment blanketing by allowing the new sediment layer to conductively warm as the model evolves through to the next timestep. Material is added or subtracted from the base of the column to maintain a constant thickness equal to a. Sediment thermal properties are readjusted for changes in porosity on compaction, lithology, and temperature and are held constant for each model timestep. This is justified because these properties change by only a small amount during a timestep. The temperature distribution at the end of one timestep becomes the initial temperature distribution for the next timestep once it has been corrected for adjustments to the finite element grid. Any sediment that remains above sea level at the end of a timestep is iteratively eroded to sea level and the timestep is repeated for the thinned layer. Temperature and subsidence predictions from the numerical model computations were compared with analytical solutions for simple cases (McKenzie, 1978) and agree to within 1%.

Assumptions and Limitations

Several assumptions are made to simplify the analysis, a major one being that the problem is considered to be one-dimensional. This means we neglect flexure of the lithosphere in response to sediment and water loads (Beaumont et al., 1982; Karner and Watts, 1982), lithospheric deflection as a result of thermoelastic bending moments (Bills, 1983; Nakiboglu and Lambeck, 1985), viscous flow in the lower lithosphere (Keen, 1985), and lateral heat transfer.

The mechanical strength of the lithosphere, the Earth's thermal boundary layer, is primarily controlled by its thickness and internal temperature distribution (Courtney and Beaumont, 1983). During rifting, the lithosphere is weakened by lithospheric thinning and fracture of the brittle upper crust. Following rifting, the lithosphere cools and thickens, and mechanical strength is restored. Our assumption of local isostatic compensation of surface loads is reasonable for the early stages of basin formation but yields increasing error in basement subsidence as the lithosphere thermally ages. However, sedimentation wanes with time and calculations indicate that these errors are less than 10% for our Labrador Shelf study by comparison with models in which flexure is included (Beaumont et al., 1982).

Lateral variations of heat flow into the base of the lithosphere can induce thermal bending moments that flex the lithosphere. However, for sufficiently broad thermal anomalies, as in the case of continental margins, surface deformations due to thermal stresses are likely small in comparison to isostatic adjustments to density changes. Nakiboglu and Lambeck (1985) conclude that surface deflections due to thermal stresses cannot be distinguished from those of local isostasy if the width of the heat source exceeds three times the effective elastic thickness of the lithosphere.

We neglect viscous flow in the lower lithosphere and assume that hotter mantle material passively upwells to replace thinned lithosphere and then cools by thermal conduction. Keen (1985) demonstrates that viscous flow can cause dynamic elevation changes in the rift zone and lead to depth-dependent thinning of the lithosphere by secondary convection in the upwelling asthenosphere. The magnitude of these effects is uncertain because they depend on the absolute value of viscosity which is poorly determined.

We assume heat transport is by vertical conduction and model rifting as an instantaneous event. Jarvis and McKenzie (1980) have shown that for rifting times less than about 20 my, instantaneous extension is a good approximation but their model considers only vertical heat conduction. Recent work (Cochran, 1983; Alvarez et al., 1984) suggests that lateral heat conduction is important during rifting over finite time intervals. For uniform extension models, the effect is to increase subsidence and decrease heat flow in the center of the basin and increase heat flow and produce uplift on the flanks. Thus, within the basin, synrift subsidence increases at the expense of postrift subsidence. These effects diminish with increased basin width and decreased length of rifting, but probably remain significant for most extensional basins. Depth-dependent extension is likely to increase these horizontal temperature gradients and further promote lateral heat flow.

We can model finite rates of extension as the superposition of a series of instantaneous stretching events but details of the rift stage, particularly its duration, are poorly constrained for the Labrador margin. It is sufficiently old, almost two lithospheric thermal time constants (time for thermal anomaly to decay by 1/e), that little evidence remains, for example, of the early subsidence and heat flux, to test for the effects of lateral heat flow and finite rates of extension. Furthermore, Alvarez et al. (1984) conclude that one-dimensional finite rifting calculations are not a substantial improvement over the instantaneous approximation if lateral conductive heat losses are significant. Therefore, the question we address is, "Are the available geological constraints sensitive to the early details of rifting at this margin, or can we account for the observations using a simple model?"

Advective heat transport by fluids during compaction should be negligible if dewatering is continuous. If instead it is episodic, thermal pulses should largely be confined to permeable zones and be of relatively short duration (Cathles and Smith, 1983). Subsurface water movement, driven by hydraulic gradients, can significantly modify the temperature distribution within basins (Garven and Freeze, 1984a, 1984b; Hitchon, 1984; Beaumont et al., 1985). This is observed in foreland basins, which are associated with high topography and have elevated water tables in recharge areas that drive regional ground-water flow. For intracratonic basins, numerical modeling of compaction-driven fluid flow suggests that heat transfer is conduction-dominated with no disturbance to the geothermal gradient because the fluid velocities are low (Bethke, 1985). Little is known about the hydrodynamics of passive continental margins. Observations from the Nova Scotian margin (Issler, 1984) and recent geochemical and modeling results (Mackenzie et al., 1985) are consistent with the assumption of conductive heat transport.

As a further simplification and for lack of reliable data, we do not include arbitrary eustatic sea-level corrections and we assume there is no preexisting basement topography prior to rifting. Thus, basement is rifted at sea level and sea level is held constant as the model evolves. Additional complications, such as erosion (Hellinger and Sclater, 1983) or crustal thickening due to underplating by basaltic melt (Beaumont et al., 1982), are ignored.

The simplistic kinematic approach to lithospheric thinning requires that deformation be distributed in a continuous manner. This is clearly not correct for the upper crust, which deforms by brittle failure, thereby making it difficult to model the early subsidence history in regions of rough basement topography. Most of the Labrador wells, for example, are drilled on the crests of rotated basement fault blocks and horsts. Therefore, most of the geological data available to test model predictions are associated with anomalous basement structures.

In summary, we use a kinematic, one-dimensional passive rifting model that assumes instantaneous extension, local isostatic compensation of surface loads, conductive heat transport, constant sea level, no preexisting basement topography, and continuous lithospheric deformation. As a result of the above assumptions, the model is limited to studying the postrift thermal subsidence of extensional basins. The model should work best for a basin with large dimensions, thick flat-lying sediments, a short rifting history, relatively thin synrift sediments, and low-lying topography. We demonstrate below, using results form the Labrador margin, that good results can be obtained even when some of these criteria are not satisfied. That good results are given by models of the Labrador margin reflects the fact that both the model and the observations are most sensitive to the present thermal regime.

Model Output, Predictions, and Tests

Computed temperature and subsidence histories can be compared with geological, geophysical, and geochemical observations to assess the validity of the model and to place constraints on model parameter values. Subsidence can be assessed by comparison with observed stratigraphy, interpreted paleobathymetry, and crustal thickness measurements. Calculated temperatures can be checked against present-day thermal measurements and products of temperature-dependent geochemical reactions.

The model should accommodate the observed sediment thicknesses and simultaneously give reasonable paleobathymetries at deposition. Sediment that does not fit into the basin at the time of deposition is removed by erosion, thereby modifying the model stratigraphy by comparison with that observed. Only qualitative comparisons can be made between computed water depths and those interpreted because paleontology cannot give precise water-depth estimates once water depths exceed several tens of meters. Faunal and floral distributions are controlled primarily by factors that are indirectly related to water depth, such as salinity, temperature, light levels, bottom substrate, and energy conditions.

We assume a prerift crustal thickness, t_c, of 35 km (Royden and Keen, 1980), which is within the range of measurements under the Precambrian Shield to the southwest of Labrador (Berry and Fuchs, 1973). Crustal thickness estimates from seismic refraction experiments (van der Linden, 1975b) can therefore be compared with postrift thicknesses, t_c/β, determined from modeling.

The model predicts the present-day temperature distribution and heat flow at points within a sedimentary basin. These predictions can be tested against heat-flow measurements and bottom-hole temperatures, corrected for the effects of drilling fluid circulation (Dowdle and Cobb, 1975; Lee, 1982; Luheshi, 1983).

Past temperature conditions are more difficult to test. The integrated temperature effects can be estimated by examining products of thermally activated organic chemical reactions. We use two approaches. The first considers unimolecular, first-order reactions involving biological marker compounds that are common constituents of most organic-rich sediments (Mackenzie and McKenzie, 1983). Second, we use the empirical method of Lopatin (1971) to compute changes in vitrinite reflectance using model-derived temperature histories.

Mackenzie and McKenzie (1983) estimated kinetic parameters for simple reactions involving the isomerization and aromatization of specific hydrocarbon molecules using data from seven sedimentary basins formed by stretching events of age 180 Ma to 15 Ma. These results were further corroborated by recent work on the Nova Scotian margin (Mackenzie et al., 1985). The reaction rate coefficient for each reaction (K) is believed to vary with temperature according to Arrhenius' Law

$$K = A\exp(-E/RT)$$

where A is the frequency factor, E the activation energy, R the ideal gas constant, and T the absolute temperature. A and E are both assumed to be constants specific to each reaction.

Figure 14.6 illustrates two of the three reactions discussed by Mackenzie and McKenzie (1983), namely, the isomerization of a sterane and the aromatizaton of steroid hydrocarbons. Steranes originate from the early diagenetic breakdown of sterols which occur in the cell membranes of algae and higher order plants (Rhead et al., 1971; Dastillung and Albrecht, 1977). In Figure 14.6, solid lines refer to carbon–carbon bonds, with carbon atoms located at line intersections. A solid (open) dot corresponds to a carbon atom above (below) the plane of the paper, having a carbon–hydrogen bond projecting out of (into) the plane of the paper. A solid triangle (dashed line) represents a methyl group above (below) the plane of the paper. Aromatic rings are indicated by an inscribed circle. The stereochemistry (three-dimensional structure) of carbon atoms not part of a ring system (e.g., C-20 in Fig. 14.6) is defined as R or S (see Gunstone, 1974).

The isomerization reaction is reversible with the forward reaction governed by the rate constant K_2 and the backward reaction controlled by K_1 (Fig. 14.6). Natural product sterols all have the 20R configuration and this characteristic is retained by the steranes of immature sediments. The 20S isomer, which has the methyl group at the C-20 site pointing out of the page rather than inward, is of similar stability to the 20R isomer. At temperatures higher than $\approx$65°C, hydrogen exchange occurs at the C-20 site until the equilibrium of 20S/(20R + 20S) = 0.54 (Van Graas et al., 1982) is steadily approached between 65°C and 130°C.

The aromatization reaction, by contrast, is thought to be irreversible. Monoaromatic steroid hydrocarbons, produced during early diagenesis,

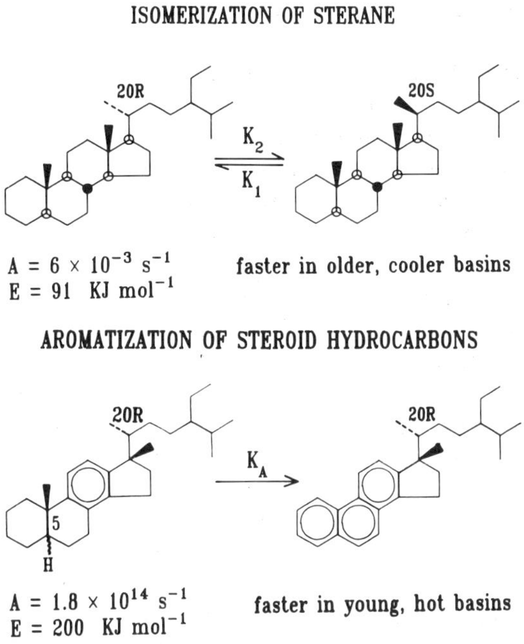

FIGURE 14.6. Diagram summarizing the reactions involving the isomerization and aromatization of steroid hydrocarbons. These reactions are assumed to obey Arrhenius' law for first-order reactions with the reaction rates primarily controlled by temperature, the frequency factor, A, and the activation energy, E. The upper part of the diagram shows the molecules of two isomeric steroid hydrocarbons that occur in sediments and are used to determine the extent of isomerization. The lines represent carbon–carbon bonds with carbon atoms located at line intersections. All carbon atoms have four bonds and where not indicated, bonds to hydrogen atoms are implied. Cyclohexane rings (hexagons) are nonplanar. A solid (open) dot represents a carbon atom above (below) the plane of the paper with a carbon–hydrogen bond projecting out of (into) the plane of the paper. A solid triangle (dashed line) corresponds to a $-CH_3$ group above (below) the plane of the paper. The isomerization reaction involves the removal and reattachment of a hydrogen atom at the site marked 20. In the lower part of the diagram, the molecule on the left represents two monoaromatic hydrocarbons that are isomeric at C-5 (shown by the sinuous C–H bond). The inscribed circle represents an aromatic ring and the numbers identify the carbon atoms. The aromatization reaction is assumed to be irreversible and with increasing temperature, hydrogen and a methyl group are lost to produce a triaromatic hydrocarbon (see Mackenzie and McKenzie (1983) for a more complete discussion of these reactions).

lose seven hydrogens and one methyl group to form triaromatic steroid hydrocarbons at temperatures above ≈90°C (Fig. 14.6).

Both of the above reactions are precursors to the main phase of oil generation and, therefore, when the reactions have gone to completion, sediments can be considered to be within the zone of significant petroleum generation. This approach potentially has several advantages over more commonly used maturation indicators such as vitrinite reflectance. These are specific reactions whose rates depend on a small number of kinetic parameters, A and E. Once these are known, the amount of reaction can be predicted for any temperature history. Vitrinite, by comparison, varies in composition and the reactions responsible for changes in its reflectivity are poorly understood. Vitrinite reflectance studies, therefore, have been limited to empirical correlations with time and temperature histories (Lopatin, 1971; Waples, 1980).

Another advantage is that aromatization-isomerization (A-I) reactions occur at different rates. The aromatization reaction will go to completion first in basins with high heat flow. Conversely, the isomerization reaction will finish first in basins having low heat flow. Cross-plots of these two reactions allow for the partial separation of time and temperature effects and the determination of heating rates. A cumulative thermal history indicator such as vitrinite cannot provide an independent estimate of heating rate when used alone.

The above technique is a recent development and is not in widespread use. We have predicted the progress of A-I reactions for the Labrador margin but the lack of sample measurements at this time forces us to use another approach to test our model temperature predictions. We use the Scotian Shelf TTI (Time-Temperature Index, Waples, 1980) calibration of Issler (1984) to compute percent vitrinite reflectance because these data are abundant for the Labrador Shelf wells.

The large scatter of data points associated with Waples' (1980) TTI-vitrinite correlation curve makes it difficult to obtain accurate estimates of percent vitrinite reflectance using his approach of a universal calibration. From a study of four sedimentary basins, Issler (1984) concluded that good local calibrations of TTI could be established for individual basins using present-day temperatures and burial history curves. If thermal conditions have remained stable over the time interval in which the vitrinite reflectance was achieved, then present-day thermal gradients can be used to approximate past temperature conditions. The Scotian Shelf, a ≈185 Ma, thermally stable continental margin that experienced continuous subsidence and has sediments presently at their maximum temperature and burial depth, is thought to satisfy this criterion. Therefore, we believe the Scotian Shelf TTI-vitrinite calibration may closely approximate the true relationship between time, temperature, and vitrinite reflectance. Some of the scatter in the universal calibration (Waples, 1980) may reflect inaccuracies in the assumed TTI values.

Labrador Shelf Examples

Use of the model is illustrated by applying it to well data from the Labrador continental margin off eastern Canada. Figure 14.7 shows the location of wells discussed in this chapter. Studies of subsidence (Keen, 1979; Royden and Keen, 1980), geology (McMillan, 1973; McWhae and Michel, 1975; Umpleby, 1979; Gradstein and Srivastava, 1980; McMillan, 1980; McWhae et al., 1980; McWhae, 1981; Srivastava et al., 1981; Hiscott, 1984), geochemistry (Bujak et al., 1977a, 1977b; Powell, 1979; Umpleby, 1979; McMillan, 1980; Rashid et al., 1980), and geophysics (van der Linden, 1975b; Srivastava, 1978; Hinz et al., 1979; Keen and Hyndman, 1979) of this region have been carried out.

The generally accepted view is that the Labrador margin developed as a result of rifting of Greenland away from North America during the Cretaceous. Grant (1980) has pointed to some apparent problems with applying plate tectonic concepts to this region, but part of this ambiguity can be attributed to uncertainties in the geological and geophysical data. Although complications do exist, we believe that most of the subsidence in the majority of Labrador Shelf wells can be explained by a simple model of a Cretaceous stretching event followed by thermal subsidence. Royden and Keen (1980) have shown, using data from four wells, that a depth-dependent extension model is necessary to explain subsidence at this margin. We go on to investigate the thermal consequences of this.

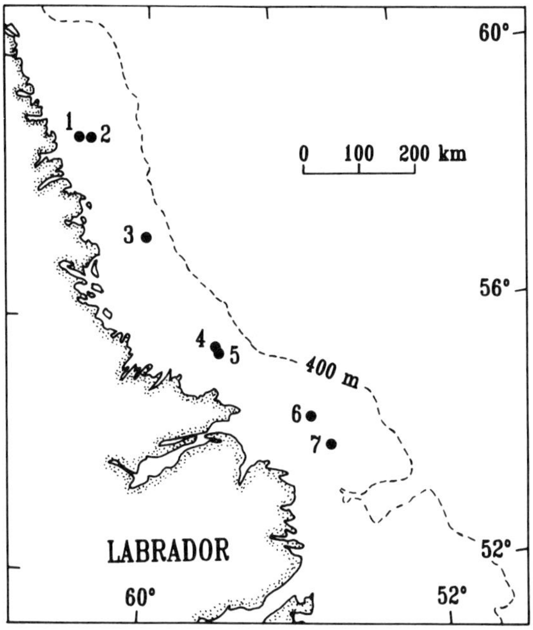

FIGURE 14.7. Location map for the Labrador Shelf wells discussed in this chapter. The dashed line is the 400-m isobath that marks most of the shelf region. The names of the wells are as follows: 1, Gilbert F-53; 2, Karlsefni A-13; 3, Snorri J-90; 4, Herjolf M-92; 5, Bjarni H-81; 6, North Leif I-05; 7, Freydis B-87.

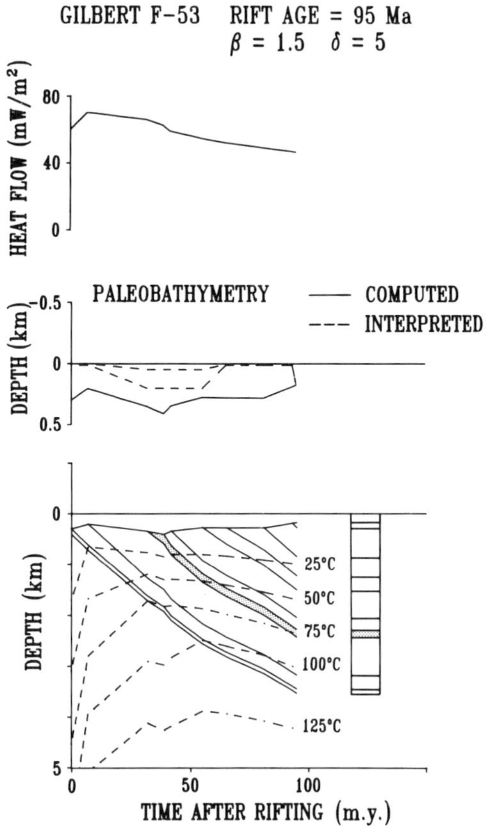

FIGURE 14.8. Heat-flow, paleobathymetry, and subsidence results for the Gilbert F-53 well. The upper diagram shows the variation of heat flow with time after rifting. The middle figure shows a comparison between computed and interpreted paleobathymetry. The lower diagram illustrates predicted subsidence, paleobathymetries, and temperatures for times after the end of rifting. The stippled layer represents the burial history of a Paleocene sediment unit and its present-day position in the stratigraphic column is shown to the right of the subsidence curves (see text for more details).

Results from the Gilbert F-53 Well

Model results for the Gilbert F-53 well (Fig. 14.7) are illustrated in Figures 14.8 to 14.10. The lower part of Figure 14.8 shows subsidence versus time after rifting, assuming instantaneous stretching at 95 Ma, with $\beta = 1.5$ and $\delta = 5$. The horizontal line at 0 km depth is the constant sea-level datum line and the dashed lines are isotherms spaced at 25°C intervals with the sea-bottom temperature held constant at 0°C. Solid curves represent the top of individual sediment units with the lowest one being top of basement, and the solid line joining these curves corresponds to sea bottom. Synrift sediments are present at the end of rifting (0 my).

We can trace the thermal evolution of a particular sediment unit as it undergoes burial and heating with time using the lower part of Figure 14.8. For example, Paleocene age sediment (stippled layer) is predicted to have been deposited between 32 and 39 my after rifting in about 400 m of water. With further subsidence and sedimentation, it conductively warms to a present-day temperature of about 75°C at approximately 2.3 km below sea level. The present-day stratigraphic column is shown to the right of the subsidence curves.

The middle part of Figure 14.8 shows a comparison between computed (solid line) and interpreted (dashed line bounds) paleobathymetry for different times after rifting. As stated above, only qualitative comparisons can be made because of our assumptions of constant sea level, local isostasy, and no preexisting basement topography. In addi-

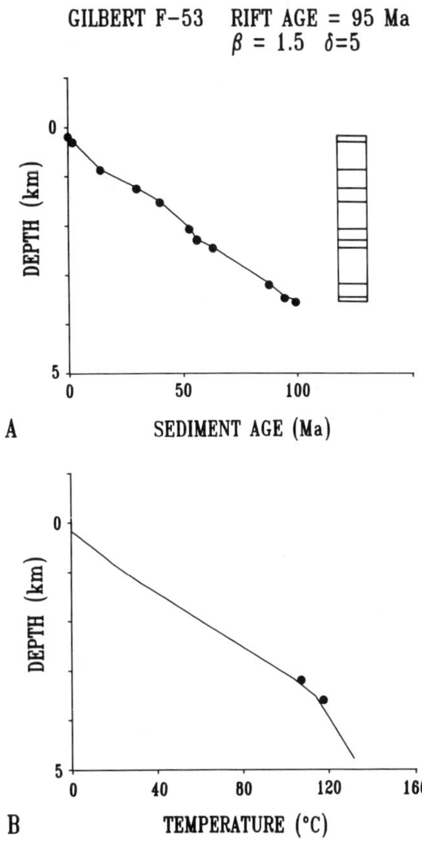

FIGURE 14.9. A, Comparison between present age-depth relationships (dots) and the stratigraphy accepted by the model. The present-day stratigraphic column is shown to the right of the curve. B, Comparison between corrected bottom-hole temperatures (dots) and the computed present-day temperature distribution.

tion, our choice of porosity-depth function will affect computed water depths. For example, if there is more shale compaction than we allow, computed water depths for the Cretaceous will decrease somewhat because the decompacted section will be correspondingly thicker. Also, depth assignments from paleontology can vary considerably depending on the diagnostic criteria used. Reports of different consultants can show a wide range of paleoenvironmental interpretations for the same well (Petro-Canada, personal communication, 1983). Many of the Labrador wells show a marked increase in water depth during the Plio-Pleistocene. Seismic reflection profiles suggest this may be partly the result of large-scale downward tilting of the margin in a landward direction (Hugh Balkwill, personal communication, 1983). We neglect this recent phase of subsidence because it is not thermal in origin and should not affect the computed temperature histories.

On stretching, there is a rapid rise in temperature and heat flow (top of Fig. 14.8), which is followed by slower thermal decay. The delay in the maximum surface heat flow following extension occurs because $\delta > \beta$ and the crust undergoes early conductive heating. With uniform stretching models, $\beta = \delta$, and the maximum surface heat flow is achieved on stretching. The exponential decay in heat flow with time implies that the margin is still cooling but that most of the original thermal anomaly has dissipated. The curve is not smooth because of the discrete nature of model timesteps and changes in the sediment properties with time. Present-day heat flow is predicted to be approximately 47 mW/m^2.

Figure 14.9A, a plot of present sediment depth versus age in the Gilbert F-53 well, confirms that the model and observed stratigraphies agree and that there was no sediment erosion. The dots represent observed sediment age versus depth from present-day sea level, and the solid curve represents model sediment age versus depth. The curve is slightly offset from the points because the computed water depth is less than the observed water depth. To the right of the curve is the present-day stratigraphic column.

Figure 14.9B is a plot of present-day temperature versus depth from sea level for the Gilbert F-53 well. The solid curve is the computed temperature distribution and the dots are corrected bottom-hole temperatures (Dowdle and Cobb, 1975; Fertl and Wichmann, 1977). The sharp bend in the temperature curve coincides with the thermal conductivity contrast at the sediment-basement interface. Bottom-hole temperatures are unreliable and must be corrected for the cooling effects of drilling fluid circulation. These corrections are generally large ($\approx 15°C$) and uncertain due to both theoretical constraints (Luheshi, 1983) and the quality of the available data. Therefore, bottom-hole temperatures by themselves are not necessarily a good check on computed temperatures.

Figure 14.10A shows the progress of A-I reactions for the Gilbert F-53 well with the shaded area (diagonal pattern) approximately marking the zone of reaction. The upper part of this zone tends to follow the 75°C isotherm and below this zone A-I reactions have gone to completion. The model predicts that, at present, A-I reactions are occurring in Paleocene (56 Ma) to Turonian age (90 Ma) sediments and that the reactions have gone to completion in the underlying thin layer of Cenomanian-Albian age sediments. If we trace the burial history of the Paleocene layer (stippled), we see that it has only recently entered the reaction zone. Model results indicate that reactions began about 40 my after rifting in the thin Albian layer overlying basement and continued for approximately 50 my. The aromatization reaction went to completion well before the isomerization reaction was finished ($\approx 75\%$ complete). Unfortunately, sample measurements are not available to test the predictions given in Figure 14.10A.

Figure 14.10B is a plot of measured (dots) and computed (solid curve) percent vitrinite reflectance (percent R_o) versus depth. Considering the uncertainties involved with both computing and measuring vitrinite reflectance (Héroux et al., 1979), the agreement is excellent. The predicted A-I reaction zone interval corresponds approximately to the range 0.40 to 0.70% R_o in this well.

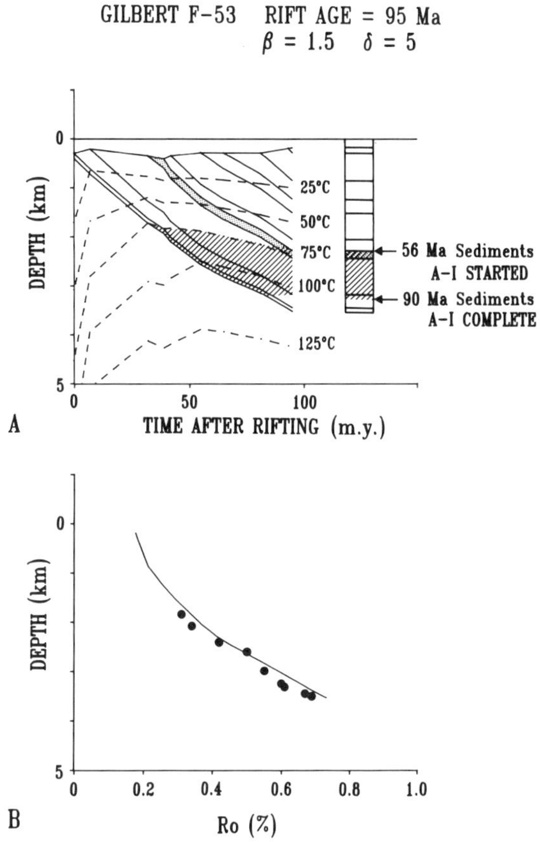

FIGURE 14.10. A, Same subsidence diagram as shown in Figure 14.8 but with the A-I reaction zone superimposed on it (see text for more details). B, Comparison between measured (dots) and computed percent vitrinite reflectance for the Gilbert F-53 well (see text for more details).

Results from Other Labrador Wells

Figures 14.11 to 14.16 show preliminary results from six other Labrador wells, listed in order from north to south along the margin. The wells were grouped according to rifting age, with rifting time set equal to 95 Ma (Cenomanian) in the north (Figs. 14.11 and 14.12), 105 Ma (Albian) for the central portion (Figs. 14.13 and 14.14), and 113 Ma (Aptian-Albian) for the southern part of the margin (Figs. 14.15 and 14.16). For each of these figures, Panel A shows subsidence and the position of the A-I reaction zone for times after rifting and Panel B shows computed (solid curve) and interpreted (dashed line) paleobathymetry. Panel C is a plot of both computed present-day temperatures (solid curve) and corrected bottom-hole temperatures (dots), and Panel D shows computed (solid curve) and measured (dots) percent vitrinite reflectance. Corrected temperatures based on poor data are labeled with a question mark in Panel C.

All models accepted the observed sediment budget without erosion. Rifting was assumed to be instantaneous and four wells (Figs. 14.12 to 14.15) had synrift sediments present at the time of rifting. Generally, computed paleobathymetries are greater than those inferred from paleontology. Uncertainties in paleobathymetry estimates are evident by the broad ranges given in Figures 14.11 to 14.13 and 14.15. Herjolf (Fig. 14.13) and Bjarni (Fig. 14.14) were drilled in close proximity to the same basement structure and exhibit similar subsidence yet have quite different interpreted paleobathymetries. For each of the wells (Figs. 14.11 to 14.16), computed water depths increase following rifting to some maximum value and then gradually shoal to present depths. Maximum computed water depths occur during late Paleocene-early Eocene

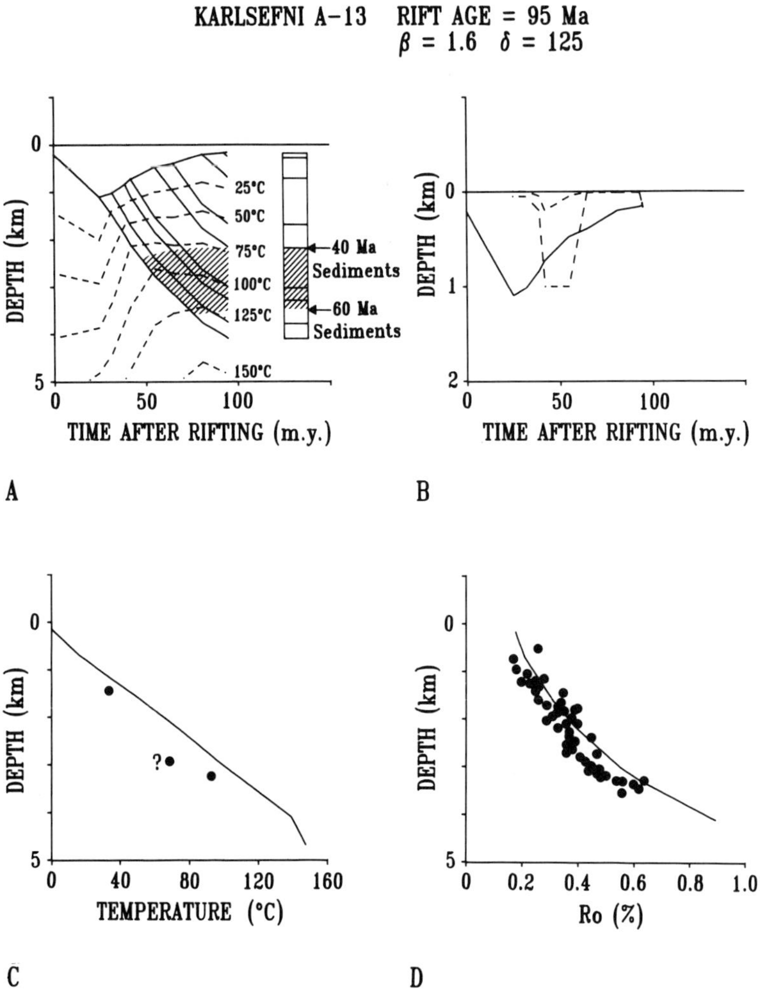

FIGURE 14.11. Diagram showing a comparison between model predictions and observations from the Karlsefni A-13 well. A, Shows predicted subsidence and temperature histories along with the development of the A-I reaction zone (see text for more details); B, compares computed (solid line) and interpreted (dashed line) paleobathymetry; C, compares corrected bottom-hole temperatures with computed temperatures; D, compares measured and computed vitrinite reflectance (see text for greater details).

except at Karlsefni (Fig. 14.11) where they reach a maximum during the Maastrichtian.

The poorest agreement between computed (>1,000 m) and interpreted (<200 m) paleobathymetry occurs with the Karlsefni A-13 model during the Cretaceous. This well was drilled near the crest of a pronounced basement high and received no sediment until the Maastrichtian when shales overstepped the high. For a rift age of 95 Ma and $\beta = 1.6$, the instantaneous stretching model predicts rapid early subsidence even for extreme values of subcrustal lithospheric attenuation. A lower value for δ would predict even greater initial subsidence. A much younger rift age for the instantaneous model, say 70 to 75 Ma, would fit the interpreted water depths better but appears incompatible with seismic-stratigraphic data. It is most likely, however, that rough basement topography that the model cannot reproduce is the reason for the apparent error in the predicted water depths. It is also possible that the paleobathymetry interpretations are incorrect because divergent water depth

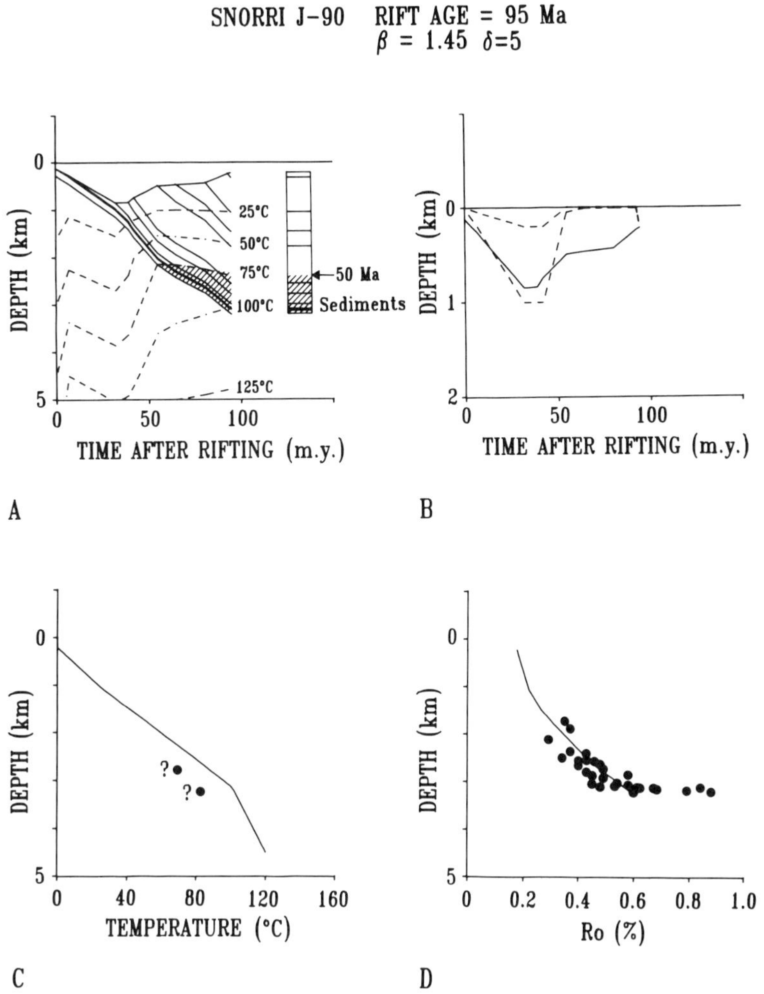

FIGURE 14.12. The same composite diagram as Figure 14.11 except that it is for the Snorri J-90 well (see text and Fig. 14.11 caption).

interpretations are given for some intervals in this well (Petro-Canada, personal communication, 1983).

Most of the models show reasonable agreement between both computed and measured vitrinite reflectance and temperatures. Computed present-day temperatures appear to be somewhat high for Karlsefni (Fig. 14.11) and Snorri (Fig. 14.12), but in the case of Snorri, bottom-hole temperature data are of very poor quality and probably significantly underestimate true formation temperatures. Computed percent vitrinite reflectance seems to be low for Herjolf (Fig. 14.13) and North Leif (Fig. 14.15). Vitrinite reflectance data for North Leif are at odds with other geochemical data that imply lower maturity levels and a mud additive and/or reworked material is suspected to have influenced these measurements (Petro-Canada, personal communication, 1983). Higher reflectance values occur in Lower Cretaceous sediments at the base of the Snorri well (Fig. 14.12) that have undergone extensive erosion prior to the onset of thermal subsidence.

Panel A in Figures 14.11 to 14.16 shows the computed progress of A-I reactions. Calculations suggest that these reactions have gone to completion in Maastrichtian-lower Paleocene sediments at the base of Karlsefni (Fig. 14.11) and are near

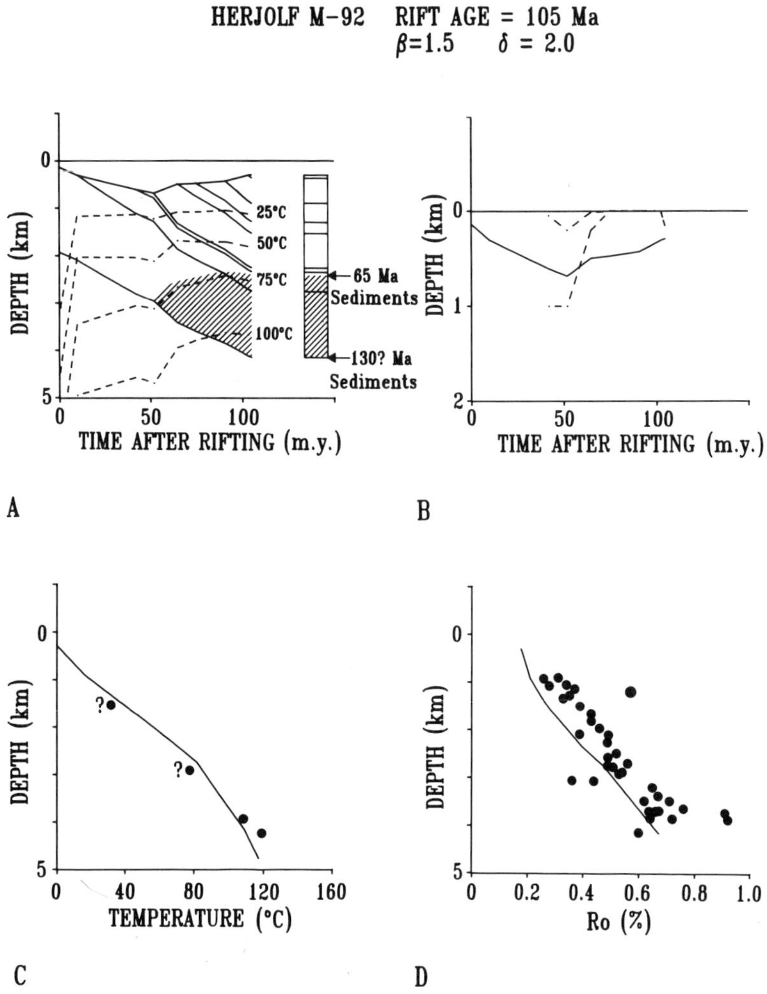

FIGURE 14.13. The same composite diagram as Figure 14.11 except that it is for the Herjolf M-92 well (see text and Fig. 14.11 caption).

completion in Lower Cretaceous sediments of Herjolf (Fig. 14.13) and North Leif (Fig. 14.15). Both the timing of A-I reactions and the thickness of the reaction zone vary with the amount of extension, sedimentation rate, and, to a lesser degree, rifting age. For example, at Freydis (Fig. 14.16) with $\beta = 1.25$, A-I reactions have not even started, whereas at Karlsefni (Fig. 14.11) with $\beta = 1.6$, they have gone to completion, Karlsefni was drilled on a basement high and only penetrated a thin veneer of Upper Cretaceous sediments.

A-I reactions appear to have started during early Miocene at Bjarni (Fig. 14.14), early Oligocene at Snorri (Fig. 14.12), middle Eocene at Karlsefni (Fig. 14.11), and early Eocene at both Herjolf (Fig. 14.13) and North Leif (Fig. 14.15). The stratigraphic column to the right of the subsidence curves in Panel A of each figure shows the location of the present-day computed A-I reaction zone. Lower Cretaceous ages are uncertain and are indicated with a question mark. Hydrocarbon generation from sediments as a function of temperature and time depends strongly on the type and amount of organic matter present. If the onset of maturity is taken to be 0.60% R_o, model results indicate that four of the seven wells in Figure 14.7 have penetrated the upper part of the petroleum generation zone.

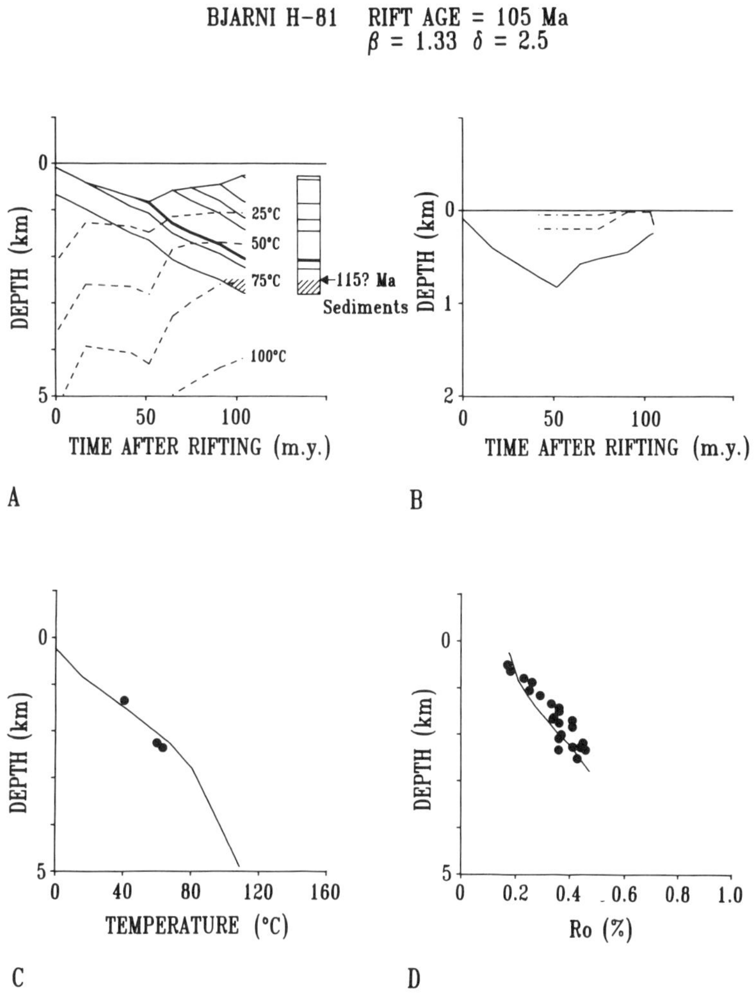

FIGURE 14.14. The same composite diagram as Figure 14.11 except that it is for the Bjarni H-81 well (see text and Fig. 14.11 caption). Total drilled stratigraphic section (2364 m) was extrapolated to a basement depth of 2550 m below sea floor.

Discussion

Although the results of Figures 14.8 to 14.16 are encouraging, several outstanding problems remain and the sensitivity of model results to parameter values remains to be tested. Controversy exists as to the timing and duration of rifting. Estimates for the onset of sea-floor spreading range from Campanian (Srivastava, 1978) to late Jurassic-Early Cretaceous (van der Linden, 1975a). Rifting had begun by at least the earliest Cretaceous as indicated by Neocomian age volcanics of the Alexis Formation overlying basement in several Labrador wells and the presence of small basic dykes at Ford's Bight, Labrador dated as 145 ± 6 Ma and 129 ± 6 Ma (McWhae et al., 1980). Evidence for the earliest possible onset of rifting is the occurrence of basic Jurassic (162 ± 5 Ma) dykes parallel to the coast of southern Greenland (Watt, 1969; McMillan, 1980).

Srivastava's (1978) interpreted magnetic anomaly pattern suggests that sea-floor spreading occurred in two stages, with spreading beginning in the southern part of the margin during the Campanian (75 Ma) and in the northern part during the Maastrichtian. A change in ridge axis orientation occurred during the Paleocene (60 Ma) and spreading continued until the late Eocene (40 Ma) when

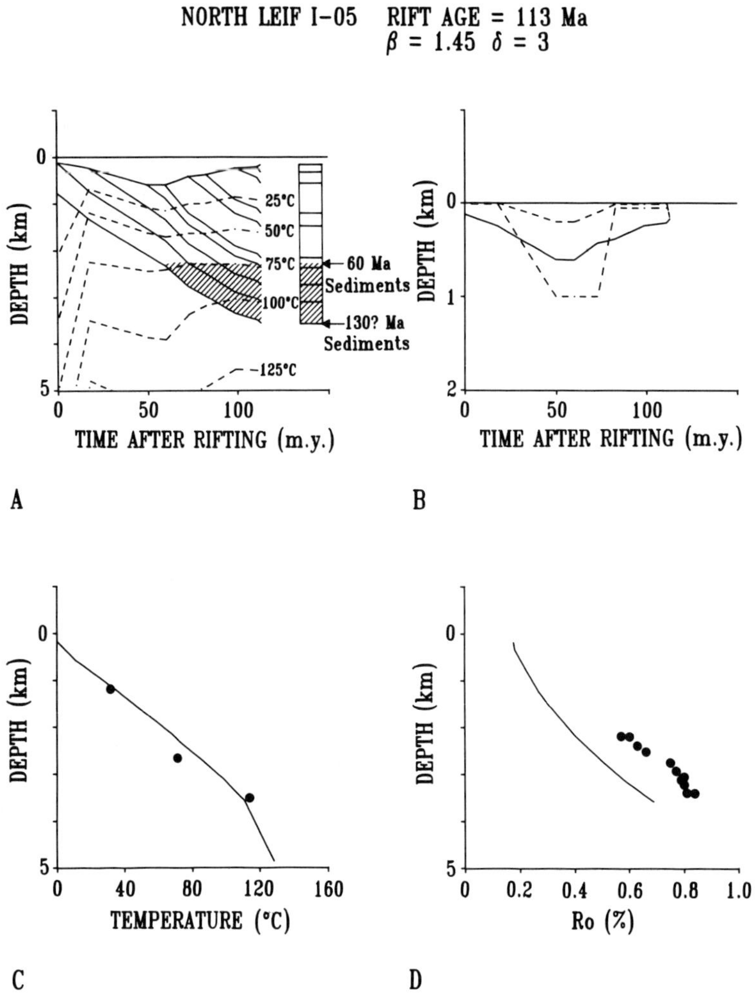

FIGURE 14.15. The same composite diagram as Figure 14.11 except that it is for the North Leif I-05 well (see text and Fig. 14.11 caption).

Greenland began to move with the North American plate. Problems with interpreting the sea-floor spreading history include the fact that magnetic anomalies are not well developed everywhere, the complex pattern of fracture zones makes correlation difficult, and the ocean-continent boundary cannot be accurately determined (Srivastava, 1978; Hinz et al., 1979). Schult and Gordon (1984) place the initiation of Labrador sea-floor spreading in the Turonian (90 Ma) using Srivastava's reconstructions and the interpretation of Cande and Kristofferson (1977) that the anomalies that Srivastava identifies as anomalies 31 and 32 are actually anomalies 33 and 34.

Royden and Keen (1980) model the onset of thermal subsidence as progressing from south to north, choosing the end of rifting as 80 Ma for the Bjarni and Herjolf wells (Fig. 14.7) and 70 Ma for Snorri and Karlsefni based on plate tectonic reconstructions (Srivastava, 1978; Gradstein and Srivastava, 1980). C.R. Tapscott and R.C. Vierbuchen (personal communication, 1983) also model thermal subsidence as commencing from south to north from about 90 Ma to 75 Ma. The onset of thermal subsidence cannot be determined strictly from well data because the wells were drilled on basement highs and did not penetrate old marine sediments.

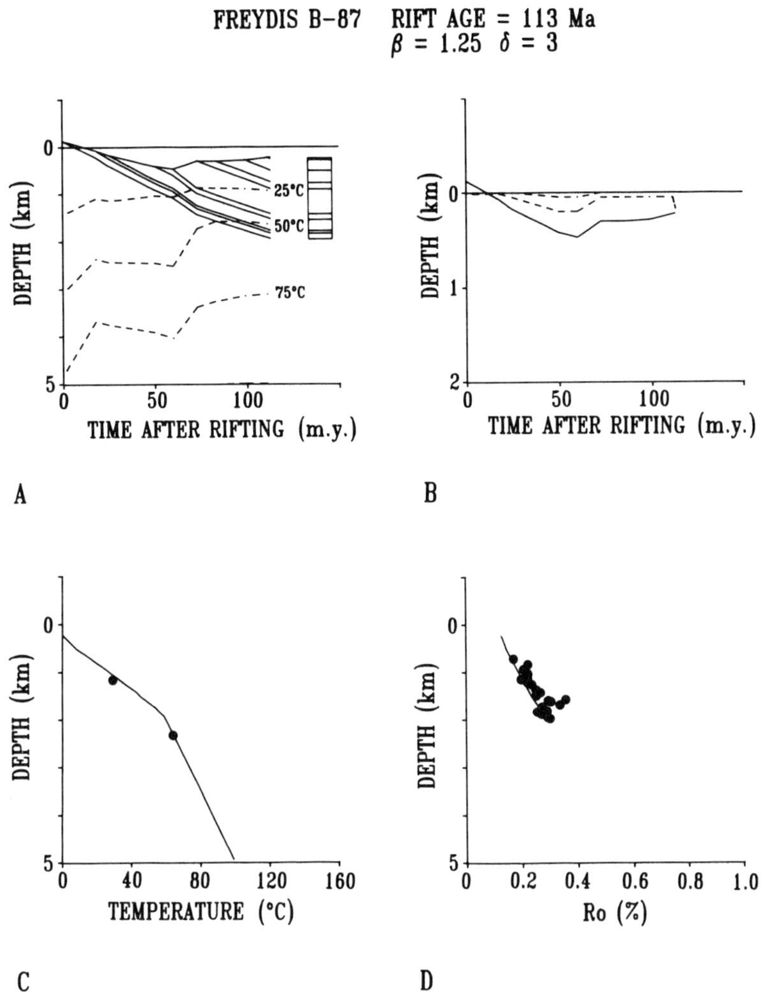

FIGURE 14.16. The same composite diagram as Figure 14.11 except that it is for the Freydis B-87 well (see text and Fig. 14.11 caption).

Seismic-stratigraphic studies suggest that the onset of thermal subsidence was nearly synchronous along the margin and of an earlier age. An Aptian-Albian (113 Ma) unconformity appears to mark the change from dominantly syntectonic deposition in half-grabens to overstepping of the grabens by fluvial, transitional, and marine facies and is interpreted as the breakup unconformity (Hugh Balkwill, personal communication, 1983).

On the basis of the available geological information, we favor an older age for the end of rifting, occurring within the range Aptian-Cenomanian, although better geological control is needed to resolve this problem. Preliminary calculations suggest that the margin is of sufficient age that variations of rift age within this range do not significantly affect the final computed results. Therefore, the model cannot be used to constrain the exact timing for the onset of thermal subsidence, although we can narrow down the most probable range. For our preliminary calculations, we have chosen to model rifting as commencing from south to north with effective instantaneous rifting ages of 113 Ma to 95 Ma. To have better constraints on the rifting history, we need more accurate age determinations on synrift and early postrift sediments and volcanic rocks, better age calibrations for magnetic anomalies in the Labrador Sea, and a clearer understanding of the ocean-continent transition.

From the evidence given above, it appears that rifting occurred over a time span of many tens of millions of years, perhaps Late Jurassic to Late Cretaceous. This puts into question our assumption of instantaneous extension. We can include stretching over a finite time interval but the results cannot be tested due to lack of biostratigraphic resolution and the absence of a complete synrift sequence in most of the Labrador wells.

Our results (Figs. 14.8 to 14.16) appear to be reasonable using the instantaneous stretching approximation, although refinement of model paleobathymetries may require adjustment of the rifting model or lithospheric parameters. The margin has a fading memory of the details of rifting and the thermal effects of rifting are not recorded or are obscured in the wells examined. This occurs because most of the Labrador wells penetrate a thin synrift sequence and much of the original thermal anomaly had decayed before substantial thicknesses of postrift sediments accumulated. The exponential dependence of reaction rate on temperature for vitrinite means that reflectivity is strongly influenced by maximum temperature conditions (e.g., Hood et al., 1975) that presently exist on the Labrador margin. These results are compatible with the conclusions of McKenzie (1981) that show that vitrinite is not a sensitive temperature history indicator for sufficiently old basins.

Another problem concerns paleontological age determinations. The scarcity of "indigenous species," reworking, borehole cavings, dilution of faunas in coarse clastic sediments, and the apparent incompatibility of palynological and micropaleontological criteria for age assignments renders biostratigraphic dating uncertain (Barss et al., 1979; Gradstein and Srivastava, 1980; Hugh Balkwill, personal communication, 1983). We use a chronostratigraphic chart that ties biostratigraphic ages to the stratigraphic geometry displayed in seismic sections and date sediments using formation tops determined from well logging. On the basis of well data, we see no evidence for a second phase of basement subsidence beginning in the Miocene as noted by others (Hinz et al., 1979; Umpleby, 1979; Grant, 1980). Although several wells show an increase in sedimentation rate during this time period, backstripped curves generally show smoothly declining subsidence. Uncertainties in post-Eocene dating may explain some of the variability observed in subsidence curves. Further offshore, however, a series of listric faults along the middle part of the shelf and possible late Eocene dykes and flows in the northern part of the margin suggest late Eocene extension and renewed subsidence (Hugh Balkwill, personal communication, 1983).

Our choice of the Harland et al. (1982) time scale to convert biostratigraphic ages into absolute ages should not seriously bias our results. Other recent time scales do not differ substantially for the Cretaceous and Cenozoic. Differences in absolute age may affect the detailed shape of subsidence curves but are unlikely to have a major effect on the final computed results.

Radiogenic heat production is an uncertain quantity that can contribute significantly to computed heat flow. For example, calculations for the Gilbert well (Fig. 14.7) indicate that radiogenic heat production in the crust and sediments accounts for about 15% of the present-day heat flow at this site. This has a major effect on computed organic maturity, which is highly sensitive to temperature, but has only a minor effect on computed subsidence. Errors of even 10°C in computed temperature can alter depth estimates for mature sediments by hundreds of meters. The sediments contribute approximately a third of the total heat production in the Gilbert model. As mentioned above, the crustal component is poorly determined due to lack of information on the thickness of the heat-producing layer and the distribution of radioactive elements.

A final problem involves uncertainties in the values of lithospheric parameters used in the model. Parsons and Sclater (1977) estimated these quantities by inverting data from observations of ocean-floor depth and heat flow as a function of age. Unfortunately, not all the parameters could be independently determined using the data because they occur as combinations in the governing equations. By assuming that densities and the thermal transport properties of the lithosphere were known, they derived values for lithospheric thickness (125 km), its thermal expansion coefficient (3.2×10^{-5} $°C^{-1}$), and its basal temperature (1350°C). Using these parameters and other values given in Appendix 14.A, we obtained reasonably good results for the thermal calculations in the models presented (Figs. 14.8 to 14.16).

The question arises whether results acquired from studies of oceanic lithosphere are applicable to continental lithosphere. A clue to this might be found by examining subsidence and paleobathymetry data. Model assumptions and interpretational uncertainties restrict us to making qualitative assessments of paleobathymetry. Nevertheless, computed water depths generally exceed those inferred from fossil evidence (Figs. 14.8, 14.11 to 14.16). Moreover, these wells were drilled on basement highs and they should have experienced minimum water depths during subsidence. Also, wells in the northern part of the margin (Figs. 14.8, 14.11, 14.12) have high values of δ relative to β. These are necessary to keep water depths as shallow as possible during the early phases of subsidence. Wells drilled on basement highs will be biased towards low values of β and high values of δ. When taken together, this evidence indicates that our models overestimate syn- and early postrift paleobathymetry.

Recent work (T.H. Jordan, 1981; Davies and Strebek, 1982; Peltier, 1984, for example) supports the concept of a thicker continental lithosphere, on the order of 200 km. The Labrador margin formed from the rifting of a Precambrian craton, and if these ideas are correct, it may be appropriate to use a thicker lithosphere in the model. Preliminary calculations using a 200-km-thick lithosphere indicate that subsidence is improved, giving shallower paleobathymetries and reducing the disparity between β and δ. Heat flow is lower but this is expected because a change in one of the lithospheric parameters will affect the values for the other parameters. From a detailed study of continental heat-flow data, Morgan and Sass (1984) conclude that mantle (reduced) heat flow is approximately constant in stable continental areas and that variations in surface heat-flow measurements are primarily the result of variations in crustal heat production. Probably our best option for constraining prerift lithospheric properties is to ensure that the reduced heat flow for the chosen lithospheric thickness agrees with observations by adjusting the thermal conductivity of the lithosphere.

Crustal thickness estimates from seismic refraction measurements near Snorri (Fig. 14.7) fall in the range 17 to 20 km (Royden and Keen, 1980), whereas the model predicts a crustal thickness of about 24 km. Model crustal thicknesses are in closer agreement with the refraction results when the thicker lithosphere is used, because greater crustal extension ($\beta \approx 2$) is necessary to reproduce subsidence at Snorri. This result is, however, dependent on our choice of prerift crustal thickness.

Despite the above uncertainties, we believe that a useful model has been developed for studying temperature and subsidence histories in extensional basins, particularly continental margins. Using measured thermal properties of crust and sediment, we achieve model results that are consistent with present-day temperature and organic maturity data from wells on the Labrador margin. The virtue of the model is its relative simplicity, the minimal number of input variables, and its ability to account for a variety of geological observations. More elaborate models may offer some improvement to the calculations presented here, primarily with regard to computed paleobathymetries, but from a practical viewpoint, the additional sophistication may not be justified, given the resolution of the available geological data.

Conclusions

We present a one-dimensional, depth-dependent stretching model for studying the postrift thermal evolution of certain types of extensional basins. Heat transport is assumed to be by vertical conduction alone, and, therefore, the model is applicable to basins of sufficient horizontal dimensions so that we can neglect lateral heat transfer. To keep the model simple, it was constructed to require a minimum number of input variables. These include a rifting age, the lithospheric stretching parameters β and δ, and a sediment budget.

The model can be tested against a variety of geological, geophysical, and geochemical observations. Qualitative comparisons can be made between computed and interpreted paleobathymetry and model and observed stratigraphy. The amount of crustal thinning determined from modeling can be checked with crustal thickness estimates from seismic refraction experiments. Constraints on thermal history calculations include corrected bottom-hole temperatures, heat-flow measurements, measured organic maturity, and other temperature history indicators.

To use the model, it must be properly calibrated to the area of interest because properties such as thermal conductivity and radiogenic heat production can show significant variation from basin to basin. Computed temperature histories, and consequently predictions of organic maturity, are strongly dependent on the thermal characteristics of sediment and crust.

Using well data from the Labrador continental margin, we demonstrate that good agreement between theoretical predictions and observations is achieved when measured thermal properties of the crust and sediments are used in the model. Although rifting was probably a prolonged event on this margin, it appears that an instantaneous stretching model can account for subsidence, organic maturity, and the present-day temperature distribution. The thermal effects of rifting cannot be resolved from the available geological data in the wells studied. An examination of interpreted paleobathymetry data suggests that a thicker continental lithosphere ($\approx$200 km) would improve subsidence calculations over those using the commonly quoted value (125 km) derived from studies of oceanic lithosphere. The model generally predicts water depths in excess of those interpreted from micropaleontology.

Numerical studies of sedimentary basin evolution have important practical applications. Models, such as the one presented here, can be used as an aid in assessing the hydrocarbon potential of a sedimentary basin, to estimate the timing of petroleum generation and its relation to trap development, and for intercomparisons of various thermal history indicators.

Acknowledgments. We thank Petro-Canada and other members of the Labrador Group of Companies (AGIP Canada Ltd.; Amerada Minerals of Canada Ltd.; Canterra Energy Ltd.; Gulf Canada Resources Inc.; Ranchman's Resources (1976) Ltd.; and Suncor Inc.) for providing financial support, geological data, and borehole samples. We are indebted to Hugh Balkwill of Petro-Canada for his help in initiating the study and explaining the regional geology to us. Trevor Lewis of the Pacific Geoscience Centre kindly provided the use of his laboratory and assisted us in obtaining thermal property measurements on samples from the Labrador Shelf. The paper was improved by the useful comments of Dale Sawyer and an anonymous reviewer.

References

Alvarez, F., Virieux, J., and Le Pichon, X. 1984. Thermal consequences of lithosphere extension over continental margins: The initial stretching phase. Geophysical Journal of the Royal Astronomical Society 78:389–411.

Angevine, C.L., and Turcotte, D.L. 1983. Porosity reduction by pressure solution: A theoretical model for quartz arenites. Geological Society of America Bulletin 94:1129–1134.

Barss, M.S., Bujak, J.P., and Williams, G.L. 1979. Palynological zonation and correlation of sixty-seven wells, eastern Canada. Geological Survey of Canada Paper 78-24, 118 pp.

Beaumont, C. 1981. Foreland basins. Geophysical Journal of the Royal Astronomical Society 65:291–329.

Beaumont, C., Boutilier, R., Mackenzie, A.S., and Rullkötter, J. 1985. Isomerization and aromatization of hydrocarbons and the paleothermometry and burial history of the Alberta foreland basin. American Association of Petroleum Geologists Bulletin 69:546–566.

Beaumont, C., Keen, C.E., and Boutilier, R. 1982. On the evolution of rifted continental margins: Comparison of models and observations for the Nova Scotian margin. Geophysical Journal of the Royal Astronomical Society 70:667–715.

Berry, M.J., and Fuchs, K. 1973. Crustal structure of the Superior and Grenville Provinces of the northeastern Canadian Shield. Bulletin of the Seismological Society of America 63:1393–1432.

Bethke, C.M. 1985. A numerical model of compaction-driven groundwater flow and heat transfer and its application to the paleohydrology of intracratonic sedimentary basins. Journal of Geophysical Research 90: 6817–6828.

Bills, B.G. 1983. Thermoelastic bending of the lithosphere: Implications for basin subsidence. Geophysical Journal of the Royal Astronomical Society 75: 169–200.

Birch, F., and Clark, H. 1940. The thermal conductivity of rocks and its dependence upon temperature and composition. American Journal of Science 238:529–558.

Brace, W.F., and Kohlstedt, D.L. 1980. Limits on lithospheric stress imposed by laboratory experiments. Journal of Geophysical Research 85:6248–6252.

Brown, C., and Girdler, R.W. 1980. Interpretation of African gravity and its implication for the breakup of continents. Journal of Geophysical Research 85: 6443–6455.

Budiansky, B. 1970. Thermal and thermoelastic properties of isotropic composites. Journal of Composite Materials 4:286–295.

Bujak, J.P., Barss, M.S., and Williams, G.L. 1977a. Offshore east Canada's organic type and colour and hydrocarbon potential: Part 1. Oil and Gas Journal 75(14): 198–202.

Bujak, J.P., Barss, M.S., and Williams, G.L. 1977b. Offshore east Canada's organic type and colour and hydrocarbon potential: Part 2. Oil and Gas Journal 75(15): 96–100.

Cande, S.C., and Kristofferson, Y. 1977. Late Cretaceous magnetic anomalies in the North Atlantic. Earth and Planetary Science Letters 35:215–224.

Cathles, L.M., and Smith, A.T. 1983. Thermal constraints on the formation of Mississippi Valley-type lead-zinc deposits and their implications for episodic basin dewatering and deposit genesis. Economic Geology 78:983–1002.

Chen, W.-P., and Molnar, P. 1983. Focal depths of intracontinental and intraplate earthquakes and their implications for the thermal and mechanical properties of the lithosphere. Journal of Geophysical Research 88: 4183–4214.

Cochran, J.R. 1983. Effects of finite rifting times on the development of sedimentary basins. Earth and Planetary Science Letters 66:289–302.

Courtney, R.C., and Beaumont, C. 1983. Thermally-activated creep and flexure of the oceanic lithosphere. Nature 305:201–204.

Dastillung, M., and Albrecht, P. 1977. Δ2 steranes as diagenetic intermediates in sediments. Nature 269: 678–679.

Davies, G.F., and Strebeck, J.W. 1982. Old continental geotherms: Constraints on heat production and thickness of continental plates. Geophysical Journal of the Royal Astronomical Society 69:623–634.

Dokka, R.K., and Pilger, R.H., Jr. 1983. A non-uniform extension model for continental rifting. Geological Society of America Abstracts 15:559.

Dowdle, W.L., and Cobb, W.M. 1975. Static formation temperature from well logs: An empirical method. Journal of Petroleum Technology 27:1326–1330.

Dzevanshir, R.D., Buryakovskiy, L.A., and Chilingarian, G.V. 1986. Simple quantitative evaluation of porosity of argillaceous sediments at various depths of burial. Sedimentary Geology 46:169–175.

Fairhead, J.D., and Reeves, C.V. 1977. Teleseismic delay times, bouger anomalies and inferred thickness of the African lithosphere. Earth and Planetary Science Letters 36:63–76.

Falvey, D.A., and Middleton, M.F. 1981. Passive continental margins: Evidence for a prebreakup deep crustal metamorphic subsidence mechanism. In: Colloquium on Geology of Continental Margins (C3). Oceanologica Acta 4 (suppl.):103–114.

Fertl, W.H., and Wichmann, P.A. 1977. How to determine static BHT from well log data. World Oil 184: 105–106.

Foucher, J.-P., Le Pichon, X., and Sibuet, J.-C. 1982. The ocean-continent transition in the uniform lithospheric stretching model: Role of partial melting in the mantle. Philosophical Transactions of the Royal Society of London, Series A 305:27–43.

Gans, P.B., Miller, E.L., McCarthy, J., and Ouldcott, M.L., 1985. Tertiary extensional faulting and evolving ductile-brittle transition zones in the northern Snake Range and vicinity: New insights from seismic data. Geology 13:189–193.

Garven, G., and Freeze, R.A. 1984a. Theoretical analysis of the role of groundwater flow in the genesis of stratabound ore deposits: Part 1. Mathematical and numerical model. American Journal of Science 284: 1085–1124.

Garven, G., and Freeze, R.A. 1984b. Theoretical analysis of the role of groundwater flow in the genesis of stratabound ore deposits: Part 2. Quantitative results. American Journal of Science 284:1125–1174.

Gough, D.I. 1986. Mantle upflow tectonics in the Canadian Cordillera. Journal of Geophysical Research 91: 1909–1919.

Gradstein, F.M., and Srivastava, S.P. 1980. Aspects of Cenozoic stratigraphy and paleoceanography of the Labrador Sea and Baffin Bay. Palaeogeography, Palaeoclimatology, Palaeoecology 30:261–295.

Grant, A.C. 1980. Problems with plate tectonics: The Labrador Sea. Bulletin of Canadian Petroleum Geology 28:252–278.

Gunstone, F.D. 1974. Basic Stereochemistry. London, English University Press, 78 pp.

Harland, W.B., Cox, A.V., Llewellyn, P.G., Pickton, C.A.G., Smith, A.G., and Walters, R. 1982. A Geologic Time Scale. New York, Cambridge University Press, 128 pp.

Hedberg, H.D. 1936. Gravitational compaction of clays and shales. American Journal of Science 31:241–287.

Hellinger, S.J., and Sclater, J.G. 1983. Some comments on two-layer extensional models for the evolution of sedimentary basins. Journal of Geophysical Research 88:8251–8269.

Héroux, Y., Chagnon, A., and Bertrand, R. 1979. Compilation and correlation of major thermal maturation indicators. American Association of Petroleum Geologists Bulletin 63:2128–2144.

Hinz, K., Schlüter, H.-U., Grant, A.C., Srivastava, S.P., Umpleby, D.C., and Woodside, J. 1979. Geophysical transects of the Labrador Sea: Labrador to southwest Greenland. Tectonophysics 59:151–183.

Hiscott, R.N. 1984. Clay mineralogy and clay-mineral provenance of Cretaceous and Paleogene strata, Labrador and Baffin Shelves. Bulletin of Canadian Petroleum Geology 32:272–280.

Hitchon, B. 1984. Geothermal gradients, hydrodynamics and hydrocarbon occurrences, Alberta, Canada. American Association of Petroleum Geologists Bulletin 68:713–743.

Hood, A., Gutjahr, C.C.M., and Heacock, R.L. 1975. Organic metamorphism and the generation of petroleum. American Association of Petroleum Geologists Bulletin 59:986–996.

Hurtig, E., and Brugger, H. 1970. Heat conductivity measurements under uniaxial pressure (in German). Tectonophysics 10:67–77.

Hyndman, R.D., Jessop, A.M., Judge, A.S., and Rankin, D.S. 1979. Heat flow in the maritime provinces of Canada. Canadian Journal of Earth Sciences 16: 1154–1165.

Illies, J.H., and Greiner, G. 1978. Rhinegraben and the Alpine system. Geological Society of America Bulletin 89:770–782.

Issler, D.R. 1984. Calculation of organic maturation levels for offshore eastern Canada: Implications for general application of Lopatin's method. Canadian Journal of Earth Sciences 21:477–488.

Jarvis, G.T., and McKenzie, D.P. 1980. Sedimentary basin formation with finite extension rates. Earth and Planetary Science Letters 48:42–52.

Jordan, T.E. 1981. Thrust loads and foreland basin evolution, Cretaceous, western United States. American Association of Petroleum Geologists Bulletin 65: 2506–2520.

Jordan, T.H. 1981. Continents as a chemical boundary layer. Philosophical Transactions of the Royal Society of London, Series A 301:359–373.

Kappelmeyer, O., and Haenel, R. 1974. Geothermics: With special reference to application. Geoexploration Monograph 4. Berlin, Gebrüder Borntraeger, 238 pp.

Karner, G.D., and Watts, A.B. 1982. On isostasy at Atlantic-type continental margins. Journal of Geophysical Research 87:2923–2948.

Keen, C.E. 1979. Thermal history and subsidence of rifted continental margins: Evidence from wells on the Nova Scotian and Labrador Shelves. Canadian Journal of Earth Sciences 16:505–522.

Keen, C.E. 1985. The dynamics of rifting: Deformation of the lithosphere by active and passive driving forces. Geophysical Journal of the Royal Astronomical Society 80:95–120.

Keen, C.E., and Hyndman, R.D. 1979. Geophysical review of the continental margins of eastern and western Canada. Canadian Journal of Earth Sciences 16: 712–747.

Keen, C.E., and Lewis, T. 1982. Radiogenic heat production in sediments from the continental margin of eastern North America: Implications for petroleum generation. American Association of Petroleum Geologists Bulletin 66:1402–1407.

Keenan, J.H., Keyes, F.G., Hill, P.G., and Moore, J.G. 1978. Steam Tables: Thermodynamic Properties of Water including Vapour, Liquid and Solid Phases. New York, Wiley, 156 pp.

Kligfield, R., Crespi, J., Naruk, S., and Davis, G.H. 1984. Displacement and strain patterns of extensional orogens. Tectonics 3:577–609.

Lee, T.-C. 1982. Estimation of formation temperature and thermal property from dissipation of heat generated by drilling. Geophysics 47:1577–1584.

Lewis, T. 1974. Heat production measurements in rocks using a Gamma Ray Spectrometer with a solid state detector. Canadian Journal of Earth Sciences 11: 526–532.

Lopatin, N.V. 1971. Temperature and geologic time as factors in coalification. Akademiia Nauk SSSR Izvestiia Seriia Geologicheskaia, no. 3, pp. 95–106 (in Russian) (English translation by N.H. Bostick, 1972, Illinois State Geological Survey).

Lucazeau, F., and Le Douaran, S. 1985. The blanketing effect of sediments in basins formed by extension: A numerical model. Application to the Gulf of Lion and Viking Graben. Earth and Planetary Science Letters 74:92–102.

Luheshi, M.N. 1983. Estimation of formation temperature from borehole measurements. Geophysical Journal of the Royal Astronomical Society 74:747–776.

Mackenzie, A.S., Beaumont, C., Boutilier, R., and Rullkötter, J. 1985. The aromatization and isomerization of hydrocarbons and the thermal and subsidence history of the Nova Scotia margin. Philosophical Transactions of the Royal Society of London, Series A 315:203–232.

Mackenzie, A.S., and McKenzie, D.P. 1983. Isomerization and aromatization of hydrocarbons in sedimentary basins formed by extension. Geological Magazine 120:417–528.

Magara, K. 1968. Compaction and migration of fluids in Miocene mudstone, Nagaoka Plain, Japan. American Association of Petroleum Geologists Bulletin 52: 2466–2501.

Magara, K. 1976. Thickness of removed sedimentary rocks, paleopore pressure and paleotemperature, southwestern part of Western Canada Basin. American Association of Petroleum Geologists Bulletin 60: 554–565.

Magara, K. 1980. Comparison of porosity-depth relationships of shale and sandstone. Journal of Petroleum Geology 3:175–185.

Maxwell, J.C. 1964. Influence of depth, temperature and geologic age on porosity of quartzose sandstone. American Association of Petroleum Geologists Bulletin 48:697–709.

McKenzie, D.P. 1978. Some remarks on the development of sedimentary basins. Earth and Planetary Science Letters 40:25–32.

McKenzie, D.P. 1981. The variation of temperature with time and hydrocarbon maturation in sedimentary basins formed by extension. Earth and Planetary Science Letters 55:87–98.

McMillan, N.J. 1973. Shelves of Labrador Sea and Baffin Bay, Canada. In: McCrossan, R.G. (ed.): Future Petroleum Provinces of Canada: Their Geology and Potential. Canadian Society of Petroleum Geologists Memoir 1, pp. 473–517.

McMillan, N.J. 1980. Geology of the Labrador Sea and its petroleum potential. In: Proceedings of the Tenth World Petroleum Congress: Vol. 2. Exploration: Supply and Demand. London, Heyden, pp. 165–175.

McWhae, J.R.H. 1981. Structure and spreading history of the northwestern Atlantic region from the Scotian Shelf to Baffin Bay. In: Kerr, J.W., and Fergusson, A.J. (eds.): Geology of the North Atlantic Borderlands. Canadian Society of Petroleum Geologists Memoir 7, pp. 299–332.

McWhae, J.R.H., Elie, R., Laughton, K.C., and Gunther, P.R. 1980. Stratigraphy and petroleum prospects of the Labrador Shelf. Bulletin of Canadian Petroleum Geology 28:460–488.

McWhae, J.R.H., and Michel, W.F.E. 1975. Stratigraphy of Bjarni H-81 and Leif M-48, Labrador Shelf. Bulletin of Canadian Petroleum Geology 23:361–382.

Montadert, L., De Charpal, O., Roberts, D., Guennoc, P., and Sibuet, J.-C. 1979. Northeast Atlantic passive continental margins: Rifting and subsidence processes. In: Talwani, M., Hay, W., and Ryan, W.B.F. (eds.): Deep Drilling Results in the Atlantic Ocean: Continental Margins and Paleoenvironment. Maurice Ewing Series, Vol. 3, Washington, DC, American Geophysical Union, pp. 154–186.

Morgan, P. 1983. Constraints on rift thermal processes from heat flow and uplift. Tectonophysics 94:277–298.

Morgan, P., and Sass, J.H. 1984. Thermal regime of the continental lithosphere. Journal of Geodynamics 1: 143–166.

Nakiboglu, S.M., and Lambeck, K. 1985. Comments on thermal isostasy. Journal of Geodynamics 2:51–65.

Panza, G.F., Mueller, St., and Calcagnile, G. 1980. The gross features of the lithosphere-asthenosphere system in Europe from seismic surface waves and body waves. Pure and Applied Geophysics 118:1209–1213.

Parsons, B., and Sclater, J.G. 1977. An analysis of the variation of ocean floor bathymetry and heat flow with age. Journal of Geophysical Research 82:803–827.

Peltier, W.R. 1984. The thickness of the continental lithosphere. Journal of Geophysical Research 89: 11303–11316.

Perrier, R., and Quiblier, J. 1974. Thickness changes in sedimentary layers during compaction history: Methods for quantitative evaluation. American Association of Petroleum Geologists Bulletin 58:507–520.

Powell, T.G. 1979. Geochemistry of Snorri and Gudrid condensates, Labrador Shelf: Implications for future exploration. Geological Survey of Canada Paper 79-1C, pp. 91–95.

Rashid, M.A., Purcell, L.P., and Hardy, I.A. 1980. Source rock potential for oil and gas of the east Newfoundland and Labrador Shelf areas. In: Miall, A.D. (ed.): Facts and Principles of World Petroleum Occurrence. Canadian Society of Petroleum Geologists Memoir 6, pp. 589–608.

Raymer, L.L., Hunt, E.R., and Gardner, J.S. 1980. An improved sonic transit-time-to-porosity transform. Transactions, Society of Professional Well Log Analysts Twenty-First Annual Logging Symposium, Lafayette, LA, pp. P1-P13.

Rehrig, W.A., and Reynolds, S.J. 1980. Geologic and geochronologic reconnaissance of a northwest-trending zone of metamorphic complexes in southern Arizona. In: Crittenden, M.D., Jr., Coney, P.J., and Davis, G.H. (eds.): Cordilleran Metamorphic Core Complexes. Geological Society of America Memoir 153, pp. 131–157.

Rhead, M.M., Eglinton, G., and Draffen, G.H. 1971. Hydrocarbons produced by the thermal alteration of cholesterol under conditions simulating the maturation of sediments. Chemical Geology 8:277–297.

Rieke, H.H., III, and Chilingarian, G.V. 1974. Compaction of Argillaceous Sediments. Developments in Sedimentology 16, Amsterdam, Elsevier, 424 pp.

Rowley, D.B., and Sahagian, D. 1986. Depth-dependent stretching: A different approach. Geology 14:32–35.

Roy, R.F., Beck, A.E., and Touloukian, Y.S. 1981. Thermophysical properties of rocks. In: Touloukian, Y.S., Judd, W.R., and Roy, R.F. (eds.): Physical Properties of Rocks and Minerals. New York, McGraw-Hill, pp. 409–502.

Roy, R.F., Blackwell, D.D., and Birch, F. 1968. Heat generation of plutonic rocks and continental heat flow provinces. Earth and Planetary Science Letters 5:1–12.

Royden, L., and Keen, C.E. 1980. Rifting process and thermal evolution of the continental margin of eastern Canada determined from subsidence curves. Earth and Planetary Science Letters 51:343–361.

Royden, L., Sclater, J.G., and Von Herzen, R.P. 1980. Continental margin subsidence and heat flow: Important parameters in formation of petroleum hydrocarbons. American Association of Petroleum Geologists Bulletin 64:173-187

Sass, J.H., Lachenbruch, A.H., and Munroe, R.J. 1971. Thermal conductivity of rocks from measurements on fragments and its application to heat-flow determinations. Journal of Geophysical Research 76:3391-3401.

Sass, J.H., Stone, C., and Munroe, R.J. 1984. Thermal conductivity on solid rock: A comparison between a steady-state divided-bar apparatus and a commercial transient line-source device. Journal of Volcanology and Geothermal Research 20:145-153.

Sawyer, D.S. 1985. Brittle failure in the upper mantle during extension of continental lithosphere. Journal of Geophysical Research 90:3021-3025.

Sawyer, D.S., Toksöz, M.N., Sclater, J.G., and Swift, B.A. 1983. Thermal evolution of the Baltimore Canyon Trough and Georges Bank Basin. In: Watkins, J.S., and Drake, C.L. (eds.): Studies in Continental Margin Geology. American Association of Petroleum Geologists Memoir 34, pp. 743-762.

Schult, F.R., and Gordon, R.G. 1984. Root mean square velocities of the continents with respect to the hot spots since the early Jurassic. Journal of Geophysical Research 89:1789-1800.

Sclater, J.G., and Christie, P.A.F. 1980. Continental stretching: An explanation of the post-mid-Cretaceous subsidence of the central North Sea Basin. Journal of Geophysical Research 85:3711-3739.

Sclater, J.G., Royden, L., Horvath, F., Burchfiel, B.C., Semken, S., and Stegena, L. 1980. The formation of the intra-Carpathian basins as determined from subsidence data. Earth and Planetary Science Letters 51: 139-162.

Selley, R.C. 1978. Porosity gradients in North Sea oil-bearing sandstones. Journal of the Geological Society of London 135:119-132.

Sewell, G. 1981. Twodepep, a small general purpose finite element program. International Mathematical and Statistical Library Technical Report 8102, 11 pp.

Smith, R.B., and Bruhn, R.L. 1984. Intraplate extensional tectonics of the eastern Basin-Range: Inferences on structural style from seismic reflection data, regional tectonics and thermal-mechanical models of brittle-ductile deformation. Journal of Geophysical Research 89:5733-5762.

Srivastava, S.P. 1978. Evolution of the Labrador Sea and its bearing on the early evolution of the North Atlantic. Geophysical Journal of the Royal Astronomical Society 52:313-357.

Srivastava, S.P., Falconer, R.K.H., and Maclean, B. 1981. Labrador Sea, Davis Strait, Baffin Bay: Geology and geophysics: A review. In: Kerr, J.W., and Fergusson, A.J. (eds.): Geology of the North Atlantic Borderlands. Canadian Society of Petroleum Geologists Memoir 7, pp. 333-398.

Steckler, M.S. 1981. The thermal and mechanical evolution of Atlantic-type continental margins. Ph.D. thesis, Columbia University, New York, 266 pp.

Steckler, M.S. 1985. Uplift and extension at the Gulf of Suez: Indications of induced mantle convection. Nature 317:135-139.

Steckler, M.S., and Watts, A.B. 1978. Subsidence of the Atlantic-type continental margin off New York. Earth and Planetary Science Letters 41:1-13.

Strang, G., and Fix, G.J. 1973. An Analysis of the Finite Element Method. Englewood Cliffs, NJ, Prentice-Hall, 306 pp.

Turcotte, D.L., and McAdoo, D.C. 1979. Thermal subsidence and petroleum generation in the southwestern block of the Los Angeles Basin, California. Journal of Geophysical Research 84:3460-3464.

Umpleby, D.C. 1979. Geology of the Labrador Shelf. Geological Survey of Canada Paper 79-13, 34 pp.

van der Linden, W.J.M. 1975a. Mesozoic and Cainozoic opening of the Labrador Sea, the North Atlantic and the Bay of Biscay. Nature 253:320-324.

van der Linden, W.J.M. 1975b. Crustal attenuation and seafloor spreading in the Labrador Sea. Earth and Planetary Science Letters 27:409-423.

Van Graas, G., Baas, J.M.A., Van Der Graaf, B., and de Leeuw, J.W. 1982. Theoretical organic geochemistry: 1. The thermodynamic stability of several cholestane isomers calculated from molecular dynamics. Geochimica et Cosmochimica Acta 46:2399-2402.

Von Driska, P.M. 1982. Double diffusive convection across a single thermohaline interface at high density stability ratios. M.Sc. thesis, University of Illinois, Urbana, IL 98 pp.

Walsh, J.B., and Decker, E.R. 1966. Effect of pressure and saturating fluid on the thermal conductivity of compact rock. Journal of Geophysical Research 71: 3053-3061.

Waples, D.W. 1980. Time and temperature in petroleum formation: Application of Lopatin's method to petroleum exploration: American Association of Petroleum Geologists Bulletin 64:916-926.

Watt, W.S. 1969. The coast-parallel dike swarm of southwest Greenland in relation to the opening of the Labrador Sea. Canadian Journal of Earth Sciences 6: 1320-1321.

Weast, R.C., Astle, M.J., and Beyer, W.H. 1983. CRC Handbook of Chemistry and Physics (64th ed.): Boca Raton, FL, Chemical Rubber Company Press, 2303 pp.

Wood, R.J. 1982. Subsidence in the North Sea. Ph.D. thesis, University of Cambridge, Cambridge, England, 94 pp.

Wright, J.A., Jessop, A.M., Judge, A.S., and Lewis, T.J. 1980. Geothermal measurements in Newfoundland. Canadian Journal of Earth Sciences 17:1370–1376.

Zorin, Y.A. 1981. The Baikal Rift: An example of the intrusion of asthenospheric material into the lithosphere as the cause of disruption of lithospheric plates. Tectonophysics 73:91–104.

Appendix 14.A

Material property values for models

1. Water layer			
Density of seawater	1,030 kg m^{-3}		
2. Sediments and volcanic rocks			
	Sandstone	Shale	Basalt
Matrix density (kg m^{-3})	2,650	2,700	2,800
Compaction constant (m^{-1})	9.884×10^{-3}	1.652×10^{-4}	0
Surface porosity (%)	48.5	47.3	5.0
Thermal conductivity at 0°C (W m^{-1} °C^{-1})	3.1	1.7	1.8
Specific heat at 0°C (J kg^{-1} °C^{-1})	708	748	847
Radioactive heat production (μW m^{-3})	.57	1.29	.16
3. Crust			
Prerift thickness (t_c)	35 km		
Density at 0°C	2,820 kg m^{-3}		
Thermal conductivity	3.1 W m^{-1} °C^{-1}		
Specific heat at constant volume × density	3.877×10^6 J m^{-3} °C^{-1}		
Radioactive heat production	1.0 μW m^{-3}		
Volume coefficient of thermal expansion	3.2×10^{-5} °C^{-1}		
Depth of radioactive layer of uniform density	7.5 km		
4. Subcrustal lithosphere			
Prerift thickness ($a - t_c$)	125 − 35 = 90 km		
Density at 0°C	3,330 kg m^{-3}		
Thermal conductivity	3.1 W m^{-1} °C^{-1}		
Specific heat at constant volume × density	3.877×10^6 J m^{-3} °C^{-1}		
Volume coefficient of thermal expansion	3.2×10^{-5} °C^{-1}		
5. Asthenosphere			
Density at T_m	3,186 kg m^{-3}		
Temperature (T_m)	1350°C		
Thermal conductivity	3.1 W m^{-1} °C^{-1}		
Specific heat at constant volume × density	3.877×10^6 J m^{-3} °C^{-1}		
Volume coefficient of thermal expansion	3.2×10^{-5} °C^{-1}		

15
A Simulator for the Computation of Paleotemperatures During Basin Evolution

A.E. McDonald, D.U. von Rosenberg, W.R. Jines, W.H. Burke, Jr., and L.M. Uhler, Jr.

Abstract

The stages of hydrocarbon generation in the various source rocks of a sedimentary basin can be estimated from maturation indices. Computation of maturation index requires the time-temperature history of the formations in the basin. Processes that occur during the formation and development of a basin are quite complex and include subsidence, sedimentation, uplift/erosion, faulting, magmatism, and surface temperature variations during changes in ocean level and climate. These processes have been modeled in our company for a number of years using GOLIATH, a two-dimensional, transient, finite-difference, geological thermal simulator.

Introduction

The stages of hydrocarbon generation in the various source rocks of a sedimentary basin can be estimated from maturation indices. The computation of these indices requires the time-temperature history of the formations in the basin (Hood et al., 1975; Waples, 1980). A mathematical model of the various processes that govern this history has been developed, and a computer solution has been implemented. The resulting program is identified as GOLIATH, an acronym for geological thermal simulator. This chapter describes the simulator.

A complete listing of the FORTRAN program, including comment statements, requires more than 100 pages. Obviously, details of all aspects of the program cannot be covered in this short chapter. An outline for the solution of the heat-flow equation is given in Appendix 15.A.

Physical Processes Simulated

The processes that occur during the formation and development of a basin are quite complex. They include subsidence, sedimentation, uplift and erosion, compaction, faulting, magmatism, water movement by convection, and variations in surface temperature during changes of ocean level and climate. Subsidence rates vary throughout the basin as a function of geological time, and a starved basin is created when subsidence is faster than sedimentation.

To model these processes in detail, it is convenient to have a computer program such as GOLIATH, a two-dimensional (one horizontal and one vertical), transient model that can be used to study a number of cycles of sedimentation/subsidence and uplift/erosion. Physical processes that are simulated in the model include the following:

Subsidence The basement sinks due to tectonic forces and loading.

Sedimentation Sediments are added concurrently with or after subsidence.

Uplift/erosion Parts of the basin rise, and the surface is eroded.

Internal heat generation Radioactivity or chemical reactions produce heat.

Surface temperature variations with geological time Climate or other determinants change surface temperature. Subsurface temperatures slowly respond.

Magmatic activity Rock temperatures respond to occurrences of dikes, sills, stocks, and so forth.

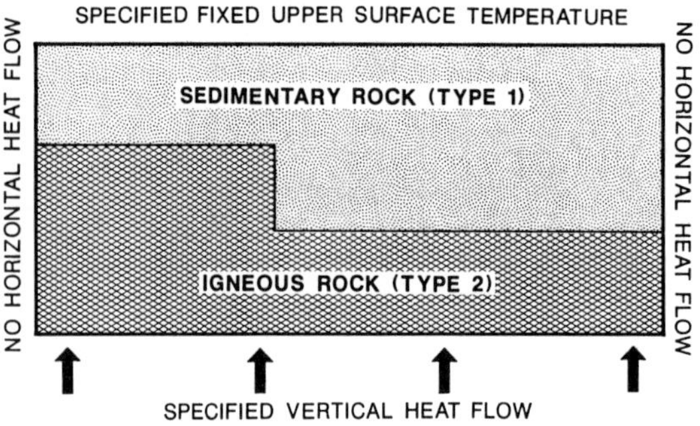

FIGURE 15.1. A simple geothermal model in which heat flows in at the bottom and out at the top.

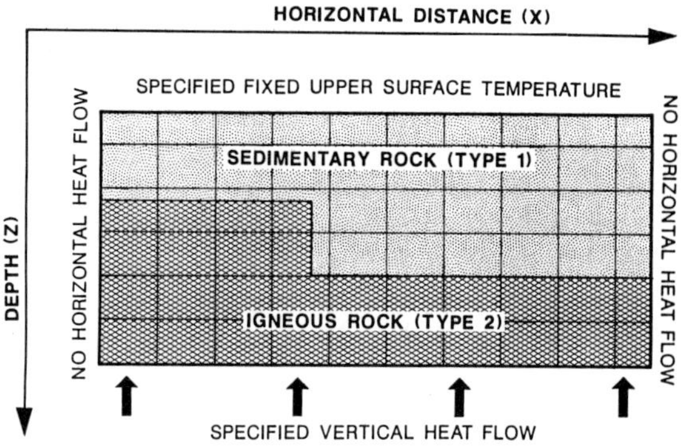

FIGURE 15.2. Rectangular grid superimposed on the simple geothermal model shown in Figure 15.1.

Vertical faulting Two horizontally adjacent blocks move at different vertical velocities.

Change in properties with burial Thermal conductivity of sediments increases due to loss of porosity with burial.

Starved basin Subsidence is faster than sedimentation so that the basin becomes "starved" for sediments.

Model Description

Mathematical equations describing conductive heat flow in the earth are given in Appendix 15.A. A simple example of the type of problem that can be modeled is shown in Figure 15.1. Heat flows in at the bottom from rocks below those included in the modeled rectangle. Heat is lost through the upper edge. Heat flows both laterally and vertically in the interior of the model. At the edges vertical flow occurs, but lateral flow is not permitted.

To model this simple example, a rectangular grid is superimposed on Figure 15.1, as shown in Figure 15.2. Thermal properties of rock Type 1 are used in the stipled blocks. Properties of rock Type 2 are used in the cross-hatched blocks. Blocks containing both rock types use integrated average thermal properties. Heat flow into the model is specified at the lower boundary as a function of horizontal dimension (x) and time (t). Surface temperatures are specified at the upper boundary as a function of x and t. Optionally, heat flows can be specified at the surface.

While mathematically the modeled portion of the earth is required to be a rectangle with horizontal top and bottom and vertical sides, in practice the upper surface can have a complex shape. This shape is obtained by specifying a fixed temperature surface below the top edge of the rectangular grid. The shape of this surface can be fitted to the shape of the topographic surface.

The grid of Figure 15.2 remains fixed, and rocks move vertically, but not horizontally, through it. New sediments enter through the top boundary. As rocks move through the bottom boundary, they are eliminated from the solution. However, if uplift occurs they will move back into the grid. This bottom boundary is placed far enough below areas of major interest that events at the bottom will not influence temperature changes within the area of interest. As rocks move through the grid, they carry their thermal properties with them. When the sedimentation and subsidence rates are equal, the upper or topographical surface stays fixed. When subsidence is at a greater rate than sedimentation, a starved basin will develop; when it is at a lesser rate, the basin will fill in.

During an uplift event, the rocks move upward, and some of them will move through the topographic surface. Usually these are considered to be eroded away; in this case, they are eliminated from the simulation and will not appear again. If a period of sedimentation follows the uplift-erosion event, new sediments will be deposited upon the surface.

A Nontrivial Case Study of Geological Thermal Behavior

Additional concepts will be illustrated through analysis of a two-dimensional slice from a hypothetical Late Cenozoic basin. The basin is ideal-

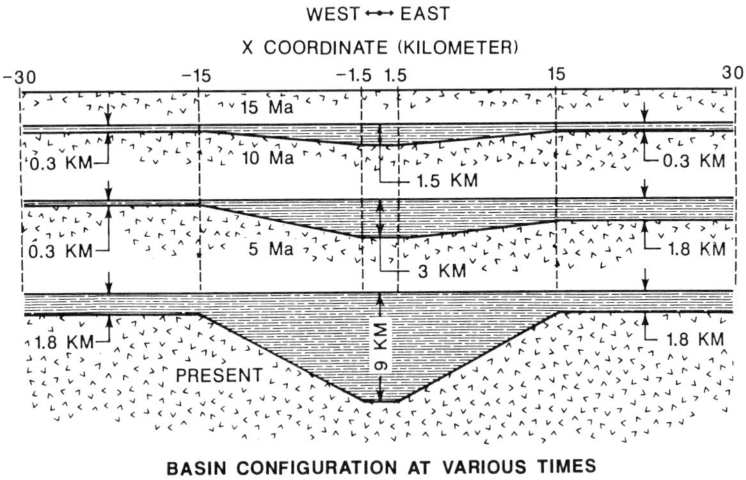

FIGURE 15.3. Configuration of the modeled hypothetical Late Cenozoic basin at various times from 15 Ma to present. On both flanks, the upper and lower sediment surfaces remain horizontal out to x values of -60 and $+60$ km, where the boundary condition of no lateral heat flow is imposed.

ized, but the example illustrates several effects of geological history on thermal gradients in various parts of the basin.

Figure 15.3 shows the shape of the basin at several geological times. Sedimentation began in the Miocene at 15 Ma (million years before present). It continued for 5 my at a rate of 300 m/my in the depocenter and at a rate of 60 m/my on each flank. Deposition rates at locations between the depocenter and the flanks are assumed to be intermediate to these. The resulting basin configuration is identified in Figure 15.3 by the label "10 Ma." During this 5-my period, and throughout the modeled basin history, depositional rate is assumed to have equaled subsidence rate. Hence the sediment-water interface remained flat.

From 10 Ma to 5 Ma, the depocenter sedimentation rate continued at 300 m/my. During this time there was no deposition on the western flank, while sediments were deposited on the eastern flank at the increased rate of 300 m/my.

From 5 Ma to the present, the center of the basin received sediments at the rapid rate of 1,200 m/my. The flank rates were reversed with the western flank now at 300 m/my and with no sedimentation on the eastern flank.

The thermal conductivity of the basement is assumed spatially and temporally constant at 7.5×10^{-3} cal/cm s C°, which is a representative value for igneous rocks. Sedimentary rock conductivity is assumed to be independent of the horizontal coordinate (x). The conductivity of these rocks is increased as they are buried more deeply. This increase in conductivity is in response to the decrease in porosity as the sediments are compacted. Change in dimensions of the sediments under burial-induced compaction is not included in the model. Conductivity of the sediments as a function of depth of burial as used in this example is given in Table 15.1. Volumetric heat capacity is assumed everywhere constant at 0.5 cal/cm^3 C°.

Boundary conditions are:

1. fixed temperature at upper surface,
2. no horizontal heat flow at $x = -60$ km and at $x = +60$ km, and
3. vertical heat flow $= 1.792 \times 10^{-6}$ cal/cm^{2}s, upward, at a depth of 60 km, independent of x and t.

Initial temperatures, just before sedimentation began at 15 Ma, were obtained from a steady-state solution of the heat-flow equation. The all-basement configuration shown at 15 Ma in Figure 15.3 was used to compute the steady-state solution.

TABLE 15.1. Example of typical variation in sediment conductivity with depth in the modeled basin.

Depth (km)	Conductivity (cal/cm s C°)
0	0.0020
0.3	0.0034
0.6	0.0039
1.2	0.0044
3	0.0051
6	0.0057
12	0.0060

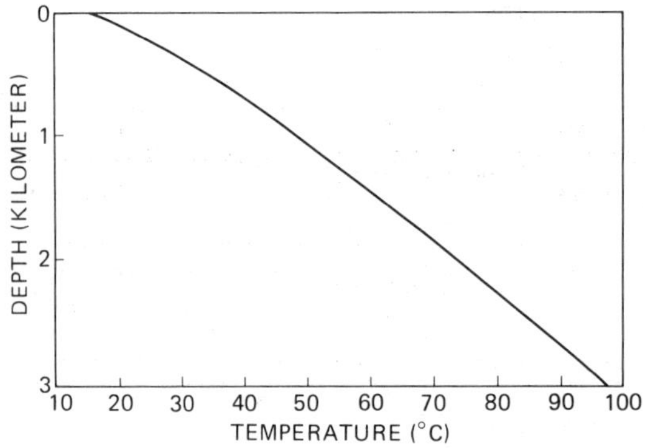

FIGURE 15.4. Present depocenter geothermal profile in the modeled Late Cenozoic basin (Fig. 15.3).

Figures 15.4 to 15.6 illustrate some of the results from this study. Figure 15.4 shows the present-day geothermal profile at the depocenter ($x = 0$) in Figure 15.3. Variations in the thermal gradient with horizontal location and with time are shown in Figures 15.5 and 15.6. For these figures, the thermal gradient, in C°/km, is defined as the difference in the temperature at a depth of 1.5 km and that at the surface, divided by 1.5. The present-day gradient at the depocenter can be computed from Figure 15.4 as (60° − 15.6°)/1.5, which yields 29.6 C°/km. This is the value shown in both Figures 15.5 and 15.6. Figure 15.5 is a comparison of the present-day gradient across the basin, as obtained from the transient solution, with the gradient for the steady-state solution of the basin in its present configuration. The *steady-state* gradient is symmetric about $x = 0$ since the basin is symmetric in its present-day configuration, as can be seen in Figure 15.3. The gradient from the *transient solution* is not symmetric since the flanks subsided at different rates. The transient gradient is also much lower than the steady-state gradient because of the rapid sedimentation rate during the last 5 my. Figure 15.6 shows the change in the gradient at the depocenter during the last 6 my. The gradient decreased rapidly at 5 Ma when rapid deposition began; as heat began to flow into the central portion of the basin from the flanks, the rate of decrease lessened, and a steady-state condition was approached.

Further analysis of these figures reveals the following:

1. Rapid deposition of sediments reduces the thermal gradient.
2. The more rapid and prolonged the deposition, the greater the depression of the gradient.
3. The more recent the deposition, the greater the present gradient depression.
4. Vertical heat flow in the depocenter is less than on the flanks because the flow of heat follows the path of least resistance. The basement provides less resistance (higher thermal conductivity) than the sediments. The curve of final steady-state gradient versus x in Figure 15.5 shows this effect clearly. The variation of steady-state gradient with x is entirely due to variation of vertical heat flow.
5. The present thermal gradient in the depocenter is less than on the flanks. There are two reasons for this: faster deposition in the center and bending of heat flow away from the center.
6. Gradients, especially in the depocenter, were significantly higher in the past than at present. The source rock maturation is significantly influenced by this effect.

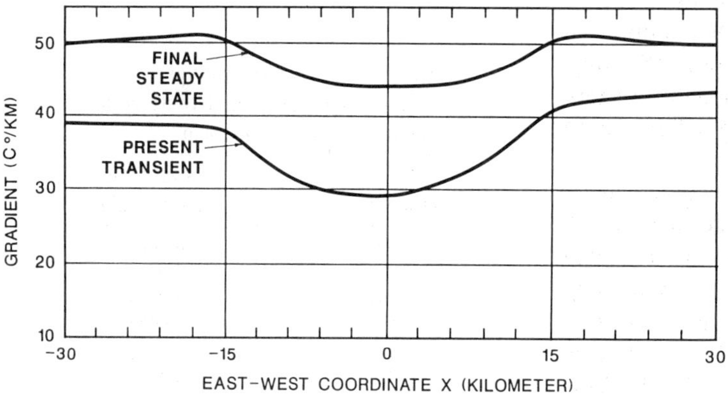

FIGURE 15.5. Final steady-state and present transient gradients at various locations in the modeled Late Cenozoic basin (Fig. 15.3).

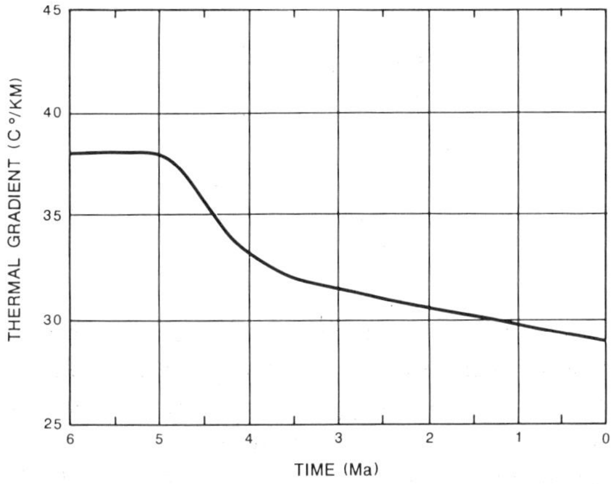

FIGURE 15.6. Plot of depocenter thermal gradient versus time in the modeled Late Cenozoic basin (Fig. 15.3).

Additional Remarks

Many capabilities of GOLIATH were not included in the above Late Cenozoic basin study to maintain simplicity of presentation. Other processes that can be simulated, such as uplift/erosion and internal heat generation, are listed above in the section entitled "Physical Processes Simulated." GOLIATH can be used in either transient or steady-state mode. In the transient case, a steady-state initial temperature distribution may be generated, if desired.

References

Hood, A., Gutjahr, C.C.M., and Heacock, R.L. 1975. Organic metamorphism and the generation of petroleum. American Association of Petroleum Geologists Bulletin 59:986–996.

McDonald, A.E. 1985. Vector computer applications in reservoir simulation. Proceedings, 1985 Cray Science and Engineering Symposium, Mendota Heights, MN.

Waples, D.W. 1980. Time and temperature in petroleum formation: Application of Lopatin's method to petroleum exploration. American Association of Petroleum Geologists Bulletin 64:916–926.

Watts, J.W. 1971. An iterative matrix inversion method suitable for anisotropic problems. Society of Petroleum Engineers Journal 11:47–51.

Appendix 15.A. Mathematical Considerations

Problem Statement

The heat-flow equation in two space dimensions x and z, including (vertical) geological subsidence, is:

$$\frac{\partial}{\partial x}\left[K\frac{\partial T}{\partial x}\right] + \frac{\partial}{\partial z}\left[K\frac{\partial T}{\partial z}\right] - v_z\frac{\partial(cT)}{\partial z} + Q(x,z) = \frac{\partial}{\partial t}(cT) \qquad \text{(A1)}$$

where:

c = heat capacity per unit volume
K = thermal conductivity
Q = heat generation rate per unit volume per unit time
T = temperature
t = time
v_z = subsidence velocity
x = lateral direction
z = depth

Initial and boundary conditions are:

$$T(x,z,0) = T_0(x,z).$$

$\frac{\partial T}{\partial x} = 0$ at left and right boundaries.

$\frac{\partial T}{\partial z} = F(x,t)$ at bottom boundary.

$\frac{\partial T}{\partial z}$ or T specified at top boundary, as functions of time.

In addition, temperature may be specified as a function of time at any (x,z) location. A finite-difference equation analogous to Equation A1 is written between two adjacent time levels denoted as t^n and t^{n+1}. In the following equations, a value of temperature at t^n is denoted as T^n; one at t^{n+1} is written simply as T with no superscript.

Difference Equations

Denote a space-dependent finite difference grid node by (x_i, z_j), or simply by (i,j). The usual 5-point difference operator at (i,j) for Equation A1 can be written as

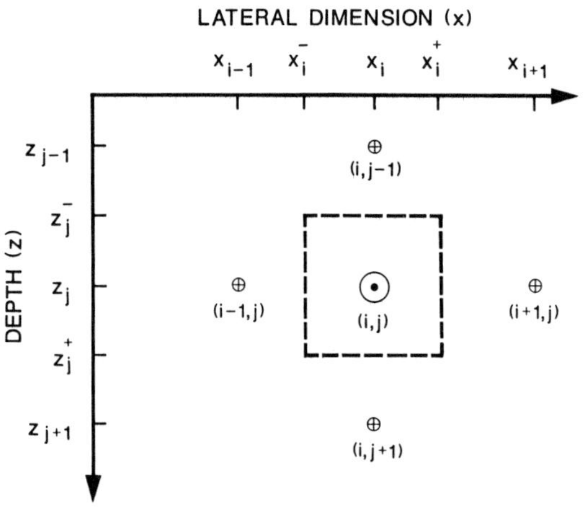

FIGURE 15.A. Cell (i,j) is outlined by the dashed lines enclosing node (i,j). Cell boundaries are midway between adjacent nodes.

$$W_{ij}T_{i-1,j} + N_{ij}T_{i,j-1} + E_{ij}T_{i+1,j} + S_{ij}T_{i,j+1} + C_{ij}T_{ij} = R_{ij} \quad \text{(A2)}$$

where T_{ij} denotes the average temperature within a rock volume (cell) "centered" at (x_i, z_j) and bounded by the rectangle shown in Figure 15.A. W, E, N, S, C, and R are defined as follows:

$$W_{ij} = \frac{2K_{i-1/2,j}}{(x_i - x_{i-1})(x_{i+1} - x_{i-1})}$$

$$E_{ij} = \frac{2K_{i+1/2,j}}{(x_{i+1} - x_i)(x_{i+1} - x_{i-1})}$$

$$N_{ij} = \frac{2K_{i,j-1/2}}{(z_j - z_{j-1})(z_{j+1} - z_{j-1})} + \frac{C_{i,j-1}v_{z_{ij}}}{z_{j+1} - z_{j-1}}$$

$$S_{ij} = \frac{2K_{i,j+1/2}}{(z_{j+1} - z_j)(z_{j+1} - z_{j-1})} - \frac{C_{i,j+1}v_{z_{ij}}}{(z_{j+1} - z_{j-1})}$$

$$C_{ij} = -(W + E + N + S)_{ij} - \frac{c_{ij}}{\Delta t}$$

$$R_{ij} = -\frac{c_{ij}}{\Delta t}T_{ij}^n - Q_{ij} + \text{boundary condition (if adjacent to boundary).}$$

Values such as $K_{i-1/2,j}$, $K_{i,j-1/2}$, and so forth, are obtained by integral average, e.g.,

$$K_{i,j-1/2} = \frac{\int_{z_{j-1}}^{z_j} K(z,t)dz}{\int_{z_{j-1}}^{z_j} dz} = \frac{\int_{z_{j-1}}^{z_j} K(z,t)dz}{(z_j - z_{j-1})}$$

Note that K varies with both z and t, since sediment position (and hence property) depends on t. Variation with t does not appear in the integral, for it is evaluated at a definite point in time (e.g., either at the end or middle of a time step). Alternatively, K could be computed using a harmonic average.

In a similar fashion, c_{ij} denotes an integrated average over the rectangle with corners at

$$\frac{1}{2}(x_{i-1} + x_i, z_{j-1} + z_j), \frac{1}{2}(x_{i+1} + x_i, z_{j-1} + z_j),$$

$$\frac{1}{2}(x_{i-1} + x_i, z_{j+1} + z_j), \frac{1}{2}(x_{i+1} + x_i, z_{j+1} + z_j).$$

Solution of Difference Equations

Equation A2 is linear. It is easily solved with variants of line successive over-relaxation (LSOR) (McDonald, 1985). This method is well known and widely published. Using LSOR as a base, we have added three features to improve performance:

1. Starting values are obtained by extrapolating temperature changes from a previous time step.
2. Starting values are improved using the correction of Watts (1971).
3. Lines are ordered according to decreasing magnitude of residual, and the LSOR solution is iterated to treat the lines according to this ordering.

Parameter and Tolerance

Default values are available for iteration parameter and convergence tolerance. The user may elect to override these values. Iteration parameter ω must be such that $1 \leq \omega < 2$. The default value is given by

$$\omega = \frac{2}{1+\sqrt{1-\mu^2}}$$

where μ depends on mesh spacings. To compute μ, let

$$\nu_x = \operatorname*{Min}_i \{x_{i+1} - x_i\} ,$$

$$\nu_z = \operatorname*{Min}_j \{z_{j+1} - z_j\} ,$$

L_x = length of top of rectangle

L_z = length of side of rectangle

Then

$$\mu = \frac{\mathrm{Cos}(\pi \nu_z/L_z)}{2 - \mathrm{Cos}(\pi \nu_x/L_x)} .$$

The convergence tolerance τ is chosen to ensure adequate energy balance. Iteration continues until LHS and RHS of Equation A2 satisfy

$$|\mathrm{RHS} - \mathrm{LHS}| < \tau$$

at each node of the grid. The default value is $\tau = 0.01$.

16

Thermal Evolution of Laramide-Style Basins: Constraints from the Northern Bighorn Basin, Wyoming and Montana

E. Sven Hagen and Ronald C. Surdam

Abstract

The Laramide-style basins of the central Rocky Mountain region are deep asymmetric structural depressions containing thick sequences of Upper Cretaceous and Tertiary strata. The combined effects of tectonics and sedimentation have contributed to the thermal evolution of the basins and to the maturation history of Cretaceous hydrocarbon source rocks.

Thermal parameters, from Cretaceous rocks and oil well temperature data, provide the primary thermal constraints for the proposed two-dimensional, finite-difference, numerical model. Input parameters for the numerical model include the geometry of the basin, thermal conductivity, heat flow, and a constant surface temperature. By integrating this model with time-temperature reconstructions, the temperature histories for Cretaceous source rocks can be determined. The Bighorn basin reached maximum diagenetic temperatures in Early Miocene time (≈ 20 Ma) prior to regional epeirogenic uplift and erosion of the basin to its present geomorphic form. Within the center or asymmetric portion of the basin, average thermal conductivities for the overburden above the Mowry Formation are higher, resulting in lower geothermal gradients ($\approx 25\,°C/km$). Along the basin margin, average thermal conductivities are lower resulting in higher geothermal gradients ($\approx 32\,°C/km$).

Ultimately, integrating basin-specific time-temperature histories with a kinetic geochemical model will allow the determination of the areal extent and timing of oil and gas generation in a particular hydrocarbon source rock, as well as potential migration pathways and hydrocarbon traps throughout the geological history of the basin.

Introduction

The thermal origin and evolution of most sedimentary basins can be divided into two types: transient and steady state (Fig. 16.1). The most common example of a transient thermal system is an extensional tectonic regime (rift or pull-apart basin) related to divergent or convergent (transform faults) plate boundaries. Numerous workers have defined the thermal histories of rift basins (Keen, 1979; Sclater and Christie, 1980; Angevine and Turcotte, 1981; McKenzie, 1981) and extensional basins (Turcotte and McAdoo, 1979; Zandt and Furlong, 1982; Heasler and Surdam, 1985). In general, tectonic processes involved include crustal attenuation by thinning (rift) or creation of a transform plate boundary (extensional), upwelling of hot asthenosphere, and subsequent cooling and thermal subsidence.

In contrast, steady-state thermal regimes occur in regions where thermal perturbations of the earth's crust are second-order when compared to extensional tectonic regimes. Common examples may include intracratonic basins and foreland basins adjacent to fold-thrust belts. Variations in the temperature history are due primarily to the structural and stratigraphic geometry (i.e., thermal conductivity structure) of the basin rather than to significant variations in regional heat flow through time.

The general scheme for determining the temperature history of a basin is threefold (Fig. 16.1). First, the geological history of a basin must be

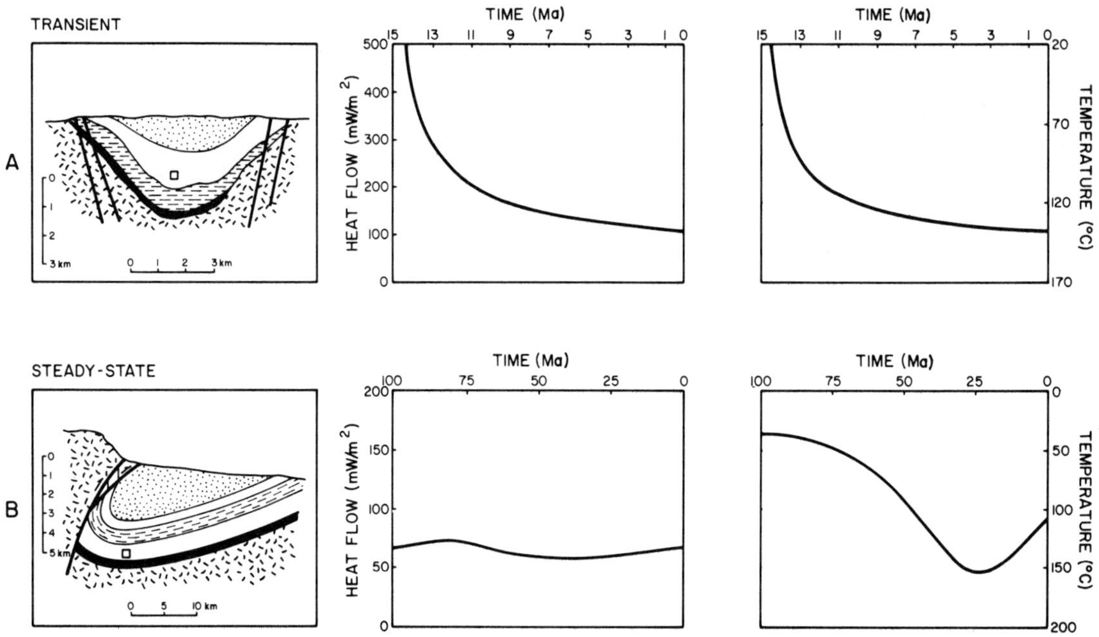

FIGURE 16.1. General thermal modeling scheme for determining the temperature histories for a transient and steady-state thermal regime; A, pull-apart (extensional); B, Laramide-style basin. The modeling scheme is divided into three aspects: geological history, heat-flow determination, and time-temperature plots. Heat-flow and time-temperature plots are constructed in reference to the box shown on the geological cross section. Diagrams for the pull-apart basin (A) are modified from Heasler and Surdam (1983).

determined. Geological parameters may include: 1) regional tectonic regime (i.e., passive margin, intracratonic); 2) structural history and basin geometry; 3) mode and history of basin subsidence (i.e., flexural, thermal); and 4) sedimentary history including erosional phases. The tectonic and geological history provide the initial constraints on the regional thermal regime.

The second phase involves constructing a thermal model for the basin of interest. The thermal model is based on structural and stratigraphic reconstructions integrated with the thermal history constrained by parameters that reflect the degree of thermal exposure: clay diagenesis (Hower et al., 1976), organic geochemical parameters (Hunt, 1979), vitrinite reflectance (Lerche et al., 1984), and fisson-track dating (Naeser, 1979). A product of the thermal model is paleo-heat flow and the basin's temperature field. Integrating these data with the geological history of the basin defines a time-temperature history for the basin of interest.

The general schemes for determining time-temperature histories for extensional (Fig. 16.1A; pull-apart) and Laramide-style (Fig. 16.1B) basins are compared. These basins display markedly different geometries as well as tectonic, thermal, and subsidence histories. Extensional basins display high heat flow in the early phases of their geological history because of lithospheric thinning. As the heat flow decays through time, the region cools and subsides. In contrast, Laramide-style basins, which are underlain by thermally mature lithosphere, may have a more uniform background heat flow through time because of the regional tectonic setting. For each basin, the combination of geological and heat-flow histories yields significantly different temperature histories. For an extensional basin, the temperature history of the sedimentary section is dependent on the transient heat-flow history and conductivity structure, whereas a Laramide-style basin is mainly dependent on the thermal conductivity structure of the basin (Fig. 16.1).

The purpose of this investigation is to: 1) develop a thermal model for a Laramide-style basin by incorporating a synthesis of the tectonic and subsidence histories; and 2) constrain this model by various thermal parameters. The thermal model is incorporated in a two-dimensional, finite-difference, numerical model that is used to define the temperature field in the basin at various times during the geological past. This model is dependent on the basin geometry, thermal conductivity structure, and regional heat flow. After integrating temperature data with the geological history of the basin, a composite time-temperature diagram is developed. The geological similarities of most Laramide-style basins in the Wyoming foreland provincc allow this model to be applied throughout the central Rocky Mountains. Specifically, we have chosen the Bighorn basin, Wyoming and Montana, as a case study.

Laramide-Style Basins

The Laramide-style basins of the Wyoming foreland province are deep, asymmetric, structural depressions bounded in part by large-scale basement-involved uplifts that have several kilometers of structural relief. Although similar style basins occur in Soviet central Asia, northern Afghanistan, northern Iran, Venezuela, and the Columbian Andes, the central and southern Rocky Mountains represent the best example of a foreland area disrupted by blockfaulted structures (Bally and Snelson, 1980). The term *Laramide-style basins* refers to those basins formed by basement-involved structural deformation of the Wyoming foreland province during the Laramide orogeny, which occurred from Campanian to Eocene time (Hagen et al., 1985).

Laramide-style basins have numerous geological features in common. Most basins in the central Rocky Mountains are bordered, on at least one side, by uplifts cored by Precambrian basement with as much as 7 km of structural relief. These basins are asymmetric adjacent to major bounding thrust faults and contain thick sequences of Upper Cretaceous and Tertiary strata.

The structural origin of Laramide-style basins has been the subject of controversy by two distinct schools of thought: vertical and horizontal tectonism. Brown (1984) gives a complete discussion of the wide range of structural styles present with basement-involved tectonics. Vertical tectonism requires vertical movement of basement blocks along high-angle faults resulting in drape folds and subsidiary thrusts in overlying strata (Prucha et al., 1965; Stearns, 1971; Woodward, 1976; Palmquist, 1978; Chapin and Cather, 1983). Horizontal tectonism requires horizontally directed compression as a mechanism for uplift and thrusting of basement blocks along moderate- to low-angel faults (Berg, 1962; Lowell, 1974; Smithson et al., 1978; Blackstone, 1983). In addition, those describing the importance of wrench faulting are Bell (1955), Sales (1968), Stone (1969), and Chapin (1983). Although the mechanics of deformation are of little importance to the thermal model presented herein, the geometry of the basement uplift (vertical versus low-angle thrusting) and rate of emplacement may have important effects on the temperature field within Laramide-style basins.

The nature of basin subsidence (Keefer and Love, 1963; Keefer, 1965; Hagen et al., 1985) and associated Tertiary sedimentation (Bradley, 1964; Keefer, 1965, 1970; Anderson and Picard, 1974; Hickey, 1980; Gingerich, 1983) are similar in most Laramide-style basins with variations because of the timing of deformation (Gries, 1983), variation in structural style (thrust versus wrench-faulting), and subsidence rates.

In general, most Laramide-style basins contain thick sequences of latest Cretaceous, Paleocene, and Eocene fluvial, lacustrine, and paludal deposits within the asymmetric portions of the basin (Fig. 16.2). Facies may vary laterally away from the basement-involved uplift as conglomerates and fluvial sandstones grade to mudstones, shales, and carbonates (Hickey, 1980). The thermal conductivity and temperature fields of each Laramide-style basin can be a function of heat flow, basin geometry, facies distribution of the Tertiary fill, and porosity and fluid composition. Variations in the timing of individual basement-block uplifts (Gries, 1983) may cause variations in sedimentary response during basin fill. Therefore, the thermal conductivity structure, basin geometry, and convective transport of fluids will change throughout the geological history of the basin. The end result will be variations in the temperature field through time.

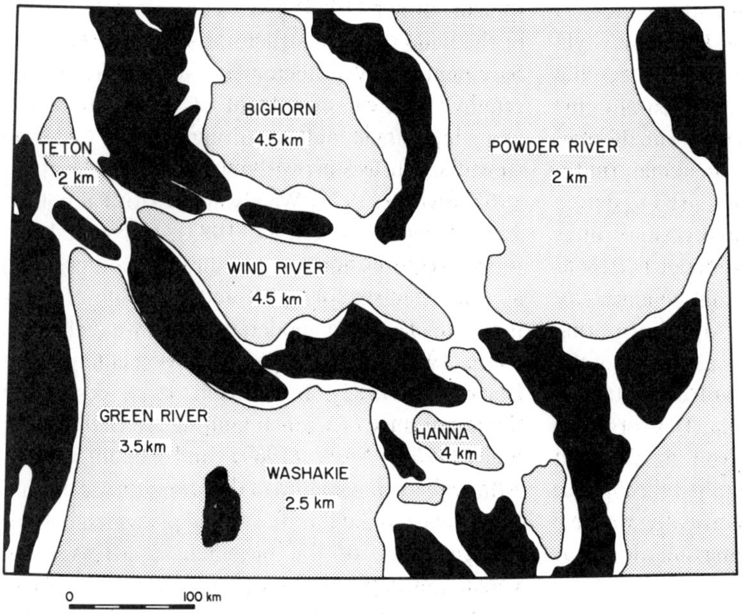

FIGURE 16.2. Distribution of Laramide-style basins in the central Rocky Mountain region illustrating the maximum Tertiary fill in each basin. Bounding uplifts are shown in solid and basins are illustrated in a stippled pattern.

Most Laramide-style basins in the Wyoming foreland province have experienced some degree of erosion to their present geomorphic form (Van Houten, 1952; Love, 1960). Regional epeirogenic uplift in Late Oligocene or Early Miocene initiated erosion of most basins in the central Rocky Mountains (Bown, 1980). Removal of Tertiary strata may range from a few hundred meters (Hanna basin) to kilometers (Bighorn basin). Therefore, in most cases maximum burial and temperatures will occur prior to epeirogenic uplift and erosion. As will be shown, this will have important implications on the thermal parameters used to constrain a thermal model for Laramide-style basins.

Bighorn Basin

Geological Constraints

The Bighorn basin, Wyoming and Montana, is both a structural and topographic basin that is asymmetric to the west adjacent to the Oregon basin-Beartooth thrust fault where a maximum vertical displacement of up to 6 km is observed (Fig. 16.3). The basin is presently bounded by the Owl Creek Mountains to the south, the Bighorn Mountains to the east, and the Absaroka Volcanics and Beartooth Mountains to the west. These elements define the margins of the basin both structurally, stratigraphically, and topographically.

The general structural geometry of the basin adjacent to the Beartooth uplift is illustrated by cross section (Fig. 16.4). The structural axis of the Bighorn basin passes under the Beartooth thrust front near this line of section where Precambrian basement rock is thrust on the order of 3 km (Foose et al., 1961) to as much as 12 km (Bonini and Kinard, 1983). The Beartooth block probably controlled both the structural configuration and post-Laramide sedimentation (i.e., Fort Union Formation) of this region (Fig. 16.4). Isopach patterns of Tertiary strata reflect rapid basin downwarping and sedimentation during Laramide deformation (Gingerich, 1983). This deformation reached the Bighorn basin area from the west by late Campanian time but maximum structural uplift of the Beartooth block did not occur until Paleocene and Early Eocene time (Bown, 1980). The structural and stratigraphic relationships (Fig. 16.4) are typical throughout the asymmetric portions of the Bighorn basin (Blackstone, 1985; Stone, 1985) where large thicknesses of Tertiary Fort Union and Willwood Formations have accumulated with thinning toward the eastern flank of the basin. Sand/shale ratios within Tertiary sediments varied

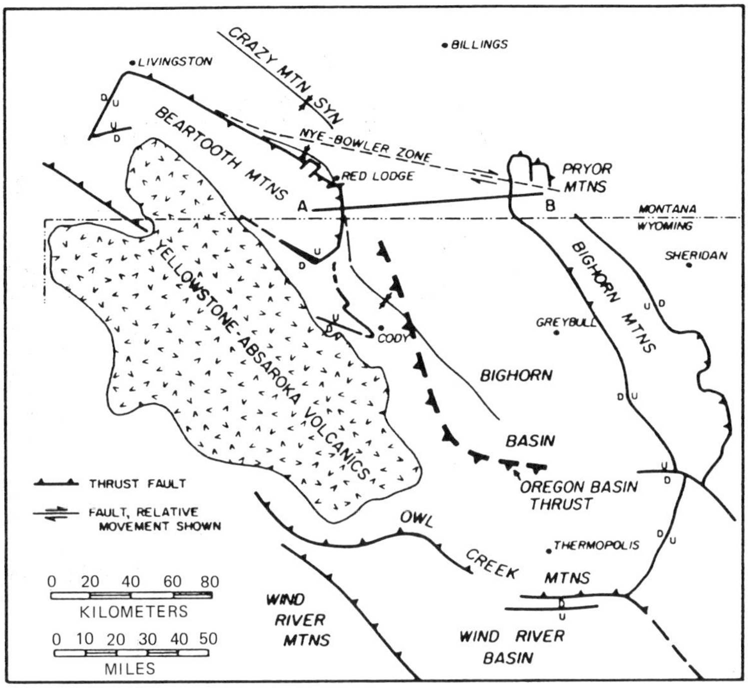

FIGURE 16.3. Location and major geological features of the Bighorn basin, Wyoming and Montana (modified from Foose, 1973). Location of the proposed thermal model and geological cross section is shown as profile A-B.

throughout the Bighorn basin depending on the proximity and timing of basement-involved uplifts (Bown, 1980).

Along the structural cross section (Fig. 16.4) more than 3 km of fluvial Fort Union Formation were deposited as conglomerates, sandstones, shales, coals, and limestones. Subsidence rates often exceeded or equaled sedimentation rates as evidenced by interbedded coal and lacustrine deposits and the dominance of mudstone over sandstone in various parts of the formation (Allen, 1978).

The thermal evolution of the Bighorn basin is highly dependent on the maximum sediment thickness obtained during and after Laramide deformation. In general, Tertiary sedimentation was considered to be continuous until Pliocene time in northwestern Wyoming (Love, 1960; McKenna and Love, 1972; McKenna, 1980). The "basin-fill" model requires net aggradation through Early Pliocene time to present-day elevations of approximately 2.7 km. Since that time, regional epeirogenic uplift initiated continuous excavation of the Bighorn basin removing virtually all of the middle to late Tertiary rocks except for scattered remnants on the crest of the Bighorn Mountains and in the southwestern Absaroka Range (McKenna and Love, 1972; Love et al., 1976) at elevations ranging from 2.3 to 2.7 km (Sundell, 1985). This model proposes that more than 1.5 km of sediment fill has been removed from most of the Bighorn basin in the past 3 my—approximately 35,000 km^3 of sedimentary rock (Sundell, 1985).

Thermal Constraints

Any thermal model for a Laramide-style basin must take into account the past and present geological history including the structural and stratigraphic framework of the basin. This includes: 1) thermal refraction from Precambrian basement uplifts (Gretener, 1981); 2) sedimentation rates during and after Laramide deformation (Kappelmeyer and Hanel, 1974); 3) basin-wide variations in lithology and thermal conductivities; 4) variations in heat flow throughout the geological history of the basin; and 5) the hydrological system of aquifers present within the stratigraphic section.

The Bighorn basin is surrounded by basement-involved structures with substantial vertical relief. The probable Precambrian overhang of these major thrust faults ranges from 1 km to 19 km

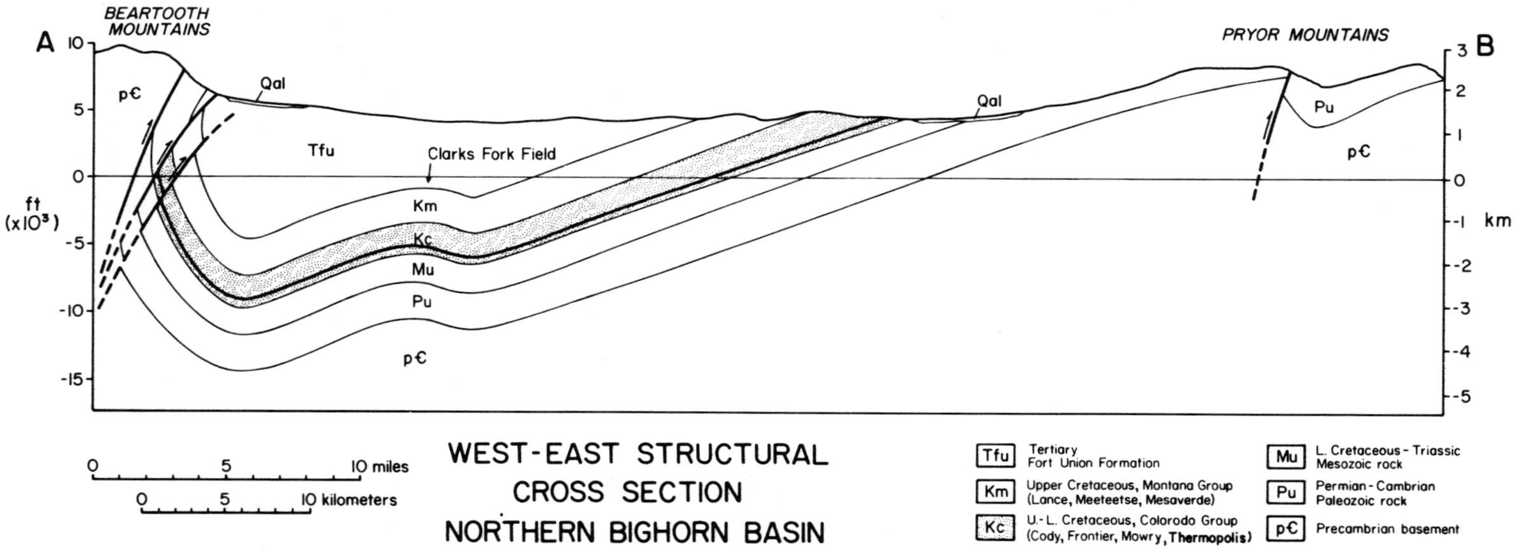

FIGURE 16.4. Generalized structural cross section through the northern Bighorn basin. Structural cross sections were integrated into the thermal model presented in this investigation to define the temperature field throughout the basin. The Mowry Formation is shown as a solid black layer within the stratigraphic sequence of Cretaceous hydrocarbon source rocks (stippled) (Hagen and Surdam, 1984). Location of cross section A-B is shown in Figure 16.3 (modified from Hagen and Surdam, 1984).

(Blackstone, 1985). Therefore, pronounced temperature anomalies due to abrupt lateral variations in rock type (i.e., thermal conductivities) may occur. The thermal model presented in this report evaluates the effects of basement involved thrusting on the temperature history of the Bighorn basin. Cross section A-B (Fig. 16.4) was chosen because it represents the narrowest part of the basin where thermal refraction may be greatest.

During Laramide deformation several kilometers of Paleocene Fort Union Formation and Eocene Willwood Formation were deposited in the deep, asymmetric portions of the basin. In the northern Bighorn basin, sedimentation rates for the Fort Union Formation reached a maximum of 280 m/my (Gingerich, 1983) or 0.3 mm/yr. Calculations for thermal disequilibrium due to rapid burial by sediments suggest that a noticeable perturbation in the thermal equilibrium of a basin occurs only after sedimentation rates exceed 2 mm/yr (Kappelmeyer and Hanel, 1974). Therefore, the cooling effect of rapid burial was neglected.

A thrust-depositional model for the Beartooth Mountains is reconstructed in two block diagrams (Fig. 16.5). Thrusting of the Beartooth block began in Paleocene time and the Fort Union Formation was deposited as thick fluvial sandstones with a coarser conglomeratic facies proximal to the mountain front (Jobling, 1974; Hickey, 1980). Maximum deformation occurred in Late Paleocene and Early Eocene time with rapid structural elevation of the basin margins and creation of a highly asymmetric basin (Bown, 1980). In general, conglomeratic facies are adjacent to the mountain front (Hickey, 1980) and grade laterally into predominantly fluvial sandstones and shales to the east. Lacustrine and paludal facies also occur within the asymmetric portion of the basin (Gingerich, 1983). The model (Fig. 16.5) illustrates that, basin-wide, there may be dramatic changes in the lithological characteristics and thicknesses of the Tertiary stratigraphic section. In effect, the center of the basin adjacent to the uplift will have a higher proportion of sandstone than the opposite basin margin (Fig. 16.5). Therefore the overall thermal conductivity for the entire stratigraphic sequence, present-day surface to base of the hydrocarbon source rock (Fig. 16.4), will be higher in the basin center than the margin. These variations in thermal conductivity will cause basin-wide variations in conductive heat transfer and, therefore, variations in the thermal gradients.

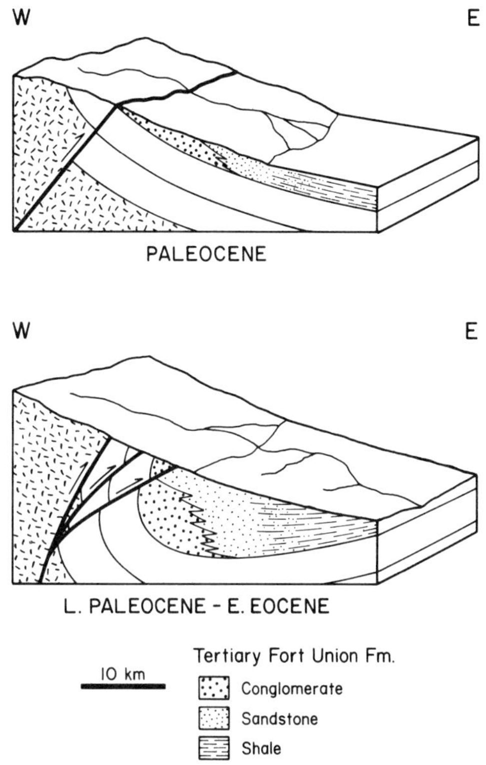

FIGURE 16.5. A thrust-depositional model for the northern Bighorn basin during Laramide deformation and deposition of the Tertiary sedimentary sequences. The model is constructed along profile A-B (Fig. 16.3) (modified from Jobling, 1974).

An additional thermal perturbation may occur in a basin because of migration of fluids in porous rock during compaction or normal groundwater transport. Heat flows more slowly through rock by conduction than if there is additional convective heat transport associated with migrating fluid (Buntebarth, 1980). Heasler and Hinckley (1985) demonstrated that most thermally anomalous areas in the Bighorn basin are associated with local geological structures where deep-heated waters are brought closer to the surface along basin-flanking anticlines. Also, Meyer and McGee (1985) demonstrated that positive geothermal anomalies were defined in the Elk basin anticline (south of profile A-B; Fig. 16.4) by lateral and upward movement of fluids through a porous medium and fracture sys-

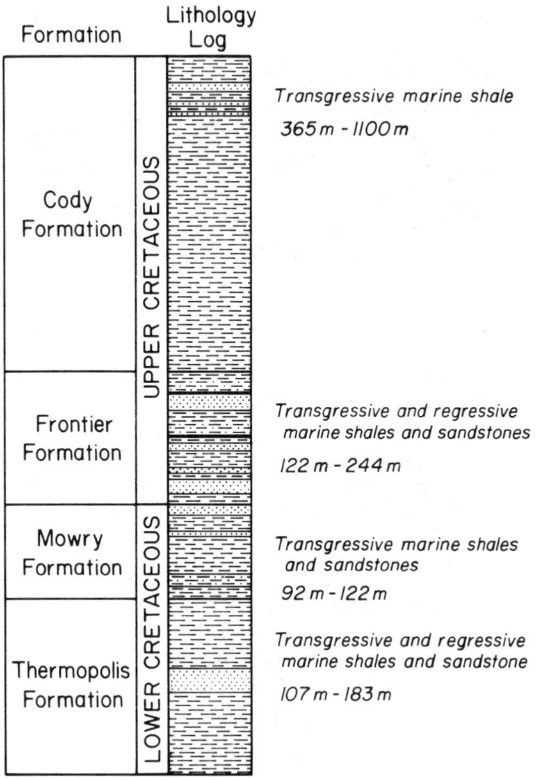

FIGURE 16.6. Generalized stratigraphic section for the Cretaceous Cody, Frontier, Mowry, and Thermopolis Formations, Bighorn basin, Wyoming and Montana. Lithology log is diagrammatic (modified from Hagen and Surdam, 1984).

tem. These geothermal anomalies manifest themselves as geothermal gradients that are much higher than regional geothermal gradients. On a basin-wide scale regional geothermal anomalies are difficult to demonstrate because of lack of geological data (e.g., past and present hydrological regime, paleo-heat flow, aquifer conditions) and the anisotropic nature of most basins. However, Hitchon (1984) demonstrated a genetic link between topography (water-table elevation), hydrodynamic regime, and geothermal gradient patterns in Alberta, Canada. Unfortunately, little published hydrological data exist for the northern Bighorn basin. In addition, Heasler and Hinckley (1985) noted no regional geothermal anomalies in the basin other than local structurally controlled geothermal systems.

Thermal Parameters

The thermal parameters presented in this investigation are two of several parameters derived from Cretaceous hydrocarbon source rocks (stippled region, Fig. 16.4; Hagen and Surdam, 1984) and oil well data. A thermal parameter is defined as an index of thermal maturation, or thermal exposure, that is obtained through geochemical analysis of a sediment. These parameters include mixed-layer illite/smectite clays, vitrinite reflectance, elemental analysis of kerogen, and pyrolysis data. Illite/smectite clays and vitrinite reflectance provide information on the degree of thermal stress or maximum temperature experienced by a sediment. In contrast, elemental analysis and pyrolysis provide information on the degree of hydrocarbon generation in a source rock. Parameters presented in this investigation are those that have been correlated to temperature: illite/smectite clays for maximum temperature and oil well bottom-hole temperatures for present-day formation temperatures.

In the Bighorn basin, the Cody, Frontier, Mowry, and Thermopolis Formations consist of over 1.2 km of marine shales, mudstones, and sandstones deposited in nearshore and offshore environments. These units were deposited in the Western Interior foreland basin during frequent transgressions of the epicontinental seaway. This created conditions favorable for both the production and preservation of marine organic matter. The combination of restricted circulation (Kauffman, 1977), high rates of organic matter accumulation, and periodic stratification in the water column creating anaerobic conditions (Arthur et al., 1981) produced organic-rich rocks. A general stratigraphic column illustrates the hydrocarbon source rocks for this sequence (Fig. 16.6).

These Cretaceous rocks yield useful thermal parameters for several reasons. They contain an abundance of volcanically derived mixed-layer illite/smectite clays. Additionally, they were deposited in a marginal marine setting accounting for the influx of terrestrial organic debris including vitrinite particles, and contain sufficient organic material to be considered effective hydrocarbon source rocks (Nixon, 1973; Hagen and Surdam, 1984). The present stratigraphic and structural position of these sequences is depicted in Figure 16.4. All thermal maturation data presented herein are from the Mowry Formation (Fig. 16.6).

Clay Diagenesis

The diagenesis of interstratified illite/smectite (I/S) clays during progressive burial of a sedimentary sequence is a widely recognized reaction and is considered to be an important diagenetic geothermometer (Burst, 1969; Hower et al., 1976; Boles and Franks, 1979). Previous investigations (Hower et al., 1976; Boles and Franks, 1979) suggest a general reaction for the conversion of I/S clays during progressive burial:

$$K^{+} + (Al^{+3}) + \text{smectite} = \text{illite} + Si^{+4}$$

Recent studies (Nadeau et al., 1984), however, suggest that interstratified clays are aggregates of fundamental clay particles (smectite or illite) whose crystallite sizes change by progressive dissolution of smectite and reconstitution to thin illite particles. Although the nature of the I/S transformation may be different than originally proposed, the principles of diagenetic trend identification by X-ray diffraction are essentially the same.

Mixed-layer clays in the Mowry Formation exhibit random ordering from 80 to 25% expandability and an ordered interlayer structure at low expandabilities (<25% expandable layers; Hower et al., 1976; Hagen and Surdam, 1984). All I/S clays examined in the Bighorn basin show some degree of diagenetic alteration. Factors controlling this diagenetic trend include temperature, pressure, pore fluid composition, bulk chemistry of the shale, composition of starting smectite, sources of potassium, and time (Hower, 1981).

Numerous studies (Hower et al., 1976; Boles and Franks, 1979; Hoffman and Hower, 1979; Hower, 1981) suggest that the conversion of random to ordered I/S clays takes place at approximately 100°C, with the conversion of smectite to illite beginning at approximately 60°C. In addition, Hoffman and Hower (1979), in a study of Cretaceous rocks from the Montana thrust belt, estimate diagenetic temperatures for the entire range of I/S conversion applicable to upper Mesozoic strata. Application of these temperatures to

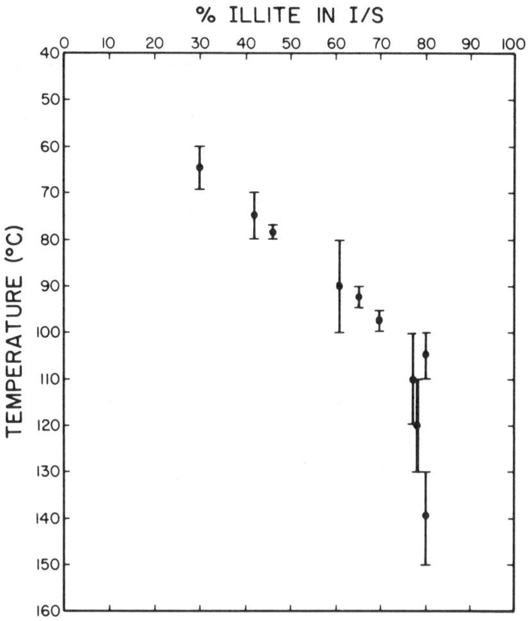

FIGURE 16.7. Mixed-layer illite/smectite clay compositions and estimated diagenetic temperatures (Hoffman and Hower, 1979). Clay compositions are estimated by X-ray diffraction techniques (Hower, 1981) for the entire depth range of the Mowry Formation along profile A-B (Fig. 16.4). Error bars illustrate the range in temperature for any particular clay composition.

I/S clay compositions in the Bighorn basin (Fig. 16.7) yields maximum diagenetic temperatures at maximum burial prior to regional epeirogenic uplift in Late Oligocene or Early Miocene (Bown, 1980) and subsequent erosion. This will be further illustrated by comparing present-day formation temperatures to maximum diagenetic temperatures used to constrain the proposed thermal model.

The variation in temperatures (100° to 150°C) at 80% illite in the I/S (Fig. 16.7) occurs because the rocks have reached burial temperatures of 100°C, sufficient to cause ordering in the illite/smectite clays. With increasing burial along cross section A-B (Fig. 16.4), there is an increase in the degree of ordering (Hower, 1981) or crystallinity of the I/S clay. Temperatures estimated for the various I/S compositions (±10°C) are consistent with temperatures estimated by Burtner and Warner (1984) for I/S clays in the Powder River basin.

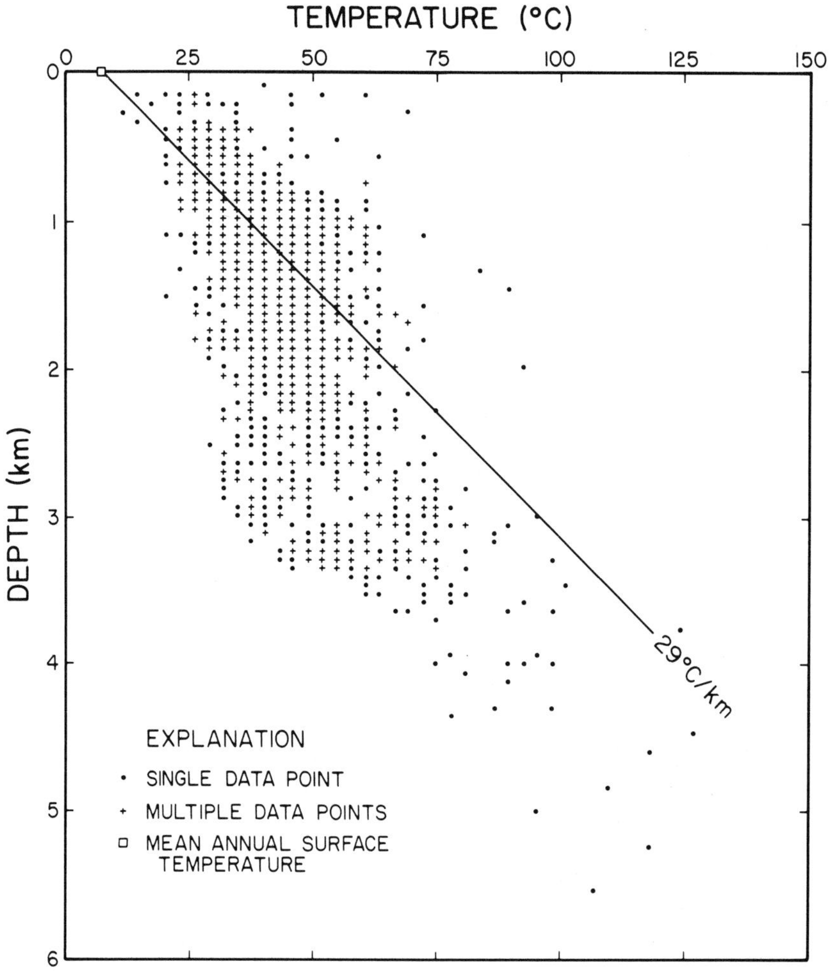

FIGURE 16.8. Temperature-depth plot of bottom-hole temperatures in the Bighorn basin. (Modified from Heasler and Hinckley, 1985.)

Temperature Data

The most prolific source of temperature information for deep subsurface temperatures is oil wells. Sources of subsurface temperature data include: 1) thermal logs from wells, and 2) oil and gas well bottom-hole temperatures (BHT). For a complete discussion of the uses and limitations of each method, the reader is referred to Gretener (1981). To summarize, thermal logs and BHT can give accurate temperature data for thermal modeling if: 1) logged holes have equilibrated to true rock temperatures; 2) there are no large fluctuations in surface temperature, especially for shallower holes; 3) temperature intervals are small; and 4) geological conditions such as faulting, overpressuring, and hydrodynamics are known. Bottom-hole temperatures are the most abundant subsurface data and represent a single temperature point at the total depth of an oil or gas well. Those temperatures reflect the thermal conductivity of the entire rock sequence, the geological conditions of the area, regional heat flow, and thermal perturbations due to drilling. As bottom-hole temperatures are normally measured a few hours after drilling has stopped, the measured temperatures are usually lower than the actual formation temperature. The well, therefore, is in a state of thermal disequilibrium due to the cooling and heating effect of drilling mud, variations in subsurface temperature, and mud circulation rates.

Temperature data for the Bighorn basin have been presented in detail by Heasler and Hinckley (1985) and results of approximately 2,035 bottom-hole temperature measurements are presented (Fig. 16.8). Gradients derived from shallow bottom-hole temperatures tend to be higher than average, and gradients derived from deep bottom-hole temperatures are lower due to drilling effects (Heasler and Hinckley, 1985). These authors have also calculated background thermal gradients based on thermal logs, thermal conductivities of basin sedimentary sequences, and heat flow (Fig. 16.8). A background thermal gradient of 29°C/km has been calculated for the Bighorn basin. In some cases the higher temperatures of shallower wells (basin flank) and lower temperatures for deeper wells (basin center) may be a result of basin geometry and lateral variations in the thermal conductivity structure of the basin.

Thermal Model

To evaluate the temperature history of the Bighorn basin, a steady-state, two-dimensional, finite-difference model is proposed. The equation:

$$\frac{\delta}{\delta x}\left[k\,\frac{\partial T}{\partial x}\right]+\frac{\delta}{\partial y}\left[k\,\frac{\partial T}{\partial y}\right]=0$$

describes the steady-state conductive transport of heat where k is the thermal conductivity and T represents temperature. The thermal model was used to compute temperatures in the basin both at times of maximum and present-day fill. Computations were performed on the University of Wyoming Cyber CDC 760 computer using methods described by Heasler (1984).

Input parameters for the steady-state finite-difference model include basin geometry and geological conditions. These parameters include basal heat flow, thermal conductivity of formations, basin geometry, and a ground-surface temperature. In effect, basin geometry will control the regional thermal conductivity structure of the basin in two dimensions, including Precambrian basement.

In a steady-state thermal regime, heat flow is considered to be constant throughout geological time. Present-day heat-flow values have been measured in various areas of the Bighorn basin and range from 54 to 75 mW/m^2 (Heasler et al., 1982). A heat-flow value of 67 mW/m^2 was used in thermal modeling based on data from Blackwell (1969), Heasler (1978), and Heasler et al. (1982).

The model presented in this report is a first approximation of the temperature history for a Laramide-style basin. Thermal conductivity values are therefore averages for four stratigraphic units (Fig. 16.9): 1) Tertiary and Upper Cretaceous, 2) Cretaceous hydrocarbon source rocks, 3) Mesozoic and Paleozoic, and 4) Precambrian. Thermal conductivity values were taken from Heasler (1978) and Hinckley et al. (1982) and from estimates based on observed rock lithologies. Because thermal conductivities have not been measured for the Upper Cretaceous and Tertiary fill in the basin, estimates were based on composite electric-log lithologies along the cross section (Fig. 16.4) and published lithological descriptions. Thermal conductivities used in each model are shown in Figure 16.9.

Given the input parameters, the finite-difference model can be used to calculate the temperatures within the basin due to the vertical and horizontal conductive transport of heat. Because the model is constrained by thermal parameters discussed earlier, the model was run at maximum basin fill (Early Miocene), which records maximum diagenetic temperatures (Fig. 16.9). A 2.7-km basin-fill model was used for the northern Bighorn basin as described earlier. A model was also run for the present-day basin configuration to compare unpublished and published (Heasler and Hinckley, 1985) present-day temperature data to modeled temperatures.

From the numerical model (Fig. 16.9), several points are worth noting: 1) because there is little contrast in thermal conductivities, juxtaposition of basement uplifts causes very little refraction ($\approx 10\%$) of the isotherms within the basin; 2) isotherms are refracted by lower conductivity strata; 3) in areas of high structural relief, such as the western edge of the basin, isotherms are more severely curved; and 4) geothermal gradients vary from the basin center to the margin. Geothermal gradients vary from 25°C/km for the basin center to 31°C/km for the basin margin (Fig. 16.9). The

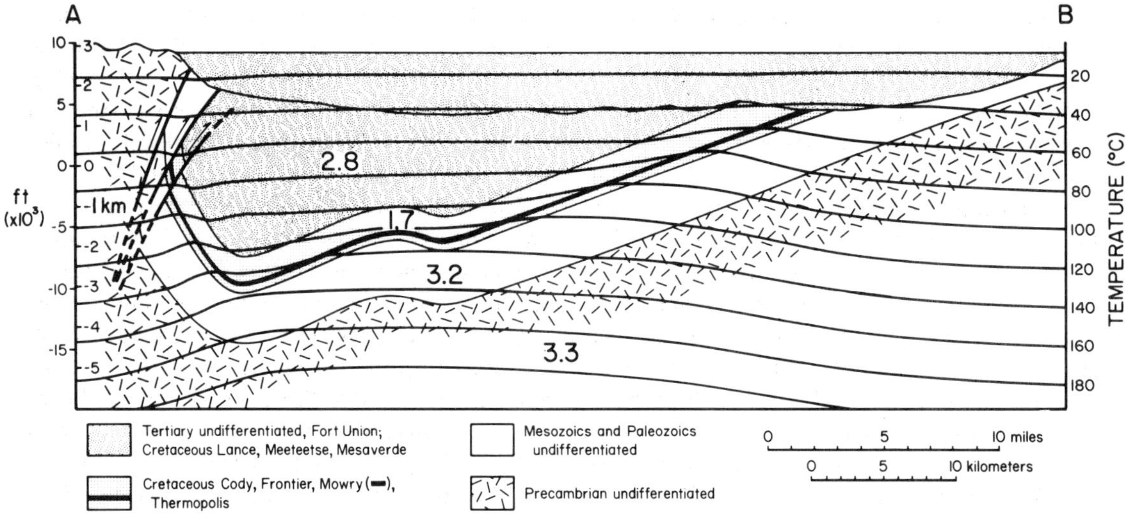

FIGURE 16.9. Results of the steady-state, finite-difference, numerical model along cross section A-B (Fig. 16.4). Numbers represent the thermal conductivities (W/m°K) used for the four units and heat flow remained constant at 67 mW/m^2. The basin is modeled at maximum fill (2.7 km) during Early Miocene time. A ground-surface temperature of 7°C was also used (Lowers, 1960).

thermal gradients may vary depending on the various combinations of thermal conductivities and heat-flow values used in the model (Hagen and Surdam, 1984).

Results of two-dimensional, finite-difference modeling were compared with temperature results of a one-dimensional model. A one-dimensional model calculates temperatures according to Fourier's Law: $dT = q/k\ dx$, where k is thermal conductivity, q is basal heat flow, dx is thickness, and dT is temperature increase. A one-dimensional model is valid for the central and eastern portions of the basin, but along the extreme western margin of the basin, two-dimensional transport of heat begins to affect temperatures by about 10%. The effect of the two-dimensional transport of heat will be less in the past during basin deformation. Consequently, a one-dimensional model can be used to approximate the temperature history of the Bighorn basin using the same input parameters as the two-dimensional thermal model.

To check the validity of the two-dimensional thermal model, calculated temperatures from the thermal model were compared with measured temperatures. Figure 16.10 illustrates calculated temperature data for the Mowry Formation with depth at present-day and maximum basin fill along profile A-B (Fig. 16.4). Data plot on two distinct linear trends representing the thermal conductivities of the Tertiary and Upper Cretaceous overburden (solid line) and Cretaceous hydrocarbon source rocks (dashed line). Both the present-day and maximum fill modeled trends overlap because only 1.5 km of the Tertiary-Upper Cretaceous overburden are removed while the basin geometry remains the same.

Mowry Formation exhibits progressively less Tertiary-Upper Cretaceous overburden moving from the basin center to the flank, which is reflected in a decrease in plotted temperature data with decreasing depth (Fig. 16.10). If D_1 (inset, Fig. 16.10) represents the basin center and D_3 the basin flank, then all predicted temperature data should plot along a line connecting D_1 and D_3. This occurs until only lower thermal conductivity rocks (Cretaceous source rocks) are present at the outcrop for the present-day fill model (dashed line, Fig. 16.10). The slope of the line connecting points D_1, D_2, and D_3 represents an "apparent gradient" in the basin and should be the same as the thermal gradient calculated for the Tertiary-Upper Cretaceous section. Calculated thermal gradients for

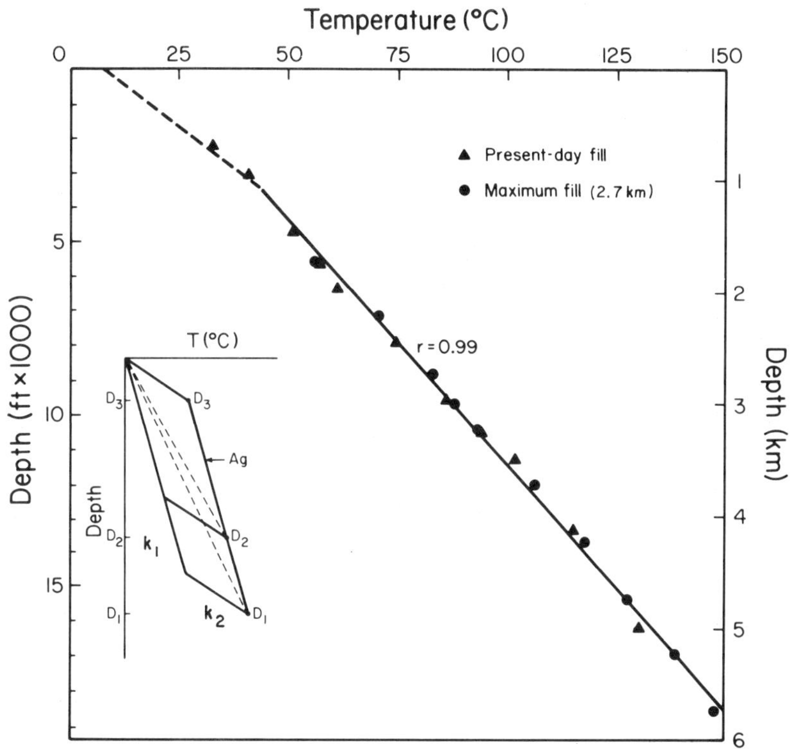

FIGURE 16.10. Calculated temperature data for the Mowry Formation with depth using the profile and data described in Figure 16.9. The solid regression line (n = 0.99) is calculated from data representing varying amounts of Tertiary-Upper Cretaceous overburden. Dashed line represents the temperature profile for Cretaceous rocks near the surface, without Tertiary-Upper Cretaceous overburden. Inset diagram illustrates schematically the computation of these temperature data at depths D_1, D_2, and D_3 where k_1 is thermal conductivity of Tertiary-Upper Cretaceous sediments and k_2 is thermal conductivity of Cretaceous source rocks. Average thermal gradients are represented by dashed lines (inset) to each depth point (i.e., D_1, D_2, and D_3). Apparent gradient (Ag) is the slope of a line connecting D_1, D_2, and D_3 and is equivalent to the thermal gradient through the Tertiary-Upper Cretaceous section.

data in Figure 16.10 and the Tertiary-Upper Cretaceous section (k = 2.8 W/m°K) are 23°C/km and 24°C/km, respectively.

The average thermal gradient down to the Mowry Formation at each depth is represented by the dashed lines (inset) in Figure 16.10. The thermal gradient progressively increases toward the flanks of the basin as suggested by the results in Figure 16.9.

An excellent correlation exists between predicted and observed temperature data. In Figure 16.11, both present-day formation temperatures and maximum diagenetic temperatures for the Mowry Formation are plotted along with the linear regression lines illustrated in Figure 16.10. Present-day formation temperatures have been calculated using corrected bottom-hole temperatures. These data were then compared to bottom-hole temperature data given in Heasler and Hinckley (1985) in order to delineate any spurious temperatures. Maximum diagenetic temperatures were computed for illite/smectite clay data and are considered to have a ±10°C error.

Present-day formation temperatures show an excellent correlation with predicted temperature trends (Fig. 16.11) with an average error of ±5%. Clay temperatures show good correlation to the predicted temperature trend (Fig. 16.11) with scat-

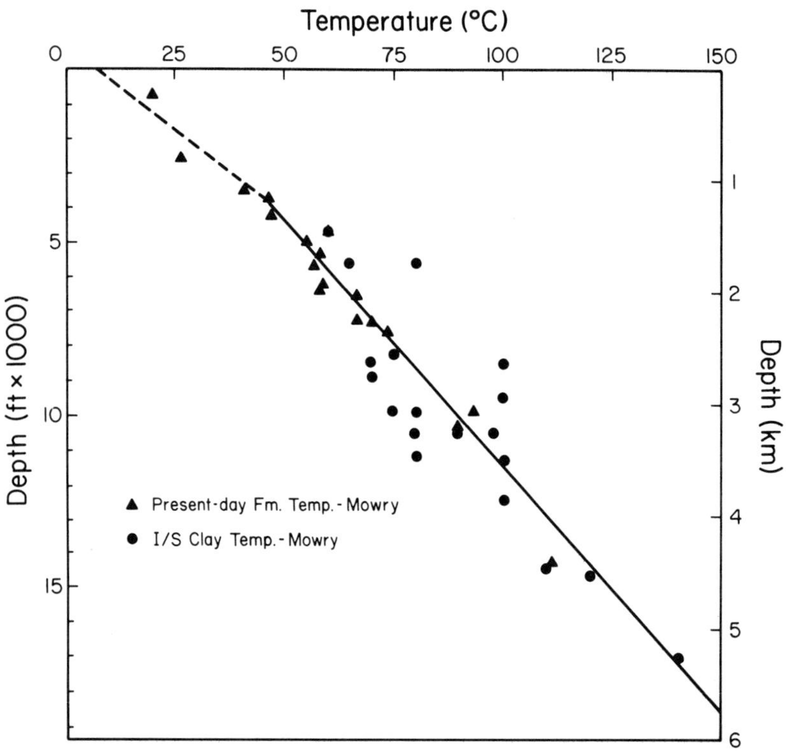

FIGURE 16.11. Observed temperature data plotted on the regression lines computed in Figure 16.10. Present-day formation temperatures for the Mowry are calculated from corrected bottom-hole temperatures. Maximum diagenetic temperatures are calculated from illite/smectite clays and represent temperatures at maximum basin-fill.

ter due to uncertainties in clay composition and/or temperature calculations. The average error between predicted and observed temperatures at maximum basin fill is $\pm 10\%$. It is therefore concluded that the thermal model and input parameters as well as the basin-fill model (2.7 km) used in this investigation adequately describe observed temperature data for the northern Bighorn basin.

Thermal conductivity will generally increase with a decrease in porosity. Therefore, in the Bighorn basin, rapid burial, compaction, and diagenesis (cementation or decementation) may drastically alter the conductivity structure from the basin center to the flank. For example, a typical Fort Union graywacke (Jobling, 1974) with an initial porosity of 50% and a thermal conductivity of 1.7 W/m°K compacting to a final porosity of 25% ($\approx$3 km of burial) will have a final conductivity of 3.0 W/m°K, if filled with water. In addition, zones of early carbonate cementation or framework grain dissolution may have local effects on the porosity and thermal conductivity of the rock.

Figure 16.12 illustrates the difference between using a constant thermal conductivity (this investigation) for the Tertiary and Upper Cretaceous section, and a varying conductivity due to compaction. Variations in thermal conductivity with compaction were calculated using the compaction curves of Sclater and Christie (1980) for porosity and conductivity calculations using the methodology described by Woodside and Messmer (1961).

The maximum error incurred by using a constant thermal conductivity is approximately +10°C at 1.5 km of burial. This is within the error of measured and predicted thermal data used in the proposed thermal model; therefore thermal conductivity variations due to compaction were neglected. It is interesting to note that this temperature error may have a significant effect on thermal gradients along the basin margin, causing the overall gradi-

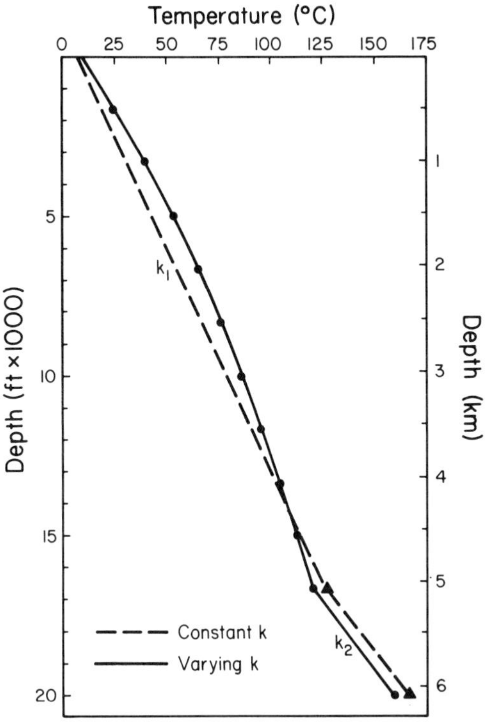

FIGURE 16.12. Temperature-versus-depth plot illustrating the effect of varying thermal conductivity with depth. The constant thermal conductivity plot was computed using a thermal conductivity of 2.8 W/m°K (k_1) for the Tertiary-Upper Cretaceous sequence and 1.7 W/m°K (k_2) for the Cretaceous source rock sequence. The varying conductivity was computed using the method of Woodside and Messmer (1961) after calculating the change in porosity with a 50:50 sand-shale sequence using compaction curves from Sclater and Christie (1980).

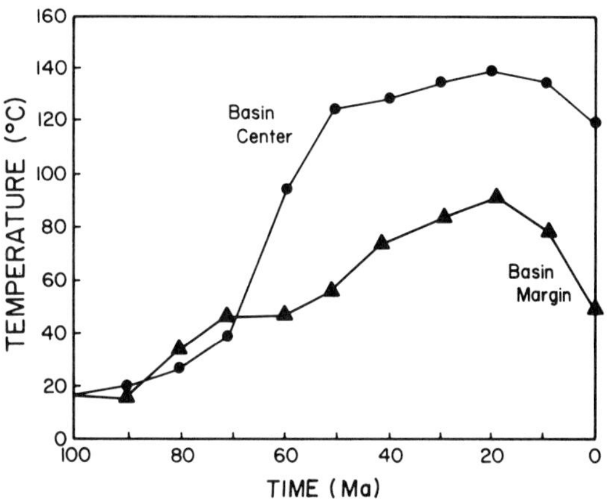

FIGURE 16.13. Temperature histories for the basin center and margin for the Mowry Formation derived from one-dimensional thermal modeling. Differences in the temperature histories are due to the basin geometry and maximum burial at each location.

ent along the margin to be higher. Using a constant thermal conductivity for approximately 1 km of burial by Tertiary and Upper Cretaceous sediment yields an overall thermal gradient for the Mowry Formation of 31°C/km. Using a compaction corrected thermal conductivity yields an overall thermal gradient of 37°C/km for the Mowry Formation.

Conclusions

The simple two-dimensional numerical model presented in this investigation adequately describes the thermal history of Cretaceous hydrocarbon source rocks in the Bighorn basin. Because of the geological similarities of most Laramide-style basins in the Wyoming foreland province, this thermal model may be applied to Laramide-style basins other than the Bighorn basin. These geological similarities include: 1) the presence of bounding, basement-involved uplifts; 2) asymmetric geometry due to thrusting of basement-involved uplifts; 3) large thicknesses of Upper Cretaceous and Tertiary sedimentary fill within the asymmetric portions of the basin; and 4) underlain by thermally mature lithosphere. Thermal histories may vary from basin to basin because of variations in the timing of basement-block uplifts and the distribution of resulting sedimentary facies.

By reconstructing the burial and tectonic history of Cretaceous rocks in the Bighorn basin and using thermal gradients defined by the thermal model, a time-temperature history can be developed for the basin center and margin. This method is similar to that proposed by Waples (1980) except that Lopatin's time-temperature index of maturity (TTI) is not determined. Instead, geothermal gradients are employed for the entire geological history of a formation of interest and the temperature history is determined (Fig. 16.13). Figure 16.13 illustrates the simplicity of the thermal model that has been presented. The curves are very similar, with each reaching maximum diagenetic temperatures at approximately 20 Ma during maximum burial by Tertiary sediments. The rapid increase in temperature between 70 and 50 Ma for the basin center is due to rapid burial over a very short time period.

Once a temperature history is determined, geochemical modeling may be used to calculate the level of thermal maturity relative to hydrocarbon generation as well as the timing of oil and gas generation for a particular hydrocarbon source rock. A theoretical maturation index, time-temperature index (Waples, 1980) or transformation ratio (Tissot and Espitalié, 1975), is determined by this process and may be compared to observed maturation parameters to provide additional constraints for a basin's thermal history. These maturation parameters include elemental analysis of kerogen and pyrolysis data.

Ultimately, geochemical modeling can be applied to frontier regions within a similar geological setting to determine the areal extent and timing of oil and gas generation in a particular hydrocarbon source rock, as well as potential migration pathways and hydrocarbon traps throughout the geological history of the basin.

Acknowledgments. Cooper Land (Golden and Land Oil Company, Bismarck, North Dakota) provided much of the stratigraphic data and valuable field discussion for the northern Bighorn basin. We gratefully acknowledge the assistance of the Montana Oil and Gas Commission (Billings, Montana) in providing access to vital drill core material. Hagen would like to thank Mobil Oil Company (Dallas, Texas) for basic research support and Amoco Production (Denver, Colorado) for financial aid to support field work; Surdam thanks Texaco Oil Company (Houston, Texas) for basic research support. We also thank the following individuals in the Geology Department of the University of Wyoming: Steve Boese (analytical chemistry) and Laurie Crossey (time-temperature diagrams). These individuals provided significant discussions and analytical support for our work. Henry Heasler provided the numerical modeling and exchange of ideas related to this project and to thermal modeling. Lastly we thank Donald Blackstone for his invaluable discussions and information concerning the structure of the Bighorn basin.

References

Allen, J.R.L. 1978. Studies in fluviatile sedimentation: An exploratory quantitative model for the architecture of avulsion-controlled alluvial suites. Sedimentary Geology 21:129–147.

Anderson, D.W., and Picard, D.M. 1974. Evolution of synorogenic clastic deposits in the intermontane Uinta Basin of Utah. In: Dickinson, W.R. (ed.): Tectonics and Sedimentation. Society of Economic Paleontologists and Mineralogists Special Publication 22, pp. 167–189.

Angevine, C.L., and Turcotte, D.L. 1981. Thermal subsidence and compaction in sedimentary basins: Applications to Baltimore Canyon Trough. American Association of Petroleum Geologists Bulletin 65: 219–225.

Arthur, M.A., Scholle, P.A., Pollastro, R.M., Barker, C.E., and Claypool, G.E. 1981. Geochemical and sedimentologic indicators of depositional environments and diagenetic trends in the Niobrara Formation (upper Cretaceous), Colorado and Kansas. Geological Society of America Abstracts with Programs 13: 398–399.

Bally, A.W., and Snelson, S. 1980. Facts and principles of world petroleum occurrence: Realms of subsidence. In: Miall, A.D. (ed.): Facts and Principles of World Petroleum Occurrence. Canadian Society of Petroleum Geologists Memoir 6, pp. 9–94.

Bell, W.G. 1955. Geology of the southeastern flank of the Wind River Mountains, Fremont County, Wyoming. Ph.D. thesis, University of Wyoming, Laramie, 204 pp.

Berg, R.R. 1962. Mountain flank thrusting in Rocky Mountains foreland, Wyoming and Colorado. American Association of Petroleum Geologists Bulletin 46: 2019–2032.

Blackstone, D.L. 1983. Laramide compressional tectonics southeastern Wyoming. University of Wyoming Contributions to Geology 22(1):1–38.

Blackstone, D.L. 1985. Foreland compressional tectonics: Southern Bighorn Basin, Wyoming. Wyoming Geological Survey Open File Report 85-3, 27 pp.

Blackwell, D.D. 1969. Heat flow determinations in the northwestern United States. Journal of Geophysical Research 74:992–1007.

Boles, J.R., and Franks, S.G. 1979. Clay diagenesis in Wilcox sandstones of southwest Texas: Implications of smectite diagenesis on sandstone cementation. Journal of Sedimentary Petrology 49:55–70.

Bonini, W.E., and Kinard, R.E. 1983. Gravity anomalies along the Beartooth front, Montana. Wyoming Geological Association Thirty-Fourth Annual Field Conference Guidebook, pp. 89–95.

Bown, T.M. 1980. Summary of latest Cretaceous and Cenozoic sedimentary, tectonic, and erosional events, Bighorn Basin, Wyoming. In: Gingerich, P.D. (ed.): Early Cenozoic Paleontology and Stratigraphy of the Bighorn Basin. University of Michigan Papers on Paleontology 24, pp. 25–32.

Bradley, W.H. 1964. Geology of Green River Formation and associated Eocene rocks in southwestern Wyoming, and adjacent parts of Colorado and Utah. U.S. Geological Survey Professional Paper 496A, 86 pp.

Brown, W.G. 1984. Basement involved tectonics, foreland areas. American Association of Petroleum Geologists Continuing Education Course Note Series 26, 92 pp.

Buntebarth, G. 1980. Geothermics, an Introduction. Berlin, Springer-Verlag, 144 pp.

Burst, J.R., Jr. 1969. Diagenesis of Gulf Coast clayey sediments and its possible relationships to petroleum migration. American Association of Petroleum Geologists Bulletin 53:487–502.

Burtner, R.L., and Warner, M.A. 1984. Hydrocarbon generation in lower Cretaceous Mowry and Skull Creek shales of the northern Rocky Mountain area. In: Woodward, J., Meissner, F.F., and Clayton, J.L. (eds.): Hydrocarbon Source Rocks of the Greater Rocky Mountain Region. Denver, Rocky Mountain Association of Geologists, pp. 449–468.

Chapin, C.E. 1983. An overview of Laramide wrench faulting in the southern Rocky Mountains with emphasis on petroleum exploration. In: Lowell, J.D. (ed.): Rocky Mountain Foreland Basins and Uplifts. Denver, Rocky Mountain Association of Geologists, pp. 169–179.

Chapin, C.E., and Cather, S.M. 1983. Eocene tectonics and sedimentation in the Colorado Plateau-Rocky Mountain area. In: Lowell, J.D. (ed.): Rocky Mountain Foreland Basins and Uplifts. Denver, Rocky Mountain Association of Geologists, pp. 33–56.

Foose, R.M. 1973. Vertical tectonism and gravity in the Bighorn Basin and surrounding ranges of the middle Rocky Mountains. In: De-Jong, K.A., and Scholten, R. (eds.): Gravity and Tectonics. New York, Wiley, pp. 443–455.

Foose, R.M., Wise, D.U., and Garbarini, G.S. 1961. Structural geology of the Beartooth Mountains, Montana and Wyoming. Geological Society of America Bulletin 72:1143–1172.

Gingerich, P.D. 1983. Paleocene-Eocene faunal zones and a preliminary analysis of Laramide structure deformation in the Clark's Fork Basin, Wyoming. Wyoming Geological Association Thirty-Fourth Annual Field Conference Guidebook, pp. 185–196.

Gretener, P.E. 1981. Geothermics: Using temperature in hydrocarbon exploration. American Association of Petroleum Geologists Education Course Note Series 17, 170 pp.

Gries, R. 1983. North-south compression of Rocky Mountain foreland structures. In: Lowell, J.D. (ed.): Rocky Mountain Foreland Basins and Uplifts. Denver, Rocky Mountain Association of Geologists, pp. 9–32.

Hagen, E.S., Shuster, M.W., and Furlong, K.P. 1985. Tectonic loading and subsidence of intermontane basins. Wyoming foreland province. Geology 13: 585–588.

Hagen, E.S., and Surdam, R.C. 1984. Maturation history and thermal evolution of Cretaceous source rocks of the Bighorn basin, Wyoming and Montana. In: Woodward, J., Meisser, F.F., and Clayton, J.L. (eds.): Hydrocarbon Source Rocks of the Greater Rocky Mountain Region. Denver, Rocky Mountain Association of Geologists, pp. 321–338..

Heasler, H.P. 1978. Heat flow in Elk Basin oil field, northwestern Wyoming. M.S. thesis, University of Wyoming, Laramie, 168 pp.

Heasler, H.P. 1984. Thermal evolution of coastal California with application to hydrocarbon maturation. Ph.D. thesis, University of Wyoming, Laramie, 76 pp.

Heasler, H.P., Decker, E.R., and Buelow, K.L. 1982. Heat flow studies in Wyoming, 1979 to 1982. In: Ruscetta, C.A. (ed.): Geothermal Direct Heat Program Roundup Technical Conference Proceedings: Vol. 1, State Coupled Resource Assessment Program. Salt Lake City, UT, University of Utah Research Institute Earth Science Laboratory Division, pp. 290–312.

Heasler, H.P., and Hinckley, B.S. 1985. Geothermal resources of the Bighorn basin, Wyoming. Geological Survey of Wyoming Report of Investigations 29, 28 pp.

Heasler, H.P., and Surdam, R.C. 1983. A thermally subsiding basin model for the maturation of hydrocarbons in the Pismo Basin, California. In: Isaacs, C.M., and Garrison, R.E. (eds.): Petroleum Generation and Occurrence in the Miocene Monterey Formation, California. Los Angeles, Pacific Section, Society of Economic Paleontologists and Mineralogists, pp. 69–74.

Heasler, H.P., and Surdam, R.C. 1985. Thermal evolution of coastal California with application to hydrocarbon maturation. American Association of Petroleum Geologists Bulletin 69:1386–1400.

Hickey, L.J. 1980. Paleocene stratigraphy and flora of the Clark's Fork Basin. In: Gingerich, P.D. (ed.): Early Cenozoic Paleontology and Stratigraphy of the Bighorn Basin, Wyoming. University of Michigan Paper on Paleontology 24, pp. 33–49.

Hinckley, B.S., Heasler, H.P., and King, J.K. 1982. The Thermopolis hydrothermal system with an analysis of Hot Springs State Park. Geological Survey of Wyoming Preliminary Report 20, 42 pp.

Hitchon, B. 1984. Geothermal gradients, hydrodynamics and hydrocarbon occurrences, Alberta, Canada. American Association of Petroleum Geologists Bulletin 68:713–743.

Hoffman, J., and Hower, J. 1979. Clay mineral assemblages as low grade metamorphic geothermometers: Application to the thrust faulted disturbed belt of Montana, U.S.A. In: Scholle, P.A., and Schluger, P.R. (eds.): Aspects of Diagenesis. Society of Economic Paleontologists and Mineralogists Special Publication 26, pp. 55–80.

Hower, J. 1981. Shale diagenesis. In: Longstaff, F.J. (ed.): Clays and the Resource Geologist. Mineralogical Association of Canada Short Course Handbook 7, pp. 60–80.

Hower, J., Eslinger, E.V., Hower, M.E., and Perry, E.A. 1976. Mechanism of burial and metamorphism of argillaceous sediments: 1. Mineralogical and chemical evidence. Geological Society of America Bulletin 87: 725–737.

Hunt, J.M. 1979. Petroleum Geochemistry and Geology. San Francisco, Freeman, 617 pp.

Jobling, J.L. 1974. Stratigraphy, petrology, and structure of Laramide (Paleocene) sediments marginal to the Beartooth Mountains, Montana. Ph.D. thesis, Pennsylvania State University, University Park, 102 pp.

Kappelmeyer, D., and Hanel, R. 1974. Geothermics. Geoexploration Monograph 4. Berlin, Gebruter Borntraeger, 239 pp.

Kauffman, E.G. 1977. Geological and biological overview: Western Interior Cretaceous basin. The Mountain Geologist 14:75–99.

Keefer, W.R. 1965. Stratigraphic and geologic history of the uppermost Cretaceous, Paleocene, and lower Eocene rocks in the Wind River Basin, Wyoming. U.S. Geological Survey Professional Paper 495A, 77 pp.

Keefer, W.R. 1970. Structural geology of the Wind River Basin. U.S. Geological Survey Professional Paper 495D, 35 pp.

Keefer, W.R., and Love, J.D. 1963. Laramide vertical movements in central Wyoming. University of Wyoming Contributions to Geology 2:47–54.

Keen, C.E. 1979. Thermal history of subsidence of rifted continental margins: Evidence from wells on the Nova Scotia and Labrador shelves. Canadian Journal of Earth Sciences 16:502–522.

Lerche, I., Kendall, C.G. St. C., and Yarzab, R.F. 1984. The determination of paleoheat flux from vitrinite reflectance data. American Association of Petroleum Geologists Bulletin 68:1704–1717.

Love, J.D. 1960. Cenozoic sedimentation and crustal movement in Wyoming. American Journal of Science 258A:204–214.

Love, J.D., McKenna, M.C., and Dawson, M.R. 1976. Eocene, Oligocene and Miocene rocks and vertebrate fossil at the Emerald Lake locality, 3 miles south of Yellowstone National Park, Wyoming. U.S. Geological Survey Professional Paper 932D, 28 pp.

Lowell, J.D. 1974. Plate tectonics and foreland basement deformation. Geology 2:274–278.

Lowers, A.R. 1960. Climate of the United States—Wyoming. U.S. Weather Bureau Climatography of the United States 60–48, pp. 116, 1128.

McKenna, M.C. 1980. Remaining evidence of Oligocene sedimentary rocks previously present across the Bighorn basin, Wyoming. In: Gingerich, P.D. (ed.): Early Cenozoic Paleontology, and Stratigraphy of the Bighorn Basin. University of Michigan Papers on Paleontology 24, pp. 143–146.

McKenna, M.C., and Love, J.D. 1972. High-level strata containing Early Miocene mammals on the Bighorn Mountains, Wyoming. American Museum Novitates 2490, pp. 1–31.

McKenzie, D. 1981. The variation of temperature with time and hydrocarbon maturation in sedimentary basins formed by extension. Earth and Planetary Science Letters 55:87–98.

Meyer, H.J., and McGee, H.W. 1985. Oil and gas fields accompanied by geothermal anomalies in Rocky Mountain region. American Association of Petroleum Geologists Bulletin 69:933–945.

Nadeau, P.H., Wilson, M.J., McHardy, W.J., and Tait, J.M. 1984. Interstratified clays as fundamental particles. Science 225:923–925.

Naeser, C.W. 1979. Thermal history of sedimentary basins: Fisson-track dating of subsurface rocks. In: Scholle, P.A., and Schluger, P.R. (eds.): Aspects of Diagenesis. Society of Economic Paleontologists and Mineralogists Special Publication 26, pp. 109–112.

Nixon, R.P. 1973. Oil source beds in Cretaceous Mowry Shale of northwestern interior United States. American Association of Petroleum Geologists Bulletin 57: 136–161.

Palmquist, J.L. 1978. Laramide structures and basement block faulting: Two examples from the Big Horn Mountains, Wyoming. In: Matthews, V., III (ed.): Laramide Folding Associated with Basement Faulting in the Western United States. Geological Society of America Memoir 151, pp. 125–138.

Prucha, J.J., Graham, J.A., and Nickelsen, R.P. 1965. Basement-controlled deformation in Wyoming province Rocky Mountain foreland. American Association of Petroleum Geologists Bulletin 49:966–992.

Sales, J.K. 1968. Crustal mechanics of Cordilleran foreland deformation: A regional and scale model approach. American Association of Petroleum Geologists Bulletin 52:2016–2044.

Sclater, J.G., and Christie, P.A.F. 1980. Continental stretching: An exploration of the post-Mid Cretaceous subsidence of the central North Sea Basin. Journal of Geophysical Research 85:3711-3739.

Smithson, S.B., Brewer, J.A., Kaufman, S., Oliver, J., and Hurich, C. 1978. Nature of the Wind River thrust, Wyoming, from COCORP deep reflection data and gravity data. Geology 6:648-652.

Stearns, D.W. 1971. Mechanisms of drape folding in the Wyoming province. Wyoming Geological Association Twenty-Third Annual Field Conference Guidebook, pp. 149–152.

Stone, D.S. 1969. Wrench faulting and Rocky Mountains tectonics. The Mountain Geologist 6:67–79.

Stone, D.S. 1985. Geologic interpretation of seismic profiles, Bighorn basin, Wyoming: Part II. West flank. In: Gries, R.R., and Dyer, R.C. (eds.): Seismic Exploration of the Rocky Mountain Region. Denver, Rocky Mountain Association of Geologists and the Denver Geophysical Society, pp. 175–186.

Sundell, K.A. 1985. The Castle Rock Chaos: A gigantic Eocene landslide-debris flow within the southwestern Absaroka Range, Wyoming. Ph.D. thesis, University of California at Santa Barbara, Santa Barbara, 236 pp.

Tissot, B.P., and Espitalié, J. 1975. L'evolution thermique de la matière organique des sèdiments: Applications d'une simulation mathematique potentiel petrolier des bassins sedimentaires et reconstitution de l'histoire thermique des sediments. Revue de l'Institut Francais du Pètrole 30:743–777.

Turcotte, D.L., and McAdoo, D.C. 1979. Thermal subsidence and petroleum generation in the southwestern block of the Los Angeles Basin, California. Journal of Geophysical Research 84:3460–3464.

Van Houten, F.B. 1952. Sedimentary record of Cenozoic orogenic and erosional events, Bighorn basin, Wyoming. Wyoming Geological Association Seventh Annual Field Conference Guidebook, pp. 74-79.

Waples, D.W. 1980. Time and temperature in petroleum formation: Application of Lopatin's methods to petroleum exploration. American Association of Petroleum Geologists Bulletin 64:916-926.

Woodside, W., and Messmer, J.H. 1961. Thermal conductivity of porous media. Journal of Applied Physics 32:1688-1699.

Woodward, L.A. 1976. Laramide deformation of the Rocky Mountain foreland: Geometry and mechanics. In: Woodward, L.A., and Northrop, S.A. (eds.): Tectonics and Mineral Resources of Southwestern North America. New Mexico Geological Society Special Publication 6, pp. 11–17.

Zandt, G., and Furlong, K.P. 1982. Evolution and thickness of the lithosphere beneath coastal California. Geology 10:376–381.

17
Thermal and Hydrocarbon Maturation Modeling of the Pismo and Santa Maria Basins, Coastal California

Henry P. Heasler and Ronald C. Surdam

Abstract

Thermal and hydrocarbon maturation models have been developed for the Pismo and Santa Maria basins of coastal California. The thermal models derived for the temperature history of Miocene Monterey Formation source rocks include the effects of subduction, triple junction migration with consequent asthenospheric upwelling, thermal refraction, and temporal changes in thermal conductivities due to diagenesis and compaction. Hydrocarbon maturation models use the technique described by Tissot and Espitalié in 1975 and the derived temperature history to predict the timing of hydrocarbon maturation.

In an attempt to predict API gravity of the generated oils, we use a two kinetic component mixture of kerogen. We assume that relatively sulfur-poor kerogen (5 weight % sulfur or less) will tend to generate high API gravity oils and that relatively sulfur-rich kerogen (9 weight % sulfur or greater) will tend to generate low API gravity oils.

Using our two kinetic component assumption and kinetic values given by Lewan in 1985 for sulfur-rich and relatively sulfur-poor type II kerogens, we are able to predict the general pattern of low API gravity oils found in the Pismo basin and both the high and low API gravity oils found in the Santa Maria basin.

Introduction

One of the most difficult exploration tasks relative to the Monterey Formation of California is the temporal prediction of generative patterns of crude oils, particularly with respect to API gravity. If the distribution of high API gravity crude oil (API greater than 20) could be successfully predicted, the exploration risk associated with the Monterey Formation would be significantly reduced. Thus, the essential problem addressed in this chapter is the determination of the level of thermal exposure necessary to generate either high or low API gravity crude oils within the Monterey Formation.

The thermal history of coastal California is a crucial parameter in our hydrocarbon maturation models. Many authors (Tissot, 1969; Lopatin, 1971; Tissot and Espitalié, 1975; Waples, 1980) stress the importance of both time and temperature in developing hydrocarbon maturation models. Thus, a maturation model must include an adequate representation of the passage of potential source rocks through time-temperature space.

Along coastal California, much of the petroleum production occurs in basins formed since the Oligocene (California Division of Oil and Gas, 1974; Hall, 1981; Isaacs, 1984). Plate tectonic interactions must be considered in this region because of their significant affect on the geological and thermal histories both before and after the Oligocene.

In this chapter we review our model for the thermal evolution of coastal California (Heasler and Surdam, 1985) and apply the thermal model to the maturation of hydrocarbons in the Pismo and Santa Maria basins of California. We refine the predictive capabilities of the hydrocarbon maturation model by using basin-specific thermal models and a maturation model that uses two different kinetic values to represent the kerogen contained in potential source rocks in the Monterey Formation.

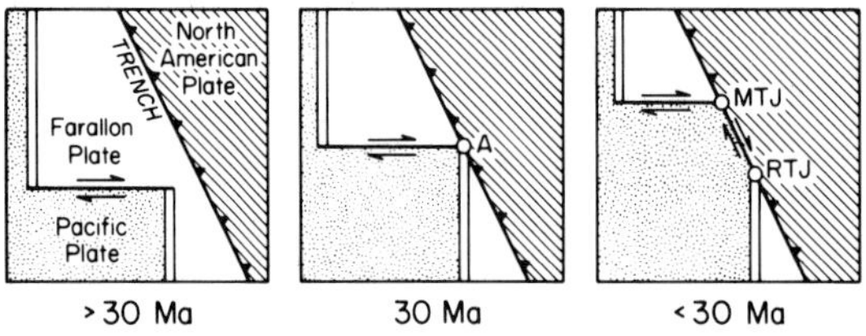

FIGURE 17.1. Generalized map showing relationship between North American, Farallon, and Pacific plates through time. Subduction of Farallon plate continued until approximately 30 Ma, when ancestral East Pacific rise crest (double line) impinged on North American plate at Point A. Continued relative motions of plates generated San Andreas transform between northward migrating Mendocino triple junction (MTJ) and southward migrating Rivera triple junction (RTJ). Subduction of remnants of Farallon plate continued north of MTJ and south of RTJ. (Modified from Dickinson and Snyder, 1979b, Fig. 1.)

Thermal Modeling

The thermal model summarized in this chapter is essentially that of Heasler and Surdam (1985). For a more complete discussion of the thermal model, the reader is referred to that paper.

The thermal and tectonic history of coastal California has been dominated for the last 30 my by interactions between the North American, Farallon, and Pacific plates. Plate interactions, as described by Dickinson and Snyder (1979a, 1979b), that must be thermally modeled in order to predict hydrocarbon maturation patterns include subduction, migrating triple junctions, and the opening of a slabless window.

The Mendocino triple junction (MTJ) and Rivera triple junction (RTJ) were formed in the Oligocene when the ancestral East Pacific rise first encountered the North American plate (Dickinson and Snyder, 1979a, 1979b). Until that time, the Farallon plate had been subducted beneath the North American plate (Fig. 17.1). After the impingement of the East Pacific Rise, there was a simultaneous migration of a fault-fault-trench triple junction (the MTJ) northward and a rise-trench-fault triple junction (the RTJ) southward along the Pacific-North American plate boundary (Fig. 17.1). The broad boundary between the Mendocino and Rivera triple junctions, with its associated wrench tectonism, has been named the San Andreas transform (Wilson, 1965). Positions of the MTJ and the East Pacific rise crest since 30 Ma are shown in Figure 17.2.

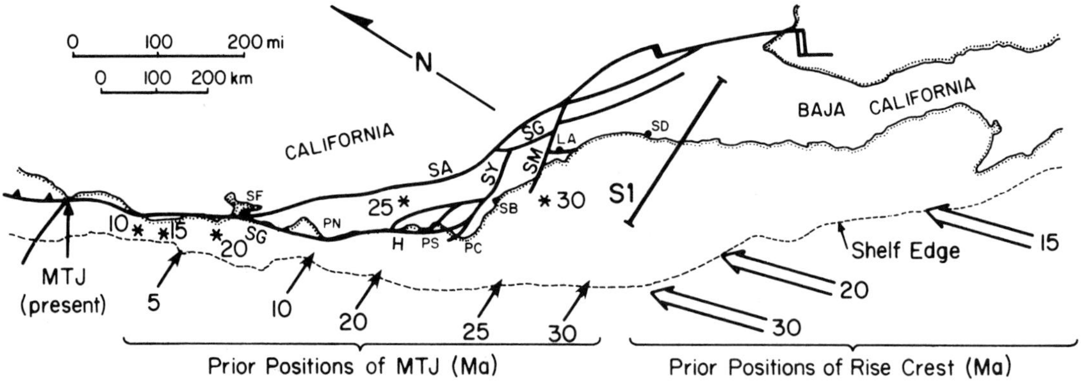

FIGURE 17.2. Map of coastal California showing positions of MTJ and East Pacific rise crest through time (after Dickinson and Snyder, 1979a, Figs. 8 and 9). Asterisks represent estimated locations at various times of MTJ passage. Northward displacement of asterisks from prior positions of MTJ along continental slope is thought to represent Neogene slip of coastal slices between San Andreas fault, San Gregorio-Hosgri fault, and continental slope (Dickinson and Snyder, 1979a). Location used for thermal modeling of subduction near San Diego at 30 Ma is shown by S1. Reference points are San Francisco (SF), Point Sal (PS), Point Conception (PC), Santa Barbara (SB), Los Angeles (LA), San Diego (SD), and Pinnacles (PN). Faults are San Andreas (SA), Santa Ynez (SY), San Gabriel (SG), Santa Monica-Malibu Coast (SM), Hosgri (H), and San Gregorio (SG). (After Hall, 1981.)

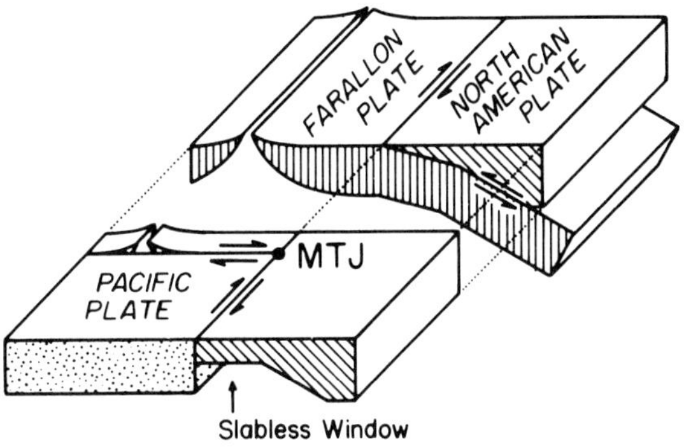

FIGURE 17.3. Diagrammatic representation of geometry of slabless window created by relative motions between North American, Farallon, and Pacific plates. (Modified from Zandt and Furlong, 1982, Fig. 2.) Slabless window was created by right-lateral movement of North American plate off Farallon plate in area of Mendocino triple junction (MTJ) (Dickinson and Snyder, 1979b).

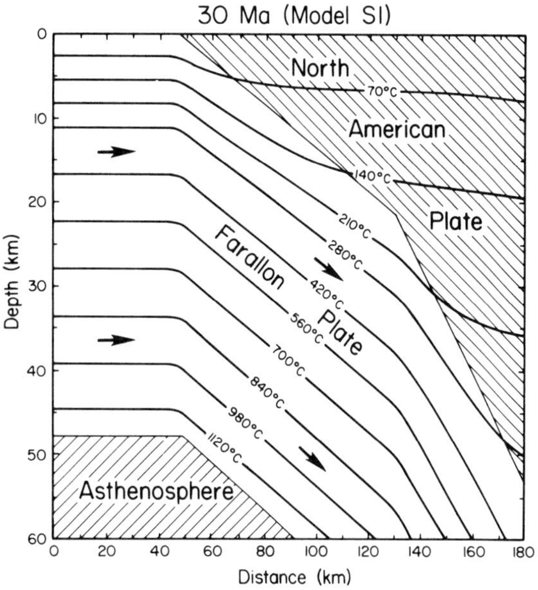

FIGURE 17.4. Result of numerical subduction model near San Diego 30 Ma. Subduction rate was 8 cm/yr (Dickinson and Snyder, 1979a). Note vertical exaggeration.

Associated with the migration of the triple junctions is the formation of a slabless window beneath the North American plate (Dickinson and Snyder, 1979b). Subduction ceased beneath the North American plate at the location where the ancestral East Pacific rise first impinged on the North American plate. Subduction of remnants of the Farallon plate (Juan de Fuca plate to the north, Rivera and Cocos plates to the south) continued north and south of the triple junctions. The relative motion of the plates resulted in the development of an area beneath the North American plate with no subducted oceanic lithosphere (Fig. 17.3).

Other authors have considered the thermal effects of these plate tectonic interactions (Lachenbruch and Sass, 1980; Zandt and Furlong, 1982; Furlong, 1984). Our approach to estimating the thermal effects of the plate tectonic interactions was to solve conductive and convective heat transport partial differential equations. Published geological and thermal literature were used to establish boundary conditions and constraints of the differential equations. The differential equations were solved using finite difference numerical techniques (Heasler, 1984).

Results of the thermal modeling of subduction give the magnitude of temperature suppression in the North American plate due to convective and conductive transport of heat. Figure 17.4 shows a result of these calculations for a region near San

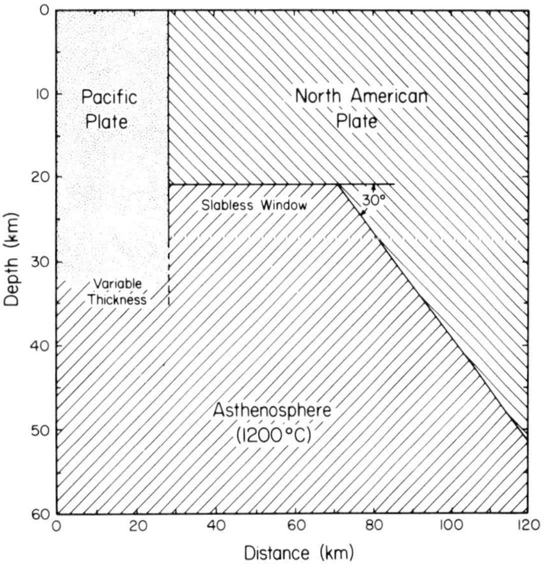

FIGURE 17.5. Initial geometry of North American plate, Pacific plate, and slabless window used in time-dependent numerical models. Note vertical exaggeration.

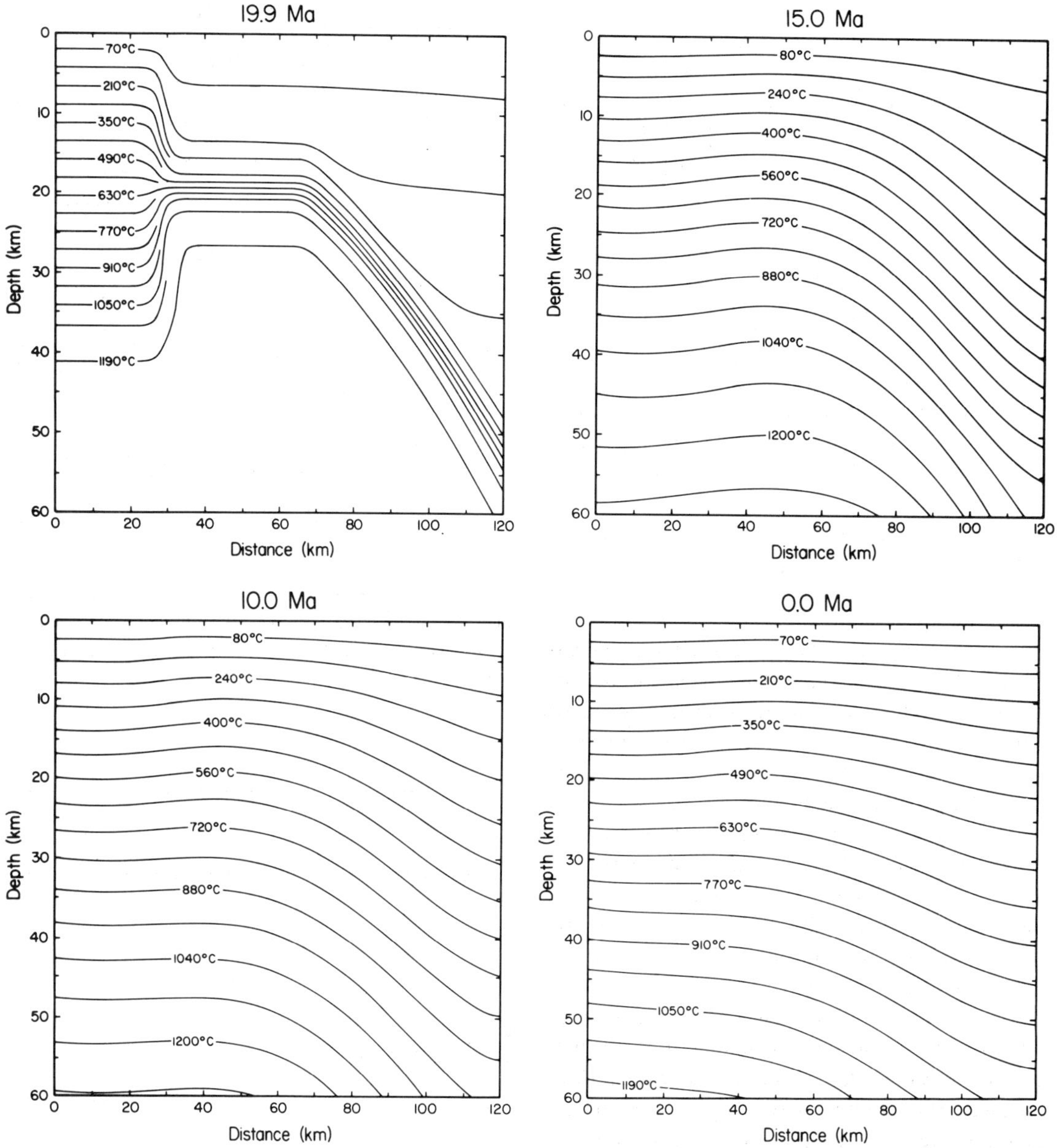

FIGURE 17.6. Results of time-dependent numerical model for region near Point Conception. Passage of Mendocino triple junction is assumed to have occurred approximately 20 Ma. Note vertical exaggeration.

Diego. The magnitude of the temperature suppression in the arc-trench gap between the North American plate and subducted oceanic lithosphere will depend on such variables as rate of subduction, angle of subduction, age of oceanic lithosphere subducted, thermal conductivities of the North American and ocean plates, and radiogenic heat production in both plates (see Heasler and Surdam, 1985).

After triple junction migration, the thermal effects of upwelling asthenosphere (i.e., the formation of a slabless window) must be approximated.

Variables that effect this solution are the geometry of the asthenospheric upwelling, the initial temperature configuration in the North American and oceanic plates, the assumed temperature of the lithosphere, and the thermal diffusivities of the North American and oceanic plates. The geometry of the model and modeled results for the region near Point Conception are shown in Figures 17.5 and 17.6. For additional details on the thermal model, see Heasler and Surdam (1985).

The described thermal modeling techniques allow the heat flow into a given region to be calculated both as a function of time and position. Figure 17.7 illustrates the variation of heat flow across the Pacific plate-North American plate boundary near Point Conception at four different times. As shown in Figure 17.7, basins that are relatively further inland and younger have experienced less thermal exposure. This is because the thermal event caused by the lithospheric upwelling decreases in strength away from the slabless window and the magnitude of the thermal anomaly decreases with time. Data such as that contained in Figure 17.7 are important because they can be used to calculate the basal heat flow into forming sedimentary basins both as a function of time and position.

Once the temporal variation of heat flow into a basin is known, the geometry of the basin through time can be estimated in order to calculate the effect of thermal refraction within the basin. Thermal refraction is due to two- or three-dimensional transport of heat. Its magnitude is a function of the geometry of the basin and the relative contrasts in thermal conductivities between rock units. Thermal refraction effects were estimated using two-dimensional finite-difference models for the conduction of heat as discussed in Heasler and Surdam (1985). Since the two basins investigated (the Pismo and Santa Maria basins) are several times longer than they are wide, we feel that a two-dimensional approximation is valid for the central areas of the basins that were modeled for hydrocarbon maturation. Examples of the calculated effects of thermal refraction for the Pismo basin are shown in Figure 17.8.

When considering individual basins such as that shown in Figure 17.8, the thermal conductivities of the rock units must either be measured or estimated. The thermal conductivity of a porous rock is dependent on the porosity value (Woodside and Messmer, 1961), which will change as a function of burial and hence time for most rocks (Steckler and Watts, 1978). However, in some rocks such as the highly siliceous Monterey Formation, porosity changes are temperature dependent due to diagenetic reactions (Murata et al., 1977; Keller and Isaacs, 1985). Thus, the temperature structure of Figure 17.8 incorporates thermal conductivity changes due to lithological differences, compaction, and temperature-dependent diagenetic reactions (see Heasler and Surdam, 1985, for details).

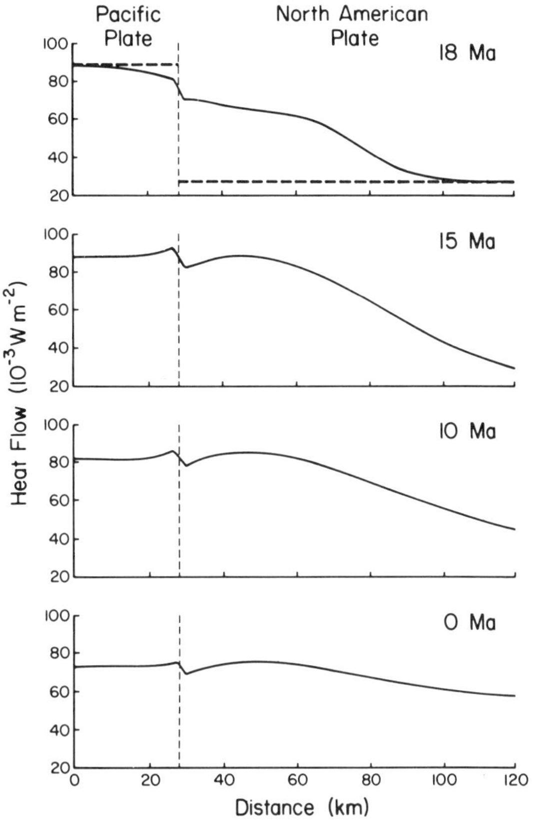

FIGURE 17.7. Surface heat flow as a function of time and position for region near Point Conception. Initial heat flow at 20 Ma is shown by heavy dashed lines on 18 Ma graph. Change in heat flow at Pacific-North American plate boundary results from thermal conductivity differences between plates. Temperatures are continuous across boundary (see Figs. 17.5 and 17.6).

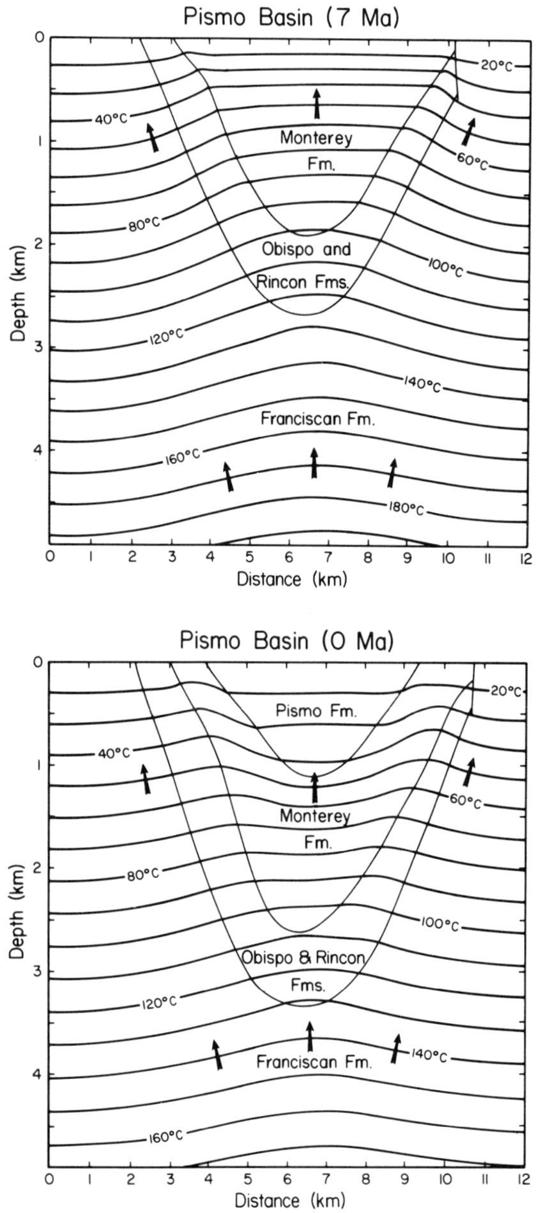

FIGURE 17.8. Examples of numerically modeled effects of thermal refraction on sediment temperatures in Pismo basin. Arrows indicated direction of heat transport. Note vertical exaggeration.

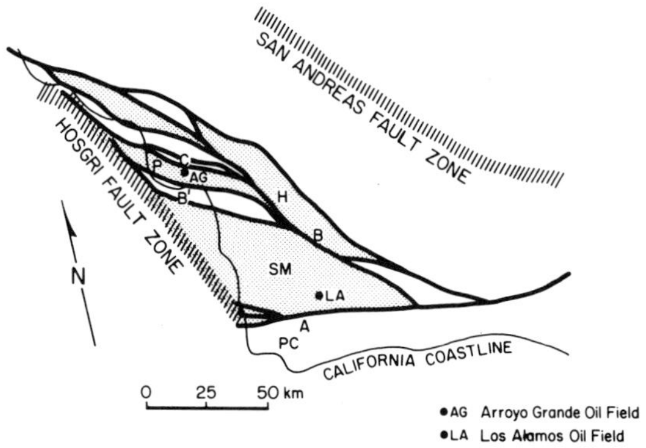

FIGURE 17.9. Simplified map showing tectonic setting of Pismo (P) and Santa Maria (SM) basins. Pismo basin is separated from Huasna basin (H) by West Huasna fault and from Santa Maria basin by Santa Maria River fault. Major tectonic element to west is Hosgri fault system. Stippled areas are depositional basins. PC refers to Point Conception. (Modified from Hall, 1973a, 1973b, 1981.) Cross section locations (Fig. 17.10) shown by AB and B′C.

General Geology of the Pismo and Santa Maria Basins

To achieve the objectives of this chapter two areas within different basins were selected for detailed modeling. First, the Arroyo Grande oil field within the Pismo basin (Figs. 17.9 and 17.10) was chosen because it produces low API gravity crude oil (API 13 to 15) generated within the Monterey Formation from the central portion of a syncline (California Division of Oil and Gas, 1974). Stratigraphic reconstructions suggest that the Monterey Formation has been buried to present-day depths no greater than 2.1 km (7,000 ft) and that the removal by erosion of overburden has been minimal (Hall, 1973a; Surdam and Stanley, 1981). Thus, the generative source rocks in the Pismo basin are at or near their maximum burial depth. Second, the Los Alamos oil field in the onshore Santa Maria (Figs. 17.9 and 17.10) basin was chosen because it produces high API gravity crude oil (API 34) generated within the Monterey Formation from structural closure at a present-day depth of approximately 2.8 km (9,300 ft) (California Division of Oil and Gas, 1974). Again, stratigraphic reconstructions suggest that erosion of overburden has been minimal (California Division of Oil and Gas, 1974). Thus, the generative source rocks responsible for the crude oil in the Los Alamos field have been subjected to burial depths of *at least* 2.8 km (9,300 ft). Lastly, these two fields were chosen because of the availability of a time-temperature

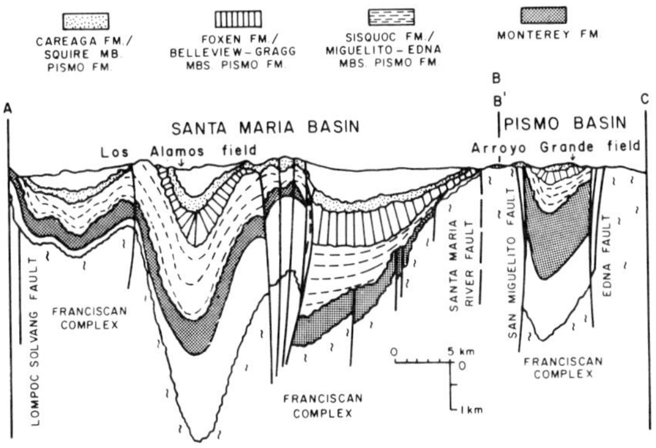

FIGURE 17.10. Generalized cross sections of the Pismo and Santa Maria basins showing approximate locations of the Los Alamos and Arroyo Grande oil fields (California Division of Oil and Gas, 1974) and the relative location of the Monterey Formation source rocks. Cross sections are modified from Magoon and Isaacs (1983) and Stanley and Surdam (1984).

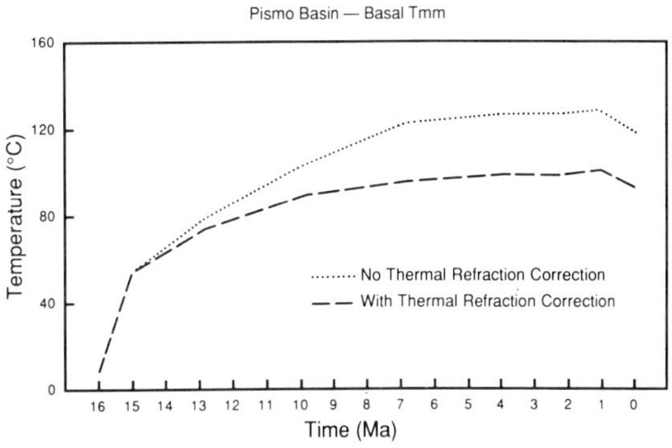

FIGURE 17.11. Temperature history of the basal Monterey Formation (Tmm) in the Pismo basin. The dotted line is the calculated temperature history without any thermal refraction correction. The dashed line represents the temperature history corrected for thermal refraction. See text for additional discussion.

framework for both areas, including essential heat flow and temperature profiles (Heasler and Surdam, 1985).

Hall (1981) has shown that both the Pismo and Santa Maria basins are pull-apart structures. Both structures extend offshore. The source rocks in both the Pismo and Santa Maria basins are contained in the Miocene Monterey Formation. Total organic carbon values for the Monterey Formation in these basins typically average from 1 to 5 weight % (Isaacs, 1981; Surdam and Stanley, 1981).

Hydrocarbon Maturation Modeling

The time-temperature reconstructions as shown in Figures 17.11 and 17.12 for the Pismo and Santa Maria basins illustrate the difference in thermal exposure experienced by potential source rocks. The differences between the time-temperature histories of the two basins may or may not be significant with regard to the generation of hydrocarbons depending on the kinetic parameters of the kerogens involved and the type of maturation model used. However, whichever maturation model is used must be able to predict the hydrocarbon generation patterns observed in a basin or else the maturation model, or the thermal model, or both are incorrect.

The hydrocarbon maturation model of Lopatin as described by Waples (1980) and the maturation model as proposed by Tissot and Espitalié (1975) both consider the interrelation of time and temperature in the generation of hydrocarbons. However, the two models differ in their predictive capabilities. Lopatin's maturation model assumes all kerogens react such that for every 10°C temperature

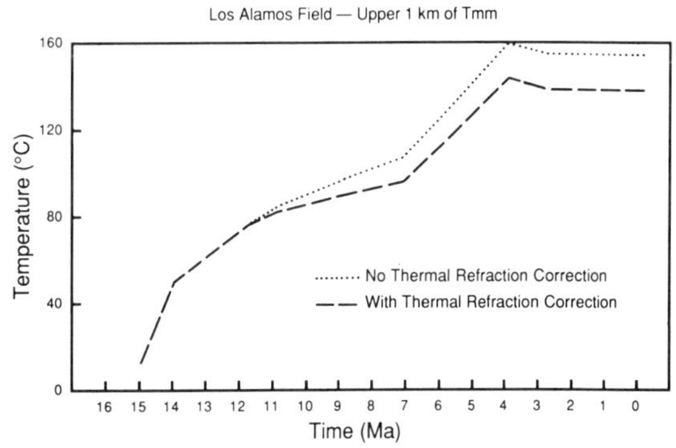

FIGURE 17.12. Temperature history of the upper 1 km of the Monterey Formation (Tmm) at the Los Alamos oil field in the Santa Maria basin. The dotted line is the calculated temperature history without any thermal refraction correction. The dash line represents the temperature history corrected for thermal refraction. See text for additional details.

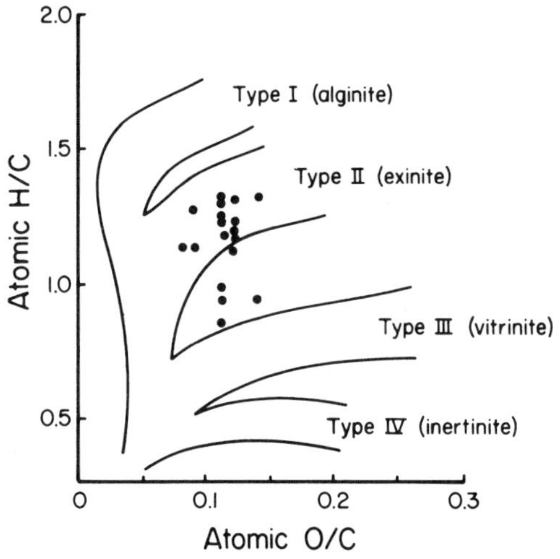

FIGURE 17.13. H/C versus O/C plot (after Waples, 1981) for kerogen for Monterey Formation in the Pismo basin. (Data are from Heasler and Surdam, 1985, Table 1.)

increase, the maturation rate increases by a factor of two. This assumption implies that the kinetic parameters for all kerogens are identical and independent of source rock type. In the Tissot and Espitalié maturation model the kinetic parameters of the kerogens can be varied. This allows the timing of maturation of different kerogens to be computed.

Lopatin's maturation model results in a value called the Time-Temperature Index (TTI). A TTI of 15 has been generally correlated with the onset of oil generation (Waples, 1981). However such a correlation may not be valid for areas that have experienced rapidly changing thermal histories over relatively short time intervals or for kerogens that have different kinetic parameters than those assumed by Lopatin.

The type of maturation model used in this work will be that of Tissot (1969) and Tissot and Espitalié (1975) in which they propose that the conversion of kerogen to hydrocarbons proceeds as a series of first-order chemical reactions. Mathematically this is stated as

$$dx_i = -x_i A_i \exp\left[-E_i/RT\right]dt$$

where x_i is the mass fraction of kerogen, E_i is the activation energy, R is the universal gas constant, A is the preexponential constant, T is the absolute temperature, and t is the time. Given time-temperature histories such as in Figures 17.11 and 17.12, the above equation may be numerically integrated to determine the mass fraction of kerogen matured at a given time. From this data a transformation ratio may be computed (see Tissot and Welte, 1978, for details). The transformation ratio is a relative measure of the maturity of a kerogen. A totally immature kerogen would have a transformation ratio of zero whereas a totally mature kerogen would have a transformation ratio of one.

TABLE 17.1. Kerogen maturation kinetics.

	Preexponential constant (10^6 yr)$^{-1}$	Activation energy (kcal/mol)
Phosphoria shale* (sulfur-rich)	4.31×10^{23}	42.7
Woodford Shale* (sulfur-poor)	2.70×10^{26}	52.2
Type II Kerogen†	1.27×10^{5} to 2.80×10^{35}	10 to 80

*Kinetic values from Lewan (1985).
†Kinetic values from Tissot and Espitalié (1975).

The computed transformation ratio is similar to the production index measured in the laboratory by pyrolysis. Thus, the transformation ratio that results from the thermal and maturation models can be compared with laboratory measurements of source rock maturity. However, laboratory pyrolysis data should be carefully interpreted with respect to the Monterey Formation, as discussed by Kablanow and Surdam (1983), due to potential problems associated with migration and other effects.

Given the Tissot and Espitalié (1975) maturation model, kinetic values must be chosen in order to estimate hydrocarbon maturation. Data from the Pismo basin indicate the kerogen to be Type II (Fig. 17.13). Orr (1984) reports Monterey kerogens containing high organic sulfur (9 to 13 weight %) and low or moderate sulfur content (4 to 6%). In addition, Lewan (1986) has shown that the organic sulfur content of pregenerative kerogen from a variety of depositional environments found in the Monterey Formation varies from 4.6 to 11.2 weight %. Both Lewan (1985) and Orr (1986) state that the greater the amount of sulfur incorporated into the matrix of a Type II kerogen, the lower the thermal stress necessary for hydrocarbon generation. Consequently, we have used kinetic values reported by Lewan (1985) for a source rock from the Phosphoria shale that contains Type II kerogen with 9 weight % sulfur and for a source rock from the Woodford Shale that contains Type II kerogen with 5 weight % sulfur (see Table 17.1).

For purposes of hydrocarbon maturation modeling we are assuming that the kerogen in the Monterey Formation consists of two kinetic components. The two kinetic components of the Monterey Formation could exist either as separate relatively sulfur-rich (9 weight % or greater) and sulfur-poor (5 weight % or less) kerogens or both kinetic components could exist in a single kerogen. However, the maturation histories for either case will be similar.

Orr (1974) has suggested that for Paleozoic crude oils in the Bighorn basin, Wyoming, thermal maturation of the crude oils and not the kerogen could change the sulfur content of the oils. He also noted that oil gravities (ranging from 12 to 55 API) varied inversely with the sulfur content of the oils (ranging from about 4 to less than 0.3%). If Orr's (1974) hypothesis regarding the thermal maturation of a single low API gravity oil in the Bighorn basin to produce high API gravity oil is applied to California basins, then our maturation analyses would have to include the determination of kinetic rate constants for such oil-oil reactions. Then, by using our thermal and maturation modeling techniques as presented in this chapter, we would be able to determine the thermal exposure necessary to mature specific oils in addition to specific source rocks. For purposes of this chapter we are considering only the kerogen-to-oil reaction.

Assuming a two-kerogen kinetic model for source rocks in the Monterey Formation allows the analysis of the timing and amount of hydrocarbon generation. The type of hydrocarbons generated by the sulfur-rich versus sulfur-poor kerogen may also be different. As stated by Orr (1984, 1986) the sulfur-poor kerogen will tend to generate high API gravity oils (low-sulfur crude oils), whereas the sulfur-rich kerogen will tend to generate low API gravity oil (high-sulfur crude oils). This inverse relationship between API gravity and percent sulfur for oils from the Santa Maria basin is shown in Magoon and Isaacs (1983, Fig. 5) and Orr (1986, Fig. 11).

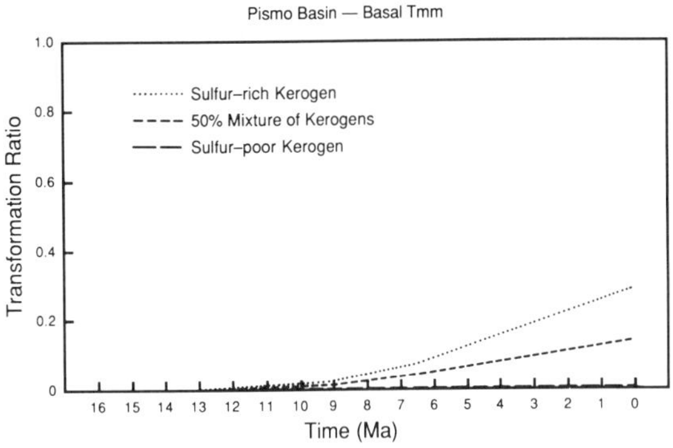

FIGURE 17.14. Calculated transformation ratio through time for the basal Monterey Formation in the Pismo basin. Corrected temperatures from Figure 17.11 were used to calculate the transformation ratio. The dotted line represents the transformation ratio calculated using kinetic values for a sulfur-rich kerogen. The long-dashed line represents the transformation ratio calculated using kinetic values for a sulfur-poor kerogen. The medium-dashed line represents the transformation ratio calculated for a kerogen composed of 50% sulfur-rich kerogen and 50% sulfur-poor kerogen.

It is important to note that in this modeling we are assuming that high-sulfur kerogen will generate high-sulfur crude oils that typically are characterized by low API gravity values and that low-sulfur kerogen will generate lower sulfur crude oils that typically are characterized by high API gravity values. In addition, crude oil fractionation effects due to migration are not part of the model. Basically, our objective is to provide the explorationist with a maturation model for the Monterey Formation that will determine the level of thermal exposure necessary to generate high API gravity oil from kerogens.

Applying the two-kerogen model to the time-temperature history for the basal Monterey Formation in the Pismo basin results in the maturation history shown in Figure 17.14. Figure 17.14 shows that if the basal Monterey Formation in the Pismo basin were composed of 100% sulfur-rich kerogen, the computed transformation ratio would be about 0.3. However, if the Monterey Formation in the Pismo basin were composed of 100% relatively sulfur-poor kerogen, the computed transformation ratio would be zero.

The maturation and thermal models suggest that only the sulfur-rich kerogen has begun to mature in

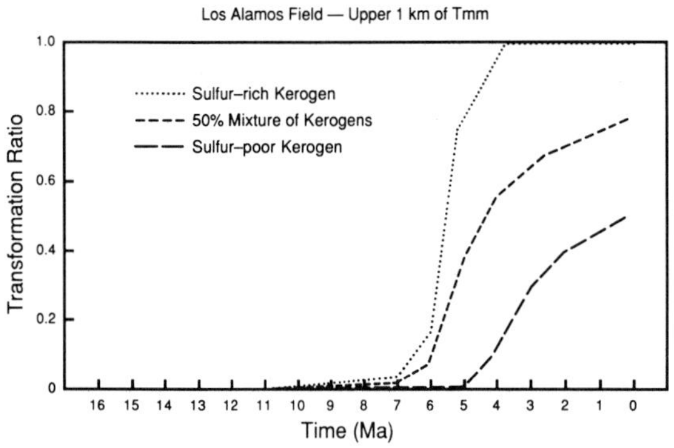

FIGURE 17.15. Calculated transformation ratio through time for the upper 1 km of Monterey Formation in Los Alamos oil field in the Santa Maria basin. Corrected temperatures from Figure 17.12 were used to calculate the transformation ratio. The dotted line represents the transformation ratio calculated using kinetic values for a sulfur-rich kerogen. The long-dashed line represents the transformation ratio calculated using kinetic values for a sulfur-poor kerogen. The medium-dashed line represents the transformation ratio calculated for a 50% sulfur-rich and 50% sulfur-poor kerogen.

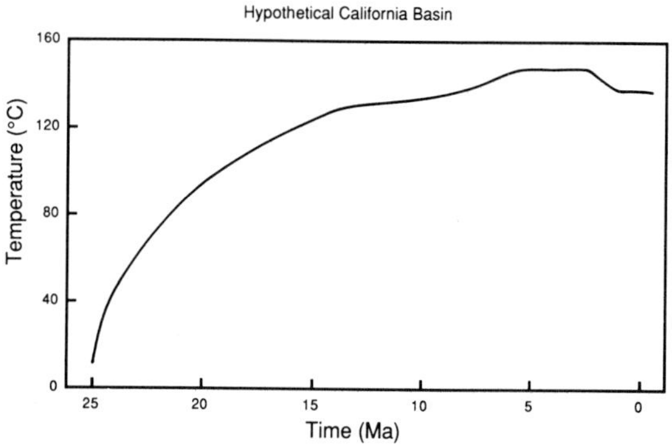

FIGURE 17.16. Temperature history of the Monterey Formation for a hypothetical coastal California basin.

the Pismo basin and consequently only low API gravity oils will be found in the basin. This agrees with data from the Arroyo Grande oil field, which produces 13 to 15 API gravity oil from Monterey Formation in the synclinal axis of the Pismo basin (California Division of Oil and Gas, 1974).

If the kerogen in the Pismo basin were a mixture of 50% sulfur-rich kerogen and 50% sulfur-poor kerogen, the computed transformation ratio would be about 0.14. This agrees roughly with 14 measurements of pyrolysis data from the Pismo basin that ranged from 0.05 to 0.31 and averaged 0.1 ± 0.08 (Heasler and Surdam, 1985). Note that with an even mixture of the two kerogen types, the sulfur-poor kerogen is still totally immature. The computed transformation ratio of 0.14 is due solely to the gradual maturation of the sulfur-rich kerogen over the last 8 my (Fig. 17.14).

Applying the two-kerogen model to the Los Alamos oil field in the Santa Maria basin results in the maturation history for the upper 1 km of the Monterey Formation as shown in Figure 17.15. The lower portion of the Monterey Formation would be subjected to higher levels of thermal exposure. Figure 17.15 shows that if the Monterey Formation in the Santa Maria basin contained 100% sulfur-rich kerogen, the kerogen would have totally matured between 7 and 5 Ma. If the Monterey Formation contained 100% relatively sulfur-poor kerogen, the kerogen would have passed through peak oil generation and would presently be leaving the oil generative window as defined by a transformation ratio of 0.45 (Tissot and Welte, 1978). If the kerogen were a mixture of 50% sulfur-rich kerogen and 50% sulfur-poor kerogen, the computed transformation ratio would be 0.73.

The maturation and thermal models for the Santa Maria basin indicate that both the sulfur-rich and relatively sulfur-poor kerogens have matured but at different times (7 to 5 Ma for the sulfur-rich kerogen versus 4 to 0 Ma for the sulfur-poor kerogen). The Los Alamos oil field produces 34 API gravity oil from the Monterey Formation from structural closure in the Santa Maria basin at a present-day depth of approximately 2.8 km (9,300 ft) (California Division of Oil and Gas, 1974). Our modeling predicts the maturity of both sulfur-rich and sulfur-poor kerogen in the Santa Maria basin and consequently the generation of both high and low API gravity crude oils. Within other oil fields in the Santa Maria basin (the Santa Maria Valley, Casmalia, Orcutt, Lompoc, and Cat Canyon oil fields) API gravities range from 9 to 27.3 for oils with sulfur values of 2.67 to 4.73% (Magoon and Isaacs, 1983).

The difference in maturation between a sulfur-rich and a sulfur-poor Type II kerogen can be illustrated by considering a hypothetical California basin whose source rocks have a time-temperature history as shown in Figure 17.16. Application of the two-kerogen kinetic model to this hypothetical basin can determine the thermal exposure necessary to mature different API gravity oils and the timing of the generation of those oils. Figure 17.17 shows the calculated transformation ratio through time for the hypothetical basin assuming a mixture of 50% sulfur-rich kerogen and 50% sulfur-poor kerogen.

For this hypothetical basin, the sulfur-rich kerogen (low API gravity oil) matures primarily from 18 to 14 Ma. The temperature range for this time interval is 105° to 130°C. The sulfur-poor kerogen (high API gravity oil) matures from 14 to 0 Ma with most of the maturation occurring from about 9 to 3 Ma. The temperature range for 9 to 3 Ma is 135° to 145°C. Thus, the maturation model for this hypothetical basin predicts that low API gravity oil would have formed 5 to 9 my before the maturation of high API gravity oil. The model also predicts the absolute dates of maturation, which could be useful in the determination of oil migration into potential structural traps.

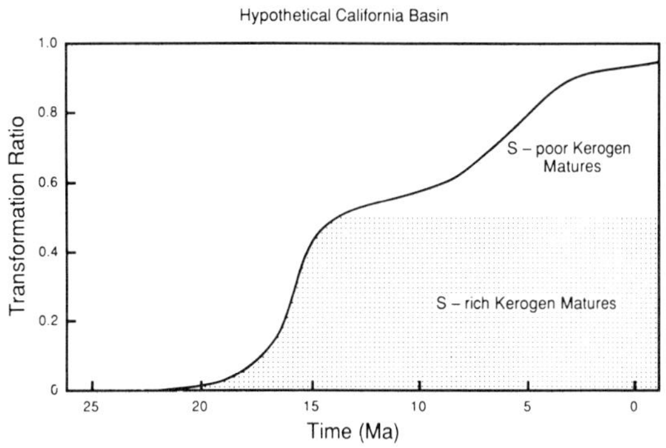

FIGURE 17.17. Calculated transformation ratio for the temperature history of the hypothetical basin shown in Figure 17.16. The transformation ratio is calculated for a 50% sulfur-rich and 50% sulfur-poor kerogen.

Thermal Model Synopsis

Thermal models developed for a basin must consider regional tectonic history, geological history of the basin, geometry of the basin, and types of sediments in the basin. The regional tectonic history helps define the temporal variation of heat flow into the area of interest. The geological history of the basin determines the burial rate of sediments and whether other thermal effects should be considered. Other thermal effects would include such things as rapid uplift and erosion, thrust faulting, localized magmatic intrusions, and rapid sediment burial. The geometry of the basin through time is important in defining the amount of heat that will be refracted into or out of the basin. The types of sediment in the basin are important in defining the thermal conductivity structure of the basin and the change in thermal conductivity of the sediments through time due to compaction or diagenetic changes. The sediment type and geological history of a basin are also important in determining if regional water flow patterns should be considered as an important convective heat transport mechanism.

The developed thermal model for a basin must be able to predict present-day measured heat-flow values and measured temperatures in the basin. Another important constraint on a thermal model for a basin is the prediction of observed hydrocarbon maturation patterns in the basin. This is important because maturation of oil depends not only on exposure of source rocks to temperature but also on the length of thermal exposure. Thus, oil maturation models place an additional constraint on the temperature history of a basin.

Given a reasonable thermal model for an area, differing maturation models may be tested in an effort to both verify the validity of the thermal model and to illuminate the subtleties of maturation patterns. In the Pismo and Santa Maria basins of southern coastal California, a sulfur-rich and sulfur-poor kerogen maturation model yields results consistent with observed data. This model predicts the presence of low API gravity oil in the Pismo basin and both high and low API gravity oil in the Santa Maria basin.

However, the maturation model may be used for more than the prediction of oil gravity. The maturation model also computes the maturation history of source rocks. The maturation history may be an important tool for explorationists to use in determining time of migration and potential traps for the generated oil.

Conclusion

Based on the thermal framework of the central California coast provided by Heasler and Surdam (1985), the kinetic values for source rocks from the Phosphoria (sulfur-rich Type II kerogen) and the Woodford (sulfur-poor Type II kerogen) shales determined by Lewan (1985), and the suggestion of Orr (1984) that high-sulfur kerogens tend to generate low API gravity crude oils, a model can be developed that predicts the level of thermal exposure necessary to generate high API gravity crude oils in the Monterey Formation. Simplifying assumptions, such as ignoring hydrocarbon migration effects, have been used, but the model does correctly predict the observed distribution of API gravity values in both the Pismo and Santa Maria basins. Lastly, the maturation model of coastal California presented herein provides the explorationist with a possible way to evaluate the level of thermal exposure necessary to generate high and low API gravity crude oils in the unconventional, but prolific Monterey Formation of California.

Acknowledgments. We wish to thank Steve Boese for chemical analyses and many helpful discussions; Mike Lewan of Amoco Production Company and Wilson Orr of Mobil Oil Corporation for discussions of kerogen kinetic parameters; and R.I. Kablanow, K.O. Stanley, and C.A. Hall for discussions of the Monterey Formation. This work was sponsored by NSF Grant EAR-8207570 and basic research grants from Texaco Incorporated and Mobil Oil Corporation.

References

California Division of Oil and Gas. 1974. California Oil and Gas Fields: Vol. 2. East-Central California. Sacramento, California Division of Oil and Gas.

Dickinson, W.R., and Snyder, W.S. 1979a. Geometry of triple junctions related to San Andreas transform. Journal of Geophysical Research 84:561–572.

Dickinson, W.R., and Snyder, W.S. 1979b. Geometry of subducted slabs related to San Andreas transform. Journal of Geology 87:609–627.

Furlong, K.P. 1984. Lithospheric behavior with triple junction migration: An example based on the Mendocino triple junction. Physics of the Earth and Planetary Interiors 36:213–223.

Hall, C.A. 1973a. Geology of the Arroyo Grande 15 minute quadrangle, San Luis Obispo County, California. California Division of Mines and Geology, Map Sheet 24, scale 1:48,000.

Hall, C.A. 1973b. Geologic map of the Morro Bay south and Point San Luis quadrangles, San Luis Obispo County, California. U.S. Geological Survey Miscellaneous Field Studies Map MF-511, scale 1:24,000.

Hall, C.A. 1981. San Luis Obispo transform fault and middle Miocene rotation of the western Transverse Ranges, California. Journal of Geophysical Research 86:1015–1031.

Heasler, H.P. 1984. Thermal evolution of coastal California with implications for hydrocarbon maturation. Ph.D. thesis, University of Wyoming, Laramie, 76 pp.

Heasler, H.P., and Surdam, R.C. 1985. Thermal evolution of coastal California with application to hydrocarbon maturation. American Association of Petroleum Geologists Bulletin 69:1386–1400.

Isaacs, C.M. 1981. Porosity reduction during diagenesis of the Monterey Formation, Santa Barbara coastal area, California. In: Garrison, R.E., and Douglas, R.G. (eds.): The Monterey Formation and Related Siliceous Rocks of California. Los Angeles, Pacific Section, Society of Economic Paleontologists and Mineralogists, pp. 257–271.

Isaacs, C.M. 1984. The Monterey: Key to offshore California boom. Oil & Gas Journal 82 (January 9):75–81.

Kablanow, R.I., and Surdam, R.C. 1983. Diagenesis and hydrocarbon generation in the Monterey Formation, Huasna basin, California. In: Isaacs, C.M., and Garrison, R.E. (eds.): Petroleum Generation and Occurrence in the Miocene Monterey Formation, California. Los Angeles, Pacific Section, Society of Economic Paleontologists and Mineralogists, pp. 53–68.

Keller, M.A., and Isaacs, C.M. 1985. An evaluation of temperature scales for silica diagenesis in diatomaceous sequences including a new approach based on the Miocene Monterey Formation, California. Geomarine Letters 5:31–35.

Lachenbruch, A.H., and Sass, J.H. 1980. Heat flow and energetics of the San Andreas fault zone. Journal of Geophysical Research 85:6185–6223.

Lewan, M.D. 1985. Evaluation of petroleum generation by hydrous pyrolysis experimentation. Philosophical Transactions of the Royal Society of London, Series A315, pp. 123–134.

Lewan, M.D. 1986. Organic sulfur in kerogens from different lithofacies of the Monterey Formation. Abstracts, Division of Geochemistry, One hundred ninety-second American Chemical Society National Meeting, Anaheim, CA (abstract No. 94).

Lopatin, N.V. 1971. Temperature and geologic time as factors in coalification (in Russian). Akademiya Nauk

SSSR Series Geologicheskaya Izvestiya, no. 3, pp. 95–106.

Magoon, L.B., and Isaacs, C.M. 1983. Chemical characteristics of some crude oils from the Santa Maria basin, California. In: Isaacs, C.M., and Garrison, R.E. (eds.): Petroleum Generation and Occurrence in the Miocene Monterey Formation, California. Los Angeles, Pacific Section, Society of Economic Paleontologists and Mineralogists, pp. 201–211.

Murata, K.J., Friedman, I., and Gleason, J.D. 1977. Oxygen isotope relations between diagenetic silica minerals in the Monterey Shale, Temblor Range, California. American Journal of Science 277:259–272.

Orr, W.L. 1974. Changes in sulfur content and isotopic ratios of sulfur during petroleum maturation: Study of Big Horn Basin Paleozoic oils. American Association of Petroleum Geologists Bulletin 58:2295–2318.

Orr, W.L. 1984. Sulfur and sulfur isotope ratios in Monterey oils of the Santa Maria River basin and Santa Barbara Channel area. Abstracts, Society of Economic Paleontologists and Mineralogists Annual Midyear Meeting, San Jose, CA, p. 62.

Orr, W.L. 1986. Kerogen/asphaltene/sulfur relationships in sulfur-rich Monterey oils. Organic Geochemistry 10:499–516.

Stanley, K.O., and Surdam, R.C. 1984. The role of wrench fault tectonics and relative changes of sea level on deposition of upper Miocene-Pliocene Pismo Formation, Pismo Syncline, California. In: Surdam, R.C. (ed.): Stratigraphic, Tectonic, Thermal and Diagenetic Histories of the Monterey Formation, Pismo and Huasna Basins, California. Society of Economic Paleontologists and Mineralogists Guidebook 2, August 13–16, 1984, pp. 21–37.

Steckler, M.S., and Watts, A.B. 1978. Subsidence of the Atlantic-type continental margin off New York. Earth and Planetary Science Letters 41:1–13.

Surdam, R.C., and Stanley, K.O. 1981. Diagenesis and migration of hydrocarbons in the Monterey Formation, Pismo syncline, California. In: Garrison, R.E., and Douglas, R.G. (eds.): The Monterey Formation and Related Siliceous Rocks of California. Los Angeles, Pacific Section, Society of Economic Paleontologists and Mineralogists, pp. 317–327.

Tissot, B.P. 1969. First data on the mechanisms and kinetics of the formation of petroleum in sediments (in French). Revue de l'Institut Francais du Petrole 24: 470–501.

Tissot, B.P., and Espitalié, J. 1975. Thermal evolution of organic matter in sediments: Application of a mathematical simulation (in French). Revue de l'Institut Francais du Petrole 30:743–777.

Tissot, B.P., and Welte, D.H. 1978. Petroleum Formation and Occurrence: A New Approach to Oil and Gas Exploration. New York, Springer-Verlag, 538 pp.

Waples, D.W. 1980. Time and temperature in petroleum formation: Application of Lopatin's method to petroleum exploration. American Association of Petroleum Geologists Bulletin 64:916–926.

Waples, D.W. 1981. Organic Geochemistry for Exploration Geologists. Minneapolis, Burgess, 151 pp.

Wilson, J.T. 1965. A new class of faults and their bearing on continental drift. Nature 201:343–347.

Woodside, W., and Messmer, J.H. 1961. Thermal conductivity of porous media. Journal of Applied Physics 32:1688–1699.

Zandt, G., and Furlong, K.P. 1982. Evolution and thickness of the lithosphere beneath coastal California. Geology 10:376–381.

Index